KU-164-355

# THE PRACTICAL SKYWATCHER'S HANDBOOK

ROBERT BURNHAM, ALAN DYER,
ROBERT A. GARFINKLE, MARTIN GEORGE,
JEFF KANIPE, DAVID H. LEVY

CONSULTANT EDITOR
DR. JOHN O'BYRNE

B L O O M S B U R Y
LONDON · BERLIN · NEW YORK · SYDNEY

Published by Adlard Coles Nautical
an imprint of Bloomsbury Publishing Plc
50 Bedford Square, London, WC1B 3DP
www.adlardcoles.com

Copyright © 2011 Weldon Owen Pty Ltd

ISBN 978-1-4081-5746-6

All rights reserved. No part of this publication may be
reproduced in any form or by any means – graphic, electronic
or mechanical, including photocopying, recording, taping or
information storage and retrieval systems – without the prior
permission in writing of the publishers.

The right of the author to be identified as the author
of this work has been asserted by him in accordance with
the Copyright, Designs and Patents Act, 1988.

A CIP catalogue record for this book is available
from the British Library.

This book is produced using paper that is made from
wood grown in managed, sustainable forests. It is natural,
renewable and recyclable. The logging and manufacturing
processes conform to the environmental regulations
of the country of origin.

Printed and bound in China by 1010
Printing Group Limited

Note: while all reasonable care has been taken in the publication
of this book, the publisher takes no responsibility for the use of
the methods or products described in the book

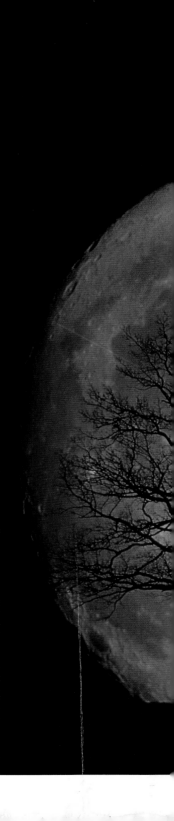

# THE PRACTICAL
# SKYWATCHER'S
# HANDBOOK

700038928306

I open the scuttle at night and see the far-sprinkled systems,
And all I see multiplied as high as I can cipher
  edge but the rim of the farther systems.

Wider and wider they spread, expanding, always expanding,
Outward and outward and forever outward.

*Song of Myself,*
WALT WHITMAN (1819–92), American poet

# CONTENTS

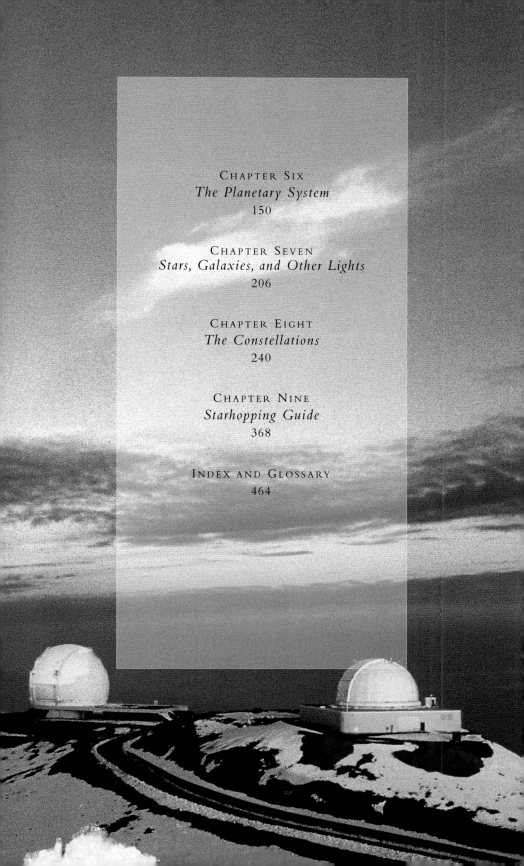

# INTRODUCTION

H ow many backyard naturalists know the sky as well as they know their local trees, birds, or animals? It's surprising how few nature-minded people do more than glance occasionally at the sky overhead. This is a great pity, because they're missing a truly beautiful part of the natural world. But perhaps more important, they are losing sight of the vastly bigger realm of nature out of which planet Earth was born.

In a sense, the universe is merely an extension of our neighborhood. From the streets and parks of our youth, we can go outward still, to the planets, moons, comets, and asteroids that make up our Solar System. The Milky Way galaxy, which is relatively nearby, lies within the Local Group of galaxies, itself a part of an immense confederation called the Local Supercluster. When we move on into the space between the superclusters, we truly leave our home and proceed to explore the great beyond—a universe where the stars grandly exceed all the grains of sand on Earth.

Einstein wrote that the most incomprehensible thing about our universe is that it is so comprehensible. The purpose of *Practical Skywatching* is to bring home that understanding, with the help of superb color photographs and illustrations, as well as engaging text. The authors are experienced in their branches of astronomy and skilled at conveying the latest ideas and discoveries through easily understood explanations.

*Practical Skywatching* is a perfect introduction to the sky and all its marvels. Use it to begin your personal voyage out into space, back through time, to an understanding of the great universe of which you are a part.

*One thing in any case is certain: man is neither the oldest nor the most constant problem that has been posed for human knowledge.*

*The Order of Things,*
MICHEL FOUCAULT (1926–84), French philosopher

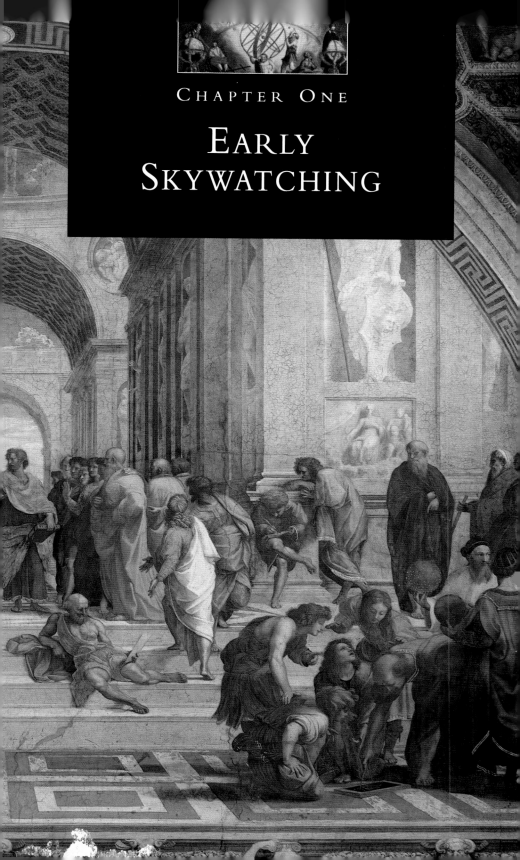

# CHAPTER ONE

# EARLY
# SKYWATCHING

# THE FIRST ASTRONOMERS

*From the time we first connected the motions of
the heavens with the changing of day into night,
we have been fascinated by astronomy.*

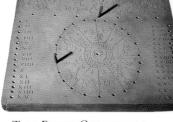

**BABYLONIAN TABLET** *recording
astronomical information, dating from
around 550 BC.*

**THE ROMAN CALENDAR** *featured a
7 day week and 12 months, each of
around 30 days. Days, weeks, and
months can be counted off using
the holes in this stone block.*

In early times, skywatchers saw the heavens in religious terms. The sky was the home of the gods as they controlled day and night, storms, and the great eclipses of the Sun and Moon. In many cultures, personalities were attributed to celestial patterns and objects and were thought to influence people's everyday lives. Astronomer–priests studied the sky, kept records, compiled calendars, and acted as custodians of the legends relating to the sky. There was no clear distinction between astronomy and astrology.

### THE BABYLONIANS
Among the first people known to have kept astronomical records were the Akkadians who lived some 4,500 years ago in the northern part of what later became Babylonia. There is some evidence that their ideas about the motions of the Sun, the Moon, and the planets were later codified by the Babylonians. From their observations, the astronomer–priests of Babylonia could predict the courses of wandering objects in the sky.

### THE FIRST CALENDARS
With a calendar that dates back to 1300 BC, the Chinese are believed to be the earliest calendar makers. The Babylonians and the ancient Egyptians also developed accurate calendars from their studies of the heavens.

Having a calendar meant people had a record of the seasons and thus knew when to plant and harvest their crops. For the Egyptians, whose economy depended on agriculture, a calendar meant

**STONEHENGE,** *in southwestern
England, is one of Europe's most striking
megalithic monuments. Initial building
has been dated at about 1800 BC, and
today's stone circles were constructed in
several stages over the next 400 years.
It is aligned to receive the rays of the
midsummer sunrise, and may have been
used to predict the motions of the Sun
and the Moon, including their eclipses.*

**NUT**, the Egyptian goddess of the sky, is usually depicted arched over the reclining Earth god, Geb, her husband. The Sun god, Ra, was thought to travel across the sky each day in his boat.

**NAVIGATION** by the stars requires precise knowledge of star positions and motions. Here, a medieval navigator is making his measurements on shore.

that they could predict when the Nile's annual flood would irrigate their fields. The priests would wait for the morning that Sirius, the brightest star in the sky, first appeared after being lost in the Sun's glare. They would then use this "heliacal rising" to predict the annual floods.

From earliest times, there has been a 7 day week, to match each quarter phase of the Moon, and the 12 months of the year reflect the Moon's completion of its cycle of phases 12 times a year.

## CHINESE ASTRONOMERS

The Chinese have a long history of dedicated sky-watching, and there is evidence that they recorded a close grouping of the bright planets in about 2500 BC.

In the fourth century BC they produced the earliest

known atlas of comets, the *Book of Silk*. A silk ribbon about 5 feet (1.5 m) long, it illustrates 29 forms of comets and lists the various types of catastrophes they heralded. The work was discovered in a tomb in 1973.

## NAVIGATION

Ever since they first took to the water in boats, sailors have had a close relationship with the heavens. Out of sight of land, they used the positions of the stars to guide them. Polynesian islanders knew how to navigate across the vast stretches of the Pacific between, for example, Tahiti and Hawaii, by plotting their course by the stars. They learned the positions of the stars and the prevailing wind patterns through poetry that they learnt by heart, passed down from generation to generation.

**NAVIGATIONAL INSTRUMENT** of a kind used for thousands of years by the inhabitants of the Marshall Islands in the Pacific.

**STAR CLOAK** worn by a Pawnee Indian and photographed in the nineteenth century. The Pawnee regarded some of the stars as gods and sought their favors through elaborate rituals.

# SHAPES *in the* SKY

*From the dawn of time, skywatchers have projected images of their own invention onto the stars, filling the heavens with gods, animals, and fantastic creatures.*

From earliest times, people have grouped stars into constellations, turning the outlines they formed into figures and animals relating to local customs and creating elaborate stories around them. Sometimes, of course, they were thought of as deities; sometimes they were seen as something more down-to-earth—for instance, the stars of Corvus, the Crow, were seen by the Arabs as a tent. Some star groupings are so striking that throughout the world there are myths associated with them.

A good example is the constellation of Ursa Major, the Great Bear, which is said to be the Callisto of Greek legend, who was transformed by Zeus or, according to another account, his wife Hera, into a bear. This constellation was also seen as a bear by many

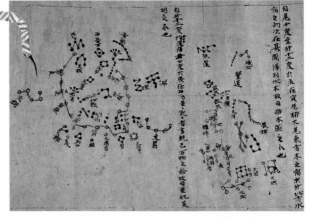

**A SNAKE** *such as this (above left), from an Australian Aboriginal bark painting, is commonly seen in star patterns.*

**CHINESE STAR MAP** *This star map from Tunhuang, China, dates from around AD 940. Except for the simple star maps on astrolabes, this is the oldest known portable star map. Ursa Major is plainly visible near the bottom of the left half of the map.*

civilizations, including several North American Indian tribes. It contains seven bright stars, best known as the Big Dipper or the Plough, and its distinctive shape has been noted by poets such as Homer, Shakespeare, and Tennyson. In Hindu mythology these seven stars represent the homes of the seven great sages; the Egyptians saw them as the thigh of a bull; the ancient Chinese saw them as the masters of the reality of heavenly influences; and Europeans saw them forming a wagon, the Anglo-Saxons associating the wagon with the legendary King Arthur.

**INDIAN STAR CHART** *This beautiful rendition of the constellations, featuring ancient Indian and Islamic patterns, accompanied a horoscope commissioned by an Indian monarch for his son in 1840.*

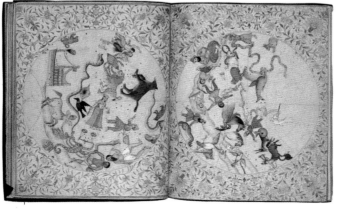

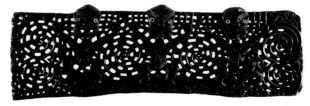

**ZODIACAL SIGNS** *(below) ring the world, which is composed of four basic elements—earth, air, fire, and water— in this encyclopedia dating from the fifteenth century.*

**MAORI SKYLORE** *Depicted in this tribal carving (above) is the separation of Ranginui (the sky father) and Papatuanuku (the Earth mother) by their sons who, living in darkness, longed to experience light and day (represented by the spirals between the figures).*

## PEGASUS IN A NEW FORM

The constellation of Pegasus is supposed to represent a winged horse, but to most people it looks more like a square, so it is generally known as the Great Square. But since it is so prominent in the evening sky during fall in North America, why not call it Pegasus, the Baseball Diamond? There are stars for each of the four bases, one each for pitcher, batter, and catcher, and the nearby Milky Way can represent the fans in the stadium!

## THE ZODIAC AND ASTROLOGY

The zodiac is a belt of 12 constellations that straddle the sky.

Largely within this belt lie the apparent paths of the Sun, the Moon, and the bright planets. The zodiacal signs are the most ancient of the 88 constellations recognized today and act as markers in astrology —the belief that the stars and planets influence human affairs.

Although there is no scientific evidence that the way the planets line up has ever affected anyone, at one time it would have been commonplace for a ruler to consult his astrologer before going into battle, or a businessman to check whether the planets favored a particular deal. Most people today are interested in astrology just for fun, but quite a few still take it seriously. Just remember that if you are worried about the gravitational pull that was exerted by Mars on your character when you were born, you can be sure that the pull exerted by your obstetrician was considerably stronger!

**THE TWELVE APOSTLES** *replace the traditional zodiacal constellations in this star map produced by Julius Schiller, a devout Catholic, for his 1627 book* Coelum Stellatum Christianum.

# CHARTING *the* UNIVERSE

*Ptolemaic theory placed Earth at the center of the universe, and it took until the sixteenth century for Copernicus to dethrone Earth and prove that Aristarchus had been right all along.*

**CLAUDIUS PTOLEMAEUS**, *better known as Ptolemy, was a brilliant and prolific author on a diverse range of topics, and made the first truly scientific maps of the heavens and Earth.*

The ancient Greeks were the first people to try to explain why natural events occurred without reference to supernatural causes and, in so doing, they changed astronomy from a mystical cult into a science. Greek thinkers came to realize that the prevailing astrological beliefs did not correspond with the "rules" of the universe that they were beginning to uncover.

Thales, who lived in the sixth century BC, was the first of the great Greek philosophers and a keen traveler. He brought back to Greece the knowledge and records of the Babylonians and the Egyptians and put forward theories that

**ARISTOTLE**, *whose theory of an Earth-centered universe dominated thinking for 1,800 years, is portrayed in this painting by Rembrandt (1606–69) contemplating a bust of the Greek poet Homer.*

**HIPPARCHUS** *(right), a Greek astronomer and mathematician of the second century BC, produced the first known star catalogue and discovered the precession of the equinoxes (see p. 113).*

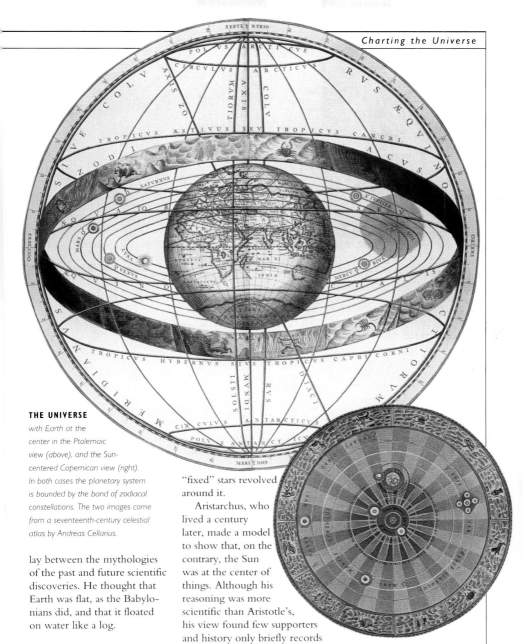

**THE UNIVERSE**
*with Earth at the center in the Ptolemaic view (above), and the Sun-centered Copernican view (right). In both cases the planetary system is bounded by the band of zodiacal constellations. The two images come from a seventeenth-century celestial atlas by Andreas Cellarius.*

lay between the mythologies of the past and future scientific discoveries. He thought that Earth was flat, as the Babylonians did, and that it floated on water like a log.

## CONFLICTING THEORIES

Aristotle, who lived from 384 to 322 BC, was one of the most influential of all the Greek philosophers, and he put forward three experimental proofs to explain that Earth was round. He also thought that Earth formed the center of the universe, and that the Sun, the Moon, the planets, and a sphere containing all the "fixed" stars revolved around it.

Aristarchus, who lived a century later, made a model to show that, on the contrary, the Sun was at the center of things. Although his reasoning was more scientific than Aristotle's, his view found few supporters and history only briefly records his heliocentric ideas.

## PTOLEMY

Ptolemy of Alexandria, another gifted Greek astronomer and scholar, published his *Almagest* in about AD 140. In this remarkable encyclopedia of ancient science he used centuries of Babylonian observations of the motions of the planets to buttress his argument for an Earth-centered universe. His intricate system of circles within circles proved to be a marvelous mathematical method for predicting the motions of the planets.

Ptolemy's "system of the world," known as the Ptolemaic system, embodied ideas that were to rule the world of astronomy for about 1,500 years. His death marks the end of the classical era of astronomy.

19

## COPERNICUS

The process of breaking down Ptolemy's system began with the work of a Polish churchman, Nicolaus Copernicus, who was born in 1473. From his early years, Copernicus thought that the Sun lay at the center of the system of planets and stars, rather than Earth, but he did not complete his formal work on the subject until he was an old man.

In 1543, just before his death, he published his masterpiece *On the Revolution of the Celestial Spheres*, which was to change humanity's view of the cosmos. Not surprisingly, it met with great hostility from the Church, which held that God had created a universe with Earth at its heart.

As a system of mathematical prediction, the Copernican view was no more successful than the Ptolemaic, but two other events made the Copernican revolution

**NICOLAUS COPERNICUS**
*(1473–1543) is the Latin name by which the Polish astronomer Niklas Koppernick is generally known. His position as Canon of the cathedral at Frauenberg allowed him to pursue his studies of mathematics, astronomy, medicine, and theology.*

inevitable: Tycho Brahe's amazingly precise observations of the sky, and Galileo's use of a simple spyglass.

## TYCHO BRAHE

One evening in 1572, Danish astronomer Tycho Brahe saw a brilliant new star on the constellation of Cassiopeia. He was so surprised, it is said, that he asked his neighbor to hit him to make sure he wasn't dreaming. We now know that this new star was a supernova —the explosion of a dying star that is so violent that it outshines the combined light of all the stars in our galaxy. In 1604, a second bright supernova blazed forth in the sky. These discoveries shattered a cornerstone of Ptolemaic thinking:

that the outermost sphere, containing all the stars, was unchanging. It was as though the heavens themselves had joined the Renaissance in Europe.

## KEPLER

Tycho's other major contribution was to pass on to his assistant, Johannes Kepler, the records of his observations

**SUPERNOVA OF 1572** *is marked as 'I' in this engraving from Tycho Brahe's book De Stella Nova.*

**URANIBORG**, *Tycho Brahe's (1546–1601) observatory on the Danish island of Hven from where he made the precise observations that were his greatest contribution to astronomy.*

of the motions of the planets that he had made from 1576 to 1597. These represent a pinnacle of achievement of naked-eye astronomy. Kepler took this data and labored for years to produce his three laws of planetary motion, laws that enabled him to predict the positions of the planets more effectively than Ptolemy or any of his predecessors.

## GALILEO AND THE TELESCOPE

In 1609, an Italian scientist named Galileo Galilei heard of an astonishing invention: by looking through two glass lenses held at a fixed distance from each other and the eye, distant objects could be magnified. Making a simple telescope for himself, he turned it toward the sky, and among his many discoveries he found that the giant planet Jupiter had four moons clearly revolving about it.

Galileo's telescope revealed a miniature version of Copernicus's Solar System, with the moons circling Jupiter in simple, nearly circular orbits.

**JOHANNES KEPLER** *(1571–1630) joined Tycho Brahe's research staff in Prague and succeeded him as Imperial Mathematician when Tycho died.*

When Galileo published his discoveries it was clear that he had abandoned the Ptolemaic system, but it was not until 1616 that the Church quietly warned him to change his views. In 1632, however, he published his book *Dialogue Concerning the Two Great World Systems*, in which three characters discuss the nature of the universe. One of them (with the suggestive name of Simplicio) supports Ptolemaic theory. The Pope, believing that Galileo was ridiculing him, put Galileo at the mercy of the Holy Office of the Inquisition and he was accused of heresy.

Forced to "abandon the false opinion that the Sun is the center of the world," Galileo lived out his days under house arrest. But the Inquisition could do nothing to stem the influence of his discoveries, which changed the face of astronomy. The

Catholic Church reconsidered the case and absolved Galileo from any wrongdoing in 1992. In 1989, a spacecraft named after him was launched to study Jupiter and its satellites—the moons that he was the first to see.

**GALILEO'S** *first telescopes were not as effective as an inexpensive telescope of today, but they changed our view of the universe.*

**GALILEO'S TRIAL** *in Rome in 1633 was marked by special indulgence and he was never jailed. A guilty verdict still resulted for having "held and taught" the Copernican doctrine.*

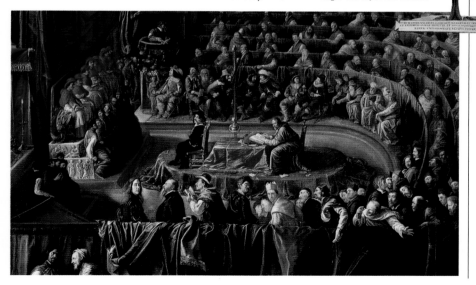

# THE BIRTH *of* ASTROPHYSICS

*The heavens have been a source of wonder since time immemorial. With the development of new tools, they began to give up their secrets at an unprecedented rate.*

Just as Galileo's telescope prompted renewed enthusiasm for observational astronomy in the early seventeenth century, the development of the spectrometer two centuries later spawned a vigorous new science that focused on the dynamics, chemical properties, and evolution of celestial bodies.

The study of astrophysics, as it was called, began modestly in Munich in 1814, when physicist and telescope-maker Joseph von Fraunhofer passed the light of the Sun and stars through a spectrometer—essentially a prism that broke the light down into the colors of the spectrum. He noticed that the spectra had hundreds of dark lines, like fence slats, running across them. Although Fraunhofer realized that the lines were important, their meaning eluded him and other scientists for 45 years.

## READING THE SIGNALS

In 1859, German physicist Gustav Kirchhoff showed that each chemical element pro - duced characteristic patterns of lines in the spectrum of the Sun. He correctly interpreted the bright spectral lines as the emission of light by particular chemical elements, and the dark lines as the absorption of light by the same elements. This meant spectral analysis

**SECRETS REVEALED**
*Gustav Kirchhoff (right) correctly interpreted spectroscopic data to reveal the composition of the Sun, while Edwin Hubble's work on galactic motions led to the Big Bang theory (left).*

could reveal, for instance, that the Sun contains sodium.

By 1863, spectroscopy had advanced enough to allow English astronomer William Huggins to publish lists of stellar spectral lines. He confirmed that the Great Nebula in Orion was gaseous, and detected hydrogen, which had already been noted in the Sun, in the spectrum of a nova—the first evidence that all stars contained hydrogen. Huggins also used spectroscopy to determine the speed and direction of the star Sirius's movement (see p. 35).

Progress in photography brought a further leap in astronomical research. Until the early 1800s, cameras were primitive pinhole types, but by the 1840s, lenses, mirrors, and better film emulsions were available. Successful photo - graphs of the Moon galvanized astronomers into photographing the Sun, which became a routine matter by the 1870s.

## DISTANT NEIGHBORS

As powerful as spectroscopy and photography were, they could not help astronomers work out the dis - tances to the stars. Industrial advances in the early 1800s, however, made possible more accurate instrumentation and optics. The German astronomer Friedrich Bessel was

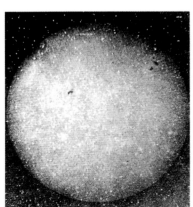

**SUNSPOTS** *can be seen on this daguerreotype (a kind of early photograph) of the Sun, taken in 1845 by Foucault and Fizeau.*

the first to capitalize on these advances. In 1838, by measuring the slight positional shift of the star 61 Cygni with respect to more distant background stars—an effect known as parallax—he estimated a distance for that star of 65 trillion miles (100 trillion km), or about 11 light-years.

Within two years, the distances to two other stars, Alpha Centauri and Vega, were determined using the parallax effect. In the decades that followed, many more distances were calculated and scientists realized that the universe was far more vast than once thought. No one, however, was prepared for how big it would "grow."

## THE INCREDIBLE EXPANDING UNIVERSE

The first decade of the twentieth century saw several discoveries that would profoundly affect astrophysics. In 1905, Danish astronomer Ejnar Hertzsprung and American astronomer Henry Norris Russell investigated the relationship between a star's color and temperature and its brightness. Russell's work on the evolution of stars led to the notion of red giants and white dwarfs. Charts that plot a star's color against its brightness are now called Hertzsprung-Russell, or H-R, diagrams in their honor (see p. 208).

In 1912, Henrietta Leavitt at the Harvard Observatory discerned a relationship between the period of brightness fluctuations and the true brightness of a class of stars called Cepheid variables (see p. 221). She noticed that Cepheids with slow cycles were brighter than those with fast cycles. Using Leavitt's period–luminosity law, Harlow

**THE MILKY WAY** *(illustrated above) is a spiral galaxy, with a short bar (yellow-white) at the center. The spiral pattern is 100,000 light-years across.*

Shapley calculated the distances to Cepheids in globular clusters. His results proved that the Milky Way was far larger than previously thought and that the Sun was not at its center.

Still under debate were the distances to, and the nature of, the "spiral nebulas," which most astronomers believed belonged to our galaxy. A series of observations, begun in 1912, showed that most spiral nebulas were receding from us. In 1924, American astronomer Edwin Hubble, scrutinizing the Great "Nebula" in Andromeda (M31), was able not only to make out individual stars in the nebula, but to determine that some of them were Cepheid variables. He then calculated that the nebula lay almost a million light-years away—a vast underestimate, as it turns out, but unquestionably beyond the bounds of our own galaxy.

But the biggest surprise came when Hubble used spectroscopy to analyze the velocities of galaxies, finding that the farther away a galaxy lay, the faster it was receding. This could best be explained, Hubble concluded, by presuming that the universe was expanding like a balloon. As time progresses, the universe's

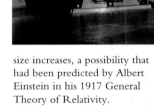

**ASTRONOMER** *Edwin Hubble (below) revised our ideas of the universe's scale.*

size increases, a possibility that had been predicted by Albert Einstein in his 1917 General Theory of Relativity.

## AGE-OLD QUESTIONS

In an expanding universe, astronomers realized, it should be possible to determine how and when the expansion began. Hubble's initial calculations indicated that the universe was only two billion years old, younger than the age of Earth as determined by geologists. Once discrepancies in Hubble's calculations had been adjusted, a more geologically agreeable age of six billion years emerged.

Astronomers now believe the universe probably began 13.7 billion years ago in an immense "explosion" of matter and energy known as the Big Bang. Since then, the expansion of space itself has carried the galaxies away from each other, in accord with Hubble's law of expansion.

# THE BIG BANG THEORY

*The "big questions" about the universe are the hardest to answer, but recent years have produced remarkable observations to match some of the amazing theories.*

**M**ost astronomers now accept that the universe can be broadly explained in terms of the Big Bang theory. This theory says that all matter and space were once compressed into a state of incredibly high temperature and pressure, from which the universe has been expanding ever since.

**THE GRAND DESIGN** *God, the "Great Architect" (above), in an illumination from a thirteenth-century French Bible.*

With the help of computers we can hypothesize that, just a fraction of a second after the Big Bang, the universe was a hot, seething mass of radiation and exotic particles. As it expanded, it cooled and more familiar sorts of matter formed. Eventually hydrogen and helium were produced, which are the main constituents of the universe today.

## MICROWAVE BACK-GROUND RADIATION

In 1965, Arno Penzias and Robert Wilson of Bell Telephone Labs were tracking sources of static in the receiver system of a radio telescope. Even after they had tightened the telescope's connections and removed some nesting pigeons from the antenna, there was one source of static that they were unable to eliminate. This turned out to be primordial background radiation coming from the entire sky at an extremely cold −455 degrees Fahrenheit (−270° C or −2.7° C above absolute zero)—none other than the echo of the Big Bang itself. The discovery won Penzias and Wilson the 1978 Nobel Prize in Physics.

In April 1992, NASA investigators revealed that their Cosmic Background Explorer (COBE) satellite telescope had detected small variations in the temperature of the background radiation. More detailed views by the WMAP satellite and other instruments reveal the seeds of the universe's later structure as early as 300,000 years after the Big Bang. From these small irregularities grew the mighty clusters of galaxies, separated by vast voids, which we see today.

**THE BIG BANG?** *How can we graphically portray the Big Bang? The flash of explosion shown here is clearly not a suitable representation, since we could never be outside the beginning of the entire universe. The Big Bang was an "explosion" of space, not one in space.*

CREATION OF THE UNIVERSE *The Big Bang is widely accepted as having been the origin of all the planets, stars, and galaxies that we see today.*

## GRAVITY

What will the universe be like in the distant future? Will it expand forever or will it slow and stop for a split second before starting to contract, ending in a Big Crunch and perhaps another Big Bang?

Until 1988, we thought the single key to this question was gravity. The gravitational effect of visible matter and unseen dark matter is important, acting to slow the expansion. But we now know that mysterious dark energy is also crucial.

## DARK MATTER

It is clear that more mass is present in the outer parts of our own galaxy than we can see. The stars orbit the galactic center too fast for it to be otherwise. The largest of the nearby clusters of galaxies, the cluster in Virgo, must also have more mass than we can see or gravity would be insufficient to keep the member galaxies from flying apart from each other.

Where is this dark matter? Hidden in black holes, perhaps, or in the tiny mass that sub-atomic particles called neutrinos may possess? Somewhere else? We don't know.

In 1993, two groups of astronomers independently reported a new type of invisible object which may form part of the answer—a Massive Compact Halo Object (MACHO). Far away in the outskirts, or halo, of our galaxy, MACHOs are likely to be large, non-shining stars or planets.

Whether or not MACHOs exist, astronomers have not yet detected nearly enough dark matter to conclude that someday the universe will stop expanding.

## DARK ENERGY

In 1998, astronomers studying extremely distant supernovae reported that the expansion of the universe was not slowing down with time. In fact, it is accelerating! Astronomers were forced to rethink their view of the universe. How can this be?

The anti-gravity effect that drives the acceleration is due to mysterious dark energy. Exactly what this is remains unclear, but dark energy is believed to make up almost three-quarters of the contents of the universe.

In an endlessly expanding universe, the galaxies would eventually be filled with ancient stars—black dwarfs that do not shine. After a time vastly more than the current age of the universe, the black dwarfs would decay, leaving nothing behind them but a mist of particles.

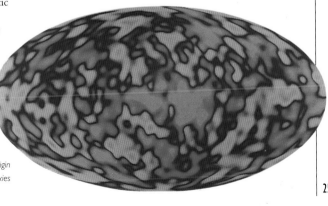

RIPPLES IN THE BACK-GROUND OF SPACE *The COBE satellite produced this image of the microwave back-ground radiation—the glow of the Big Bang spread across the sky. The pink and blue ripples represent the origin of the structure we see today as galaxies and clusters of galaxies.*

# THE EMPTINESS *of* SPACE

*By looking at the following images and comparing their relative sizes,*
*we can step out into space and begin to conceive of*
*the sheer vastness of our universe.*

**LANDSAT VIEW OF WASHINGTON DC** *The Potomac River can be seen here dividing Maryland from Virginia. The city areas appear blue, while vegetation has been colored pink. To convey an idea of the scale involved, the width of this image covers approximately 30 miles (50 km) in each direction.*

**EARTH** *as photographed by the European meteorological satellite, Meteosat. The view of Washington DC (above, left) would occupy less than ¹⁄₆₄ inch (0.3 mm) of this image of our 7,927 mile (12,760 km) diameter globe. The color in this picture has been added to simulate Earth's natural appearance. The first people to be far enough away to see the whole Earth from space were the astronauts of the Apollo 8 mission, in 1968.*

**EARTH AND MOON**, *as viewed by the Galileo spacecraft on its way to Jupiter. On this scale, the Moon would be 60 inches (1.5 m) behind Earth. Other planets have moons, but only Earth and Pluto have moons that are a large fraction of the planet's size. Contrast and color have been enhanced, especially for the Moon, to improve visibility.*

**THE SOLAR SYSTEM** *The Sun and its family of planets are featured in this portrait (orbits and planetary sizes not to scale). The blue ring near the center is Earth's orbit, 93 million miles (150 million km) from the Sun in the center. Pluto's angled orbit, crossing Neptune's, is about 40 times larger, while more distant Kuiper Belt objects are not shown.*

**NEAREST STARS** *(below) These stars, all lying within 20 light-years of the Sun, make up our solar neighborhood. On the scale of this illustration, the stars would really be infinitesimal points the size of atoms.*

*I Proxima Centauri, Alpha Centauri A, Alpha Centauri B; **2** Barnard's Star; **3** Wolf 359; **4** Sirius A, Sirius B; **5** Epsilon Eridani; **6** Epsilon Indi; **7** 61 Cygni A, 61 Cygni B; **8** Tau Ceti; **9** Procyon A, Procyon B.*

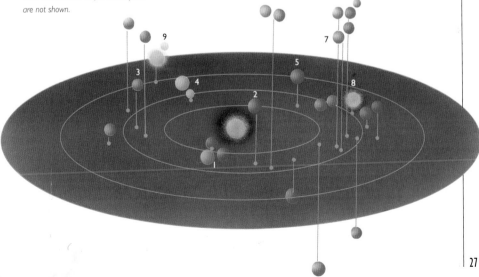

> *Put three grains of sand inside a vast cathedral, and the cathedral will be more closely packed with sand than space is with stars.*
>
> SIR JAMES JEANS (1877–1946),
> English astronomer

**THE MILKY WAY GALAXY,** seen here as a glowing maelstrom of some 200 billion suns, is about 150,000 light-years across. Our Sun is no more than a minute speck in one of the galaxy's great spiral arms. The solar neighborhood shown in the previous image would be a tiny patch just ¹/₁,₀₀₀ of an inch (0.03 mm) across. At the speed of light, it would take 28,000 years to get from the local neighborhood to the center of the galaxy.

## DISTANCE SCALES

While a mile or a kilometer is a suitable unit of distance here on Earth, when we say that the Andromeda Galaxy (M31), the nearest large galaxy to the Milky Way, is 17 quintillion miles (27 quintillion km) away—that's 17,000,000,000,000,000,000 —the number is so large that it becomes meaningless.

### ASTRONOMICAL UNITS

A useful measure in our Solar System is the Astronomical Unit (AU), the average distance between the Earth and Sun—approximately 93 million miles (150 million km). Mercury is ⅓ of an AU from the Sun, while Pluto averages 40 AU from the Sun.

### LIGHT-YEARS

For measuring distances further afield, we use the light-year—the distance light travels in a vacuum in a year. Traveling at about 186,000 miles (300,000 km) a second, light can travel the equivalent of seven times around the Earth in a single second. A light-year is about 6 trillion miles (10 trillion km).

### PARSECS

Astronomers also use a parsec, equivalent to about 3.3 light-years or 206,000 AU, as a unit of distance. If an object were only one parsec away, it would have an annual parallax (see below) of one second of arc (about ¹/₁,₈₀₀ the diameter of the Moon).

### PARALLAX

In astronomy, the term parallax refers to the apparent displacement of a nearby star against the background of more distant stars. This is not caused by the motion of the star itself, but by the Earth's motion as we observe the sky from different parts of our orbit around the Sun.

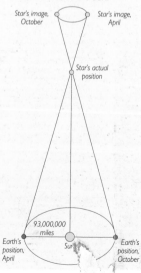

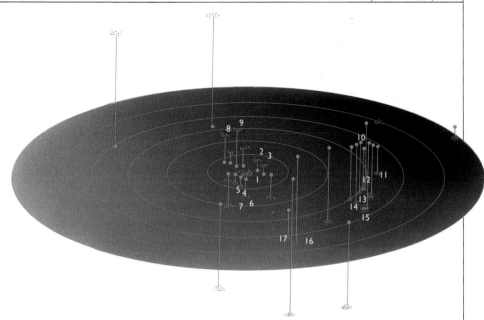

**THE LOCAL GROUP** *is a small cluster of some 40 galaxies within 3 million light-years of the Milky Way. The Milky Way and Andromeda are the dominant galaxies within it. The galaxies are shown at about double correct size relative to their distance from one another.*

*1 Milky Way; 2 Draco; 3 Ursa Minor; 4 Small Magellanic Cloud; 5 Large Magellanic Cloud; 6 Sculptor; 7 Fornax; 8 Leo I; 9 Leo II; 10 NGC 185; 11 NGC 147; 12 NGC 205; 13 M32; 14 Andromeda (M31); 15 M33; 16 IC 1613; 17 NGC 6822*

**UNIVERSE OF GALAXIES** *This view from the Hubble Space Telescope shows a cluster of galaxies of all types, one of thousands of such clusters scattered across the sky. From this far out, our entire Milky Way Galaxy is hardly more than a small blob of light.*

# CHAPTER TWO
# MODERN ASTRONOMY

*Nothing puzzles me more than time and space;*
*and yet nothing troubles me less.*

Letter to Thomas Manning, 2 January 1810,
CHARLES LAMB (1775–1834), English essayist

# OUR VIEW
## *of the* UNIVERSE

*The technological progress made in just one century has opened wide our windows on the universe.*

For millennia, humans have gazed at the stars and planets and wondered how far away they lay, what they were made of, and how they got there. Skywatchers have charted and analyzed these celestial bodies—first with quadrants and armillaries, then using telescopes, spectroscopes, and spacecraft. The Moon and planets have been mapped, and millions of star clusters, nebulas, and galaxies have been cataloged.

Today, huge telescopes and sensitive satellites examine the stars, and probes have been sent to some planets, but the goal remains the same: to discover what the universe is and why it is the way it is.

### THE STARS FROM EARTH
Astronomy has developed exponentially since the mid-1800s, when the universe was a fairly small and uncomplicated place, and the view of the cosmos was Sun-centric. The planets were simply little-known cousins of Earth. The

stars were thought to be immutable, and galaxies did not exist. Today, we know the planets as separate worlds, each with a unique geology and meteorology. Each star, too, displays a complex evolutionary history, and galaxies are recognized as giant stellar systems independent of our own.

In the 1860s, the development of spectroscopy—the breaking down of light into its constituent colors—started to expand astronomy from the

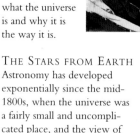

**OBSERVING** *in 1925 at the Royal Greenwich Observatory (left) was still mostly concerned with determining the positions of objects.*

study of the positions of celestial objects to the study of their physical properties. For the first time, astronomers could extract information from starlight. This allowed them to classify the stars and chart them from birth to death.

Giant reflecting telescopes at California's Mount Wilson and Palomar observatories were constructed in 1917 and 1947, and for decades they were the largest optical telescopes in the world. Now, each of the Keck telescopes on Mauna Kea, Hawaii, has a mirror 400 inches (10 m) in diameter. Grander instruments up to 1,650 inches (42 m) across are planned.

Just as importantly, photographic cameras have now

**TRUTH AND FICTION** *In the nineteenth century, Jules Verne imagined this train of capsules (top) on their way to the Moon. A hundred years later, astronauts were actually floating in space (right).*

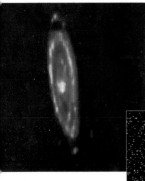

**NON-VISUAL RAYS** *Infrared image of Andromeda Galaxy (left). Cygnus X-1 (below), photographed by ROSAT, is a powerful X-ray source. It is a binary star system that probably hides a black hole.*

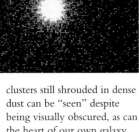

been replaced by electronic detectors called charge-coupled devices (see p. 94). CCDs capture images quickly, and are sensitive over a wide range of visible light.

## INVISIBLE SIGNALS

Visible light waves, however, reveal only a fraction of the big picture. Just after World War II, developments in radio technology enabled scientists to measure energetic radio waves emitted by the Milky Way, other galaxies, and nebulas. Advances in electronics and solid-state physics in the 1960s and 1970s made radio astronomy one of the most powerful means of surveying the cosmos. It has been especially effective in probing distant galaxies, particularly their active centers, which are thought to harbor supermassive black holes.

Astronomers who were eager to explore the universe at other invisible wavelengths (see p. 37) had to get above the distorting effects of Earth's atmosphere. Since the late 1950s, Earth-orbiting satellites have allowed them to do this.

On the low-energy end, infrared wavelengths reveal the warm-to-cold universe. With infrared cameras, star

clusters still shrouded in dense dust can be "seen" despite being visually obscured, as can the heart of our own galaxy. On the more energetic end of the spectrum, ultraviolet and X-ray wavelengths show the hot, active universe. Here, astronomers study novas and supernovas, quasars and the cores of active galaxies, and the disks of gas around black holes and neutron stars.

All-sky surveys beginning in the 1970s and 1980s revealed thousands of new objects, including new classes of infrared-bright galaxies, high-energy binary stars, black-hole candidates, and enigmatic objects that emit bursts of gamma-ray energy.

Nothing better illustrates how rapidly the frontiers are advancing than the stream of discoveries from the Hubble Space Telescope (see p. 40). Its detailed imaging and spectroscopic observations have been of incalculable value.

The last fifty years have also seen amazing leaps in the study of the universe from interplanetary travel. The dream of venturing into space is an old one. As far back as AD 160, a Greek named Lucian of Samosata wrote a story about traveling to the Moon. In the early seventeenth century, a journey to the Moon and speculations about weightlessness figured prominently in a story called *Somnium* ("Dream") by the German astronomer Johannes Kepler. Imagine what such visionaries might say about the Apollo missions to the Moon, the rovers on Mars, or the New Horizons mission to remote Pluto and beyond.

**THE HUBBLE SPACE TELESCOPE**
*looked deep into space to produce this image—each speck of light is a galaxy.*

# FROM *the* GROUND

*Our view of space from the ground is improving all the time as telescopes get bigger and more sophisticated, and computers provide new ways to gather and interpret astronomical data.*

Visible light is radiation that the human eye can see. Compared with the total range of the electromagnetic spectrum, visible light constitutes a very narrow span (see p. 37). Yet much of what we know about the universe has come to us through this slim window.

The faintest star visible to the naked eye is about 6th magnitude, but binoculars and telescopes greatly extend this reach. Even a 6 inch (150 mm) scope gathers about 500 times more light than the human eye can—no wonder astronomers always want larger telescopes.

## BIGGER AND BETTER

One of the first giant telescopes was built in 1845, by William Parsons on his Irish estate (see Box p. 97). Its mirror spanned 72 inches (1.8 m), and until 1908, it was the largest telescope in the world. In 1917, the 100 inch (2.5 m) Hooker Telescope was built by Ameri - can astronomer George Hale on Mount Wilson, California. Hale started work on an even bigger telescope, but died before it was finished in 1948. The 200 inch (5 m) Hale Telescope on Mount Palomar, California, was the world's largest for nearly 30 years.

In 1977, the Russians completed the 240 inch (6 m) Bolshoi Telescope in the Caucasus Mountains. Although larger than the Hale, design flaws prevented it realizing its full potential. In 1992, the 400 inch (10 m) Keck I Telescope on Mauna Kea, Hawaii, became the world's biggest. It was followed by Keck II (also 400 inches) in 1995. Such large telescope mirrors are very difficult to make in one piece, so each was created from 36 hexagonal segments.

Other telescope makers used thin monolithic mirrors in their giant new telescopes. They rely on computers to keep the mirror in the right shape to produce sharp images. Several telescopes in the 320 to 400 inch (8 to 10 m) range have been built. The multi-national Gemini twins, for example, are 328 inches (8.2 m) in diameter—one in Hawaii and one in Chile.

Already even larger ground-based telescopes are imagined, ranging up to the massive disk of the Overwhelming Large (OWL) telescope which is 330 feet (100 m) in diameter.

The resolution capabilities of ground-based telescopes have been limited by the effects

**BIG SCOPES** *Getting ready to photograph with the Hale Telescope (top). The Anglo-Australian Telescope (left) is quite different to new telescopes such as the VLT, Paranal Observatory, Chile (below).*

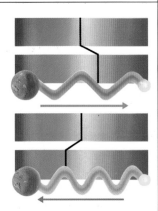

**SPECTRAL SHIFT** *William Huggins (left) based much of his work on the idea that if a star's spectral lines shift toward the red end of the spectrum (right, top), the star is moving away from Earth. If it shifts toward the blue end (right, bottom), the star is approaching.*

of atmospheric seeing—the disturbing effects of Earth's atmosphere. This, however, is becoming less of a problem. Computer-controlled adaptive optics systems can rapidly adjust the shape of a telescope's optics to reduce the turbulent effects of Earth's atmosphere and produce sharper images. This technique is being applied to all the new generations of large telescopes.

Even better resolution is promised by linking telescopes to form "interferometers." This technique has been pioneered by specialized facilities in several countries but is now being extended to some of the giant new telescopes: the twin 336 inch (8.4 m) mirrors of the Large Binocular Telescope (LBT) in Arizona, the Keck telescopes in Hawaii and the four 328 inch (8.2 m) telescopes of the European Southern Observatory (ESO) Very Large Telescope (VLT) in Chile.

## CCD TECHNOLOGY

In the 1970s, a light-sensitive electronic charge-coupled device, known as the CCD, increased the telescope's light-gathering capability a hundredfold and revolutionized astronomical imaging (see p. 94). Today, CCDs are the tools of choice for amateurs and professionals alike. They are used in all aspects of astronomical

research, from studying variations in starlight to imaging the centers of active galaxies. They are far more efficient then photographic film at gathering light and allow telescopes to peer deeper into space.

## SPECTROSCOPY

Telescopes and CCDs may gather and record starlight, but only a spectrograph can extract its secrets. A spectrograph separates white light into its component wavelengths— a spectrum—crossed by numerous dark and bright spectral lines. It then records these lines as a kind of bar code of the object's physical properties. Spectroscopy is the basis of modern astrophysics— most of what we know about the chemical composition, temperature, and pressure in any astronomical object is encoded in its spectral lines.

One of spectroscopy's most useful applications is in determining an object's motion toward or away from Earth. This was discovered in 1868 by William Huggins, who found that dark lines in the star Sirius were shifted toward the red end of the spectrum, implying a decrease in frequency. Based on work by Austrian physicist Christian Doppler, who theorized that light from a source exhibits a change in frequency if it is moving toward or away from the

observer, Huggins determined that Sirius was moving away from the Sun at about 25 miles per second (40 km/s). Had the spectral lines been blueshifted, increasing in frequency, it would have indicated that Sirius was approaching Earth.

Edwin Hubble later applied this method to galaxies, demonstrating that most galaxies are redshifted and their recessional velocities increase with distance (see p. 23).

Fiber-optic technology is improving the efficiency of spectroscopy. Many optical fibers can be arranged so that each receives light simultaneously from a different star or galaxy. The opposite ends of the fibers can then be aligned so that they feed into the spectograph at once, allowing hundreds of spectra to be gathered at the same time.

**DOUBLE VISION** *The Sydney University Stellar Interferometer (SUSI), Australia, uses pairs of mirrors up to 2,110 feet (640 m) apart to mimic a large telescope.*

35

# OTHER WINDOWS
## *on the* UNIVERSE

*There is a wealth of information about the physical universe
that lies beyond the realm of visible light.*

**THE VERY LARGE ARRAY** *Telescope
in New Mexico consists of 27 radio
dishes linked to a central receiver.*

The physical universe is represented by more than our eyes are able to see. Radiation spanning the full electromagnetic spectrum (see diagram, facing page) can tell us a great deal about celestial objects.

Signals from space—such as visible light, X-rays, and ultraviolet, gamma, and infrared rays—all reach Earth as waves, and the various "wavelengths" indicate the distances between the crests of the waves.

The features in these other regions of the spectrum are picked up by telescopes designed specifically to "see" them. The air above us absorbs and scatters particular wavelengths; to pick these up, some telescopes are placed beyond Earth's atmosphere.

### THE LOW END

The low-frequency, low-energy part of the spectrum is the domain of infrared, milli - meter, and radio radiation.

Infrared rays are the radia - tion we feel as heat. Most infrared sources radiate at temperatures between −430 and 1,800 degrees Fahrenheit (−260° and 1,000° C). Infra - red observations can pierce dense dust clouds to reveal young stars that are invisible at optical wavelengths,

accentuate dusty disks around stars, and even pinpoint remote dusty galaxies. Some near- and mid-infrared observations can be ground-based, but longer-wavelength infrared observations must be conducted from balloons, rockets, or satellites.

**Millimeter Waves** Millimeter and submillimeter telescopes replace optical lenses and mirrors with radio "mirrors"—radio dishes and receivers specifically tuned to receive short-wavelength radio signals. One of the largest millimeter-wave telescopes, with a dish 50 feet (15 m) in diameter, was the James Clerk Maxwell Telescope (JCMT) on Mauna Kea, Hawaii. The array of 66 44-foot (12-m) and 23-foot (7-m) millimeter and

submillimeter telescopes in Chile, known as ALMA, is set to revolutionize this branch of astronomy. The millimeter window is ideal for observing giant molecular clouds in which stars are likely to form.

**Radio Signals** Longer-wavelength radio radiation reveals the processes occurring in gaseous nebulas, pulsars (spinning neutron stars), supernova remnants, and the active cores of distant galaxies.

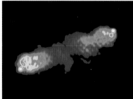

**SINGLE TELESCOPES** *like
the Green Bank Radio Telescope
(above right) and the JCMT
Millimeter Telescope (right)
have poor resolution, but several
together can produce images of
galaxies like Cygnus A (above).*

## KARL JANSKY'S STATIC FROM SPACE

In 1928, just a year after graduating from the University of Wisconsin, radio engineer Karl Jansky (1905–50) accepted a job with Bell Telephone Laboratories. One of his first assignments was to track down the source of static that interfered with radio reception and ship-to-shore communications. Although much of the static could be attributed to storms, aircraft, or local electrical equipment, Jansky detected a fainter type of background static that appeared to rise and set with the Sun. As he continued to track the radio noise with his directional radio aerial system (above), he noted that the noise moved ahead of the Sun by about four minutes each day.

Realizing that this is the same amount that the stars gain on the Sun each day, Janksy reasoned that the source of the static lay beyond the Solar System. By 1932, he had determined that its source lay in the constellation Sagittarius, toward the center of the Milky Way.

Although Jansky did not pursue the science of celestial static any further, his discovery marked the birth of radio astronomy, which by the 1970s, had taken its place alongside optical astronomy as a highly effective means of studying the universe.

A radio telescope consists of an antenna, or antenna-array, connected to a receiver. The antenna may be as simple as a dipole-type antenna, which looks something like a TV antenna. More often, though, it is a parabolic reflector dish, or an array of dishes. The radio waves are focused and collected by a secondary antenna, known as a feed. They are then transferred to a receiver for amplification and analysis. As with an optical telescope, the sensitivity of the radio tele – scope increases with the surface area of the antenna.

When two or more radio telescopes are linked so that their signals are fed to a common receiver, they are jointly termed an interferometer. The greater the distance between the radio telescopes, called the baseline, the finer the detail that can be resolved. The Very Large Array, near Socorro in New Mexico, has baselines up to 22 miles (36 km) long, while the Very Long Baseline Array consists of 10 dishes spread over about 5,000 miles (8,000 km)—from Hawaii across the USA to the Virgin Islands in the Caribbean Sea. The next step will be the SKA—a billion dollar multi-national project to build a radio telescope with 3,300 square feet (1 square km) of collecting area.

### THE HIGH END

At the high-frequency, high-energy end of the spectrum, ultraviolet, X-ray, and gamma-ray radiation unveil the physical and chemical properties of objects that are incredibly hot and energetic. These wavelength regimes are particularly useful for studying the hot gas swirling around black holes and neutron stars; galactic haloes; interactions between cosmic rays and in - terstellar gas; and the rarefied outer atmospheres of stars.

### ELECTROMAGNETIC SPECTRUM

*This illustration shows the range of radiation wavelengths and some of the instruments designed to detect them. Only optical, radio, and some infrared rays can pass through Earth's atmosphere; satellite-based telescopes are used for detecting shorter wavelengths.*

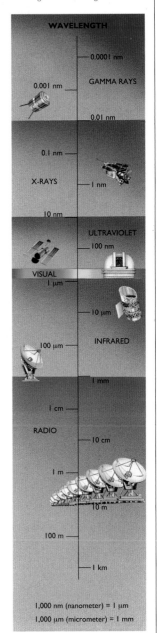

WAVELENGTH

0.0001 nm

0.001 nm — GAMMA RAYS

0.01 nm

0.1 nm

X-RAYS — 1 nm

10 nm

ULTRAVIOLET — 100 nm

VISUAL
1 μm

10 μm

100 μm — INFRARED

1 mm

1 cm

RADIO — 10 cm

1 m

10 m

100 m

1 km

1,000 nm (nanometer) = 1 μm
1,000 μm (micrometer) = 1 mm

# AROUND *the* EARTH

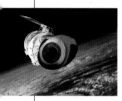

*Infrared, ultraviolet, X-ray, and gamma-ray signals have always been there, but until the late 1950s most of them were hidden behind the curtain of Earth's atmosphere.*

Observing the universe from space became a real possibility on 4 October 1957, when Russia launched Sputnik 1, the first artificial satellite. The United States lofted Explorer 1 into orbit on 31 January 1958, and between then and 1984, 65 Explorer-class satellites were successfully launched. These missions studied such phenomena as Earth's magnetic fields, solar wind (see p. 125), and ultraviolet (UV) radiation from stars and galaxies.

## GETTING INTO SPACE

Overcoming Earth's gravity and putting a payload into space requires a rocket to accelerate its cargo to about 17,500 miles per hour (28,000 km/h). The rocket's thrust must be greater than its weight, so even a small cargo uses a lot of fuel. One way of reducing the fuel required is to launch the rocket in an easterly direction near the Equator. Earth's rotation acts like a sling, providing extra lift. Cape Canaveral, Florida, is an ideal location, boosting a rocket's velocity by almost 900 miles per hour (1,450 km/h).

With enough speed, a rocket can place satellites

or other spacecraft into one of three types of orbit: equatorial, polar, or geostationary.

Equatorial satellites usually orbit at low altitude—about 100 miles (160 km) above Earth—and, because they are launched eastward, they travel from west to east.

Polar-orbiting satellites pass over the entire Earth in high-altitude orbits at a minimum of about 1,000 miles (1,600 km). Most of these scan geological or weather phenomena, but some are dedicated to military reconnaissance. They travel north to south, or south to north. From Earth, they appear to move more slowly because of their greater altitude.

Geostationary, or geosyn-chronous, satellites circle Earth once a day at an altitude of 22,000 miles (36,000 km). Because their speed matches that of Earth's rotation, they remain over a fixed point. This makes them ideal for gathering weather-forecasting data and for communications such as television transmissions.

## WAVELENGTH WINDOWS

Earth's atmosphere often interferes with the reception of radiation, so once receivers could be positioned beyond the atmosphere, a new era in the study of the universe began.

The first cosmic X-rays were picked up in 1962 by a

**A PRE-LAUNCH CHECK** *of the ROSAT X-ray and extreme ultraviolet telescope (right). A model of IRAS, the Infrared Astronomical Satellite (top).*

**SPACE EYE** *The Astro Observatory aboard the shuttle (below). A burst of gamma radiation recorded by the Compton Observatory (bottom). Repairing Syncom, a communications satellite (left).*

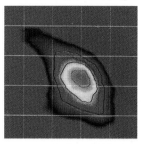

detector aboard a small rocket. Their source was later found to be Scorpius X-1, a binary-star system that is the brightest X-ray source in the sky.

Satellites, including Japan's ASCA, the European Space Agency's (ESA) EXOSAT, and the German–British ROSAT, have provided intriguing insights into X-ray sources, such as black holes, neutron stars, and supernova remnants. This work continues using NASA's Chandra X-ray observatory and ESA's XMM-Newton telescope.

UV astronomy began in earnest in December 1968 with the Orbiting Astronomical Observatory-2, which returned data covering one-sixth of the sky. From this, a catalog of bright UV sources was compiled. Subsequent missions glimpsed sources of extreme UV radiation—hot stars, supernovas, the active cores of galaxies, and quasars (the intensely energetic centers of far-off galaxies)—and produced more than 90,000 UV spectra.

In December 1990, an instrument package aboard the space shuttle *Columbia* provided spectroscopic observations in the far- and extreme-UV wavebands. The Astro mission also made images of UV sources and assessed how UV radiation is scattered by cosmic dust and strong magnetic fields. The Extreme Ultraviolet Explorer

(EUVE) was launched in June 1992 and surveyed the entire sky until 2001, studying white and red dwarf stars, eruptive variable stars, and thin inter-stellar gas. More recently, the GALEX ultraviolet telescope has surveyed the history of star formation in galaxies across the visible universe.

EXPLORING FURTHER
Seeking and investigating sources of gamma rays, the most energetic form of radiation, was the objective of COS-B, launched in 1975 by the European Space Agency. The mission ran until 1982. From 1991 to 2000, the Compton Gamma Ray Observatory set out to explore further, particularly to identify sources of mysterious gamma-ray bursts. The multi-wavelength SWIFT satellite now patrols the sky, accompanied by the Fermi Gamma-ray Space Telescope.

The Infrared Astronomical Satellite (IRAS), a joint mission of the United States, Britain, and the Netherlands, discovered more than 200,000 infrared sources during an all-sky survey of infrared objects in 1983. The Spitzer Space Telescope, launched in 2003, has set the standard for infra-red imaging in more recent times. For all infrared missions, observations in space are limited by the supply of coolant to chill the infrared camera.

THE BIG PICTURE
One of the most significant observations of the twentieth century was made in the early 1990s by the Cosmic Background Explorer. COBE measured the feeble background glow thought to be remnant radiation from the Big Bang that created the universe.

In 1992, careful analysis of COBE data revealed slight temperature fluctuations in the background glow, which were interpreted as tiny variations in the smoothness of the universe 300,000 years after the Big Bang. These variations had been predicted by the Big Bang theory. Later missions such as WMAP have refined these initial results to help unravel the structure of the universe on the largest scales.

# THE HUBBLE SPACE TELESCOPE

*No other orbiting telescope has generated as much excitement or so astounded astronomers.*

**UP CLOSE** *The HST (left) has vastly improved our view of space. The image of the galaxy M33 (right, top) was taken with the Hale Telescope, but a small part of it, NGC 604, inset, was captured in greater detail by the HST (right, bottom).*

Launched in April 1990, the Hubble Space Telescope (HST) has provided spectacular observations in visual, near-infrared, and ultraviolet wavelengths, as well as spectroscopic studies of stars, the thin interstellar gas, and galaxies. Consisting of a 95 inch (2.4 m) mirror and a suite of sensitive scientific instruments, it is controlled by hundreds of astronomers and technicians. It is unique in that it was designed to be serviced in space with instruments that could be replaced.

The Wide Field/Planetary Camera 2 (WFPC2) was, for over a decade, the most often used of HST's instruments. It could detect objects as faint as 28th magnitude (about a billion times fainter than can be seen with the naked eye). It was replaced in 2009 by the even more sensitive Wide Field Camera 3 (WFC3). The original Faint Object Camera was replaced in 2002 by the Advanced Camera for Surveys (ACS) that images ultraviolet, visible, and infrared light.

Two earlier replacements are the Near Infrared Camera and Multi-Object Spectrometer (NICMOS), and the Space Telescope Imaging Spectrograph (STIS). NICMOS handles both imaging and spectroscopic observations of objects at near-infrared wavelengths.

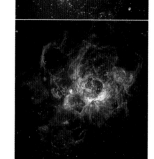

Scientists learn much from these observations about the birth of stars in dense, dusty globules, the infrared emission produced by the active centers of distant galaxies, and the nature of a class of galaxies as bright as quasars at infrared wavelengths

The STIS covers a broad wavelength range and can block out, or occult, the light of distant stars to search for black holes and Jupiter-size planets in other galaxies. It suffered an electronic failure in 2004 but was repaired in 2009.

Finally, HST's fine-guidance sensors, necessary for pointing the telescope and locking onto its target, can measure the positions of stars to 0.0003 arc-

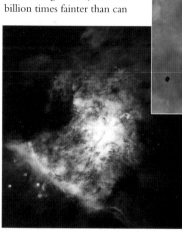

**THE GREAT NEBULA** *in Orion. The close-up (above) shows an area of dusty disks where young stars form. The larger composite image (left) is made up of 15 separate fields.*

**AN UNUSUAL EDGE-ON GALAXY**
*with a warped dusty disk, as seen by HST's Wide Field and Planetary Camera 2.*

seconds. Such precision enables astronomers to note slight wobbles in a star's position.

## WHAT HST HAS SEEN

HST has made hundreds of thousands of exposures, yielded significant insights into the formation of stars and stellar disks, disclosed important evidence for the existence of black holes in galaxies and quasars, increased our knowledge of the size and age of the universe, and detected galaxies that formed less than a billion years after the Big Bang.

Its high-resolution images of Mars, Jupiter, Saturn, and Neptune are yielding details surpassed only by space-probe photographs. The world was astonished by the spectacular images it produced in July 1994 when 21 fragments of the comet Shoemaker-Levy 9 collided with Jupiter.

HST images of star-forming regions have shown nascent stars embedded in globules of dust and gas. In the Great Nebula in Orion, dusty disks visible around protostars have been interpreted as solar systems in the making. At the

**DYING STARS** *seen by HST: NGC 6543 (above right), and Mz 3 (right) are planetary nebulas, the glowing remains of Sun-like stars.*

other end of stellar evolution, HST has produced a stunning image of Eta Carinae, a star/nebula system 8,000 light-years away (see p. 221). This is expected to explode one day as a supernova.

We can now see great disks of matter swirling around supermassive black holes at the centers of galaxies and quasars, as well as structural details in the spiral arms of nearby galaxies. Looking at the most remote corners of the universe, a series of 800 HST exposures taken over 400 orbits formed the Hubble Ultra Deep Field Image. It revealed an

assemblage of galaxies of various sizes completely filling a single speck of sky in Fornax. Many of these may date back almost to the beginning of the universe.

The HST has established itself in astronomical history and continues to make dramatic observations and discoveries. Nevertheless, astronomers are already building the next generation space telescope. The James Webb Space Telescope (JWST) will be able to look in even greater detail at a period in the universe when the primordial seeds of the galaxies began to evolve—just a few hundred million years after the Big Bang. Astronomers long to study this epoch because it may help explain the origin and fate of the universe.

The new telescope's 260-inch (6.5-m) mirror will soak up light from remote proto-galaxies as well as study nearby objects in the universe, mainly in the infrared region of the spectrum. With this new technology, we will address such burning questions as: How did galaxies form? and, What were the first generations of stars like?

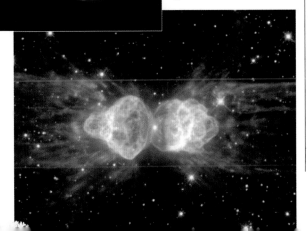

# SPACE TRAVELERS

*The fantasies of imaginative writers such as Jules Verne,*

*H. G. Wells, and Cyrano de Bergerac suddenly became reality*

*in the 1960s with the first human steps in space.*

In 1957, the steady beeping of the Soviet Union's Sputnik 1 satellite, the first to orbit Earth, flung down the gauntlet before United States policy-makers. Who would control the high ground of space? On 1 October 1958, the United States government established the National Aeronautics and Space Administration (NASA). So began the space race, which, in retrospect, may be more accurately described as a race of political ideals embodied in the technological developments achieved by the United States and the Soviet Union. At any rate, the next 12 years saw phenomenal accomplishments, illustrious glory, and tragedy.

**RECORDS** *Mercury 2 sent a chimp (top) into space. The descent module (above) of Vostok 1, in which Yuri Gagarin became the first man to orbit the Earth. The first woman in space was Valentina Tereshkova (right).*

## THE SPACE RACE

The Soviet Union leaped ahead first, with two Sputnik launches—the second of which carried Laika, a live dog. Laika survived for several days before her air supply ran out. The Soviets then placed the first human into Earth orbit, on 12 April 1961. The spacecraft, Vostok ("east"), was little more than a cabin in a cannonball, covered entirely by a heat shield. The cosmonaut, Yuri Gagarin, sat in an ejection seat, which he activated to safely parachute back to Earth.

Ostensibly, the goal of America's incipient space program, called Project Mercury, was to place astronauts into space, test their reactions and the spacecraft's maneuvering abilities, and return them safely to Earth. The long-term goal, however, as proclaimed in no uncertain terms by President John F. Kennedy in May 1961, was to land a man on the Moon by the end of the decade. It was a bold challenge, not only to the Soviets, but to the American people and their science and technology.

In contrast to Russia's daring steps into space, the Mercury program proceeded more cautiously, testing the footing each step of the way. There were 25 flights in all, only some of which were piloted. To ensure the safety of spacecraft for humans,

NASA, like Russia, placed several animals into orbit. In November 1961, a chimp named Enos rocketed into orbit and safely returned. This inhumane use of chimps and monkeys would continue into the Apollo program.

## INTO ORBIT

A month after Gagarin's historic mission, the first piloted Mercury flight was made by astronaut Alan Shepard, in space capsule Freedom 7. This was a suborbital flight, traveling beyond Earth's atmosphere without going into orbit. The second Mercury flight was also suborbital. By the time John Glenn earned the distinction of being the first American astronaut to orbit Earth, on 20 February 1962, the Soviet Union had already completed four piloted missions—one lasting four days. There were six Vostok flights in all, including one that carried Valentina Tereshkova, the first woman in space, in June 1963.

Later Mercury flights tested maneuvering systems, satellite deployment, and human performance in zero gravity.

## TWO'S COMPANY

The success of the Soviets' Vostok flights led them to embark on a brief but more ambitious program called Voskhod ("sunrise"). This entailed just two flights using a redesigned Vostok spacecraft, but each carried more than one cosmonaut at a time. In October 1964, five months before the first

**MILESTONES** *Walter Schirra is pulled from the Mercury capsule after his 1962 mission (right). Edward White spacewalks during the Gemini 4 mission (below).*

two-person Gemini mission, Vladimir Komarov, Boris Yegorov, and Konstantin Feoktistov rode Voskhod 1 into space and successfully completed 17 orbits. The first spacewalk was made by Aleksei Leonov from Voskhod 2 in March 1965.

In the same month, the first manned Gemini launch—Gemini 3—took place, piloted by Virgil Grissom and John Young. The astronauts performed orbital changes and other maneuvers that were important practice for future flights. Gemini 4, launched in June 1965, featured the first spacewalk by a United States astronaut. While pilot James McDivitt manned the spacecraft, Edward White drifted in space for 22 minutes, propelling himself on the end of a tether with a hand-held maneuvering device.

The next two years saw a succession of Gemini flights, which tested spacecraft maneuvers, refined spacewalking techniques, and studied the effects of spaceflight on humans. Perhaps the most important accomplishment of the program was the world's first docking maneuver in space, completed by Neil Armstrong and David Scott during the Gemini 8 flight in March 1966. Docking, a procedure in which one spacecraft joins another in space, was crucial to the success of future Moon missions. Other dockings followed with Geminis 10, 11, and 12.

## TRAGIC OUTCOME

The Soviet's new Soyuz ("union") program dovetailed with the conclusion of Project Gemini. The Soyuz craft was designed for docking and for taking cosmonauts to the Moon. In April 1967, Soyuz 1 was to rendezvous with a second vehicle, Soyuz 2. The flight, piloted by Vladimir Komorav, ended in tragedy when Soyuz 1 developed serious control problems early in the flight. Its landing parachute failed to open, and Vladimir Komorav was killed.

Docking attempts continued with the launch of Soyuz 2 and 3 in October 1968. This attempt failed, but both spacecraft returned safely to Earth. Not until January 1969 did Soyuz 4 and 5 achieve a successful docking and crew transfer from one spacecraft to the other. By then, however, the Americans were orbiting astronauts around the Moon.

# THE APOLLO PROGRAM

*The race to put a man on the Moon was the impetus of NASA's ambitious Apollo program, which, in just six short years, changed the boundaries of human aspirations.*

After the Gemini program came Apollo, NASA's most ambitious project yet. It began with seven unpiloted test flights, mostly involving the Saturn I and the Saturn IB rockets. These test flights led to the design and building of the much larger Saturn V rocket, which would become the heavy-lift vehicle for the Apollo program. The Saturn V was capable of hefting 285,000 pounds (130,000 kg) into Earth orbit, and 100,000 pounds (45,000 kg) to the Moon.

The cone-shaped Apollo command module, designed to carry three astronauts, sat atop the service module. While in Earth orbit, the lunar module was docked with the command module. Once in lunar orbit, two of the astronauts would enter the lunar module for the trip to the Moon's surface, leaving the command and service modules in orbit.

### REACH FOR THE MOON

Like the Soviet's Soyuz series, the Apollo program had a disastrous beginning. On 27 January 1967, an electrical fire erupted during an Apollo 1 training exercise just a month before its scheduled launch, killing astronauts Virgil Grissom, Edward White, and Roger Chaffee. The Apollo program was grounded for more than

**GIANT STEPS** *Apollo 4, the first unmanned craft in the program, on the launchpad at the Kennedy Space Center (left). Walter Schirra (above) aboard the first piloted craft, Apollo 7. The Apollo 11 logo (top left) echoes Neil Armstrong's message that "the Eagle has landed."*

a year until modifications made the spacecraft safer. Apollos 4, 5, and 6 were unpiloted flights designed to test the new systems and refine procedures for putting hardware into orbit. Their success led to the piloted Apollo 7 mission in October 1968, and the first lunar orbit by humans, in Apollo 8 in December 1968, which returned startling images of Earth from space. At least part of President Kennedy's goal had been realized.

On 3 March 1969, Apollo 9 tested rendezvous and docking procedures with the lunar module in Earth orbit. Two months later, Apollo 10 returned to lunar orbit to check the module's guidance and navigation systems. Astronauts Thomas Stafford and Eugene Cernan brought the lunar module to within 9 miles (15 km) of the first landing site, the Sea of Tranquillity, before firing their ascent engine and returning to the command module, piloted by John Young. The lunar surface was now within reach.

### MISSION ACCOMPLISHED

Less than a month later, on 20 July 1969, astronauts Neil Armstrong and Edwin "Buzz" Aldrin separated Apollo 11's lunar module, the *Eagle*, from

**APOLLO 11** *"Buzz" Aldrin steps onto the lunar surface (above) as part of the 1969 mission. The command module in orbit over the Moon (right).*

the command and service modules and began the descent to the surface of the Moon. Soon after, the *Eagle* touched down on a flat area in the Sea of Tranquillity. The two astronauts checked the equipment, then Armstrong stepped out of the lunar module onto the powdery surface of the Moon, uttering the famous words: "That's one small step for a man, one giant leap for mankind."

(A gap in transmission obscured the word "a" for most Earth-bound listeners.)

This historic mission was followed by five successful lunar landings: Apollos 12, 14, 15, 16, and 17. Apollo 13 was aborted on the way to the Moon because of an explosion that ruptured an oxygen tank and damaged other systems. Even this mission, however, was considered a victory of sorts—a catastrophe averted by the skill of the NASA team.

The Apollo Moon missions returned 844 pounds (382 kg) of rock and soil samples, amazing photographs of lunar features, and a vast quantity of data generated by instruments set up on the surface. But that was not all it accomplished. As one of the greatest techno-logical feats of our species, it elevated our view of the place of humankind in the universe to unprecedented heights.

## WEIGHTLESSNESS

Bungee jump from a bridge or stall an aircraft, and you will feel as if your weight is reduced to zero. Astronauts experience the same sense of apparent weightlessness during space missions, where it is a source of both enjoyment and discomfort. The term "zero gravity" is sometimes applied to this state. Like the term "weightless," however, it is somewhat misleading. Gravity is still present—gravity is what keeps the spacecraft in orbit. And although astronauts do not feel their own weight in space, they are still attracted by Earth's gravitational field. What causes the sense of weightlessness is the fact that the astronauts and the spacecraft are falling together toward the Earth.

Under such conditions, you do not walk, but float from place to place. Objects drift about (right) and heavy items are easily lifted. Astronauts may experience temporary nausea, called space sickness, but more serious is a loss of muscle tone and bone strength, and a decrease in white blood cells, which fight diseases. To deal with these effects, astronauts must perform special exercises during long space flights.

*Beautiful, beautiful …*

*Magnificent desolation.*

Edwin "Buzz" Aldrin (b. 1930), Apollo 11 astronaut, as he followed Neil Armstrong to the surface of the Moon.

# SHUTTLES
## *and* STATIONS

*The space race is over, replaced by a spirit
of cooperation where expertise is shared.*

With the Moon race won, NASA began work on a new type of space vehicle that could be launched conventionally via rockets, but would also be reusable. On 12 April 1981, the space shuttle *Columbia* rocketed off the launch pad at Cape Canaveral, piloted by astronauts John Young and Robert Crippen. After a 54-hour test flight, it glided safely back to Earth.

The shuttle has three components: the orbiter, the external tank, and the solid rocket boosters. The orbiter is the distinctive, delta-wing vehicle that carries the payload into orbit. Its underside is covered with 23,000 tiles that protect it from burning up during reentry. The payload bay, at the center of the orbiter, is large enough to hold a passenger bus. The external tank fuels the shuttle engines during launch and burns up in the atmosphere, while the two solid rocket boosters that help propel the shuttle off Earth's surface are ejected prior to orbit. These parachute into the Atlantic Ocean, where they are retrieved and refurbished.

Once in orbit, the shuttle is maneuvered by means of two onboard engines, which are fired at mission's end to slow the orbiter's speed. As the shuttle reenters the atmosphere, aerodynamic controls take over. Moving into a shallow glide path, it makes an unpowered landing.

**IN SERVICE** *The space shuttle (right)
has proved invaluable for making
adjustments to satellites, including the
very successful repair of the Hubble
Space Telescope in 1993 (below).*

**AT HOME IN SPACE** *The ISS, when complete, will have as much habitable space as two jumbo jets and house six or seven people at a time.*

In 2011 NASA retired the remaining shuttles—*Discovery, Atlantis,* and *Endeavor. Challenger* and *Columbia* were lost, with their crews, in accidents, reminding us of the risks of space flight (see Box). It will be replaced by a return to less expensive expendable rockets—both Russian and American.

The shuttle racked up many accomplishments. Reusable space laboratories flew aboard it many times, conducting scientific experiments and astronomical observations. The shuttle's ability to retrieve and deploy satellites from the payload bay made it a valuable asset in the launching, servicing, and repair of satellites of all types. The Hubble Space Telescope was successfully serviced five times from the shuttle's payload bay.

## SPACE STATIONS

In the early 1970s, the Soviets launched Salyut ("salute"), the first manned space station. Although rather cramped, Salyut established the viability of operating and maintaining laboratories in space. Six successful Salyut missions were made between 1971 and 1986.

The United States launched their space station Skylab on 14 May 1973. Hundreds of experiments and observations were made during its three missions, before it was allowed to burn up on re-entry over Australia in 1979.

Launched in February 1986, the Russian Mir ("peace") space station was occupied almost continuously between 1986 and 2001 by men and women from several nations. Its place has been taken by the much larger International Space Station (ISS), a joint project of five international space agencies.

## CHALLENGER AND COLUMBIA

Supposed to be a day of "firsts" for the space shuttle, 28 January 1986 is, instead, one of the American space program's darkest. *Challenger* lifted off with seven crew members, including its first civilian, Christa McAuliffe, a schoolteacher. Less than two minutes into the flight, the huge external tank exploded and *Challenger* broke into pieces. Minutes later, its cabin and crew fell into the sea. A rubber O-ring seal in one of the rocket boosters had failed.

Engineers labored for two years to rethink all critical items that went into a shuttle launch. By the time *Discovery* lifted off on 29 September 1988, safety was the number-one priority, making the shuttle a reliable, but complex and expensive, way into orbit.

However, on 1 February 2003, disaster struck the shuttle program again. A piece of foam insulation from the external fuel tank broke off and struck the left wing of *Columbia* during launch. The heat shield was damaged, causing the shuttle to disintegrate during its return to Earth two weeks later. Again the shuttle fleet was grounded for two years of investigation. *Discovery* again led the shuttles back into space on 25 July 2005.

*Flight directors in Houston Mission Control watch, horrified, as* Challenger *explodes.*

# TO *the* PLANETS

*Even before the first person was put in orbit,
space visionaries were looking to more distant and
tantalizing horizons—the planets.*

The space race of the 1960s may have had a political agenda, but it also served to inspire the exploration of an entirely new frontier. Expeditions to the planets were made a few cautious and calculated steps at a time.

## THE FIRST JOURNEYS

Venus, the closest planet and a sister world to Earth, being similar in size and mass, beckoned as a reasonable and pragmatic first objective. Success, however, was literally hit and miss. Venera 1, launched by the Soviets on 12 February 1961, failed when radio contact was lost two weeks after launch. The probe missed the planet by 60,000 miles (97,000 km) and went into orbit around the Sun. A similar fate awaited Venera 2, while Venera 3 crashed into Venus. Not until Venera 4 would the Soviets successfully rendez-vous with Venus, in October 1967, but the capsule that was supposed to make a soft landing was lost.

The Americans fared better with Mariner 2, a flyby mission to Venus, launched in August 1962. Mariner's instruments studied the planet's atmosphere and measured its lead-melting surface temperatures. The probe also discovered that Venus did not have a magnetic field or radiation belts.

**ASSIGNMENT VENUS** *The Mariner 2 spacecraft (right) made a successful flyby of Venus in 1962, five years before a Venera probe (left) reached the planet.*

The Soviets continued their efforts to land a spacecraft on Venus with Venera 7 and 8, launched in 1970 and 1972. Both capsules made it to the surface and transmitted data confirming Venus's extreme temperatures and pressures. A few years later, in 1975, Venera 9 and 10 were the first spacecraft to return images successfully from another planet's surface. Like the other kinds of data, the images were transmitted via radio.

## MARINER MISSIONS

While the Soviets were busy with Venus, United States scientists were conducting a series of flybys of the planet Mars. Mariner 4 was launched in November 1964 and yielded a total of 22 photo - graphs that revealed a heavily cratered terrain. Mariners 6 and 7 produced 201 photos, including the first close-up views of a Martian polar cap.

The most successful of the early Mars missions was Mariner 9, which entered orbit around the Red Planet on 3 November 1971, and spent the next 349 days photographing the surface.

**MARINER 9** *returned images of Mars and its moon Phobos (left).*

The probe's two video cameras returned more than 7,000 images and obtained fascinating close-up views of the planet's two moons, Phobos and Deimos.

Both Venus and Mercury were the targets of Mariner 10, launched 3 November 1973. Its cameras, equipped with ultraviolet filters, took photos of the layered cloud tops of Venus and revealed a com-plex global circulation pattern. Then the probe flew on to Mercury, where it captured more than 10,000 close-up photos of the Moon-like surface and discovered the planet's magnetic field.

48

*Nature … Doth teach us*

*to have aspiring minds …*

*And measure every*

*wandering planet's course …*

Tamburlaine the Great,
CHRISTOPHER MARLOWE
(1564–93), English dramatist

**PIONEER II** *was launched in 1973. It returned some of the first close-up images of Jupiter (left), before visiting the rings of Saturn (below) in 1979.*

## THE PIONEER SPIRIT

The United States reached farther out into the Solar System with the Pioneer launches in 1972 and 1973. Pioneer 10 made the first flyby of Jupiter on 4 Decem - ber 1973. Its closest pass was within about 80,000 miles (130,000 km) of the cloud tops. The photos it relayed showed Jupiter's churning atmosphere, including the Great Red Spot, and three of the planet's four large moons—Europa, Ganymede, and Callisto.

Almost exactly a year later, Pioneer 11 flew even closer to Jupiter—to within about 30,000 miles (48,000 km). Five years later, the probe passed Saturn, making a close approach of 18,600 miles (30,000 km) and returning 440 images and new data about the planet, its moons, and its ring system.

Today, both Pioneers 10 and 11 are heading into deep space in opposite directions. One day thousands of years from now, one or both of them may serve as an emissary from Earth to other intelligent life forms in the galaxy. Mounted on each spacecraft is a plaque that will convey information to anyone out there about our Solar System, Earth, and the human race.

## MARS AGAIN

In the middle to late 1970s, the United States embarked on a bold exploration of the outer Solar System. It also returned to Mars with the Viking probes. Each consisted

of an orbiter and a lander. The lander was designed to photograph the surface, collect and analyze soil samples, and make meteorological observations. The orbiter's duty was to image and map the planet's surface in detail. By the summer of 1980, the two Viking landers had sent back 4,200 photos of the surface and three million weather reports. The images showed a desolate rock and sand-dune terrain stretching to the horizon and a dusty pink sky. Viking 1 measured winds, with gusts of up to 30 miles per hour (50 km/h). From afternoon to night, temperatures fell from −20 to −120 degrees Fahrenheit (−30° to −85° C).

Landers returned to the surface of Mars in 1997 with the Pathfinder lander and 24-pound (11-kg) Sojourner rover. This paved the way for the larger Mars Exploration rovers Spirit and Opportunity that ranged across the Martian surface, beginning in early 2004.

**ROUGH SURFACE** *The Viking landers returned images of Mars's rocky terrain (left), while the orbiters showed features such as huge canyons (above left).*

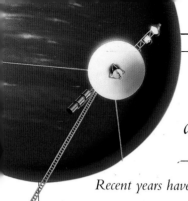

# VOYAGER
## and BEYOND

*Recent years have seen even more ambitious explorations of the Solar System, from the outer planets to the Sun at its center.*

In March 1979, Voyager 1 flew past Jupiter, sending back thousands of high-resolution photos of the cloud tops as well as intriguingly detailed images of the moons Callisto, Ganymede, Europa, and Io. Jupiter's gravity gave Voyager 1 the velocity needed to reach its ultimate destination, Saturn, in November 1980.

The planetary tour of Voyager 2 was more extensive. It reached Jupiter in July 1979, where it skimmed past the moons Callisto, Ganymede, Europa, and Amalthea, then went on to Saturn, traversing the planet's ring plane and passing closer to its moons than had Voyager 1. In January 1986, Voyager 2 reached Uranus, where it discovered several new moons. Finally, in August 1989, Voyager 2 made a close approach to Neptune, revealing an aqua-blue atmosphere punctuated by a huge dark spot and fleecy white cirrus clouds.

### GALILEO
Intrigued by the Voyager findings at Jupiter, planetary scientists once again turned their attention toward that planet in the early 1980s. They designed a mission to orbit Jupiter and monitor the planet and its satellites for several years.

The Galileo spacecraft did not have enough power for a direct flight, so a three-year

flight path was designed to "sling" Galileo to Jupiter using gravity assists from Venus and Earth. The flight path also brought it near asteroids Gaspra in 1991 and Ida in 1993. High-resolution images of Gaspra indicated that it was possibly a fragment of a larger body, while those of Ida revealed that it was orbited by a small moon.

The spacecraft finally made it to Jupiter in December 1995. On 7 December 1995, Galileo released a probe that parachuted into the upper cloud decks and returned 57 minutes of data before the

**WORTH THE WAIT** *Voyager 2 (top left) returned the first close-up images of Neptune. Since its launch from the space shuttle Atlantis in 1989 (below), the Galileo probe has returned amazing images, such as this (right) of the surface of Jupiter's moon Europa.*

signal was lost. Scientists were surprised to discover that, rather than being rich in water, oxygen, carbon, and other materials, Jupiter's mostly hydrogen atmosphere was rather primitive and unprocessed, resembling that of the Sun. Moreover, the atmosphere at this location was dry, even clear in places, with no trace of water vapor or ice.

Meanwhile, the orbiter returned tantalizing data on the Galilean satellites, discovering, among other things, that

**LONG-TERM PROJECTS** *The Ulysses spacecraft (left) on its way to the Sun; and the ESA's Giotto space probe (below) heading toward Halley's Comet.*

focus of increasing scrutiny.

In March 1986, several space probes passed near Halley's Comet as it approached Earth. Two Russian probes, Vega 1 and 2, approached within 5,520 and 4,990 miles (8,890 and 8,030 km) respectively on 6 and 9 March. Onboard spectrometers successfully measured the composition of the comet's dust.

Four days later, the European Space Agency's probe, Giotto, intercepted Halley's Comet. It penetrated the coma, or head, of the comet, passing within 370 miles (600 km) of the nucleus and sending back astonishing images of bright geyser-like jets erupting from a black potato-shaped body.

Among more recent missions, the Deep Impact spacecraft crashed a probe into the surface of Comet Tempel 1 on 5 July 2005. A few months later, the Stardust spacecraft returned a sample to Earth from Comet Wild 2.

## SOLAR PROBES

Ulysses, a joint mission by NASA and the European Space Agency (ESA) to study the Sun's environment and the solar wind, was launched in October 1990. This was the first time a spacecraft had been deliberately sent out of the ecliptic (the plane of the Solar System). After receiving a gravity assist from Jupiter in

February 1992, the Ulysses probe went on to pass over the Sun's south pole in May 1994, then over its north pole a year later. As well as gathering data about the solar wind, Ulysses' instruments returned important information about the interplanetary magnetic field, interplanetary dust, and cosmic rays.

The Japanese have sent several solar observatories designed to study extremes of solar energy. Hintori ("firebird") and Yohkoh ("sunbeam") carried X-ray and ultraviolet telescopes to image the Sun's rarefied corona and to study solar flares. Hinode ("sunrise") is the last, imaging the Sun in visible light, ultraviolet, and X-rays.

The Solar Heliospheric Observatory (SOHO) was launched in December 1996. SOHO was placed into orbit around a Lagrangian point— a region about 930,000 miles (1.5 million km) sunward from Earth where the gravitational pull of the Sun and Earth are in balance. SOHO's instruments are studying the temperature and structure of the Sun's interior, the corona, and the solar wind. It remains the main source of solar data for prediction of space weather around Earth.

Europa may have a relatively thin icy crust covering a mantle of liquid water or slush. This concoction may contain chemicals that could nurture life.

## CASSINI

A return visit to Saturn was the mission of the Cassini spacecraft, launched on 15 October 1997. It followed a complex path that involved gravitational boosts from Venus and Earth that took it to Jupiter in late 2000, before reaching Saturn in 2004.

Cassini has made repeated fly-bys of many of Saturn's major moons and discovered several new ones. Among its discoveries are plumes of water and perhaps a sub-surface ocean on the moon Enceladus.

On Christmas Day 2004, the Huygens probe separated from the Cassini orbiter and several weeks later descended in the atmosphere of Titan, Saturn's largest moon and the only moon in the Solar System with a substantial atmosphere. During the descent and for 90 minutes after landing, Huygens produced remarkable images showing dunes and surface lakes filled with liquid hydrocarbons.

## COMET MISSIONS

Although the planets have been the destinations of many space probes, minor members of the Solar System have been the

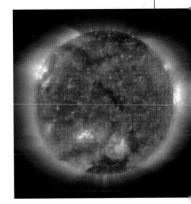

**SOHO** *returned this ultraviolet image of the Sun's corona in 1997.*

# TODAY'S EXPLORATIONS

*If just some of the dreams of the science visionaries come true, this century promises to be a bold new era of exploration and discovery in astronomy.*

With ever more sophisticated technology and mountains of data pouring in, it is hard to imagine what revelations even the next 25 years will bring. We will never understand everything about the universe because each answer poses new questions, but astronomers are closing in on some notable controversies: the location of the water that once flowed on Mars; the intriguing conditions

**IN FLIGHT** *The Mars Global Surveyor (left) on its way to the Red Planet. The Cassini orbiter (right) studying Saturn.*

on the moons of Jupiter; the ages of the oldest stars; and the presence of Earth-like planets around other stars. Still ahead lie the daunting tasks of ascertaining the physical characteristics of the earliest galaxies, and the composition of the mysterious "dark matter" and "dark energy" believed to constitute more than 90 percent of the universe.

In some respects, the future is already here. Ground-based astronomy has used new mirror-making technology to create larger, less-expensive telescopes. Even larger telescopes are planned. With

computer-controlled optics to counteract the effects of Earth's atmosphere, these instruments are almost like space telescopes on the ground. Radio telescopes linked to form the equivalent of one huge telescope are looking into the hearts of galaxies and quasars, and mapping the universe for unknown sources of radio energy.

Space missions currently underway include NASA's New Horizons spacecraft on its way to the dwarf planet Pluto and beyond. It was launched on 19 January 2006 on a mission expected to exceed 10 years. The spacecraft

**MARS MISSION** *The Mars Global Surveyor (left) reached Mars in 1997 along with the Pathfinder lander and solar-powered Sojourner rover (below), which inspected rocks around the landing site.*

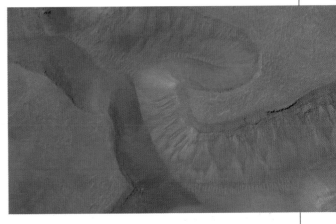

**EVIDENCE OF WATER?** *These gullies, seen from Mars Global Surveyor, may have been formed recently by groundwater seeping from the walls.*

will fly within 6,200 miles (10,000 km) of Pluto and its moons on 14 July 2015. It will then continue further into the Kuiper Belt. NASA's Messenger mission is currently studying Mercury for the first time in over 30 years.

Closer to home, in February 1996, NASA launched its Near Earth Asteroid Rendezvous (NEAR Shoemaker) mission. NEAR's trajectory took it past asteroid 253 Mathilde and on to 433 Eros. It touched down on Eros on 12 February 2001 to study the surface. The European Rosetta spacecraft is on its way to a 2014 rendezvous with comet Churyumov-Gerasimenko. It also has a lander to study the surface and interior of the comet.

Scientists have ambitious plans for Mars, too. The NASA Mars Exploration rovers are set to be followed by a variety of landers and sample return missions from the USA, Russia, Canada, and India. Along with orbiting spacecraft, Mars will be studied as no other planet apart from Earth.

## MISSIONS IN VIEW

Apart from an ambitious Mars program, many other missions are being considered, subject to available funding.

NASA's Flagship Program is a series of major exploratory missions. These may include further study of Venus, Titan, and Neptune. First off may be a joint NASA/ESA mission composed of two orbiters working together to determine if any of Jupiter's moons may support life.

Smaller and much sooner is the Juno mission to Jupiter, scheduled for launch in 2011. Like New Horizons on its way to Pluto, this is part of NASA's New Frontiers Program that may also include missions to Venus and the Moon.

Study of the Moon is increasing in interest once again, with missions proposed from several countries as part of their developing space programs. NASA plans to send astronauts back to the Moon as a possible prelude for a mission to Mars, although this remains long term and very uncertain.

## PEOPLE IN SPACE

The International Space Station is expected to remain in operation for at least a decade. It has, so far, been visited by astronauts from 15 nations. It serves as a research laboratory and a test bed for future manned space missions.

## SPACE JUNK

There are literally thousands of artificial "satellites." The US Space Command's Space Surveillance Center near Colorado Springs, Colorado, monitors the whereabouts of more than 8,500 artificial objects. A few hundred are working satellites and spacecraft, but thousands more are spent rocket boosters, shards from exploded rockets, chunks of solar panels, screws and tools from previous space missions, and several thousand defunct satellites. This space debris (depicted below as yellow dots) is becoming a big safety issue. A fleck of paint moving at speeds of 17,000 miles per hour (27,000 km/h) has the energy to chip a space shuttle window or pierce the skin of a satellite. Imagine how damaging, and dangerous, an errant screw or a large piece of metal could be moving at that velocity.

Objects in low Earth orbit reenter the atmosphere and burn up after a few months. Objects at greater altitudes could remain in orbit for centuries. Each piece of debris must be diligently tracked to avoid harm to spacecraft.

# WHERE *in the* WORLD?

*You may never travel in a spacecraft, but you can*

*visit one of the many professional observatories*

*and space centers on Earth.*

This map marks just some of the world's largest observatories and space centers. Many of these provide tours, brochures, and programs for the public. Space centers tend to be located in major cities, but functioning optical and radio observatories will be found in more remote regions, well away from city light pollution and radio noise. Old observatories in urban areas, such as the Royal Greenwich Observatory in London, are often kept as museums and are well worth a visit.

**KECK** *These domes in the W. M. Keck Observatory house the world's largest telescopes—Keck I and Keck II. They are located on the summit of Mauna Kea, Hawaii, where the steady air produces superb seeing conditions.*

▲
**THE VERY LARGE ARRAY** *(VLA) near Socorro, New Mexico, consists of 27 dish antennas, each 80 feet (25 m) across, arranged in a huge Y-shape. The antennas can be linked to form a powerful radio telescope—a technique known as interferometry. The VLA is operated by the National Radio Astronomy Observatory (NRAO).*

### WORLD OBSERVATORIES AND SPACE CENTERS

1 Keck & Gemini North Obs., Hawaii
2 Lick Observatory, California
3 Jet Propulsion Lab, California
4 Mt Wilson Observatory, California
5 Palomar Observatory, California
6 Kitt Peak & Whipple Obs., Arizona
7 Very Large Array, New Mexico

8 McDonald Observatory, Texas
9 Johnson Space Center, Texas
10 Yerkes Observatory, Wisconsin
11 Goddard Space Center, Maryland
12 Green Bank Radio Telescope, West Virginia
13 Kennedy Space Center, Florida

**ESOC** *Banks of monitors stretch across the mission-control room of the European Space Operations Centre (ESOC). Located in Darmstadt, Germany, ESOC oversees satellite operations, data reception, and processing for space flights conducted by the European Space Agency (ESA).*

**KEY**

● optical telescope

● radio telescope

● space center

**AAO** *The Australian Astronomical Observatory at Siding Spring in New South Wales, Australia, includes the UK Schmidt Telescope. This 70 inch (1.8 m) instrument is used mainly for wide-field surveys of the sky.*

**ARECIBO** *Cornell University operates a spherical 1,000 foot (305 m) radio telescope near Arecibo, Puerto Rico. The enormous dish is formed by means of wire mesh suspended above the ground in a natural hollow.*

⓮ Arecibo Telescope, Puerto Rico
⓯ ESA launch site, French Guiana
⓰ Paranal Obs., Chile
⓱ Cerro Tololo, Cerro Pachon, & La Silla Obs., Chile
⓲ La Palma Observatory, Canary Is.
⓳ Jodrell Bank, England
⓴ Effelsberg Telescope, Germany

㉑ European Space Operations Centre (ESOC), Germany
㉒ Bolshoi Telescope, Russia
㉓ Baikonur Cosmodrome, Kazakhstan
㉔ Kagoshima Space Center, Japan
㉕ South African Astronomical Obs.
㉖ Australia Telescope
㉗ Australian Astronomical Observatory

*At length, by sparing neither labor nor expense, I succeeded in constructing for myself an instrument so superior that objects seen through it appear magnified nearly a thousand times …*

GALILEO GALILEI (1564–1642),
Italian astronomer, mathematician, and physicist

# CHAPTER THREE
# SKYWATCHING TOOLS

# NAKED EYE ASTRONOMY

*For tens of thousands of years we have enjoyed the spectacle*
*of the night sky without the benefit of binoculars or a telescope,*
*simply by looking upward in the dark.*

**RURAL OBSERVERS** *These naked-eye astronomers were painted by Donato Creti (1671–1749).*

**A NAKED-EYE SPECTACLE** *(right) Venus (top) and Jupiter (center) accompany the Moon. The eagle-eyed observer would also spy Mercury within the tree, just above the band of clouds.*

**SKYWATCHING WITH CHILDREN** *(right) is especially rewarding. You don't need a telescope to take a look at the wonders of the universe.*

Without any optical aid at all, you can observe a wide range of phenomena in the sky. You can see the large dark areas on the Moon that are now known to have resulted from large objects crashing into it almost four billion years ago. So easy are they to make out that they have given rise to legends about there being a man, or a hare, in the Moon.

You can follow the nightly wanderings of five of the planets, and from one place on Earth or another you can spot the 88 constellations—the traditional star patterns. You can watch certain stars—the variables—alter in brightness over the course of days, weeks, or months. You can see star clusters like the Pleiades in Taurus and gas clouds like the Great Nebula in Orion. If the night is dark, the vast expanse of our own galaxy, the Milky Way, will be visible, winding across the sky, as will three of its neighboring galaxies.

There are also artificial satellites to be seen, and showers of tiny meteors. If you are far enough north or south, you can enjoy the fabulous fireworks display of the northern or southern lights (the aurora borealis and the aurora australis).

## THE CHANGING SKY

The sky is constantly changing. The Moon rises about 40 minutes later, on average, every night, so as well as its phase differing each time it rises, its position relative to the stars also differs.

Although they move much more slowly than the Moon, the planets trace elegant paths among the stars and you can easily follow the movements of the five that are closest to us over a season of observing.

The stars themselves rise about four minutes earlier each night, which may not sound like much, but the difference adds up to about an hour every two weeks and a day over the course of a full year.

You can even see the sky alter during the course of a single night. In only two hours, stars that were near the eastern horizon will have risen to prominence high in the eastern sky, while others will have set in the west.

## GOING OUTSIDE

For most of us, dusk marks the end of a day: for skywatchers (unless they watch the Sun), it marks the beginning. Although any clear night provides an invitation to go outside and see what is up, some nights offer special events, such as a lunar eclipse, when the Moon passes through Earth's shadow. Meteor showers (which, like eclipses, are also forecast) provide another reason for turning off the TV and turning on to the sky.

### FIRST NIGHT OUT

When skywatching for the first time, it is best not to rush outdoors to begin learning the constellations as they are plotted on the Starfinder charts.

For a beginner, especially a child, the sky is a great spur to the imagination. On your first night out, try connecting the stars to make your own patterns and figures. Let your mind roam. It should be fun, and you are unlikely to forget the positions of the stars in your personal constellations. By doing this, you will start skywatching in an entertaining and memorable way.

# BECOMING *a* BETTER OBSERVER

*Skywatchers can spend a fortune on elaborate equipment, but the sky comes free of charge.*

Armed with no more than a good astronomy guidebook, you can enjoy skywatching from any location. A beautiful grouping of the Moon and planets, the swirling display of an aurora, or the progress of stars and constellations through the seasons—these are the simple pleasures of low-cost, naked-eye astronomy.

A pair of binoculars, which you may already own, opens up new vistas—from the Milky Way resolved into thousands of stars, to a glowing comet or the shuttling moons of Jupiter.

While telescopes are seen as the entry tickets to the hobby, many people buy one far too soon. If you do not yet own a telescope, wait until you know your way around the sky. Take the opportunity at star parties and observatories to look through a variety of telescopes. When you are ready to buy one of your own, consult the guidelines in this chapter. If you already have some skywatching equipment, the advice in this chapter will help you make the most of it.

## RECORDING THE SKY

Some amateur astronomers specialize in particular kinds of observations: perhaps mapping storm features on Jupiter, tracking down an asteroid, or estimating the brightness of a variable star. Submitting formal reports of these observations to organizations can provide a great sense of accomplishment. Indeed, astronomy is one of the few sciences in which amateurs still regularly make genuine contributions.

**NOVICE SKYWATCHERS**
*(left) should learn key stars and constellations before buying a telescope. Keep a sky diary (top) with notes and sketches as your personal record.*

Even if your observations go no farther than home, you will find keeping some type of sky diary or logbook very satisfying. It can be simply a list of what you see each night out, embellished with some descriptions. Many people like to add sketches of eyepiece views. Others use cameras to record the sky (see pp. 86–95).

## CHOOSING A SITE

Your observing site will have a greater effect on your skywatching than any piece of equipment. Of course, a dark site far from city lights is ideal, but few of us have that luxury. City-bound observers may not be able to find faint galaxies, but the Moon and planets can still look splendid.

If a bright light is glaring at you, your eyes will not fully adapt to the dark and it will be difficult to see fainter targets. Try to find a site shadowed from street and yard lights. A black cloth thrown over your head, in the style of old-time photographers, can also help.

Some enthusiastic amateurs build backyard observatories (check back issues of astronomy magazines for instructions). Some are simple garden sheds with flip-top or roll-off roofs. Others have rotating domes and heated rooms. No matter how modest, the convenience of a home observatory is hard to beat, even if the sky conditions are less than ideal.

*URBAN SKIES have much to offer stargazers, with the Moon and planets unaffected by skyglow (left). However, the spread of light pollution from cities (indicated by white on the satellite image below) means many astronomers must travel farther to find dark skies.*

As with most naturalist pursuits, however, the beauty of the night sky is enhanced if you leave the urban sprawl behind. A drive of 30 to 45 minutes beyond city limits is often enough to get out from under the dome of light that covers every city and town.

Particularly for observers in the Northern Hemisphere, the best direction to drive from a city, if possible, is south. This puts the city glow to the north and places the southern sky—the most interesting part, which includes the Milky Way—above the darkest horizon. Parks are good destinations, but introduce yourself to the officials in charge or they might think you are a late-night vandal. Local astronomy clubs often maintain rural observatories on private land for their members. Observing with a group provides security and companionship.

## CLEAR SKIES

Amateur astronomers are also weather-watchers, paying close attention to daytime cloud patterns and weather maps. They soon learn that a summer sky filled with puffy cumulus clouds during the day often heralds a clear night sky.

If the sky is truly clear, the Milky Way may be seen even from suburban locations, and you are more likely to find faint objects such as nebulas.

Moisture of any kind in the air degrades the transparency of the sky. Clear nights usually come after a cold front sweeps out haze and humidity, then brings a dry high-pressure center (marked with an H on weather maps). Nights with a low dewpoint also tend to be transparent. The dewpoint is the temperature at which moisture in the air condenses, and a low dewpoint means moisture is unlikely to condense out of the atmosphere as the temperature falls during the night.

## GOOD SEEING

Memorable nights of viewing can, however, come with many kinds of weather. For the Moon and planets, humid nights with stagnant haze or fog can actually bring the sharpest views. Even though little else is visible in the sky, planet disks appear rock steady, revealing astonishing detail. This is called good seeing—it happens when the layers of Earth's atmosphere are calm and stable, and not mixed up by winds at different altitudes. Under turbulent conditions, the poor seeing turns the disks of planets into boiling blobs with, at best, fleeting moments of sharp views.

*A SKYWATCHING KIT might include anything from a hot drink in a vacuum flask to a computerized tracking system.*

At most sites, nights of good seeing are often the least transparent, and vice versa. Only the finest mountaintop observatories boast the best of both worlds. Backyard astronomers soon learn to adapt their observing priorities to the conditions of the night, perhaps pursuing a galaxy hunt on a clear night, and a planet study on a night when the seeing is superb.

# CITY *and* SUBURBAN SKIES

*Do you need a dark country sky for skywatching?*
*Not necessarily. Even in the city you can locate constellations,*
*observe the planets, and enjoy double stars.*

**BAD LIGHTING** *This is a classic example—a sign illuminated from below, directing light up into the sky.*

Compared with a dark country sky, does the illuminated sky over a city or the suburbs have much to offer? Is looking up even worthwhile?

You bet it is. In some respects a bright sky is better for a beginner than a dark one, because you do not have to contend with a confusing undertow of 3,000 faint stars. It is easier to find the outlines of the major constellations when there are just a few dozen of the brightest stars in evidence, and observing the Moon and the bright planets is just as easy in the city as it is in the country. A group called the San Francisco Sidewalk Astronomers are well known for giving the public a chance to see what can be viewed through a telescope from within the city.

For a beginner, the best place from which to make your observations would be the familiar surroundings of your own backyard, porch, or even rooftop.

Any open space will do, but it is obviously best to be away from direct lighting, if possible. Choose somewhere that is close to home, then it will be easy for you to just take a few minutes whenever you feel like it to do a little skywatching.

## LIGHT POLLUTION

Astronomer David Crawford, from the Kitt Peak National Observatory, has calculated that the United States spends about 2 billion dollars a year on lighting the underbellies of birds and airplanes!

The problem with cities, from an astronomer's point of view, is not so much the amount of lighting there is, but the direction in which it is pointed.

We need efficient lighting for road safety and security, but those who argue for this are in fact in agreement with those who long to have a darker sky. A brightly lit street can also be poorly lit, especially if the lighting is irregular, with alternating bright and dark areas. The answer lies in having well designed fixtures that direct the light downward and shield the bulb from view.

Some cities, such as Tucson, Arizona, have switched to using sodium lighting. These lights—especially the low pressure variety —emit single colors that astronomers can filter out, and they spread a diffuse light that is free from glare.

**CITY LIGHTS** *Sights such as this are not uncommon in the city. Both the buildings and the sky are equally brightly lit.*

## THE INTERNATIONAL DARK SKY ASSOCIATION

In 1987, David Crawford and Tim Hunter, an amateur astronomer from Tucson, Arizona, founded the International Dark Sky Association. This organization aims to foster an approach to lighting that promotes safety and economy, as well as preserving the beauty of the night sky.

If you would like to persuade your local authorities to adopt better lighting practices, a group of you will need to do quite a bit of lobbying, but the International Dark Sky Association can provide assistance.

**CITY STARS** *Orion's brilliant stars can be easily seen, even above the Vancouver skyline, although the photograph shows more stars than the eye alone might see.*

**GOOD LIGHTING** *The yellow light of low pressure sodium lighting is preferred by astronomers and it is also electrically efficient. The sodium-lit car park (left) is an example of such lighting, as are the factory doors (above), but the latter are better lit because well-designed fittings direct the light downward, which is where it is needed.*

# NATURAL SKYLIGHT

*Not all skylight is light pollution. The ghostly glow
of auroras and the gentle radiance of the zodiacal
light never fail to enchant skywatchers.*

The phenomena of auroras and the zodiacal light demonstrate that Earth is a planet within a system of planets. An aurora is the reaction of Earth's upper atmosphere to particles streaming outward from the Sun, while the zodiacal glow is created when sunlight reflects off countless specks of interplanetary dust.

## CURTAINS OF LIGHT

The compelling sight of the aurora borealis, or northern lights, has inspired, frightened, and otherwise transfixed people living in Arctic regions for centuries. To the Inuit of North America, the undulating gossamer curtains of light were the play of unborn children, or the torchlight held by the dead to aid the living in winter. Some say that if you whistle gently, the lights will respond by drawing nearer. In Scandinavia, it was once common to use auroras for weather forecasting, and auroras are still called "wind lights" and "weather lights."

Auroras are spawned around Earth's magnetic poles by activity on the Sun (see p. 126). Generally, the more active the Sun, the more prominent the aurora. Solar activity increases and decreases over about 11 years, with most sunspots and solar flares appearing toward the middle of the cycle. The flares produce tremendous outflows of charged particles that

**PHOTO TIPS** *Longer exposures will pick up a greater range of colors (left), while short exposures can freeze rippling curtains of light (above). A photo taken from the space shuttle shows the full extent of an aurora australis (below).*

intensify the solar wind. When the particles reach Earth, they excite, or ionize, neutral oxygen and nitrogen molecules in Earth's upper atmosphere. This ionization produces the eerie colored glow that we see as an aurora.

In both the Northern and Southern hemispheres, most auroral activity is confined to high latitudes—the region above 65 degrees. Auroras are commonly seen in Alaska, Canada, and Scandinavian countries, but in the Southern Hemisphere—where the phenomena is called the aurora australis, or southern lights— few inhabited locations are far enough south to see it often. During periods of high solar activity, auroras can extend down to latitudes of 40 degrees or less, and during extreme activity, they may even reach the tropics. One major outburst of solar activity in March 1989 produced auroras that were seen from the Caribbean Sea.

Auroras are usually visible for about an hour, but during peaks in solar activity, they

**PINK AND GREEN** *auroras appear when the solar wind ionizes oxygen molecules in our atmosphere. When nitrogen molecules dominate, blue auroras occur.*

the border of the zodiacal light. If it is not immediately visible, try using averted vision and look for the zodiacal light "out of the corner of your eye" (see p. 68).

## AN ELUSIVE LIGHT

A manifestation of the zodiacal light appears in another part of the sky, where it has a different name—the gegenschein, or counterglow. A very faint elliptical glow some 10 degrees across, this phenomenon appears opposite the Sun in what is called the antisolar region of the sky.

The best time to locate the gegenschein is around midnight in early fall or early spring. This is when it is projected into a relatively star-poor region of the sky. Dark, clear moonless nights are requisite to see this most elusive of sky glows. The technique of averted vision will also help you see the gegenschein.

can last all night. For Northern Hemisphere observers, an aurora begins as a dome of reddish or greenish light in the north. The glow eventually shapes itself into a distinctive arc that gradually creeps southward. During its peak, green, red, and blue streamers of light may extend to the zenith and beyond.

Spectacular photographs of auroras can be made using a tripod-mounted camera and a wide lens. Simply aim the camera toward the aurora and expose for 10 seconds (see also p. 89). Color photographs will often reveal subtle colors that the eye does not detect.

## THE ZODIACAL LIGHT

Of all the astronomical phenomena visible to the naked eye, the zodiacal light is one of the most delicate. It appears as a diaphanous wedge of light, no brighter than the Milky Way, extending up from the horizon well after sunset. The glow is faint and you can easily see dim stars through it. In fact, even its most "brilliant" display is often mistaken for sunset glow or uplighting from a nearby city or town.

The zodiacal light occurs when sunlight scatters off the countless dust particles that

orbit the Sun in a great disk. Like the planets, the dust particles are confined to the ecliptic—the plane of the Solar System. The constellations through which the ecliptic passes are collectively known as the zodiac, and since this is where the light is concentrated, it is called the zodiacal light.

Because the zodiacal light is aligned along the ecliptic, it reaches a higher altitude when the ecliptic is steeply inclined to the horizon. In middle-latitude locations, it is best seen in the west after nightfall during the local spring, and in the east before dawn during the local fall. In the tropics, where the ecliptic is more often nearly perpendicular to the horizon, the zodiacal light can be seen throughout the year in both the evening and morning sky.

To observe the zodiacal light, find a dark-sky location away from city lights. Give your eyes about 20 minutes to grow accustomed to the dark. Scan along the ecliptic from west to east, looking for a slight difference in brightness between the darker sky and

*THE ZODIACAL LIGHT appears here as a delicate wedge of light extending up from the western horizon after sunset.*

# SUNLIT CLOUDS

*By day and by night, the interaction of sunlight and moisture in our atmosphere creates delightful sights in the sky.*

**A MOON HALO** is depicted in this 1872 painting (left). Noctilucent clouds sometimes show a crossed pattern known as herring structure (below).

As sunlight passes through the ice crystals and water droplets suspended in the air above us, it creates an endless variety of arcs, rings, dapples of color, and splashes of light that can be seen throughout the year.

Most of these phenomena occur within Earth's lower atmosphere, which means that they are actually meteorological in nature. Noctilucent clouds, however, can claim both meteorological and astronomical ties.

Noctilucent clouds appear at night—usually an hour or so after sunset, when the Sun is more than 10 degrees below the horizon. This phenomenon is known to appear around midsummer in both the Northern and Southern hemispheres, but it can only be seen between latitudes of 45 and 60 degrees, with most sightings occurring in Canada and northern Europe.

These silvery-blue clouds form at the edge of space at an altitude of 50 miles (80 km), which is almost five times higher than most of Earth's weather-making systems. At such heights, the temperature can dip to −150 degrees Fahrenheit (−100° C). The scant amount of water vapor present at this altitude condenses onto meteoric dust and charged particles in the atmosphere, forming the clouds.

Sightings of noctilucent clouds have doubled since the 1950s. A possible explanation is that industrial pollution has increased the amount of particles at high altitudes around which the clouds can form.

## OVER THE RAINBOW

Perhaps the most familiar of all atmospheric sights is the rainbow. A rainbow begins when sunlight enters a raindrop. Most of the light passes straight through the drop, but some of it is refracted, or bent, into the colors of the spectrum. It then reflects off the back of the raindrop and is refracted again as it exits. The result of this refraction and reflection through millions of raindrops is a multicolored ring around a point opposite the Sun. From the ground, we see a beautiful arch, which is part of the ring.

## SUN COMPANIONS

Another common sight in daytime skies is the halo, sometimes called "a ring around the Sun." Haloes are caused by the refraction of sunlight through hexagonal ice crystals in cirrus clouds. As the sunlight passes through one of the six sides of the tumbling crystal, it is bent through a 22 degree angle, producing a halo with a radius of 22 degrees. Occasionally, when sunlight passes through the ends of hexagonal prisms, or through cubic ice crystals, it is bent through a greater angle, producing a second, fainter, 46 degree halo.

When the Sun is low in the sky, and when high cirrus clouds are present, look for

THE SUN *and Earth's atmosphere conspire in different ways to produce delicate sun haloes and sun dogs (left), and the spectacular green flash (below).*

two bright spots of sunlight on either side of the Sun. These spots, which will be tinged red on the inside and blue on the outside, are commonly called sun dogs or mock suns. The technical name is parhelia.

Sun dogs are produced by sunlight refracting through hexagonal ice crystals that are oriented with their bases parallel to the horizon. Sun dogs sometimes touch the outside edge of a 22 degree halo.

The safest way to observe sun dogs and haloes is to block your view of the Sun with a nearby structure such as a tree limb, chimney, or flagpole.

Yet another ice-related phenomenon, sun pillars, can be seen around sunrise or sunset. When low-angle sunlight reflects off ice crystals in cirrus clouds, a spike—or pillar—of light may be projected up from the just-set or pre-dawn Sun.

At night, moonlight can create moon haloes, moon dogs or mock moons, and moon pillars, but these effects are fainter and less common than the Sun phenomena.

*One can enjoy a rainbow without necessarily forgetting the forces that made it.*

Queen Victoria's Jubilee,
MARK TWAIN (1835–1910),
American writer and humorist

## A FLASH OF GREEN

One of the most spectacular solar-terrestrial events, as well as one of the rarest, is the sudden flash of green light seen just as the Sun rises or sets. The green flash, as it is called, is the result of Earth's atmosphere bending different colors of light by various amounts, especially near the horizon. You effectively have red, green, and blue suns that set at very slightly different times. The blue sun should be last, but the air usually blurs it out, leaving our final glimpse of the Sun as a flash of green.

To see the green flash, you will need a well-defined, unobstructed horizon and clear, haze-free conditions. An ocean panorama is ideal, which is why the green flash is most commonly seen from ships and from coastal plains and mountains.

Even when setting, the Sun can damage your eyesight, so only ever look at it for a few brief moments at a time.

## CREPUSCULAR RAYS

Sunsets are often depicted in paintings with majestic beams streaming up radially into the sky from the already set Sun. Exaggerated as the artist may make them appear, sunbeams can often be seen in clear skies 10 to 15 minutes after sunset (or before sunrise). This phenomenon is called crepuscular radiation (from the Latin word *crepusculum*, meaning "twilight").

Crepuscular rays are caused by a cloud below the horizon partially blocking the Sun. The shafts of sunlight that get past the cloud illuminate dust particles in the air above the horizon, creating spectacular rays that occasionally reach to the opposite point in the sky (below). As the cloud moves and the Sun's angle changes, the rays, too, shift across the sky. The phenomenon may last for several minutes.

# OBSERVING TECHNIQUES

*If you learn some simple measuring methods*

*and are able to take the time to prepare yourself,*

*you'll enjoy your skywatching to the full.*

When we enter a darkened room after being somewhere well lit, we have trouble seeing until our eyes have become used to the reduced light. Called dark adaptation, this process takes at least 15 or 20 minutes, but it is time well spent if you want to prepare yourself to make a thorough observation. During this time the pupils of your eyes will gradually open to their fullest extent (as much as ¼ inch [7 mm] for children; probably only ⅕ inch [5 mm] for older people), to allow in more light from the stars.

Clyde Tombaugh was a farmer in Kansas when he became interested in sky-watching. To get the most out of the Kansas sky he used to first sit in a completely dark room for an hour, simply allowing his eyes to become adapted to the dark. Taking his hobby seriously paid off, for at the age of 22 he was

taken on to work at the Lowell Observatory, and a year later he discovered the planet Pluto.

When you are adapting your eyes to the dark, avoid accidentally looking at bright lights. If you are in your backyard, turn off the porch lights and close the curtains so as to hide the light indoors.

Always protect your eyes from excessive light at other times. If you have been out in the bright Sun all day without sunglasses, your eyes will take much longer to dark adapt.

## RED LIGHTS
Many amateur astronomers use dim red lights or red-filtered flashlights to keep their eyes dark adapted during observing sessions, as any bright white

**THE RED VIEW** *A normal flashlight with a red plastic filter is an essential accessory when using the Starfinder charts in Chapter 8.*

**HOW MANY STARS?** *If you adapt your eyes to the dark, it will enable you to see many more stars.*

light will shrink the eyes' pupils again within seconds.

A piece of red plastic or cellophane fixed across the front of the flashlight is all you will need to do the job. Have the red light just bright enough to read by, but no brighter.

## AVERTED VISION
Appreciating the sky's fainter gems takes a certain amount of practice. Subtler characteristics are easier to spot if you look "through the corner of your eye," a technique known as averted vision. To do this, shift your center of attention slightly, while concentrating on the object. Fainter details will become visible since you will be using the more sensitive rods in the retina of your eye, rather than the cones at its center.

## MEASURING SIZES AND DISTANCES

Astronomers use degrees, minutes, and seconds to measure sizes and distances in the sky. For example, 90 degrees is the distance from the horizon to the zenith point directly overhead. The Moon and Sun present disks of about ½ degree, or 30 arcminutes, in size. The finest detail your eye can resolve without an optical aid is about 1 arcminute, or 60 arcseconds.

An outstretched hand, held at arm's length, will be about 20 degrees wide from the tip of the thumb to the tip of the little finger—roughly the distance between the first and last stars of the Big Dipper (the

**JUDGING DISTANCES** *in the sky, using your hand at arm's length.*

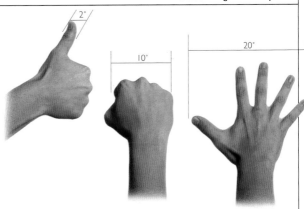

Plough) in the constellation of Ursa Major, which is visible from the Northern Hemisphere. Smaller distances can be measured with your fist at arm's length (about 10 degrees) and your thumb at arm's length (about 2 degrees). Your fist at arm's length would roughly cover the distance between the bright belt stars and brilliant blue–white Rigel in Orion, and your thumb would appear four times wider than the Moon.

After experimenting with this technique for a while, you will develop a feel for the distances between the various

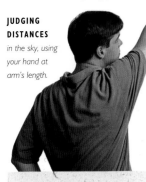

**"HANDS ON THE SKY"** *Your hand can help you measure distances in the sky.*

stars. Later on you can use this approach to determine the relative sizes of the different constellations and to find your way around much larger regions of sky. The constellation charts in Chapter 8 have symbols indicating size in terms of the outstretched hand.

**READY FOR ACTION** *A hood connected to the jacket would be a worthwhile addition to this winter observing outfit.*

## DRESSING FOR SUCCESS

For maximum skywatching enjoyment, it is wise to dress appropriately. Even in the summer you can become quite chilled if you are motionless for a long while outdoors, so always make sure you take a jacket or a sweater. And those ultra-frigid winter nights can be almost comfortable if you wear a snowsuit.

Protecting your head is most important. If you wear an insulated hat of some kind,

especially a hood connected to the rest of your outfit, you are bound to stay reasonably warm.

If you are planning to be outside for some time, take a groundsheet and a rug to sit or lie on, and maybe a sandwich and a vacuum flask of hot soup or coffee. A bag of nuts and dried fruit might also be welcome. In warm weather, the bugs are likely to be out, so you will need insect repellant.

Finally, be safety conscious. If you are on home territory, make sure you know where the obstacles are in your immediate surroundings. It's surprisingly easy

to collide in the dark with garden furniture or fall into the fish-pond in the back yard. If you decide to drive some distance from home, let someone know where you are planning to go, what roads you intend to use, and when you expect to return.

# THE BEST BINOCULARS

*Contrary to popular belief, the first piece of skywatching equipment to buy is not a telescope but a pair of binoculars.*

Anyone with an interest in the sky should own a pair of binoculars. Many people mistakenly think that binoculars cannot be as good as a telescope because they do not magnify enough. True, binoculars magnify only 10 to 20 times at best—not enough to show details on the planets or resolve small clusters of stars. But their modest magnifying and light-gathering power is more than enough to reveal many of the sky's most interesting objects.

A three-day-old Moon set in the deep blue twilight and

**BINOCULAR TESTING** *in 1939 (right). When trying out binoculars, check that the image looks sharp and bright.*

bathed in earthshine is an ideal binocular target. Using binoculars, you can see three or four of Jupiter's moons and even spot Uranus and Neptune as pale blue-green "stars." Dozens of asteroids come within reach of binoculars each year, while bright comets look spectacular.

No astronomical life is complete without a binocular tour of the fabulous dark lanes and star clouds near the center of our galaxy in Sagittarius and Scorpius. With binoculars, you can even range out past the Andromeda Galaxy to the bright galaxies of the Virgo Cluster, 50 million light-years away. Not bad for an instrument costing only about a quarter of the price of a standard beginner's telescope.

## WHY BINOCULARS?

The advantages of a pair of binoculars over a telescope are significant. Binoculars are inexpensive and readily available. Most are lightweight and require no effort to set up.

The twin barrels of binoculars are comfortable to look through. Because they are actually two small telescopes, they create a three-dimensional effect. Binoculars provide a

**THE WIDE FIELD OF VIEW** *provided by binoculars can sometimes be more suitable than a telescope view. Some galaxies, such as Andromeda Galaxy (above), and comets, such as Hyakutake (right), are large targets best seen through binoculars.*

*… the sky*

*Spreads like an ocean hung on high,*

*Bespangled with those isles of light …*

*Siege of Corinth*, LORD BYRON (1788–1824), English poet

wide field of view
that makes it easy
to find your target.
It also helps that the
images you see are right
side up, unlike the
upside-down images in
many telescopes.

## WHICH BINOCULARS?

The best binoculars for astronomy are 7 x 50 and 10 x 50
models. The 7 or 10 figure
indicates the magnification.
The 50 is the aperture—the
diameter of each front lens in
millimeters. Compared with
7 x 35 binoculars, a common
size for daytime use, models
with 50 mm lenses collect
twice as much light, yielding
brighter views of the night sky.

It is a tossup between a 7x
and a 10x pair. The higher
power of the 10x model is
better for revealing lunar
craters, star clusters, and faint
stars. High-power binoculars,
however, are harder to
hold steady and can be less
comfortable to look through
because your eyes must be
closer to the eyepieces. In
low-cost models, the extra
power may reveal flaws in
the optics that would be less
apparent in 7x binoculars.

*APERTURE is the key feature in
binoculars. Models with front lenses
50 mm across (left) are
the most popular for
astronomy, while
an aperture of
35 mm (below)
is a common size
for daytime use.*

## THE EXIT PUPIL

The exit pupil of
a pair of binoculars
is the diameter of the
beam of light leaving each eyepiece and entering each eye.
It is easy to calculate—simply
divide the aperture by the
power. For a 7 x 50 model,
the 50 mm aperture divided
by the 7 power equals a
7.1 mm exit pupil.

At night, under ideal
conditions, the pupil of a
dark-adapted human eye
opens to 7 mm. Binoculars
with a 7 mm exit pupil—
a 7 x 50, an 8 x 56, or a
9 x 63 model—produce as
wide a beam of light as the
average human eye can

accept, yielding maximum
image brightness. Such models
are commonly recommended
for astronomy—indeed, the
7 x 50 model is sometimes
called a night glass.

So why consider a 10 x 50
model? The nighttime viewing
conditions most of us contend
with are far from pitch black,
so our eyes never fully dilate
to 7 mm. Age also takes its
toll. The pupils of most
people over 30 open to only
6 mm in the dark. By the age
of 50, our pupils' maximum
aperture may be no more
than 4 to 5 mm, so the light
from binoculars with a 7 mm
exit pupil is simply wasted,
reducing the effective aperture
of the binoculars. For many
skywatchers, especially city
dwellers, binoculars with
a 5 or 6 mm exit pupil (a
10 x 50, or a lighter-weight
7 x 42) are a better choice.

*YUJI HYAKUTAKE used binoculars,
albeit very large ones, to discover the
brightest comet of 1996.*

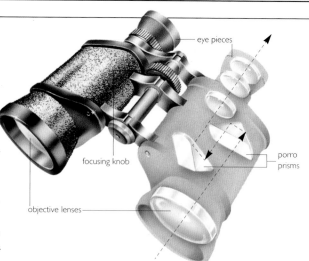

eye pieces

focusing knob

porro prisms

objective lenses

## FIELDS OF VIEW

The field of view is usually stamped on binoculars, but may be given as "feet at yards." For example, 367 feet at 1,000 yards equals a 7 degree field of view, and 262 feet at 1,000 yards equals 5 degrees.

Most 7x binoculars take in about 7 degrees of sky—roughly 14 times as big as the Moon—while 10x models see a smaller area, about 5 degrees.

Special wide-angle models provide impressive views of 8 to 10 degrees. But there are trade-offs: stars at the edge of the field tend to distort, and your eyes need to be close to the eyepieces.

Zoom binoculars sound attractive at first, but in almost all cases they are crippled by inferior optics and restricted fields of view.

## EYE RELIEF

Eye relief is the distance that your eyes must be from the binocular eyepieces to see the whole field. An eye relief of less than 9 mm makes for uncomfortable viewing.

If you wear eyeglasses, especially ones that correct for astigmatism, you will get sharper views if you keep them on when using binoculars. In this case, an extra-long eye relief of at least 15 mm and rubber eyecups that roll back are essential.

## TYPES OF PRISMS

Binoculars are made with one of two types of internal prism: porro prisms or roof prisms. Porro-prism binoculars, the most common type, have a zigzag shape. They are the best choice for skywatchers.

Roof-prism binoculars have a compact "straight-through" shape and cost more than

### PORRO-PRISM BINOCULARS

*This is the style familiar to most people. The shape of the binoculars results from the prism arrangement, which produces an upright image that is the correct*

porro-prism models. In all but the most expensive models, however, binoculars with roof prisms exhibit spikes of light on bright stars, making them unsuitable for astronomy.

## LENS COATINGS

A key difference in otherwise similar binoculars is the lens coating. Optical coatings give the lenses a blue, green, amber, or red tint. They improve image brightness and contrast, without affecting the color.

Low-cost models have only single-layer coatings, often on just the outside lenses. In better models, all optical surfaces are coated, and the best ones feature multi-layer coatings.

Some binoculars have rubber armor to cushion the optics against minor bumps. Premium models are often waterproof, which prevents dew and contaminants from seeping into the innards.

**BINOCULARS** *give great views and are easy to use, which is especially important for kids.*

*way round. Roof prism binoculars, on the other hand, are typically smaller and use a different arrangement of prisms to achieve the same result.*

## IN FOCUS

Waterproof binoculars often have two focus adjustments, one on each eyepiece. These individual-focus models are awkward for daytime use, but are fine for astronomy because every subject is located at infinity, requiring only one focus setting.

Even so, most people prefer to use center-focus models. These have a single focus adjustment that controls both

## IN-STORE TESTS

Try before you buy is a good rule with binoculars. When comparing the models available, carry out these simple tests.

- Do the images look sharp in the center? At what point do the images start to fuzz toward the edge of the field? Better models exhibit less edge distortion.
- Hold the binoculars at a distance. Look down the eyepieces. The circles of light should be evenly illuminated. Dark squared-off edges are a sign of the lower-quality BK7 prisms.
- Look into the main lenses. Lots of white reflections indicate poorly coated optics, while dark lenses with a few deep purple or green reflections indicate high-quality multi-coated optics.
- Are the binoculars comfortable to hold? Can you see the entire field without pressing your eyes right up to the eyepieces?
- Pass your hand across the front lenses so you look through just one side of the binoculars, and then the other. Does the image jump back and forth? If so, the binoculars are "cross-eyed," or out of collimation—a serious defect.

eyepieces, as well as a "set-once-and-forget" adjustment on one eyepiece that compensates for focus differences between your right and left eyes.

In recent years, some manufacturers have sold binoculars that have no focus adjustments. These fixed-focus models are unusable for astronomy. They do not focus at infinity and provide no means to compensate for the focus variations among different people's eyes.

### BINOCULAR MOUNTS

If you mount binoculars on a steady tripod, you will see much more detail in everything you look at. Many binoculars feature a centrally located threaded socket. This accepts an L-shaped bracket, which in turn bolts to any camera tripod.

**CANTILEVER STANDS,** *shown here supporting big binoculars, make it easy to look anywhere in the sky, even straight up.*

To make it easier to get under the binoculars and look through them when they are aimed high in the sky, some manufacturers offer cantilevered swing arms that suspend the binoculars away from the tripod. Some more expensive binoculars are internally "stabilized" to minimize jitter when hand-held.

Another way to steady your view is to lie back in a deck

chair or recliner, and use the chair's arms for support. What must be the ultimate luxury, however, is a rotating binocular-observer's chair.

### BIG BINOCULARS

Every skywatcher, novice or experienced, should own a pair of binoculars in the 50 mm aperture league. Serious binocular users may want to graduate to a pair in the 70 to 80 mm class. With 11 to 20 power, these big binoculars cost as much as a small telescope and their hefty weight demands tripod mounting. Binoculars in this class can provide bright, detailed views of Milky Way star fields, large nebulas, comets, and lunar eclipses unmatched by any other type of instrument.

### CARE OF BINOCULARS

With care, binoculars can last a lifetime. Avoid touching the lenses with your fingers and clean them as you would telescope eyepieces (see p. 83). Most importantly, try not to drop them. A sharp jolt can loosen the internal prisms or knock the two halves out of collimation (alignment). It can cost as much to repair binoculars as to buy a new pair of similar quality.

# TELESCOPE TYPES

*From a no-frills beginner's telescope to
an advanced computerized model, a superb selection of
tempting equipment awaits the telescope shopper.*

For the first-time buyer, the extra-ordinary variety of telescopes available can be overwhelming. There are three main optical designs. Reflecting telescopes use a mirror to gather light; refracting telescopes use a lens. A third type of telescope, the catadioptric, uses a combination of a mirror and a lens. Schmidt-Cassegrains and Maksutovs are examples of catadioptric telescopes.

Each of these three types has its selling points (see p. 76). All of them can show you details on the Moon as tiny as a mile across, the stunning rings of Saturn, the changing clouds of Jupiter, and the brightest galaxies, nebulas, and star clusters of deep space.

When shopping for a new telescope, the choice of optical design is less critical than the features listed below in our Telescope Shopper's Checklist. Most importantly, avoid

any telescope sold by how much it can magnify. Low-quality telescopes are often promoted by claims such as "450 power" or "high-power professional model." Many of the 2.4 inch (60 mm) refractors and 4.5 inch (110 mm) reflectors commonly seen in camera shops and department stores fall into this category. They feature wobbly mounts, eyepieces and finderscopes with poor optics, and plastic fittings. The views through these "toy" telescopes are disappointing, to say the least. However, there is nothing inherently wrong with smaller telescopes. A good 2.4 inch (60 mm) refractor with the features in our checklist can make a fine starter scope.

Most telescopes are sold with at least one or two eye-pieces. Try to find a telescope with quality eyepieces—such as Kellner, Orthoscopic, Modified Achromat, and Plössl types—giving no more than 75 to 100 power. Additional eyepieces can be purchased separately (see p. 82), but if the telescope provides only high-power (150x to 300x) eyepieces, it is probably of poor quality

## TELESCOPE SHOPPER'S CHECKLIST

A focuser that slides back and forth smoothly with no wobbles.

Interchangeable eyepieces—not fixed or zoom.

A good finderscope—preferably a 6 x 30 mm model, although most small telescopes come with only 5 x 24 finders (see p. 85).

All metal and wood construction, with minimal use of plastic, especially on any moving parts.

Slow-motion controls on both axes of the mount.

A tripod and mount that will not shake or bend (see p. 78).

**OUR CHECKLIST** *of features (left) is marked on a Newtonian reflector, but applies to all types of telescope. Almost any telescope you buy today will be more effective than Galileo's early telescopes, which were refractors (top).*

**BEFORE YOU GO SHOPPING**, *read this chapter carefully and compile a list of questions based on the skywatching activities you plan to pursue.*

telescope can resolve stars about 1 arcsecond apart, while an 8 inch (200 mm) telescope is capable of resolving 0.5 arc-seconds. (By comparison, the human eye can resolve only about 60 arcseconds.)

So when you are shopping, consider the telescope in your price range with the largest aperture. Then again, the biggest telescope you can afford is not necessarily the best for you. Think about where you will store it, and where and how you will use it. A smaller, portable instrument may get used more frequently than a large, unwieldy one. The best telescope for you is the one you will use most often.

and should be avoided. Many basic telescopes accept only 0.965 inch (24.5 mm) diameter eyepieces, but if you are prepared to spend a bit more, look for a telescope that takes the better grade of 1.25 inch (31.8 mm) diameter eyepieces.

## APERTURE NOT POWER

The key specification of any telescope is its aperture—the diameter of the lens or mirror that collects the light. The magnification is unimportant—by changing eyepieces it is possible to make any telescope magnify any amount. But the maximum power any telescope can deliver is equal to about 50 times its aperture in inches. A 2.4 inch (60 mm) telescope cannot operate much higher than 120 power. The limit for a 4.5 inch (110 mm) scope is 225 power. Exceed these limits and the image will indeed get bigger, but it will also look faint and fuzzy.

The critical factor is the light-gathering power of the instrument. The greater the aperture of the telescope, the more light it can collect. In fact, for every doubling of aperture, light-gathering power goes up by a factor of four. For example, a 6 inch (150 mm) mirror has four times the surface area of a

3 inch (75 mm) mirror. It collects four times the light, making the images four times brighter.

Doubling the aperture of a telescope also doubles its resolving power—the ability to see fine details on planets or split closely spaced stars. Under excellent seeing conditions, a 4 inch (100 mm)

## BUYING FOR A CHILD

With their sharp eyesight and insatiable curiosity, children make excellent skywatchers, and the gift of a telescope can lead to a lifelong hobby—or even a career.

When buying a scope for a child, avoid cheap "high-powered" instruments. Even adults have a difficult time using these shaky, fuzzy telescopes. Purchase a good-quality 2.4 inch (60 mm) refractor or, better yet, move up to a larger refractor or a 4 to 6 inch (100 to 150 mm) reflector. Choose one with an altazimuth or Dobsonian mount, rather than a complicated equatorial mount (see p. 78). It is important that the child can use the telescope unsupervised, although you will both get much more out of the purchase if you spend time skywatching together.

If you are concerned that a good-quality telescope is too expensive to start with—or if the child is still too young for a delicate instrument—purchase binoculars instead. Buy a set of star charts (see p. 96) and spend a year with the child learning the constellations and picking out binocular targets. You can sample telescopic views at star parties and public observatories.

*A 6 inch (150 mm) reflector on a Dobsonian mount.*

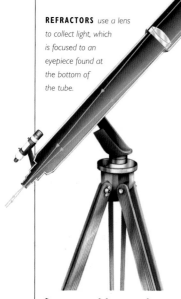

**REFRACTORS** *use a lens to collect light, which is focused to an eyepiece found at the bottom of the tube.*

## LENS OR MIRROR?

The choice of serious but budget-minded skywatchers is usually a Newtonian reflector, a design relatively unchanged since Isaac Newton invented it in 1668. As a rule, reflectors provide much greater aperture for the money than refractors do, and an economical 4 or 4.5 inch (100 or 110 mm) Newtonian can serve as a good starter scope. Several manufacturers now offer very affordable 6 inch (150 mm) reflectors on simple Dobsonian mounts (see p. 78). These provide views of planets and deep-sky targets so superior to those in smaller telescopes that it is worth making the jump up. The main drawback of a Newtonian reflector is that it requires some maintenance—the exposed and delicate mirrors need occasional cleaning and collimation.

**CATADIOPTRIC TELESCOPES**, *such as the Schmidt-Cassegrain (near right), use a mirror and a corrector lens. The light travels back and forth, then exits through a hole in the mirror to an eyepiece at the back of the telescope.*

Small refractors—in 2.4 inch (60 mm), 3.1 inch (80 mm), and 3.5 inch (90 mm) sizes—are also popular starter scopes. These rugged, maintenance-free instruments offer crisp images and are easy to aim. They can double for daytime activities, such as birding.

In larger sizes, refractors are premium instruments—a 4 inch (100 mm) refractor can cost five times as much as a 4 inch reflector. They often feature apochromatic (color-free) lenses, and are popular with optical connoisseurs and avid astrophotographers who prize their ultra-sharp images.

## HYBRID SCOPES

Catadioptric telescopes—the Schmidt-Cassegrains and Maksutovs—are hybrids, combining a reflecting mirror with a large corrector lens that eliminates optical distortions.

The Schmidt-Cassegrain design employs optical technology invented by Guillaume Cassegrain in the seventeenth century and Bernhard Schmidt in the 1930s. First introduced to amateur astronomers in the early 1970s, this telescope's key selling point is its portability: an

**NEWTONIAN REFLECTORS**
*(right) use a mirror to collect light, which is reflected and focused back up the tube to an eyepiece near the front of the telescope.*

8 inch (200 mm) model has a tube only one-third as long as an 8 inch Newtonian reflector. Extensive systems of accessories make Schmidt-Cassegrains good choices for astrophotography. They cost at least 50 percent more than reflectors—but far less than refractors of the same aperture.

The Maksutov, a similarly compact instrument, was invented independently by Bouwers and Maksutov in the 1940s. Also known as "Maks," these scopes come in apertures of up to 8 inches (200 mm), the most popular being the very portable 3.5 inch (90 mm) models. Maksutovs provide sharp, high-contrast images, making them favorites for planetary viewing.

## WHICH WAY UP?

All telescope designs produce upside-down images, but the refractors and the Schmidt-Cassegrains are usually used with an accessory called a star diagonal, which fits just in front of the eyepiece. This turns the image right way up but also swaps it left to right, so that you see a mirror image. The images in reflectors are generally upside down.

**TELESCOPE BUILDING** *can become a serious and engrossing hobby (left), but a simple Dobsonian mount (below) can be built in a couple of weekends.*

## BUILDING YOUR OWN

In the past, many amateurs ground their own telescope mirrors. Few pursue this aspect of the hobby today because quality commercial optics are now widely available.

Building a telescope at home, however, is as popular as ever. You simply buy a mirror and other parts, then place them into a thick cardboard tube. This sits on top of a Dobsonian mount, which you can construct from plywood. There are some excellent references on building your own telescope.

## SCOPE SPEED

A specification usually marked on the telescope's tube or given in its instruction manual is the focal length, almost always given in millimeters. This is the length of the light path from the lens or mirror to the focus point. With refractors and reflectors, the focal length is roughly equal to the physical length of the tube. But in Maksutovs and Schmidt-Cassegrains, the light path is folded back and forth several times, and the focal length increased, allowing long focal-length optics to reside in a shorter, more compact tube.

Dividing the focal length of a telescope by its aperture in millimeters gives its f-ratio. A telescope with a focal length of 2,000 mm and an aperture of 200 mm (or 8 inches) is an f/10 telescope. If the focal length were 1,000 mm, it would be an f/5 scope.

The smaller the f-ratio, the faster the telescope. Fast telescopes are an advantage for deep-sky photography— an f/5 telescope will record a nebula in a quarter of the exposure time required by an f/10 telescope. Hence, the term "fast" telescope. For visual use, however, there is little advantage of one f-ratio over another. Slower tele - scopes are sometimes sharper, making them better suited to high-power planetary viewing, but they often have longer, less portable tubes. Fast telescopes give lower powers and wider fields with any given eyepiece, making them better suited to deep-sky viewing, but they also tend to exaggerate flaws in the optics.

There is no single ideal telescope, but by taking into account your skywatching interests, you can certainly make an informed choice.

### ALVAN CLARK AND SONS

At the end of the nineteenth century, it was the well-heeled and fortunate amateur astronomer indeed who owned a telescope made by Alvan Clark and Sons. Based in Massachusetts, this famous company built refractors in sizes from 3 inches (75 mm) on up. Way up! In 1897, Clark's son Alvan Graham (shown here, on the left, with assistant Carl Lundin) completed the optics for the 40 inch (1 m) Yerkes refractor in Williams Bay, Wisconsin. It remains the largest working refractor ever made and is still in use. Clark and Sons' smaller refractors for amateurs provide views surpassed by few telescopes today, and are much sought after by collectors of fine optics.

# CHOOSING *a* TELESCOPE MOUNT

*The best optics in the world are worthless*

*without a sturdy mount—it is half the telescope.*

When a telescope has a poorly made mount, the image in the eyepiece jumps around—finding objects becomes a frustration, while seeing details within objects is nearly impossible. Fortunately, many telescopes sold today come with excellent mounts that move smoothly across the sky, and also remain firmly on target when released.

Small "department-store" telescopes, however, often have shaky mounts and flimsy tripods that flex and bounce with every touch and breath of wind. Do not be fooled by the high-tech dials and cables on some cheap mounts. Here is an easy showroom test: give the telescope tube a sharp rap. All telescopes will shake for a moment, but in a well-mounted telescope vibrations should die out within three to four seconds.

## ONE OR TWO MOTIONS

There are two main designs to consider—altazimuth and equatorial. The simplest and most common is the altazimuth mount, which swings up-and-down and left-to-right. Moving an altazimuth scope around the sky to find a target is easy. But as Earth rotates, the target moves. To keep it in view, you must move the telescope a little in both directions every few seconds, which can be tricky. Motorized altazimuths, operating under computer control are now becoming quite common (see p. 81).

Prior to the advent of computer control, the equatorial mount was the usual solution for tracking objects. It still offers some advantages. One of its axes— the polar axis—must be lined up so that it points to the

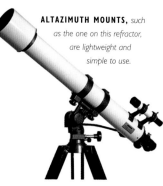

**ALTAZIMUTH MOUNTS,** *such as the one on this refractor, are lightweight and simple to use.*

celestial pole, the point in the sky around which the stars appear to turn during the night. When you rotate the polar axis, the telescope moves in the same east-to-west direction as the stars, allowing it to track a particular object with a single motion. Add a motor to the polar axis and the telescope can follow objects automatically, leaving your hands free.

## WHICH MOUNT?

Altazimuth mounts are inexpensive and quick to set up. A special type of altazimuth is now one of the most common for reflector telescopes. Popularized by the Californian telescope-maker John Dobson, this wooden mount uses simple Teflon pads to provide smooth motions in both axes. It moves

**THIS DOBSONIAN MOUNT** *(left), built by John Dobson in the late 1970s, certainly has a much simpler design than the complex mount photographed in 1912 (top left).*

effortlessly, yet stays firmly on target. A gentle nudge every minute or two is all you need to keep objects in view.

Their simplicity, economy, and stability make Dobsonian mounts an ideal choice for anyone looking for the best performance for the least money. A computer-controlled and motorized altazimuth mount certainly makes it easier to study objects such as planets under high magnification. Even with a computer-tracking system, altazimuths need some extra equipment to be suitable for most astrophotography (see p. 90).

A motorized equatorial mount is easier for tracked astrophotography. But equatorial mounts are much heavier, less portable, and more complex than altazimuth mounts. They require polar alignment (see Box) and can be confusing for beginners to move around the sky. And they are more expensive.

There are several varieties of equatorial mounts, but the most popular ones are the fork and the German equatorial mount. Fork mounts work well for short-tube telescopes such as Schmidt-Cassegrains, while German equatorials are commonly supplied with refractors and Newtonian reflectors—telescopes that have longer tubes.

## GETTING ALIGNED

For an equatorial mount to work properly, it must be polar aligned to the celestial pole. For Northern Hemisphere observers, this means aiming the polar axis toward Polaris, the northern pole star. For Southern Hemisphere observers, the pole star is the faint Sigma Octantis. You need not be too precise—unless you are taking photographs (see p. 91), getting within 5 degrees of the pole will be fine.

To polar align a fork mount, place it so the fork arms, which parallel the polar axis, aim due north (true north, not magnetic north)—or due south in the Southern Hemisphere. Level the top of the tripod, then adjust the tilt of the wedge so that the angle between the ground and the fork arms equals your latitude. For example, an observer in New York, which is at 40 degrees latitude, would tilt the wedge to a 40-degree angle. There may be a scale on the mount to help set the angle, or you can use a protractor. This angle needs to be set only once—on subsequent nights, simply aim the fork arms toward the pole star.

For a German equatorial mount, the procedure is the same, but you aim and tilt the mount's polar-axis housing (indicated below).

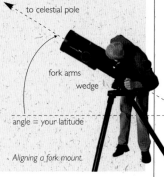

to celestial pole

fork arms

wedge

angle = your latitude

*Aligning a fork mount.*

## MOTOR DRIVES

Motor drives come in either AC or DC models. AC models run from household power, or from a power inverter connected to a car battery. DC models that have a separate battery pack can be used anywhere. Many motor drives are low voltage. They can also be used with a transformer to run from AC power.

*The heavens call to you, and circle around you, displaying to you their eternal splendors …*

The Divine Comedy, DANTE
(1265–1321), Italian poet

**THE GERMAN EQUATORIAL MOUNT**
*was invented by Joseph von Fraunhofer in the early 1800s, and is still a very popular mount. A motor drive is often available as an option.*

to celestial pole

polar axis

declination setting circle

polar-axis housing

motion in declination (north-south)

declination axis

slow-motion control

right-ascension setting circle

motion in right ascension (east-west)

angle = your latitude

# COMPUTER CONTROL

*Computer technology now makes it possible
to find even the most obscure galaxy
at the mere push of a button.*

**COMPUTER CHIPS** *are everywhere,
even inside astronomical telescopes.*

Finding objects in
the night sky is a
challenging task for
most newcomers to the
hobby. They stand in awe
as veteran skygazers aim
their telescopes at seemingly
blank areas of the sky to
reveal extraordinary views
of galaxies and nebulas.
How do the experts do it?

Starhopping is the answer—
using a good finderscope to
hop from a bright, easy-to-find
star over to fainter stars and
then to the target. Chapter 9
contains starhopping charts for
many of the sky's best and
brightest deep-sky wonders.
Conduct a starhop a few times
and you will probably remember the star pattern, making
it a snap to locate a particular
object again. You will have
graduated from a novice lost in
the stars to an expert trekking
across the sky with confidence.
But is there another way to
navigate the sky?

## SETTING CIRCLES

Most equatorial mounts have
numbered dials on each axis.
These are setting circles, and
they can be used to locate an
object using its right ascension

**COMPUTERIZED TELESCOPES**
*provide easy access to the thousands of
deep-sky objects on the Messier and
NGC lists at the push of a button.*

and declination coordinates
(see Box p. 107). However,
newcomers to astronomy
rarely find setting circles very
helpful. The dials require
careful adjustments with
every use and can quickly
lose accuracy as the sky
moves during the night.

Add-on digital setting circles
come with a pair of encoders,
one for each axis of the mount.
The encoders sense the motion
of the mount and keep track of
how far it has moved.

Digital setting circles are
easy to use and provide a

highly accurate method of
locating objects. They can be
added to most of today's popular telescopes. Remarkably,
they do not need a polar-
aligned equatorial mount,
and will work on altazimuth
or Dobsonian mounts. Most
models sell for the price of a
beginner's telescope.

## HANDS-OFF ASTRONOMY

A step up in sophistication is
a telescope with high-speed
motors on each axis and a
computerized hand controller
to drive them. The computer
is programmed with the
location of thousands of stars,
nebulas, clusters and galaxies—

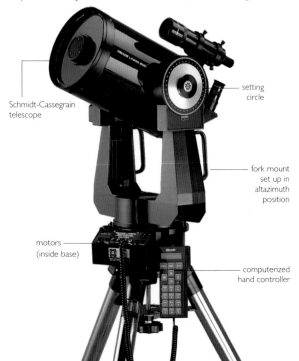

Schmidt-Cassegrain
telescope

setting
circle

fork mount
set up in
altazimuth
position

motors
(inside base)

computerized
hand controller

## EQUATORIAL OR ALTAZIMUTH?

Large observatory telescopes and amateurs wishing to photograph the sky have, in the past, needed to put their telescopes on an equatorial mount that allowed a single constant motorized motion of the telescope to track the stars as they rotated across the sky.

The Anglo-Australian Telescope (left), for example, was one of a group of 150 inch (4 m) aperture telescopes built around the world in the 1970s. They all featured massive equatorial mounts, but were the first big telescopes designed to be run by computers.

By the time the current generation of 320 to 400 inch (8 to 10 m) telescopes were designed in the 1990s the thinking had changed. Much smaller altazimuth mounts were preferred because computer control could now adjust the drive rate of two motors to keep the telescope on target. The VLT's four telescopes are an example (right). In fact, computers were used to keep the mirror itself in shape.

Amateur astronomy was not far behind. The new "Go To" telescopes employ the same principles as the giant telescopes—fortunately for a fraction of the cost.

and often also the planets. Select an object on the hand controller's display, press "Go," and the telescope automatically slews itself across the sky to the target. These have become known as "Go To" telescopes. You can also operate these telescopes from a laptop computer with a sky-charting program. Point at a galaxy on the screen, click the mouse, and away the telescope goes to find the real galaxy.

You begin by "telling" the telescope where it is on Earth—your location. Then you tell the telescope the current position of the stars by aiming the telescope at two bright stars on either side of the sky. These two positions are all the computer needs to know to guide you to any of the objects in its memory. Call up one of those objects and the telescope automatically seeks out the correct location and indicates when you are on target. Look in the eyepiece and there the object is!

This technology is most popular on fork-mounted Schmidt-Cassegrain telescopes. Remarkably, it works whether the telescope is on an equatorial or an altazimuth mount (see photo opposite). The computer will pulse each of the motors by the correct amount to track an object moving across the sky. One

catch is that the field of view rotates with time with an altazimuth mount. So the view rotates in the eyepiece during the night—not a problem when viewing by eye, but a problem for long exposure photographs. Some telescopes have de-rotators to overcome this.

The ability to find and follow objects easily without polar alignment has made computerized Schmidt-Cassegrains very popular. Of course, there is a price to pay—a computerized model will cost about twice as much

**A LAPTOP COMPUTER**

*can control a high-tech*
*telescope from*
*up to hundreds*
*of feet away.*

as a similar telescope on a manually operated mount. More recently, reduced costs have seen "Go To" technology appearing on smaller, less expensive telescopes. The computer age is having a big impact on skywatchers.

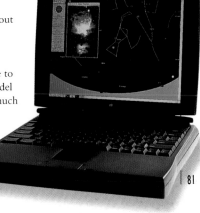

# SELECTING EYEPIECES

*Whenever you look through a telescope, you look through an eyepiece. Eyepieces are usually the first accessories telescope owners buy.*

A telescope's main lens or mirror gathers the incoming light and focuses it into an image, but it is the eyepiece that magnifies the image. To change magnifications, you need to change eyepieces. Even the best telescopes may come with only one general-purpose eyepiece as standard equipment. Two or three additional eyepieces are essential.

## EYEPIECE FOCAL LENGTHS

Eyepieces are sold not by their magnifications but by their focal lengths (always given in millimeters). Look for numbers

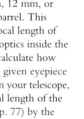
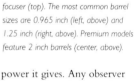

such as 25 mm, 12 mm, or 9 mm on the barrel. This indicates the focal length of the miniature optics inside the eyepiece. To calculate how much power a given eyepiece will produce on your telescope, divide the focal length of the telescope (see p. 77) by the focal length of the eyepiece.

A 20 mm eyepiece inserted into a telescope with a focal length of 2,000 mm, for instance, would give 100 power. The same eyepiece inserted into a 1,000 mm focal length scope would give 50 power.

The shorter the eyepiece's focal length, the higher the

**EYEPIECES** *slide into the telescope's focuser (top). The most common barrel sizes are 0.965 inch (left, above) and 1.25 inch (right, above). Premium models feature 2 inch barrels (center, above).*

power it gives. Any observer can use at least three eyepieces: a low power (35x to 50x), a medium power (80x to 120x), and a high power (150x to 180x) eyepiece. This means that if your scope has a focal length of 1,000 mm, for example, you might select 25 mm, 12 mm, and 6 mm eyepieces. Low power provides wide fields for locating targets and for panoramic views of star fields; medium power resolves clusters and double stars; and high power reveals details on the planets. Magnifications much higher than 180x are rarely of any use—the image gets bigger but becomes blurry in the process. Also avoid zoom eyepieces—they may promise to combine a range of powers, but in practice they deliver poor images.

**NAGLER**
(8 lens elements)

**PLÖSSL**
(4 lens elements)

**KELLNER**
(3 lens elements)

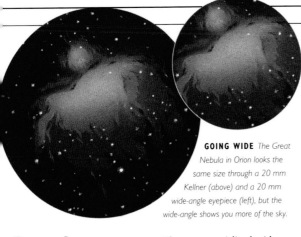

**GOING WIDE** *The Great Nebula in Orion looks the same size through a 20 mm Kellner (above) and a 20 mm wide-angle eyepiece (left), but the wide-angle shows you more of the sky.*

## BARREL SIZES

Eyepieces come in three barrel sizes: 0.965 inch (24.5 mm), 1.25 inch (31.8 mm), and 2 inch (50.8 mm). Many small starter telescopes accept only 0.965 inch eyepieces. Unfortunately, the Huygenian and Ramsden models of eyepieces usually packed with these telescopes are of mediocre quality. Better models—such as Kellners, Orthoscopics, and Modified Achromats—are available for this smaller barrel size. Upgrading to higher-quality eyepieces is one of the best things you can do to improve the performance of a low-cost telescope.

A telescope that accepts the larger 1.25 inch eyepieces is a definite plus. These eyepieces are consistently better quality and are available in a vast selection of models. A small number of eyepieces, however, have even larger, 2 inch barrels.

*Light from distant places has made the journey to earth, and it falls on these new eyes of ours, the telescopes …*

Cosmic Landscape,
MICHAEL ROWAN-ROBINSON
(b. 1942), British mathematician

These are specialized wide-angle eyepieces that fit only some of the premium telescopes on the market.

## OPTICAL DESIGNS

Eyepieces are available in various optical designs. Some manufacturers have exclusive rights to certain designs—such as Edmund Scientific's RKE models, Meade's Modified Achromats, and TeleVue's Nagler and Panoptic series. But the most common designs are sold by nearly every dealer.

Kellner, RKE, and Modified Achromat eyepieces use three lens elements for good image quality at low cost.

A step up in quality and price are the four-element Orthoscopic and Plössl models. Most manufacturers have matched sets of Plössls in their catalogs—they represent the best buys in today's marketplace. Every telescope owner can use a set of three or four Plössls.

In the premium price bracket are eyepieces known by generic names such as Erfles, and brand names such as Wide Fields, Ultra Wide Angles, Panoptics, and Naglers. These five- to eight-element models come in focal lengths from 35 to 4.7 mm. Compared with a standard Plössl, these models all show a much wider field, providing a wonderful "picture-window" view of the universe.

### KEEPING OPTICS SPARKLING

Nothing can smear your view of stars and planets better than an eyepiece coated with dust, eyelash oil, and fingerprints. To clean an eyepiece, first blow off loose dust with a lens blower. Then lightly moisten a Q-tip cotton swab with camera-lens cleaning fluid. Wipe the lens gently with the wet swab, then again with a dry one to remove the streaks. Never pour the fluid directly onto the lens—it can seep into the eyepiece, staining interior lenses. And do not try to disassemble an eyepiece—the lenses could all tumble out.

The main lenses and mirrors of telescopes should be cleaned with caution and as seldom as possible. To clean refractor lenses, blow off as much dust as you can, then moisten lens cleaning tissue with lens cleaning fluid. Wipe the lens with gentle strokes. Do not rub hard.

Reflecting mirrors are the most delicate of all. They must be cleaned only with gentle swipes of cotton balls moistened with a solution of distilled water and mild detergent, then rinsed with pure distilled water. Other methods can leave scratches, which do far more harm than dust. To keep the dust off in the first place, never store telescopes with their lenses or mirrors exposed.

# THE BEST ACCESSORIES

*A few well-chosen accessories can improve
the performance of your telescope and
increase your enjoyment of skygazing.*

Like photography, back-
yard astronomy is a
hobby filled with gadgets
and accessories. Some are frills,
but the best ones can enhance
your view of the universe.

## FILTERING OUT LIGHT
You can select from a wide
range of filters according to
your skywatching interests.
Filters usually screw into the
base of the telescope's eyepiece.

Some filters minimize
the effects of light pollution.
Known as LPR (light pol-
lution reduction) filters, they
block the green and yellow
wavelengths emitted by
street lights, but transmit the
red and blue-green colors of
nebulas. While these filters can
marginally improve views of
star clusters and galaxies, they

work best enhancing emission
nebulas (ones that produce
their own light).

LPR filters come in several
varieties. Broadband, or deep-
sky, filters transmit the widest
range of colors, providing
a modest enhancement of a
range of deep-sky objects.
Narrowband, or nebula, filters
are much more effective on
emission nebulas and are the
best choice if you are planning
to buy just one filter. Line
filters, such as Oxygen III and
Hydrogen-beta filters, transmit
only a single color, providing
excellent results on a limited
number of nebulas.

**FILTERED LIGHT** *Colored
filters (top) and LPR filters
(right) provide clearer images
of some objects. A light-polluted
view of the Veil Nebula (below left) shows
a dramatic improvement (below right)
when a narrowband LPR filter is used.*

## SOLAR SYSTEM SIGHTS
In the past, many entry-level
telescopes came with "sun
filters" that screwed into an
eyepiece. These filters are very
dangerous and should never be
used. Thankfully, they are
rarely supplied today.

The *only* safe way to view
the Sun directly through a
telescope is with a filter that
covers the entire front aperture
of the scope. Mylar filters are
the least expensive and are
available in sizes to
fit most telescopes. They produce a blue-tinted view of
the Sun. Glass filters coated
with metal show a

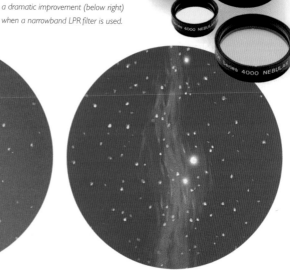

**BARLOW LENSES**
*expand the use of your eyepieces and the range of your magnification choices. They come in 2x and 3x models.*

natural-looking yellow Sun, but cost about 50 percent more. Both Mylar and glass filters give "white-light" views of features such as sunspots and faculae (see p. 126). To see towering prominences on the edge of the Sun, however, you need a Hydrogen-alpha solar filter. These can cost as much as a small telescope.

Colored filters can accentuate particular features on the planets: a green filter brings out Jupiter's Red Spot; a red filter improves views of the elusive dark markings on Mars. A set of four to six colored glass filters often costs no more than an eyepiece. An even less expensive option is to purchase colored gelatin filters from a camera store (see pp. 158, 167, and 173).

## EASY POWER
Eyepieces with a short focal length (4 to 8 mm) provide high power, but, as a rule, the higher the power of the eyepiece, the shorter its eye relief. The shorter the eye relief, the closer your eye must be to the eyepiece, making it less comfortable to use.

An alternative is a Barlow lens inserted between an eyepiece and the telescope. A 2x Barlow, which doubles the power of the eyepiece, is the most common size. A 12 mm eyepiece with a 2x Barlow

produces the same high magnification as a 6 mm eyepiece, but without giving up the comfortable eye relief of the longer 12 mm eyepiece.

## DEWCAPS
When exposed to the cool night air, the front lens of a refractor or Schmidt-Cassegrain can readily attract dew or frost. One remedy is a dewcap—a tube extending in front of the telescope. Dewcaps come with most refractors, but they must be bought separately for Schmidt-Cassegrain telescopes. Dewcaps are fine for light dew. On very humid or frosty nights, however, a

heater coil that wraps around the telescope tube is much more effective. Heater coils are also available for eyepieces and finderscopes.

## CREATURE COMFORTS
The best accessories of all are items that increase your comfort during long skywatching sessions. In the cold night air, warm boots and clothing are essential. In warm climates, you may need insect repellent. An adjustable-height stool or chair lets you sit comfortably at the eyepiece of your telescope. A folding table provides a place to lay out charts and accessories, while a foam-lined camera case is good for storing them. A flashlight with a red filter helps you find things in the dark without affecting your night vision.

## FINDING THINGS

Finderscopes are small, wide-field secondary telescopes mounted on the main telescope. They make it easy to aim the main telescope and conduct one of the starhops detailed in Chapter 9.

Many entry-level telescopes come with 5 x 24 finders—these give 5x magnification and have apertures of 24 mm. Their small lenses are fine for centering a telescope on a bright planet. If you want to locate fainter targets, however, consider upgrading to at least a 6 x 30 finder. For owners of 6 to 8 inch (150 to 200 mm) telescopes, an 8 x 50 finder is a better choice.

An accessory some people find wonderfully useful, either instead of or as well as a finderscope, is one of the reflex finders, such as the Telrad. Looking through a reflex finder, you see a naked-eye view of the sky with a red dot or bull's-eye floating among the stars (above).

# BECOMING *an* ASTROPHOTOGRAPHER

*Some astrophotography techniques require no more than the camera gear you may already own.*

Hardly anyone who has a camera and an interest in astronomy can resist taking photos of the night sky. Snapshot astrophotos will record a memorable night-sky scene just as your eye saw it. Long-exposure photos can actually capture far more than meets the eye, recording striking colors and faint objects that are invisible by eye even through a telescope. These images may serve as valuable scientific records or simply as beautiful "paintings" of celestial light.

**LONG EXPOSURES** show up colors in the Lagoon and Trifid nebulas (above) that remain invisible to the eye. With a camera adapter (left), you can easily attach a camera to a telescope.

## CAMERAS
CCD cameras (p. 94) are becoming more popular, but 35 mm is the recording medium for most first-time astrophotographers. For casual sky shooting, almost any design of camera will do, but it must have a B (for Bulb) setting for the shutter. This setting holds the shutter open for as long as the shutter button is pressed—an essential feature because most night-sky photos require exposures of several seconds, if not minutes. To hold the button down during these long exposures, you use a locking cable release, which screws into the shutter button of some cameras.

The drawback of most new cameras is that they are battery operated. Relying on batteries is risky—they may fail in the cold night air or during long exposures. This is why some astrophotographers seek out older mechanical single-lens-reflex (SLR) film cameras on the secondhand market. It might seem strange to forgo the electronic features that adorn today's cameras, but none of them is needed for astrophotography if you are one of the increasingly rare photographers using photographic film.

While a close-up of the Moon or a clear image of a faint nebula requires shooting through a telescope (see pp. 92–93), many sky subjects can be captured with standard camera lenses (see pp. 88–91). Lenses that have fixed focal lengths are usually of better quality and faster than zoom lenses. For 35 mm cameras, a set of three fixed lenses is useful: a wide-angle lens (one with a focal length of 24 to 35 mm), a normal lens (50 to 55 mm), and a short telephoto lens (85 to 200 mm). Try to choose fast lenses—these have a maximum aperture of f/2.8 or f/2.

For photographing the night sky, especially with color film, most of the filters employed by photographers and amateur astronomers are of limited value. Broadband LPR, or deep-sky, filters (see p. 84) can help darken sky

**THIS EARLY ASTROPHOTO** *of the region surrounding the Lagoon and Trifid nebulas was taken by E. E. Barnard.*

will help you learn which exposures and methods work best. Always shoot a range of exposures—a technique called bracketing. A shorter or longer exposure will often produce a better result than an estimated "best" exposure.

backgrounds but usually shift the color balance to green.

## STILL ON FILM?

With the rapid spread of electronic digital cameras, technology and techniques once restricted to professional astronomers are now available to anyone (see p. 94). However, many of the same rules and techniques still apply.

Digital cameras are much more sensitive to light than film is. Your camera determines that. For film, however, the speed of a film—its ISO rating— indicates how quickly it reacts to light. Faster films, with higher ISO rating, allow shorter exposure times, but they also tend to produce grainier images.

Twilight scenes or bright objects such as the Moon are best captured on slow, fine-grained films, which have an ISO rating of 50 to 100.

Most celestial shots require the extra light-gathering ability of at least an ISO 400 film. For ultra-high speed, film speeds up to ISO 3200 have been offered and found to be suitable for astrophotography.

In these days where film cameras are becoming increasingly rare, your photographs will be safer with a specialized professional lab.

Always shoot a daylight scene at the start of the roll so the lab can work out where the frames are on your film.

## KEEPING RECORDS

The best tip for improving your astrophotos is to keep records of the exposure data, sky conditions, and equipment used for every shot. This

*I discovered in the heavens many things that had not been seen before our own age.*

GALILEO GALILEI (1564–1642), Italian astronomer

## E. E. BARNARD

Imagine painstakingly guiding a telescope for three hours on a cold Wisconsin winter night. "How do you keep warm?" visitors to the Yerkes Observatory would ask. "We don't," replied Edward Emerson Barnard (1857–1923). For this pioneer of astrophotography, the arduous conditions were part of a normal night's work.

From the 1840s on, astronomers had been among the first to put the new technique of photography to work. Until Barnard, however, nobody had used it to map the Milky Way. From 1889 to 1895, he used an astrograph—a device with a 6 inch (150 mm), f/5 lens—and multi-hour exposures to photograph the Milky Way. By 1905, he had switched to the 10 inch (250 mm) Bruce Photographic Telescope, which used glass plates. Images taken with these two instruments formed the basis of Barnard's 1927 *Atlas of Selected Regions of the Milky Way*, a work that proved our galaxy is riddled with dark dust and nebulosity. Many of the famous star fields portrayed in this stunning atlas were first photographed, and in some cases, discovered, by Barnard. And while our optics and films have improved, few astrophotographers have surpassed the quality and quantity of E. E. Barnard's century-old images.

# CAMERA *on* TRIPOD

*With today's fast films, the only equipment
you need to start taking spectacular photographs
of the night sky is a camera and a tripod.*

Taking pictures of the night sky is back-to-basics photography. With today's automated cameras you need only point and shoot to get technically perfect photos—at least during the day. At night, especially with faint subjects such as stars, old-fashioned skill has to replace the microchip circuits of modern cameras. Light meters and autofocus sensors often will not function in the dim light.

An astrophotographer must manually set the f-stop and the shutter speed. The f-stop is the adjustable aperture of a lens, which controls how much light enters the camera; the shutter speed is the length of time the shutter remains open to the sky. Simple astrophotography—shots done with a camera on a tripod—usually employs fast apertures of f/2 to f/2.8. A lens set to an aperture of f/2.8, for example, lets in twice as much light as a lens set to f/4.

**SHORT EXPOSURES** *keep stars looking like points. This portrait of Moon, planets, and stars was taken on film with an exposure of 4 seconds.*

**LONG EXPOSURES** *allow the stars to trail. This photo shows the same scene as the portrait does, but was taken with an exposure of 90 minutes.*

With a good-quality fast lens, you can capture striking portraits of constellations and Milky Way star clouds without a tracking system. No guiding, no polar alignment, no elaborate equipment to buy or set up. Just place the camera on a sturdy tripod, focus its lens at infinity, and, with the camera in manual-exposure mode, set the lens aperture to f/2.8. Frame the scene, then lock

the shutter open on its B setting for 10 to 80 seconds.

Try a range of exposures at first. With exposures of less than 60 seconds, you can record stars far fainter than you can see with the naked eye.

## CONSTELLATIONS

With this simple technique, it is possible to capture constellation patterns, the Milky Way, and even bright nebulas such as the Eta Carinae Nebula and the Great Nebula in Orion. Only the brightest stars will record in light-polluted skies, so the darker the sky, the better.

Of course, the sky is always moving. How long an exposure you can use before the stars start to streak depends on the lens's focal

## CONSTELLATION EXPOSURES

To avoid star trailing, use exposures no longer than these.

| Lens focal length | Near 90° dec. (cel. poles) | Near 45° dec. | Near 0° dec. (cel. equator) |
|---|---|---|---|
| **28 mm** | 80 seconds | 50 seconds | 30 seconds |
| **35 mm** | 60 seconds | 35 seconds | 20 seconds |
| **50 mm** | 40 seconds | 25 seconds | 15 seconds |

**THIS LUCKY SHOT** *captures a meteor streaking across a sky that is lit up by a red and green aurora.*

### TOP CAMERA-ON-TRIPOD TARGETS

1 Total solar eclipse

2 Rare all-sky aurora

3 Crescent Moon near Venus in twilight

4 Milky Way in Sagittarius

5 Circumpolar star trails

length and on where in the sky it is pointed (see Table). Stars near the celestial equator, such as those in Orion, will trail more quickly than stars near the celestial pole, such as those in the Big Dipper.

## STAR TRAILS

But why not let the stars trail across the frame? Star-trail pictures are always impressive. For a dramatic composition, include a landscape or a fore-ground object such as a tree.

For first attempts, try exposures of 5 to 30 minutes at f/4. Check the results—if the sky is brightly lit, the exposures were too long, but if the sky is dark, your site is good enough to allow even longer exposures. At very dark sites, exposures can last all night, especially with the lens stopped down to f/8 or so.

## MOONLIGHT PORTRAITS

Astrophotography need not stop when the Moon comes up. On moonlit nights, venture out to scenic locations. During a Full Moon, exposures well under a minute at f/2.8 will record a scene that looks like daylight, complete with blue sky. And yet there will

be stars in the sky—a nightscape shown in literally a different light.

## TWILIGHT SCENES

Few images portray the beauty and tranquillity of the night sky better than a photograph of the crescent Moon at twilight, especially when a bright planet such as Venus is nearby. Use exposures of less than a second at f/2.8. With a sensitive light meter, you might be able to take an exposure reading off the twilit sky—this may be the only astrophotograph where a light meter is of any value.

## AURORAS

The Northern and Southern lights are photogenic, but elusive, subjects. During a bright display, use a fish-eye (8 mm to 16 mm), wide-angle (24 mm to 35 mm), or normal (50 mm) lens set at f/2.8 or faster. There is no single correct exposure—a shorter exposure freezes the rippling curtains of light, but may not pick up the subtle shades of red, green, blue, and purple that film can record but the eye usually cannot see.

**THIS NIGHTSCAPE** *is illuminated by a Full Moon just outside the frame.*

## METEORS

Take enough constellation and star-trail shots, and chances are a meteor will eventually streak across a frame. It is, however, much more difficult to catch a meteor deliberately. The best times are during an annual meteor shower (see p. 198). Use a 24 mm to 50 mm lens set to f/2.8, and exposures as long as your site allows. For longer meteor streaks, aim away from the shower's radiant. Keep shooting all night—with luck, one frame will grab a meteor.

# PIGGYBACKED CAMERA

*By using an equatorial mount to*
*track the stars during an exposure, you can*
*reach deep into space and record fainter targets.*

The next step up in complexity still uses the camera's own lens to shoot the sky, but now the camera rides piggyback on an equatorially mounted platform that follows the stars. The advantage is that stars will record as points, even in 10- to 30-minute exposures. Because the exposures can be so long, a piggybacked camera is able to capture stars and nebulosity far fainter than any stationary camera will record.

However, the platform must be aligned with the celestial pole. As the focal length of the lens goes up, so does the intolerance for poor alignment. For example, telephoto lenses with focal lengths of 100 to 300 mm demand alignment within ⅕ degree. Just aiming at Polaris or Sigma Octantis will not do (see Box). An altazimuth

> **TOP PIGGYBACK TARGETS**
>
> 1 **Bright comets such as Hale-Bopp**
>
> 2 **Milky Way in Sagittarius**
>
> 3 **Constellation of Orion**
>
> 4 **Nebulas and clusters of Carina/Crux area**
>
> 5 **Large Magellanic Cloud**

**A TRIPOD HEAD** *attached to a telescope allows the piggybacked camera to aim independently of the telescope.*

mount will not do either, even if computer-controlled, as the view rotates as the sky moves.

## THE PIGGYBACK PLATFORM

The simplest platform is the barn-door tracker, a home-built contraption made of two wooden boards bolted together. As you slowly turn the bolt, the "barn door" gradually opens up, swinging the camera from east to west

across the sky. Exposures of up to 10 minutes are possible.

Any motorized equatorial mount can serve as a platform for a piggybacked camera. A few manufacturers market small "scopeless" mounts that serve as dedicated camera platforms, but most people attach the camera to the side of a telescope tube. Your telescope mount may already have an attachment bolt, or you can buy an adapter bracket.

## FILM AND EXPOSURES

The ideal subjects for piggyback photos are constellations, Milky Way panoramas, large nebulas and galaxies, and comets. These are best shot with lenses set to an aperture of f/2.8 to f/4. Lenses with faster settings are available, but most perform poorly when used "wide open" at f/1.4 or f/2. Stars appear

**A CONVENTIONAL CAMERA** *riding piggyback can capture an expanse of the Milky Way. This image was taken with a wide-angle lens set at f/4.*

*... look, how the floor*

*of heaven*

*Is thick inlaid with patines*

*of bright gold.*

The Merchant of Venice,
WILLIAM SHAKESPEARE (1564–1616)

bloated, especially at the corners of the frame.

Exposure times depend on sky conditions. In a dark sky, a few minutes at f/2.8 records an astonishing amount of nebulosity.

## GUIDING OUT ERRORS

When shooting with short focal length lenses (under 135 mm), you can usually turn on the drive, open the shutter, and walk away. Longer lenses,

however, are more demanding, and slight irregularities in the speed of the drive must be guided out. To do this, you monitor a star through the main telescope during the exposure. An illuminated reticle—an eyepiece with cross hairs illuminated by a dim red light—can help you detect any wandering of the guide star.

The mount should have drive motors on both axes,

**A TELEPHOTO LENS** *works well for nebulous areas. This is a 20-minute film exposure of the Eta Carinae Nebula, taken with a 300 mm lens set at f/4.*

each with a push-button speed control. If the star wanders one way along the cross hairs, press the appropriate button to jog it back. Taking a long exposure demands constant vigilance at the eyepiece, or the use of an automatic CCD guider (see p. 93).

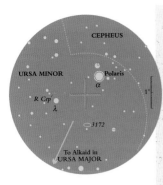

**NORTH CELESTIAL POLE**

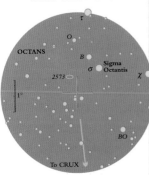

**SOUTH CELESTIAL POLE**

### PRECISE ALIGNMENT

To record pinpoint stars in long exposures on an equatorial mount, you need to zero in on the *exact* location of the north or south celestial pole. One way to do this is to use these charts (left). Each has a field of view of 5 degrees, about the same as most finderscopes.

**Step 1** Ensure that the finderscope's cross hairs aim at the same spot in the sky as your main telescope.

**Step 2** Place the mount so that its polar axis (see p. 79) aims roughly at the celestial pole.

**Step 3** Swing the tube so that it is aimed at 90 degrees declination according to the declination setting circle.

**Step 4** By moving the *whole* mount, now aim the finderscope at the pole location shown on the chart. Take care not to move the mount in declination or right ascension. Use the star patterns as a guide— while peering through the finder with one eye, watch Alkaid or Crux with the other eye to work out which way to offset the telescope.

**Trouble-shooting** This method can go awry if your declination setting circle does not read a true 90 degrees. To calibrate it, set the telescope at 90 degrees, then rotate the telescope back and forth in right ascension (east-west). If the telescope is set accurately, the stars will appear to revolve around the center of the eyepiece field. If they do not, move the telescope slightly in declination and try again. Once you have the stars revolving around the center, loosen the declination circle and turn it until it reads 90 degrees. Then lock it down. You should not need to adjust it again. Now return to Step 4.

# THROUGH *the* TELESCOPE

*By shooting through a telescope, you can add frame-filling snapshots of the Moon and close-ups of glowing nebulas to your astro-album.*

When newcomers to the hobby think about taking astro-photographs, this is what they have in mind—attaching a camera to the focus of a telescope, a technique often called prime-focus photography. The lens is now the telescope itself, so a single-lens-reflex (SLR or DSLR) camera is essential—it allows you to view and focus through the same lens that will take the photograph.

To adapt a camera to a telescope, remove the lens from the camera and the eyepiece from the telescope. Attach a T-ring designed for your camera brand onto the camera's lens mount, then screw the T-ring onto a camera adapter. The tube of the camera adapter often just slides into the focuser of the telescope in place of an eyepiece.

Any telescope mount will do for snapshots of the Moon

**MOON SHOTS** *at the prime focus of most telescopes record the entire lunar disk (above). Lunar close-ups (right) require eyepiece projection and exposures of around a second. For good deep-sky images, you may need an off-axis guider between scope and camera (top).*

(and of the Sun—with a safe filter, see p. 84). However, to take long-exposure photographs of planets and deep-sky objects such as nebulas and galaxies, you will need the tracking ability of a sturdy equatorial mount equipped with an accurate drive and variable speed controls.

## SHOOTING THE MOON

The best subject for trying out prime-focus photography is the Moon. Exposures are only a fraction of a second (see Table).

The size of the Moon in the frame depends on the telescope's focal length (see p. 77). A 2,000 mm focal length system produces a lunar disk just big enough to fill a 35 mm frame but too big for most digital cameras. With a 1,000 mm system, the Moon will be about 9 mm across—still impressive.

| LUNAR EXPOSURES | | | | |
|---|---|---|---|---|
| Assuming the use of ISO 50 film. Exposure times given in fractions of a second. | | | | |
| **f-ratio** | **Thin Crescent** | **Thick Crescent** | **Quarter Moon** | **Gibbous Moon** | **Full Moon** |
| **f/4** | $1/30$ | $1/60$ | $1/125$ | $1/250$ | $1/500$ |
| **f/5.6** | $1/15$ | $1/30$ | $1/60$ | $1/125$ | $1/250$ |
| **f/8** | $1/8$ | $1/15$ | $1/30$ | $1/60$ | $1/125$ |
| **f/11** | $1/4$ | $1/8$ | $1/15$ | $1/30$ | $1/60$ |
| **f/16** | $1/2$ | $1/4$ | $1/8$ | $1/15$ | $1/30$ |
| *Digital cameras will allow shorter exposures.* | | | | |

## TOP PRIME-FOCUS TARGETS

1 **Great Nebula in Orion**

2 **Ring Nebula**

3 **Omega Centauri**

4 **Moon with earthshine**

5 **Close-ups of First Quarter Moon**

Because the viewing screens of most cameras go dark at such long focal lengths, the principal challenge is to get the Moon in focus. A camera with an interchangeable focusing screen helps. Matte screens made for telephoto-lens work provide a brighter image, making it easier to get that sharp lunar snapshot.

### ZOOMING IN

Filling a frame with a small section of the Moon requires lots of focal length, more than most telescopes offer. The answer is eyepiece projection. Many camera adapters come with an add-on extension tube that accepts an eyepiece, although Schmidt-Cassegrain telescopes require a separate adapter called a tele-extender. The eyepiece projects a highly magnified view of the Moon onto the film.

Eyepiece projection is also essential for photos of planets. Without the extra magnification, their disks will be far too small to show any detail.

Lunar close-ups and planet portraits usually require faster exposures of up to a second. To make the exposure, hold a black card over the front of the telescope, open the camera shutter on Bulb, then flip the card away and back again. Without this "hat trick," your photographs may be blurred by the slight vibration of the camera's own shutter.

### DEEP-SKY SPLENDORS

For faint deep-sky targets such as nebulas and galaxies, faster telescopes—ones with f-ratios of f/4 to f/7—are a definite advantage because they allow shorter exposure times. Alternatively, you can speed up a slower telescope with an accessory called a tele-compressor.

Even at fast f-ratios, you will need exposures of 15 or more minutes. It would be wonderful if we could simply attach a camera at prime focus, open the shutter, and walk away. Try it at these focal lengths and the result will be horribly trailed stars, despite the most precisely aligned mount.

Errors in the drive gears are the chief culprit, but flexing mounts and tubes, atmospheric refraction, and wind all contribute to wiggly stars. Some method of guiding the telescope is called for.

Schmidt-Cassegrain owners usually turn to off-axis guiders. These devices contain a small prism that picks off a star image just outside the film frame, allowing the photographer to monitor the guide star's position while photographing through the same telescope. Owners of refractors and Newtonian reflectors often opt for a separate piggybacked guidescope—perhaps a small 2.4 or 3 inch (60 or 75 mm) refractor.

The least expensive but most difficult method of guiding is to do it by hand, tweaking speed controls to keep a star centered on the illuminated cross hairs of the guiding eyepiece. Today there is an alternative. Automatic guiders using CCD chips detect any wandering of the star's image and send pulses to the telescope's motors, which move the scope so the star returns to its centered position. The constant, unfailing corrections result in a photograph more perfectly guided than any human is capable of.

**PRIME FOCUS** *Shot through an 8 inch (200 mm) Schmidt-Cassegrain, the Great Nebula in Orion fills the frame.*

**A NORMAL EYEPIECE** *inserted into a camera adapter will magnify the image—an essential technique for planet shots.*

# VIDEO *and* CCD CAMERAS

*With CCD technology, amateurs can capture images of planets and faint galaxies that were once available only through observatory telescopes.*

**SIMPLE ASTRO-MOVIES** *can be made by holding a home camcorder above the eyepiece of your telescope.*

Professional astro-
nomers no longer use
photographic emulsions
to record the sky. Film has
been replaced by electronic
chips. This digital revolution
is now also almost complete
in amateur astronomy.

## PUTTING THE MOON ON TV

Going digital with astro-
photography can be as simple
as holding a camcorder up
to a telescope eyepiece. Aim
the telescope at the Moon,
insert a low-power eyepiece,
then hand-hold the camcorder
above the eyepiece or, better
yet, place the camcorder on a
separate tripod so that it looks
directly into the eyepiece.
Focus the camcorder's built-in
lens to infinity and zoom it
in to its telephoto position
to see highly magnified views
of small areas of the Moon.
You can also use a camcorder
in this way to capture images
of the planets.

## DIGITAL IMAGING

Most home camcorders are
not sensitive enough to pick
up images of nebulas and
galaxies. For such distant
objects, astro-imagers have
turned to specialized CCD
cameras. CCD stands for
charge-coupled device,

a type of light-sensitive chip
similar to the chips used in
home camcorders.

Instead of movies, CCD
cameras record single long
exposures. During an exposure, light falls onto an array
of pixels arranged in a grid
pattern on the chip. At the
end of the exposure, each
pixel reads out a charge that
corresponds to how much
light fell onto it. The readings
are digitized (turned into
binary numbers consisting
of 0s and 1s), then sent to a
computer that can re-create
the image on its monitor.

CCD cameras have major
advantages over photographic
film. A CCD image is digital
data, which can be manipu-
lated, stored on computer

disks, and transmitted across
the Internet. You can also see
the results immediately at the
telescope. Most importantly,
CCDs are much more sensitive
than film, requiring exposures
of only 5 minutes to record
what might take a fast film
30 to 60 minutes to pick up.
With these shorter exposures,
there is often less need for
elaborate guiding. Because the
images are digital, they can
be enhanced on a computer,
making it possible to discover
faint objects even in light-
polluted or moonlit skies.

The main drawback is the
expense of a CCD camera—
which can cost as much as a
good telescope—and also the
hardware and software that

**IMAGE-PROCESSING SOFTWARE**
*such as Photoshop can merge three
black-and-white CCD images taken
through filters and create a color image.*

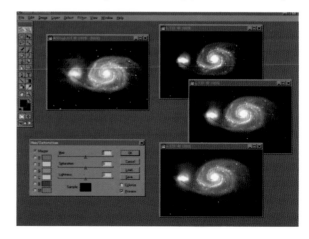

goes along with it: you will need a personal computer loaded with plenty of RAM, space on the hard drive, and image-processing programs. You may also want a laptop computer so you can easily operate the camera in the field.

## CHOOSING A CCD

The main difference between CCD cameras is the size of their chips. The lowest-priced cameras use chips with a 196 x 165 pixel array, while top-end cameras have arrays of 4,000 x 4,000 pixels.

Chips are getting bigger and cheaper each year. The advantage of a larger chip is that it can record a larger area of sky. This is critical since even large chips are much

**THIS CCD IMAGE** of the Dumbbell Nebula (M27), in Vulpecula, was captured by Jack Newton.

**A CCD SYSTEM** is complex. The heart of the system is the round CCD head, which attaches to the telescope's focus.

smaller than a 35 mm film frame, and at focal lengths of 2,000 mm or more, they "see" an area of sky only a few arcminutes wide, smaller than the disk of the Moon. Nevertheless, CCD cameras are superb devices for recording planets, galaxies, and smaller nebulas and clusters.

It is important to match the CCD camera to a telescope of the appropriate focal length. Many of today's CCD cameras have tiny pixels just 9 microns, or 0.009 mm, across. Attached to a telescope with a focal length of at least 1,000 mm, these high-resolution cameras will produce images as sharp as the atmosphere allows. When used on shorter focal length telescopes, however, the pixels will be larger than the star images, yielding odd-looking square stars.

Most CCD cameras are black and white. To produce a color image, you need to take three exposures, one each through red, green, and blue filters, and then merge them in the computer using image-processing software.

Producing a clean, noise-free image requires taking two calibration frames. A "dark frame"—an image with the same exposure time as the main exposure but taken with the lens cap on—records the background noise inherent in all cameras. A "flat field" frame—a short exposure of a blank screen or bright light—records flaws such as dead pixels and dust specks. You can then use image-processing software to electronically subtract these unwanted defects from the main image.

CCDs are not devices for the technologically challenged, but if you love computers, you will love CCD imaging.

### JACK NEWTON

If you see an astounding color CCD image, chances are it was taken by Jack Newton. From his home observatory in British Columbia, Canada, Newton pioneered color CCD techniques.

"The results I got with the first CCD cameras in 1990 blew me away," says Newton. "I couldn't believe what I was able to record. Now, with the latest cameras, it's possible to routinely take images that duplicate the best photos taken with the 200 inch Hale Telescope." Indeed, backyard telescopes can image objects never seen before from Earth. "With a 16 inch telescope, anyone can discover five or six new asteroids a night, every night." In the 1990s Newton predicted that film would be all but gone in 10 to 20 years. "CCDs will replace film just as CDs replaced vinyl records." He was right!

# THE SKYWATCHER'S LIBRARY

*Exploring the universe of information about astronomy can be almost as engrossing as exploring the universe itself.*

**FANCIFUL FIGURES** *adorn early star charts such as these celestial globes of 1630 (above). The first serious star atlas was Bayer's* Uranometria *in 1603 (top).*

Starting to learn about the universe requires no more than one or two good books. But every backyard stargazer soon discovers the need for a few standard reference works.

## STAR ATLASES AND FIELD GUIDES

A star atlas is a road map to the sky. It can help you find hundreds of telescopic targets, but you must first know how to find key constellations and bright stars. This is where a planisphere can help. These "star wheels" show the entire sky on a disk that rotates to any date and time of night: The bimonthly sky charts in Chapter 8 give a similar view on a series of charts designed for specific dates and times.

A step up in detail is one of the "field guides" to the night sky. These books feature month-by-month charts of the entire sky, supplemented by more detailed maps that are often arranged by constellation. The "Guide to the Sky" in Chapter 8 is an excellent example of this type of field guide.

More detail is provided by popular star atlases, such as *Norton's 2000.0 Star Atlas and Reference Handbook* and *The Cambridge Star Atlas*. They contain excellent star charts that plot all stars down to the naked-eye limit of magnitude 6, as well as hundreds of other clusters, nebulas, and galaxies. Both atlases and field guides contain lists of the finest telescopic targets for each area of the sky, and any one of them is a valuable addition to an astronomy library.

More detailed still is *Sky Atlas 2000.0*. This large-format atlas plots all stars down to magnitude 8 and includes 2,500 deep-sky objects—many more than the magnitude 6 atlases do. The "Starhopping Guide" in Chapter 9 covers selected highlights of the sky at the same level of detail.

One of the most comprehensive atlases in print is the *Uranometria 2000.0*, a three-volume set (including a massive catalog, the Field Guide) that divides the sky into 473 charts, plotting all stars down to magnitude 9, as well as 14,000 deep-sky objects to magnitude 15.

The *Millennium Star Atlas* goes even deeper, with over 1 million stars reaching down to magnitude 11. It is based on stellar positions, brightnesses and distances measured by the Hipparcos Satellite. It includes over 22,000 multiple stars and 10,000 non-stellar objects.

## DEEP-SKY CATALOGS

A glance at any star map reveals cryptic references to objects such as M31, NGC 4565, and IC 434. What do they mean?

In the late 1700s, the French astronomer Charles Messier compiled the *Catalogue of Nebulous Objects and Star Clusters*, which remains the most popular list of bright deep-sky wonders. Number 31 on Messier's list, for example, is the Andromeda Galaxy (M31). Modern versions of the

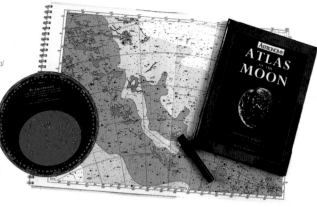

**CHARTS AND ATLASES** *of the Moon, planets, and stars are essential "tools" for skywatchers.*

list contain 110 entries, and seeing all of them is a challenge for amateur astronomers (see p. 423).

By the mid-1800s, astronomers had discovered many more fuzzy telescope targets in the night sky. In 1864, the English astronomer John Herschel published the *General Catalogue of Nebulae*, a list of more than 5,000 objects discovered mostly by him and his father, William Herschel (famous for discovering the planet Uranus).

In 1888, J. L. E. Dreyer published a master list, the *New General Catalogue of Nebulae and Clusters of Stars* (see Box). Supplementary *Index Catalogues* (IC) were added in 1895 and 1908.

Astronomers such as Wilhelm and Otto Struve and S. W. Burnham compiled catalogs of double stars in the nineteenth century. For example, stars designated with the Greek letter sigma ($\Sigma$) are from Wilhelm Struve's catalogs. Collinder, Melotte, Ruprecht, Stock, Trumpler, and many others in the twentieth century have cataloged star clusters and lent their names to clusters that Messier and the NGC authors missed. In the 1950s, George Abell cataloged clusters of galaxies—distant Abell clusters are challenging targets for owners of large telescopes.

CELESTIAL HANDBOOKS
Many of these catalogs are now out of print or inaccessible to amateurs, but some excellent encyclopedic works are still available. The three-volume *Celestial Handbook* by Robert Burnham, Jr., originally compiled in the 1960s, remains the most comprehensive. Another good reference is Rev. T. W. Webb's *Celestial Objects for Common Telescopes*, first published in 1859 but regularly updated (see p. 225).

You can keep in touch with celestial events and discoveries by subscribing to one of the popular astronomy magazines such as *Sky & Telescope* or *Astronomy*. For a bit more local flavour, British observers should look for *Astronomy Now* and Australians for *Sky & Space*. Among the pages of all these magazines are descriptions of many objects—from the bright and famous known to everyone, to the faint and obscure, which few reference books ever mention.

## THE NEW GENERAL CATALOGUE

In 1874, the largest telescope in the world was owned by an amateur astronomer. At his ancestral home of Birr Castle, Ireland, the third Earl of Rosse, William Parsons, built what became known as the "Leviathan of Parsonstown" (pictured below). For its day, it was a monster—a 72 inch (1.8 m) diameter telescope slung on cables between two massive brick walls. With this unwieldy instrument, Rosse discovered that many "nebulas" had a spiral shape. Today we know them to be spiral galaxies. Rosse himself rarely used the telescope, but many other keen observers did. Among them was Johann Louis Emil Dreyer. Dreyer served as Rosse's assistant from 1874 to 1878, recording the mysterious spiral nebulas, among others, with sketch pad and notebook.

By 1886, Dreyer and other observers had discovered so many nebulas and star clusters that a new catalog was needed. The Royal Astronomical Society assigned Dreyer the task. Published in 1888, the *New General Catalogue* (NGC) contained entries for 7,840 objects. The two *Index Catalogues* (IC) later added another 5,386 objects, some discovered with the newly invented process of photography. A century later, the 13,000-plus objects of the NGC and IC lists form the core of all of today's comprehensive databases of deep-sky wonders.

# ASTROCOMPUTING

*Computers have revolutionized the way amateurs find information, and it's getting easier every year.*

**PLANETARIUM PROGRAMS** *present more detailed views of the sky than many printed atlases.*

Personal computers and astronomy seem made for each other. One of the most useful pieces of software is a planetarium program, which is like an electronic star atlas. Most versions can be used to print out customized star charts— perhaps showing a select region of the sky or a new comet's path, or printed in mirror-image fashion to match the view through some telescopes and finderscopes.

You can set planetarium programs for any date thousands of years into the past or future. They are indispensable for figuring out details such as what the sky will look like at a certain time, when Venus will be closest to the Moon, or where a comet will appear.

At the very least, these programs plot all Messier and many NGC objects and stars down to magnitude 9. Many programs contain databases of thousands more deep-sky objects from newer catalogs, including the *Hubble Guide Star Catalog*. This database contains more than 19 million stars and non-stellar objects.

## BEYOND STAR CHARTS

Many planetarium programs go far beyond star-charting functions. For example, they may simulate flights around the Solar System, showing the sky from other planets. Many contain still images and movies of planets and deep-sky objects, or even allow you to create animated movies of celestial events.

Multimedia programs available on CD or DVD are like electronic, interactive books. You can tour the planets or fly over a landscape on Venus or Mars. Such programs are an excellent way to get your family and friends interested in astronomy.

## ASTRONOMY ONLINE

Professional astronomers were among the first to put the Internet to use, back in the days before the World Wide Web and graphical browsers made it easy for anyone to navigate "the Net." Today, astronomy-related sites on the Internet are among the most popular. Selected images

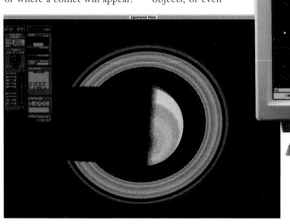

**EXPLORE A VIRTUAL UNIVERSE** *using programs that can show you various views of Saturn or help you to locate a particular star or galaxy.*

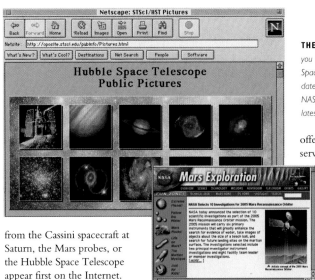

**THE WORLD WIDE WEB** *brings you the latest images from the Hubble Space Telescope and keeps you up to date on the space shuttle missions. NASA's array of web sites provides the latest US space news.*

from the Cassini spacecraft at Saturn, the Mars probes, or the Hubble Space Telescope appear first on the Internet. NASA maintains a host of sites rich in images and information.

You can do your telescope shopping online. The web sites of equipment manufacturers provide information and advice on the latest models, and allow you to order them direct. You can even download samples of new software.

Using the Internet, amateur astronomers can control telescopes halfway around the world, creating digital images that they can then download onto their own computers. Current aurora predictions, the latest pictures of a new comet, this hour's satellite weather map—all this and more are on the Web. There is so much to explore that some amateurs conduct their hobby via the computer screen, neglecting the real sky for the digital universe.

One of the more remarkable uses of computers is "SETI@home." Over three million computers around the world have SETI@home installed, usually as a screensaver. Together they form an enormously powerful computer. Each one downloads blocks of data from the radio telescope at Arecibo, looking for signs of intelligent signals.

## GETTING SOCIAL

Using the Web's newsgroups and the special-interest forums offered by commercial online services, amateur astronomers from every nation now talk to each other every day. Newcomers can post questions about what to buy or how to get started in the hobby. FAQ (Frequently Asked Questions) files provide useful advice on equipment and astrophotography. The world has become one big virtual astronomy club, although nothing can replace meeting fellow skywatchers face to face.

## FINDING OUT MORE ON THE WEB

For the keen computer user, here are some useful addresses to help you further your research.

**American Association of Variable Star Observers (AAVSO):** www.aavso.org
**Association of Lunar and Planetary Observers (ALPO):** www.lpl.arizona.edu/alpo
**Astronomical Society of the Pacific (ASP):** www.astrosociety.org
**Astronomy picture of the day:** antwrp.gsfc.nasa.gov/apod/astropix.html
**British Astronomical Association (BAA):** www.ast.cam.ac.uk/~baa
**The Central Bureau for Astronomical Telegrams (CBAT):** cfa-www.harvard.edu/cfa/ps/cbat.html
**European Space Agency (ESA):** www.esrin.esa.it
**HST images:** www.stsci.edu/pubinfo/Pictures.html
**International Astronomical Union (IAU):** www.iau.org
**International Dark-Sky Association (IDA):** www.darksky.org
**International Meteor Association (IMO):** www.imo.net
**Jet Propulsion Laboratory (JPL):** www.jpl.nasa.gov
**NASA:** www.nasa.gov
**NASA space missions:** spacescience.nasa.gov/missions
**The Planetary Society:** planetary.org
**The SETI Institute:** www.seti-inst.edu
*Sky & Telescope* **(magazine):** www.skypub.com
**Space Telescope Science Institute (STScI):** www.stsci.edu
**Visual Satellite Observer's Home Page:** www.satobs.org/satintro.html

# CONTRIBUTIONS *of* *the* AMATEUR

*A small number of amateur astronomers practice*
*their hobby with diligence and expertise*
*and produce professional results.*

Most of the discoveries in astronomy today are made by professional astronomers who work with medium to large telescopes. Nevertheless, amateurs play a valuable part in modern astronomy because they provide the continuous monitoring and broad sky coverage that professionals cannot afford.

Amateurs routinely monitor thousands of variable stars and they often detect outbursts of the more irregular variables. Many comets, novas, and supernovas have been discovered by amateurs who dedicate an enormous amount of time and experience to careful, systematic searches of the sky.

## THE CENTRAL BUREAU FOR ASTRONOMICAL TELEGRAMS

If you discover something new during your skywatching, such as a nova or a comet, the Central Bureau for Astronomical Telegrams (CBAT) will want to hear about it. Producing about 300 circulars and "electronic telegrams" a year, the CBAT announces discoveries by amateurs and professionals, publishes the predicted positions of comets and asteroids, and discusses observations made of everything from meteors to stars that emit X-rays.

Amateur astronomers are urged, however, to verify their findings before reporting them, as more than 90 percent

**AMATEUR TELESCOPE** *An 8 inch (200 mm) Schmidt-Cassegrain with the Moon, Venus, Saturn, and Jupiter.*

of the reports of new comets have proved to be false alarms. Ask an experienced amateur to repeat your observation or repeat it yourself the following night, to look for changes.

## LESLIE C. PELTIER AND VARIABLE STARS

On a May evening in 1915, Leslie Peltier, an Ohio farm boy, was walking outdoors. "Something," he wrote, "perhaps a meteor, caused me to look up for a moment. Then, literally out of that clear sky, I suddenly asked myself, 'Why do I not know a single one of those stars?'" Deciding to save money to buy a telescope, he picked 900 quarts of strawberries on the family farm, at 2 cents a quart, to raise the necessary $18. Three years later, thrilled by their antics, he started observing variable stars, an activity he was to pursue for the rest of his life.

Peltier's reputation as a skilled observer spread, and he was offered the use of a 6 inch (150 mm) refracting telescope by Henry Norris Russell of the Princeton Observatory.

Three years later, on 13 November 1925, Peltier discovered his first comet, one of 10 comets that bear his name,

# DAVID LEVY'S DISCOVERY OF A COMET

The early morning (around 3 am) of 20 May 1990, was partly cloudy but dark as I stepped into my backyard, walked a few steps and pulled the 10 foot square (3 × 3 m) roof off my garden-shed observatory. Although Comet Austin had been visible through binoculars some time earlier in the night, it had set. Besides, I was not planning to look at old comets; I was looking for new ones.

I took the cover off my 16 inch (400 mm) reflector and pointed the telescope toward the east. Then, with eye at the eyepiece and hand on the telescope, I began searching slowly through the eastern sky, moving the telescope one field of view at a time, checking each field for a faint fuzzy object that could be a comet. Finding these faint fuzzies is easy, but most of them turn out to be galaxies, nebulas, or star clusters.

About an hour after I started, the waning crescent Moon rose in the east. Since moonlight floods the sky with enough light to make faint comets disappear, I usually stop hunting when the Moon is up, but on this occasion I did not feel like stopping. I moved the telescope past Alpheratz (Alpha [α] Andromedae), one of the stars in the square of Pegasus, and moved on inside the square.

A few minutes later I stopped moving the telescope and studied the field

**COMET LEVY 1990C** *moved relative to the stars during this exposure, resulting in trailed star images. The telescope was tracking the nucleus of the comet, which is lost in the larger coma in the photograph.*

of view. There was a fuzzy spot! I knew this part of the sky so well that I suspected this to be the real thing. I contacted the CBAT by electronic mail, with information that included where the object was in the sky and how bright it was.

A day later, when the slowly moving comet had proved me right, the CBAT announced it as Comet Levy (1990c). At the moment of discovery, I had been looking at something that no one had ever seen before. I felt as though the sky had looked down on me and said, "OK, David, since you've been watching faithfully for so long, here is something special."

Within a week, several observers had reported accurate positions so that the comet's orbit could be determined. The news was good: the comet was expected to get bright enough to be visible to the naked eye. By the end of June, Comet Levy was an easy target through binoculars, and it continued to brighten steadily as the summer progressed. It was at its best in early August. I remember setting up my telescope at a dark roadside location. I was about to turn the scope toward the comet when I looked up and saw the Summer Triangle of Vega, Deneb, and Altair high in the sky. I had first seen those same three stars 30 years earlier, when I was starting as a skywatcher. But on this August night the view was different, for in the middle of that triangle was bright Comet Levy. It was a night I'll never forget.

**A LOG BOOK** *is an important part of the observer's equipment. Here, David Levy's log book records comet observations and viewing with his friends.*

**COMET IKEYA SEKI,** *named after two Japanese amateur astronomers, was a spectactular sight in daylight from some parts of the world in 1965. It came so close to the Sun that it was known as a Sungrazing comet.*

associations will supply detailed comparison charts and guidance to the amateur wanting to become serious about variable stars.

## RECORDING YOUR OBSERVATIONS

Keeping an observing diary or a log book is perhaps the best way to remember the experiences you have had with your telescope. If you wish to study variables seriously, search for comets, or just take photographs, you will probably find the systematic approach of keeping a diary extremely useful.

## OBSERVING A VARIABLE STAR

Observing variable stars is perhaps the easiest really productive work the amateur can undertake, since binoculars are all that are needed. To learn the art, you cannot do better than follow the example of famed variable star observer, Leslie Peltier (see p. 100).

For each of the 132,000 variable star observations that Peltier made during his lifetime, he followed a set procedure. After finding the variable with one of his telescopes, he would choose two nearby stars, one a bit brighter than the variable, and another that was fainter. A star chart made especially for the variable would have these "comparison stars" plotted, together with their magnitudes, which might be, say, 8.2 and 8.8. Then by interpolating, he would estimate the brightness of the variable star to be, say, 8.6.

For over 60 years Peltier reported his observations to the American Association of Variable Star Observers (AAVSO). Headquartered in Cambridge, Massachusetts, the AAVSO has followed the behavior of more than 1,000 variable stars since 1911, using reports from amateurs from all over the world. Professional astronomers recognize the AAVSO as a reliable and valuable resource.

Associated programs are run by the British Astronomical Association (BAA) and the Royal Astronomical Society of New Zealand (RASNZ). All three

---

## TEACHING THE TEACHERS

Larry Lebofsky is a planetary scientist who has had a prodigious career. He discovered Larissa, a moon circling Neptune; the existence of water on certain asteroids; and Asteroid 3439 has been named in his honor. But he now tends to use binoculars rather than large telescopes, and spends more time with teachers than with research astronomers.

This new dimension to his career arose from a visit that his wife, Nancy, made to their daughter's second grade classroom. Miranda's teacher had prepared a lesson about the phases of the Moon—first quarter, full, and "gibbons." Lebofsky began wondering why he was studying asteroids when teachers appeared not to know

what a gibbous Moon was, so he and Nancy joined forces in building a program that introduces teachers of young children to ways of teaching astronomy. Their Astronomy-Related Teacher Inservice Training (ARTIST) workshops are now held in schools throughout southern Arizona.

**OBSERVATORIES** *for the amateur come in all shapes and sizes, from the classic dome (below) to the "portable observatory" that is a modified tent (left).*

## BACKYARD OBSERVATORIES

Should you become serious about your observing, you may find it useful to build an observatory in your back-yard—somewhere you can keep your telescope set up and ready for use.

A number of companies sell prefabricated structures, ranging from modified tents to simple metal or fiberglass observatory buildings. If you are inclined to be more ambitious, you could even buy a ready-made dome.

You may prefer to make a sliding-roof observatory out of a garden shed, providing you know your way around a hammer and screwdriver. With such a design, when the roof is slid back you will be able to swing your telescope around to see the full sky rather than just a section of it. There are many other possibilities. Have a look through back issues of astronomy magazines for more ideas.

If you have a place set up solely for looking at the sky, you will be in the best possible position to begin a serious observing program.

**STAR PARTY** *One of the benefits of joining an astronomy club is the chance to go along to observing nights. You can then talk to other amateur astronomers about their telescopes and compare the views that they produce.*

## JOINING AN ASTRONOMY CLUB

After skywatching on your own for a while, you might find it helpful to join a group of people who are also fascinated by the night sky. Most large towns have astronomy clubs that host regular talks, courses, and star parties. A club can be an excellent source of advice and learning.

A favorite event is the local star party. Most clubs will host monthly observing sessions at their club site or observatory in the country. At these gatherings, a newcomer can look through a variety of instruments and talk to the owners about what they like and dislike about their telescopes. Public observing sessions—perhaps at a local school or mall—are another club activity. It can be very rewarding to take your passion for astronomy to the public.

To contact your local club, call a nearby planetarium, nature or science center, college, or university. You could also check the Internet for lists of clubs in your area.

Once a year, most regions also host an enormous star party, usually at a park or campground with dark skies. These events attract hundreds of amateur astronomers and, for many, the annual star party becomes their main chance to escape light-polluted skies. Swap-tables become popular places to pick up bargains, and vendors often set up sales booths. Check astronomy magazines or their Web sites for dates of major star parties coming up in your area.

Look at the stars! look, look up at the skies!

O look at all the fire-folk sitting in the air!

The bright boroughs, the circle-citadels there!

*The Starlight Night,*
GERARD MANLEY HOPKINS (1844–89), English poet

# CHAPTER FOUR

# UNDERSTANDING *the* CHANGING SKY

# MAPPING *the* STARS

*Although we no longer believe that Earth is at
the center of the universe, it is a view that can help us understand
why the sky changes from hour to hour and night to night.*

**BEYOND THE CELESTIAL SPHERE?**
*This colored woodcut, reminiscent of
sixteenth-century engravings, illustrates
the old belief that Earth was surrounded
by a vast star-studded dome.*

People once believed that
the stars were attached
to a huge sphere that
rotated around Earth once a
day. Today we know that the
stars and all the other heavenly
bodies lie at different distances
from us and that they only
appear to move around Earth
(from east to west) because
Earth itself is spinning on its
own axis (from west to east).
Nonetheless, when describing
the positions of stars and the
ways in which they appear
to rise and set, it is useful to
imagine Earth being enveloped by a vast celestial sphere.

## EARTH AND THE CELESTIAL SPHERE

To describe the positions of
objects in the sky, astronomers
developed a grid system of reference points and lines similar
to the one used to determine
locations on Earth. Here is
where the concept of a celestial sphere is of value to us.

Let us imagine that Earth is
a balloon that can be inflated
until it fills this imaginary
celestial sphere. If the most
important points and lines on
Earth's surface were somehow
transferred onto the celestial
sphere, the celestial equator
would correspond to Earth's
equator, the celestial poles
would correspond to
Earth's rotational north
and south poles, and the
lines of longitude and
latitude would translate
as celestial coordinates
called right ascension

and declination. Tilted at an
angle to the celestial equator
would be the ecliptic, the
apparent path of the Sun
around the celestial sphere.

## THE EQUATORIAL COORDINATE SYSTEM

Declination in the sky is
measured in the same way that
latitude on Earth is measured:
in degrees, minutes, and
seconds north and south of
the equator—in this case, the
celestial equator. It increases
from 0 degrees on the equator
to 90 degrees at the poles.
North of the celestial equator
declination is listed as positive
(+), while south is negative (-).

**THE GREENWICH MERIDIAN** *is
marked by a line at the Old Royal Greenwich Observatory, near London. Here you
can stand with one foot in each of the
Eastern and Western Hemispheres.*

**LATITUDE AND LONGITUDE**
*form a grid on Earth, defined by
the north and south poles, by which
any point may be uniquely located.*

**THE CELESTIAL SPHERE** *and its system of "equatorial" coordinates mimic the surface of Earth. The important extra feature is the ecliptic—the apparent path of the Sun across the background of the celestial sphere.*

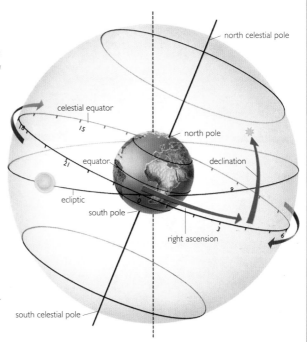

For example, Polaris has a declination of just over +89 degrees; Rigel (in Orion) lies at -8 degrees and 15 minutes.

The celestial equivalent of longitude on Earth is called right ascension (abbreviated to RA), and is measured not in degrees, as longitude is, but in units of time: hours, minutes, and seconds (where 24 hours equals 360 degrees and one hour of RA equals 15 degrees). The right ascension of Polaris is thus referred to as 1 hour, 49 minutes; and that of Rigel as 5 hours, 12 minutes.

The zero (0) line of RA (the celestial equivalent to Earth's Greenwich meridian of longitude) passes through the point where the Sun crosses the celestial equator at the first moment of spring in the northern hemisphere—the vernal equinox. Hours in right ascension are measured eastward from this point, until we reach 23 hours, 59 minutes. One minute later, of course, and we find ourselves back at 0 hours.

## POSITIONS IN THE SKY

The altazimuth coordinate system is especially useful if you are trying to explain to a fellow observer where an object is in the sky at any particular moment. This system is mainly used for navigation, but it also enables skywatchers to specify the position of a celestial body with respect to their horizon and at a particular time using coordinates called altitude and azimuth. These figures differ for each observer, depending on his or her position on Earth and the time that the observation is made.

**ALTITUDE (ALSO KNOWN AS ELEVATION)** This is the angle above the observer's horizon. The point directly overhead, at 90 degrees, is known as the zenith.

**AZIMUTH** This is the angle measured clockwise from north along the horizon to the point on the horizon that lies beneath the star. Thus, N = 0 degrees or 360 degrees; E = 90 degrees; S = 180 degrees; and W = 270 degrees.

**MERIDIAN** This is an imaginary great circle that passes through your zenith from north to south, dividing the sky in two: the eastern and the western halves. It is important to be aware of this line because when an object crosses it, it's as high in the sky as it's going to get.

The Sun crosses the line of the meridian around noon every day. We say that the Sun, or any star, culminates when it crosses the meridian.

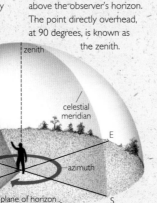

# A SPINNING EARTH

*Every hour of every day, a panorama unfolds above our heads,*
*and all because Earth is spinning.*

Our planet Earth is a roughly spherical globe which revolves around its own axis, an imaginary line that runs through the center of Earth from the north pole to the south pole. If it were not for the stars way beyond Earth's atmosphere, it would be difficult to prove that this motion actually takes place, since everything rotates along with us—the land, air, oceans—all of our immediate natural environment.

## A SHIFTING HORIZON

As Earth rotates from west to east, our perception is that the stars, the Sun, and all other heavenly objects move around us in the opposite direction—from east to west. The apparent motion of the stars in the sky, however, will depend on where you happen to be on Earth. Once again, the concept of the celestial sphere (introduced on p. 106) can help us explain what stars we can see

and how they appear to move with respect to our particular viewing location. The latitude of the observation point is the key: a skywatcher in Canada sees the stars with much the same perspective as someone in central Europe or in northern Japan; the same is true for people in Buenos Aires, Cape Town, and Adelaide. See the illustration below.

### At the North Pole, 90° N

For this observer, the celestial north pole, marked by the star Polaris, corresponds with the zenith—the point in the sky directly overhead. From this position on Earth, the celestial equator is parallel with the horizon, and because the stars move along a path parallel to this horizon, only the stars in the northern half of the celestial sphere are ever visible.

### At the Equator, 0° From this

latitude, an observer can see all the stars of the sky. Here, the

**STAR TRAIL PHOTOGRAPH** *showing circumpolar stars arcing around the north celestial pole and never setting.*

celestial equator rises up in an imaginary vertical line from the horizon and runs through the zenith. The north and south celestial poles lie exactly on the horizon. The stars in the east rise straight up and sink straight down again below the horizon in the west.

**At 40° S Latitude** For all locations lying between these extreme latitudes, there is a part of the sky that always remains invisible—that which surrounds the celestial pole of the opposite hemisphere. On the other hand, the area of the sky close to the visible pole remains in view all the time. Stars here never set, but seem to circle around the celestial pole. These are called circumpolar stars, and the closer you are to the pole, the more circumpolar stars there are.

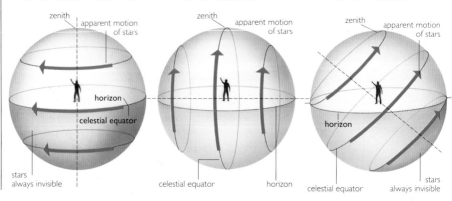

**AT THE NORTH POLE, 90° N**    **AT THE EQUATOR, 0°**    **AT 40° S LATITUDE**

zenith | apparent motion of stars | horizon | celestial equator | stars always invisible

zenith | apparent motion of stars | celestial equator | horizon

zenith | apparent motion of stars | horizon | celestial equator | stars always invisible

To persons standing alone on a hill
during a clear midnight such as
this, the roll of the world eastward
is almost a palpable movement.

*Far from the Madding Crowd,*
THOMAS HARDY (1840–1928),
English poet and novelist

# TIME *and the* SEASONS

*While the tilted Earth spins around its own axis, it is also tracing out a path around the Sun. These two motions, and where we are on Earth, determine what we see in the sky at a given time.*

THE SEASONS *personified, in this painting by Walter Crane (1845–1915).*

" Why is it hotter in summer than in winter?" In *A Private Universe,* a film from the 1990s, this question was posed to a group of Harvard graduates. Most had no idea of the correct answer—they said that it was because Earth is closer to the Sun in summer.

SEASONS *on Earth are caused by the tip of Earth's axis, not by changing distance from the Sun.*

## THE REASON FOR THE SEASONS

Earth orbits the Sun at an average distance of 93 million miles (150 million km). But because Earth's orbit is slightly elliptical Earth is, just as the Harvard graduates stated, sometimes nearer to the Sun than at other times. But why we experience different seasons has to do with the way Earth is tilted relative to the plane of its orbit around the

Sun, rather than the distance between them at different times of the year.

From Earth, the Sun is seen projected against the remote background of the celestial sphere. This apparent path of the Sun around the sky in the course of a year is known as the ecliptic. (In the Starfinder Charts, Chapter 8, the ecliptic is marked with a dotted line.) Since most of the planets revolve around the Sun in more or less a flat plane, viewed from Earth the paths of the other planets across the sky tend to stay fairly close to the ecliptic. Knowing this, ancient people attached great importance to this band of constellations on the ecliptic, known as the zodiac.

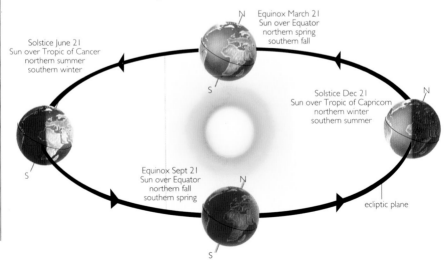

Solstice June 21
Sun over Tropic of Cancer
northern summer
southern winter

Equinox March 21
Sun over Equator
northern spring
southern fall

Solstice Dec 21
Sun over Tropic of Capricorn
northern winter
southern summer

Equinox Sept 21
Sun over Equator
northern fall
southern spring

ecliptic plane

**OLD WORLD MAP** *showing India and the East Indies, surrounded by the celestial sphere with the prominent colored band of the zodiac.*

**MIDNIGHT SUN**, *seen from Mawson Base, Antarctica, illustrating how the Sun never sets during midsummer in arctic or antarctic regions.*

However, Earth's axis is not perpendicular to the ecliptic—rather, it is tipped at an angle of 23½ degrees. The ecliptic, therefore, is tipped with respect to the celestial equator by the same angle. Because of this, Northern and Southern Hemispheres experience opposite seasons.

In June, the Southern Hemisphere is in winter because it is leaning away from the Sun, while the Northern Hemisphere is experiencing midsummer. The Sun's rays no longer reach the South Pole, and for half the year there the Sun will fail to rise, resulting in continuous night.

Six months later, in December, Earth will have gone halfway round the Sun. The Northern Hemisphere is now in midwinter, while the Southern Hemisphere is in midsummer. At the South Pole this time, the Sun will remain above the horizon for six months, resulting in a phenomenon referred to as midnight Sun. In March and September, however, both hemispheres have an equal share of day and night.

## DAY AND NIGHT

The lengthiness of summer days and the corresponding brevity of winter days increases as you move to higher latitudes. During winter, the Sun is low in the sky, which means that the days are short and the nights long. Also, the Sun's rays must pass through a greater thickness of the atmosphere; some of the heat is absorbed, and the low angle from which the rays come means that they are more spread out.

In summer, on the other hand, the Sun passes high across the sky as it rises and sets each day, yielding longer days than in winter. Also, because the Sun is high during this season, the Sun's rays are less spread out and it heats Earth more directly.

## THE SUN'S MOTION ACROSS THE SKY

Midway between sunrise and sunset the Sun reaches its highest point in the sky around noon. If, over the course of a year, you observed the shadow cast by a stick on a flat piece of ground, you would graphically see how the Sun's height in the sky at noon varies with the passing of the seasons (see the the illustration below).

During summer, the shadow is shortest at noon on the solstice, the day with the longest daylight hours. Here, the Sun reaches its highest point in the sky for the year. In winter, the Sun traces its lowest path across the sky on the day of the "winter" solstice. It is during this time that the Sun casts the longest shadows and gives us the day with the fewest daylight hours.

**PATH OF SUN ACROSS THE SKY** *at different times of the year, as viewed by a Northern Hemisphere observer. A Southern skywatcher would look north, instead of south, to see the same effects.*

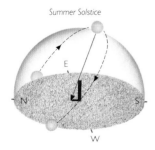

*Summer Solstice*

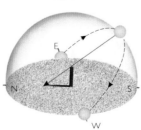

*Equinox*

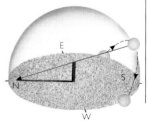

*Winter Solstice*

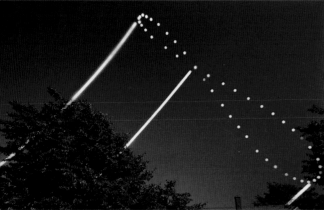

**AN ANALEMMA** *is the unusual "figure 8" symbol found in various forms on accurate sun dials. This photo illustrates its origin by recording the position of the Sun in the eastern US sky at 8.30 am every 10 days. The Sun's different positions are caused by the tilt of the ecliptic relative to the celestial equator, plus small changes in the Sun's daily speed across the sky.*

At noon on the first days of spring and fall, the length of the shadow cast by the stick is midway between its summer minimum and winter maximum. These days are the vernal and autumnal equinoxes (around March 21 and September 21 for Northern Hemisphere spring and fall, respectively), where day and night are of equal duration.

## SOLAR AND SIDEREAL TIME

The Sun's daily motion determines the most basic time cycle—day and night. Our day of 24 hours is how long it takes Earth to complete one rotation around its axis relative to the Sun. Our clocks indicate time on this basis, and we refer to this as "mean solar time."

We might expect the celestial sphere to appear to rotate around us in the same period, but it does not. Instead, throughout the year as Earth moves along its path around the Sun, the stars appear to rise about 4 minutes earlier each night—in other words, an observer will see a particular star at the same position 4 minutes earlier on successive nights. Therefore, relative to the stars, the celestial sphere rotates once around Earth in 23 hours and 56 minutes. Because (for our purposes) the stars are essentially fixed in space, this is Earth's true period of rotation and is the basis of what we call "sidereal time," the time astronomers refer to when determining an object's position in the sky.

## TIME ZONES

All the times given in this book are in mean solar time. We actually use a system of standard times which divides the 360 degrees of longitude around Earth into 24 time zones, each 15 degrees wide. Because the difference between solar and standard times is usually small, where we are in longitude makes little difference to what we see in the sky, but where we are in latitude does (see p. 245).

When we say, for instance, Orion is on the meridian at 10 pm on Jan 10, this is true for anywhere in the world. (In the case of Orion this is *literally* true, since it spans the celestial equator and can be seen, at

**ORION**, *as seen looking south in winter from Edinburgh, Scotland. These two views are at the same time of night but two weeks apart, illustrating the slow motion of the sky during the year.*

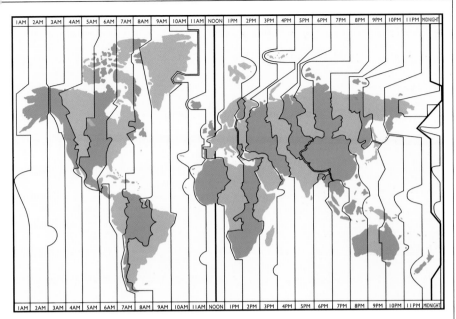

| 1AM | 2AM | 3AM | 4AM | 5AM | 6AM | 7AM | 8AM | 9AM | 10AM | 11AM | NOON | 1PM | 2PM | 3PM | 4PM | 5PM | 6PM | 7PM | 8PM | 9PM | 10PM | 11PM | MIDNIGHT |

least in part, from anywhere on Earth.) The 10 pm refers to local time.

But what happens when someone in San Francisco wants to compare notes about an eclipsing variable star in Orion with someone in New York City? Orion will indeed be on the meridian at 10 pm local time on January 10 for both of them, but the San Francisco skywatcher will see this 3 hours later in fact! If they both want to see the star at the same moment as it eclipses, they will need to allow for the time difference in making the observation—say, 10 pm in New York City and 7 pm in San Francisco.

An easier way is to adopt a time system that everyone can use: Universal Time (UT) is essentially Greenwich Mean Time (GMT), which is local time at Greenwich. So now, our two observers can both agree, without confusion, that the eclipse begins at 1 hour UT on Jan 10 (the next day!) because that's the time people in Greenwich will see it.

This simple picture, however, is complicated slightly by "daylight saving" or "summer" time (DST), when we move our local time ahead by 1 hour in order to keep the Sun from rising overly early in summer. Benjamin Franklin proposed this feature in 1784 to save on the cost of lighting.

## PRECESSION

Does Earth's axis always point in the same direction in space? Actually, it does not. Because of tidal effects of the Sun and Moon, Earth "wobbles" like a spinning top, causing the direction of the vernal equinox to shift in the sky. This wobble is called precession. The sky's entire coordinate system shifts along with that wobble over a leisurely 26,000 years. Therefore, whenever we say that an object has a particular right ascension and declination, we must specify a precise moment in time, like the year 2000, to define our celestial coordinates. That moment is called the epoch. The vernal equinox is now in Pisces, but over the full wobble it will move through all the signs in the zodiac. The celestial poles move too, so Polaris, our northern pole star, will be of as little use to us a thousand years from now, as it was to the ancient Egyptians.

**TIME ZONES** *(above) are used to keep countries, states, or provinces on a common time. As a result, some locations differ by many minutes between true local time and standard time.*

# NAMING *the* STARS

*Over thousands of years, people of countless cultures have given the stars names and projected their dreams onto them.*

C oming mainly from the Arabs and Greeks, the names we use for the stars above us are strange and entertaining. One favorite is Zuben El Genubi, the ancient Arabic name given to the brightest star in Libra. Both the Greeks and the Arabs saw this star as representing nearby Scorpius's southern claw, which is what this splendid name means. The star marking the old northern claw is called Zuben Eschamali.

Betelgeuse is another strange name. Some older texts say, incorrectly, that this should be pronounced "BEETLE-juice", perhaps leading some to wonder why such a majestic star should be diminished to liquid refreshment for bugs. "BET-el-jooze" is the correct pronunciation, which means "house of the twins" in

**PTOLEMY** (left) observing the Moon with Urania, who was the Greeks' patron of astronomy.

**BAYER'S URANOMETRIA** (1603) was the first serious star atlas, with one map devoted to each of the 48 traditional constellations. This illustration shows Centaurus, the Centaur; Lupus, the Wolf; and Crux, the Southern Cross.

Arabic, because the Arabs regarded the star as part of the neighboring constellation of

**JOHN FLAMSTEED** was the first Astronomer Royal of England and the first director of the Royal Greenwich Observatory. His great star catalogue, Historia Coelestis Britannica (1725) was published after his death.

Gemini. It now marks Orion's shoulder.

BAYER LETTERS AND FLAMSTEED NUMBERS
At the start of the seventeenth century, the German astronomer Johann Bayer set down the positions and magnitudes of all the known stars in an organized way. His star atlas *Uranometria*, published in 1603, labelled many of the stars visible to the naked eye using a system based on the Greek alphabet, whereby the brightest star was generally called alpha ($\alpha$), the next brightest beta ($\beta$), and so on. Betelgeuse is therefore also called Alpha ($\alpha$) Orionis, Orionis being

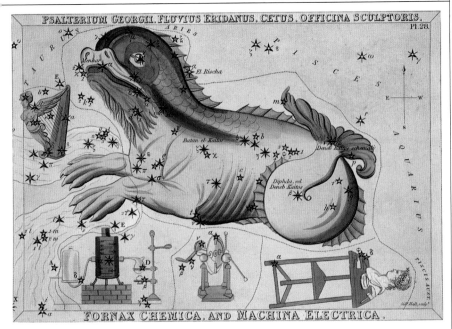

FORNAX CHEMICA, AND MACHINA ELECTRICA.

the genitive form of the constellation name.

Early in the eighteenth century, the Reverend John Flamsteed, appointed by King Charles II of England as the first Astronomer Royal, extended the list of "named" stars by adding Arabic numbers and covering many fainter stars. In his system, the stars are numbered west to east across a constellation, so Betelgeuse is also 58 Orionis.

The fainter a star is, the less interesting is its name. For example, near Zuben El Genubi in Libra is Flamsteed's 5 Librae, best seen through binoculars, and the Smithsonian Astrophysical Observatory's SAO 158846, a magnitude 9.2 star too faint to be seen without a telescope. The SAO Catalogue, which includes 258,997 stars, is another listing often encountered by amateur astronomers. Another bigger and more recent list is the Hubble Space Telescope (HST) Guide Star Catalogue which includes some 19 million stars! There are also many catalogues of star clusters, nebulas, and

**URANIA'S MIRROR** *(1825) was a set of cards depicting the constellations. Cetus, the Sea Monster, is seen here with some surrounding constellations, including the obsolete Psalterium Georgii and Machina Electrica.*

galaxies. The Messier Catalogue, however, is the most famous, followed by the NGC and IC catalogues (see below).

*The stars I know and recognize and even call by name. They are my names, of course. I don't know what others call the stars.*

OLD WOMAN QUOTED BY ROBERT COLES IN *The Old Ones of New Mexico*

## THE MESSIER CATALOGUE

As astronomers grew increasingly familiar with using telescopes and searching the sky, they found more and fainter objects.

From about 1759, French astronomer Charles Messier began listing all the fuzzy objects he encountered in his search for new comets. He listed 103 of these objects in a catalogue, published in 1781, which was to be revised and added to over several years. In its current form the Messier

inventory includes 110 objects—mainly clusters, nebulas, and distant galaxies—spread over most of the sky.

There are also a number of other lists of deep-sky objects, the best known being the New General Catalogue (NGC) of 1888. This, along with its two index catalogues (IC), lists over 13,000 objects. Thus the famous Great Nebula in Orion is known as both M42 and NGC 1976.

# STAR BRIGHTNESS *and* COLOR

*Like lights on a street, stars are fainter the farther away they are,*

*but some stars are supergiants, intrinsically more luminous than*

*mere dwarfs like the Sun and its near neighbors.*

The first thing you notice on looking skyward on a clear night is a hodgepodge of stars of varying brightness. How do we describe the brightness of stars and planets?

**APPARENT MAGNITUDE**
Our system of "magnitudes" dates back to the second century BC when Greek astronomer Hipparchus divided the stars into six brightness groups, the brightest stars being 1st magnitude and the faintest 6th magnitude.

By 1856, Norman Podgson of Radcliffe Observatory had quantified this relationship, making a 1st magnitude star 100 times brighter than the faintest star visible without a telescope. A 2nd magnitude star was then 2.5 times fainter than a 1st magnitude star and a

**APPARENT BRIGHTNESS** *of stars, like the street lights (below) depends, in part, on their distance. For example, the middle light meter shown in the illustration (right) is twice as far from the light as the front meter and therefore measures only ¼ the brightness. The meter furthest from the light source records only ⅑ of its brightness.*

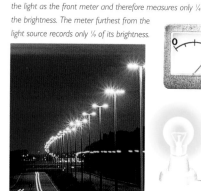

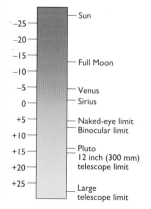

3rd magnitude star, in turn, 2.5 times fainter again.

Podgson defined Polaris, the North Star, as being 2nd magnitude. Most of the stars in the Big Dipper are also about 2nd magnitude. Vega, the brightest star in the Summer Triangle, is 0 magnitude. The brightest star in the sky, Sirius, has a minus magnitude of −1.4. The Moon and Sun are even brighter, of course, but are still accommodated on the scale at −12.6 (Full Moon) and −26.8, respectively.

**APPARENT MAGNITUDE SCALE**
*ranges over a factor of around 100 billion in brightness to encompass the brightest stars and the faintest galaxies seen in large telescopes.*

| | |
|---|---|
| −25 | Sun |
| −20 | |
| −15 | |
| −10 | Full Moon |
| −5 | |
| 0 | Venus |
| | Sirius |
| +5 | |
| | Naked-eye limit |
| +10 | Binocular limit |
| +15 | Pluto |
| | 12 inch (300 mm) |
| +20 | telescope limit |
| +25 | |
| | Large |
| | telescope limit |

**ABSOLUTE MAGNITUDE**
The stars differ in magnitude for two reasons: some are closer to us than others, and some really are brighter than others. In order to describe a star's intrinsic brightness, astronomers have defined absolute magnitude as the apparent magnitude a star would have if it were 10 parsecs or 33 light-years from us (see p. 28).

**ILLUSION AND REALITY**
Comparing apparent and absolute magnitudes is a little like looking at light fixtures on the street. Although all the street lights are the same strength, the ones closer to you appear brighter than the ones down the road. However, the lights in the

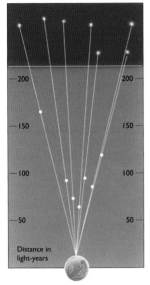

parking lot at the end of the street appear much brighter, even though they are farther away, because they are intrinsically more powerful. Most reference in this book to a star's brightness concerns its brightness in the sky as seen from Earth—its apparent magnitude. Only when we wish to establish how luminous a star actually is do we use absolute magnitudes.

**ILLUSION VERSUS REALITY** *The stars of Ursa Major form the familiar Big Dipper (or Plough) purely by chance since they actually lie at different distances from Earth.*

## STAR COLORS

As a careful look skyward will show, the stars differ not merely in brightness, but also in color. If you look at Antares in Scorpius, Aldebaran in Taurus, or Betelgeuse in Orion, you will see that all these stars are reddish—a subtle red rather than the red of traffic lights. Vega in Lyra and Rigel in Orion are a delicate blue.

A star's color gives us a clue to its nature. Blue signifies that the star is much hotter than the Sun; red means it is cooler (see p. 210).

Star colors are subtle to the eye. In color images, however, the colors appear more obvious, but they are also influenced by the color sensitivity of the camera.

The eye itself has a particular color sensitivity— we see yellow and green clearly, but infrared or ultra-violet light is invisible.

The type of magnitudes we usually encounter are more properly called visual (V) magnitudes since they reflect the brightness of stars as seen by the human eye. Yet some stars are very dim in visible light but bright in, say, the infrared (which we can't see). Our visual magnitude system provides a poor reflection of the true output of such stars.

**STAR COLOR** *tells us the temperature of the surface of a star. For many stars—the main sequence stars— increasing temperature is usually paralleled by an increase in size.*

| 20,000° C Blue-white | 10,000° C White | 6,000° C Yellow | 4,500° C Orange | 3,000° C Red |
|---|---|---|---|---|

## THE TWENTY NEAREST STAR SYSTEMS

| Name *denotes star has companion | Constellation | Apparent magnitude | Distance (light-years) |
|---|---|---|---|
| Proxima Centauri | Centaurus | +11.1 | 4.24 |
| Alpha Centauri* | Centaurus | −0.27 | 4.4 |
| Barnard's Star | Ophiuchus | +9.5 | 6.0 |
| Wolf 359 | Leo | +13.5 | 7.8 |
| Lalande 21185 | Ursa Major | +7.5 | 8.3 |
| Sirius* | Canis Major | −1.4 | 8.6 |
| Luyten 726-8* | Cetus | +12.4 | 8.7 |
| Ross 154 | Sagittarius | +10.5 | 9.7 |
| Ross 248 | Andromeda | +12.3 | 10.3 |
| Epsilon Eridani | Eridanus | +3.7 | 10.5 |
| Lacaille 9352 | Piscis Austrinus | +7.4 | 10.7 |
| Ross 128 | Virgo | +11.1 | 10.9 |
| Luyten 789-6* | Aquarius | +12.7 | 11.3 |
| 61 Cygni* | Cygnus | +5.2 | 11.4 |
| Procyon* | Canis Minor | +0.3 | 11.4 |
| Groombridge 34* | Andromeda | +8.0 | 11.6 |
| Sigma 2398* | Draco | +8.8 | 11.6 |
| Epsilon Indi | Indus | +4.7 | 11.8 |
| G 51-15 | Cancer | +14.8 | 11.8 |
| Tau Ceti | Cetus | +3.5 | 11.9 |

Source: Pasachoff, *Astronomy: From the Earth to the Universe* (Brooks/Cole, 2002)

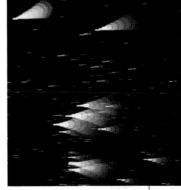

**THE COLORS OF STARS** *Even photographs do not reveal true star colors because bright stars flood the film with light and appear whitish. Changing camera focus during a star trail exposure overcomes this by spreading out the light to ensure that each star has its color well recorded at some point along its trail.*

# THE MOON *and* PLANETS *in* MOTION

*Unlike distant stars, the Moon and the planets are not far from us, so we can watch their movements across the sky every night.*

The Moon takes a month (a "moonth") to orbit Earth. As a result of the gravitational pull between Earth and the Moon, it also takes the same amount of time to rotate on its axis, so it always keeps the same face turned toward us. However, this face is presented about 50 minutes later each night as the Moon moves among the stars, because of its orbit around Earth.

**AN ORRERY** *was a mechanical device used to illustrate the motion of the planets around the Sun, seen here in a work by British painter Joseph Wright (1734–97). This orrery appears to show the positions of the planets in November 1757.*

The Moon also presents a varying amount of its lit face to us during its monthly orbit, because of the changing relationship between it, the

Sun, and Earth. These are known as its phases.

At full moon, it is opposite the Sun from our point of view, so we see the entire side of it that faces us bathed in sunlight. When it is "new," the Sun shines on the far side of the Moon, so the sunlit face is hidden from us. At other phases, we see only part of the Moon's sunlit surface. All the time, whether it is lit or not, we are only ever looking at the one face of the Moon.

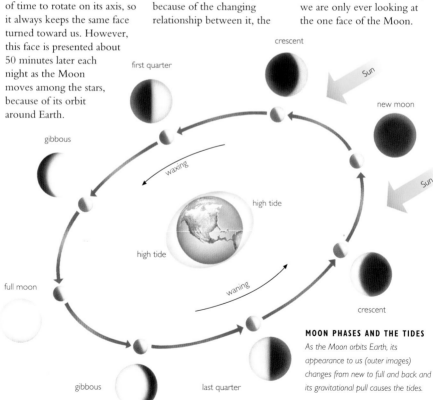

**MOON PHASES AND THE TIDES**
*As the Moon orbits Earth, its appearance to us (outer images) changes from new to full and back and its gravitational pull causes the tides.*

Here on Earth, the ebb and flow of the tides are a constant reminder of the Moon's gravitational influence.

## THE WALTZ OF THE PLANETS

Although they move much more slowly than the Moon, the planets trace elegant paths among the stars that you can easily follow during a season of observing. The farther away a planet is from us, the more slowly it appears to move against the stars at any time.

During the course of some months, you can actually watch a planet move backward as it is being overtaken by Earth in the same way that you see a car recede as you overtake it on the highway. As you pass the car, it appears to slow down, then move backward, then resume its forward movement after you have gone by.

Moving more slowly around the Sun than we do, Mars, Jupiter, and Saturn all exhibit the same type of motion, moving eastward, appearing to reverse course

as we pass them, and then resuming their eastward trek. This process is called retrograde motion.

## THE INNER PLANETS

Mercury and Venus are closer to the Sun than we are, and orbit it more quickly. Their movements appear very different from those of the outer planets.

Venus, for example, will appear close to the Sun early one evening. With each passing week, it will appear to rush away from the Sun, then it will slow down and for a few days it will seem to move very slowly—a time known as greatest elongation. From a maximum of 47 degrees from the Sun, it will then start to move closer again, becoming so close that we cannot see it. Its closest apparent position to the Sun is called conjunction. In fact, at this point it is no closer to the Sun than usual; it is simply almost in line with us and the Sun. After that, it appears in the morning sky and repeats its performance.

**RETROGRADE MOTION** *is a characteristic of the outer planets, as Earth overtakes them in its faster orbit. Here, a retrograde loop of Mars is superimposed on the stars of the constellation Taurus.*

Mercury moves in the same way, but it goes through the acts of its drama much faster than Venus, and it never reaches more than 28 degrees from the Sun.

**CRESCENT NEPTUNE**, *as seen from Voyager 2, illustrates how a planet that is closer to the Sun than the observer can show a crescent phase.*

Over the rim of the waiting earth the moon lifted with
low majesty till it swung clear of the horizon and rode off,
free of moorings ...

The Wind in the Willows,
KENNETH GRAHAME (1859–1932), British author

CHAPTER FIVE

# SUN and MOON

# OBSERVING *the* SUN

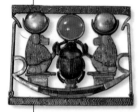

*Life-giver, and even law-giver in the past,*

*the Sun has a strong influence on the world*

*that anyone can feel on a sunny day.*

A Sun god has been a major feature in almost every human culture. Among the Hopi Indians of the arid American Southwest, for instance, the Sun is the "keeper of the ways." Each day it climbs out of its eastern kiva (an underground ceremonial chamber) and up into the sky to keep watch on the world.

Sun-watching in ancient times was inextricably mixed with astrology and divination. In the European Renaissance, however, the first telescopes were invented and studies of the Sun finally started to address its physical reality.

Amateur astronomers were responsible for many of these early studies. For example, an amateur determined that sunspots occur in cycles, and two amateurs discovered the existence of solar flares and their links to geomagnetic disturbances such as auroras.

During this century, most solar research has become "Big Science," with budgets and sophisticated instruments to match. Ambitious large-scale projects include the SOHO and TRACE satellite observatories (see p. 51).

But solar observation remains a rich and enjoyable field for the amateur, thanks in part to the Sun's ever-changing appearance. No other area of astronomy offers so many opportunities to see a celestial object change.

**SUN GOD** *An Egyptian pendant shows a Sun god in scarab-beetle form (top). The setting Sun (above). To match your telescope to an H-alpha filter, you may need a kit of extra accessories (right).*

Some amateurs also find that solar viewing fits more easily into their daily lives than does nighttime astronomy, with its late hours and long trips to find dark skies.

## THE NEAREST STAR

The Sun is our closest star—in fact, it is about 270,000 times closer than the next nearest star, Alpha Centauri, and is the only star whose surface we can see in any detail.

The Sun's proximity makes proper safety precautions essential. It is possible to view the Sun with complete safety, but you must follow the procedures described in Chapter 3 (see p. 84) and below. Cutting corners can permanently damage your eyesight—Galileo's blindness in later life was probably caused by slapdash solar-observing methods.

## THE SOLAR SCOPE

A solar telescope does not need much aperture—6 inches (150 mm) is ample, and good observations can be made with apertures of 2 to 3 inches (50 to 75 mm). Because sunlight usually creates poor seeing, you cannot use the higher resolution afforded by a larger scope.

To reduce the effects of bad daytime seeing, try observing in the early morning, before the Sun has had a chance to heat up the surroundings. Erect your telescope on a grassy area, avoiding pavement, rooftops, or any other heat-absorbing surface. Another solution is to view the Sun across a lake, where the water helps to keep the air relatively settled.

The only safe way to view the Sun directly is to use a solar filter that fits over the entire aperture of the telescope (see p. 84). An aperture filter reduces sunlight to safe levels before it enters the scope. This keeps the optics cooler and improves the viewing.

Most solar filters are made from Mylar plastic or glass; both kinds will show sunspots and faculae (see pp. 126–127). Expensive filters that isolate narrow wavelengths, such as Hydrogen-alpha (H-alpha), can show other surface details and spectacular prominences.

**TWO SAFE METHODS** *You can view the Sun directly through a telescope fitted with an aperture filter (above), or you can use a solar projection screen (right).*

### PROJECTING THE SUN

The other technique used for solar viewing is projection. It requires no costly equipment and allows a group of people to view the Sun at one time. Aim your telescope at the Sun and affix a sheet of white cardboard behind the eyepiece, focusing until the image looks sharp. Shield the projected solar image from any direct sunlight. Several observers can then gather around the telescope and study the Sun and its features. A projected solar diameter of 6 inches (150 mm) is easy to view and useful for sketching.

When aiming the telescope toward the Sun, never sight upward along the tube—instead, look at the scope's shadow on the ground. Cover the finderscope's aperture so that it cannot accidentally burn you and no one is tempted to look through it. Never leave the telescope unattended—and if your group includes children, watch them like a hawk.

The heat of the Sun can ruin the optical cement in a compound eyepiece, so do not use your best eyepiece. Opt for a cheaper design such as a Huygenian or a Ramsden. If your scope's aperture exceeds 4 inches (100 mm), place an opaque mask with a circular cutout over the aperture to reduce it to 4 inches, or even less. This reduces the heat buildup inside the instrument.

---

## PHOTOGRAPHING THE SUN

Solar astrophotography can become a simple and enjoyable matter with the right equipment—an aperture solar filter, an adapter to couple your camera to the telescope, and, preferably, a telescope mounting that can track the Sun (see pp. 86–93).

Some amateurs take a photograph of the Sun every clear day, building up a valuable archive of images. Others photograph specific solar activity, such as sunspots, or use narrow-wavelength filters, such as H-alpha filters, to capture the active surface of the Sun.

The Sun's brightness means you can use short exposures. Solar filters reduce the Sun's brightness to about that of the Full Moon—refer to the lunar exposure table on page 92 as a guide to exposure times for solar photographs, but take a lot of test images to see what works with your equipment.

Experienced solar photographers know that the toughest problem to contend with is poor seeing. Shooting in the early morning or across a body of water can help, but part of the challenge lies in discovering what works best for an individual site.

*This false-color image of the Sun's surface was taken through an H-alpha filter. Regions of activity are shown as yellow.*

# THE STRUCTURE *of the* SUN

*The Sun is a vast, layered structure, comprising all but a small fraction of the mass of the entire Solar System.*

**THE FACE OF THE SUN** *This nine-teenth-century painted woodcarving from Germany depicts "Mrs Sun."*

The Sun is a ball of gas about 865,000 miles (1,392,000 km) in diameter. If you could put Earth at its center, the Moon would orbit about halfway to the Sun's surface. It would take 109 Earths, lined up side by side, to span the Sun's diameter. The Sun contains about 333,000 times as much matter as Earth does, and comprises more than 99.99 percent of the mass in the entire Solar System. The remaining 0.01 percent makes up the rest of the Solar System—that is, the nine planets and all the moons, comets, asteroids, and dust.

The Sun is a medium-size, medium-hot star in the middle of its life, which began some 4.6 billion years ago. We can be thankful that it is such an ordinary star. If the Sun were 10 times more massive, it would already have lived its life and exploded as a supernova (see p. 222). And had the Sun been 10 times smaller, it would not produce enough light and heat for life to exist in the Solar System. A single element, hydrogen, accounts for 92.1 percent of all the atoms in the Sun.

The next most common element is helium, which amounts to 7.8 percent. All the other elements together make up the remaining 0.1 percent. The Sun is fueled by thermonuclear reactions in its core, in a process where hydrogen atoms are fused to make heavier helium atoms, which results in an enormous release of energy.

## A HOT HEART

Solar astronomers use a technique called helioseismology to look below the surface of the Sun. Using telescopes positioned around the Earth (to keep the Sun under constant watch), they study subtle pulsations in the Sun's surface. These pulsations— essentially sound waves— probe the Sun's interior in much the same way that seismic waves from earth-quakes enable geophysicists to study our planet. Such studies have shown that the Sun has a layered structure. The core extends outward to about a quarter of the Sun's radius, containing merely 7 percent of the volume, but half the mass, of the Sun. It

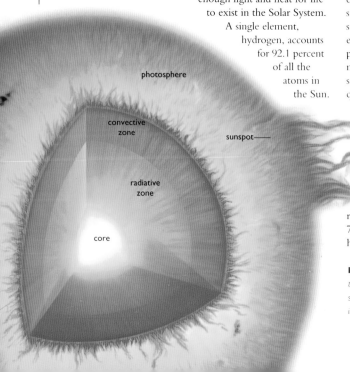

photosphere

convective zone

sunspot

radiative zone

core

**INSIDE THE SUN** *Scientists believe that the Sun has a layered structure, as shown in this diagram. Nuclear fusion in the core produces the Sun's energy.*

has a temperature of some 27 million degrees Fahrenheit (15,000,000° C). The core is the "engine room" of the Sun, where the energy released from nuclear fusion streams outward. This energy is carried by radiation in the form of photons, which are parcels of electromagnetic energy.

Three-quarters of the way to the surface, the solar gas becomes considerably cooler and undergoes a boiling, convective motion. Convection carries the core's energy along the last step outward. Just below the photosphere—the visible surface—the big upwelling cells of gas break down into smaller ones, resulting in the granules seen on the surface. By the time it reaches the photosphere, the gas has cooled to about 11,000 degrees Fahrenheit (6,000° C).

## THE PEARLY CORONA

Above the Sun's photosphere lies a thin, cool, pinkish layer called the chromosphere. It is visible only during solar eclipses or through an H-alpha filter. The top of the chromosphere merges into the corona, the Sun's vast outer atmosphere, which reaches outward at least 10 times the solar radius. It, too, can usually be seen from Earth only during solar eclipses, when its pearly light

**THE SUN'S CORONA**—*its outer atmosphere*—*extends far into space in this infrared image (above), taken with a telescope that created an artificial eclipse.*

is a beautiful sight to behold. The coronal gas is very thinly spread but extremely hot—about 2 million degrees Fahrenheit (1,000,000° C). It is heated by processes that remain unclear.

Beyond the corona, the Sun sends out a stream of charged particles—mostly negatively charged electrons and positively charged protons—into space in what is called the solar wind. Near the Sun, this wind blows at 2,500 miles per second (4,000 km/s), but at Earth's distance, the velocity is a tenth of that figure.

The solar wind sometimes interacts with Earth's atmosphere and creates auroras. It also controls the shape and direction of a comet's ionized gas tail (see p. 193).

## THE FUTURE OF OUR STAR

Astronomers predict the future of the Sun by looking at what has happened to similar stars. Its life is now about half over. As it consumes hydrogen in its core, leaving an "ash" of helium behind, the Sun will grow hotter and brighter. When its hydrogen begins to run

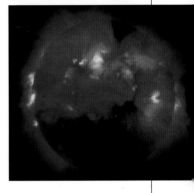

**THIS X-RAY VIEW** *shows the hot gases of the inner corona above the photosphere, which looks dark in X-rays.*

low, about five billion years from now, the Sun will grow into a red giant, expanding its diameter a hundredfold and swallowing the planet Mercury in the process.

The Sun will spend a few hundred million years as a red giant. Then, as nuclear reactions in its core start to sporadically fuse the helium "ash" into carbon, the Sun may begin to pulsate like the variable star Mira. Finally, as it exhausts its reserves, the Sun will throw off its outer layers to form a planetary nebula. This shell of gas will endure for a few tens of thousands of years before it dissipates into space. Left behind will be the white-hot core of the Sun, a white dwarf that will slowly cool and fade over billions of years.

---

### SUN FACTS

**Distance from Earth** 92,960,000 miles (149,600,000 km)

**Mass** 333,000 × Earth's mass

**Radius** 109 × Earth's radius

**Apparent size** 32 arcminutes

**Apparent magnitude** −26.7

**Rotation (relative to stars)** 25.4 days at equator to 35 days near poles

# SOLAR ACTIVITY

*The surface of the Sun is a spectacular, turbulent place,*

*and signs of intense solar activity, such as solar*

*flares and sunspots, can be seen from Earth.*

The visible surface of the Sun is called the photosphere ("sphere of light"). It looks solid, but in reality it is a transitional shell of gas about 250 miles (400 km) thick. At about 11,000 degrees Fahrenheit (6,000° C), this outer layer is much cooler than the gas deep beneath it.

The edge of the photosphere looks dark compared with the center of the disk. This limb-darkening effect occurs because at the center of the disk we see deeper into the Sun where temperatures are higher and the gas shines more brightly. At the limb we are seeing cooler gas at higher elevations and it appears darker.

Viewed at high magnification, the photosphere shows many features. The smallest of these are individual cells of gas called granules, which average about 700 miles (1,100 km) across. Granules can last from several minutes to half an

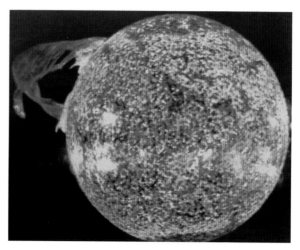

**THIS SPECTACULAR PROMINENCE** *is erupting more than 370,000 miles (600,000 km) into space.*

hour, but are usually difficult to observe because of the poor seeing during daytime.

The photosphere also has large, irregular patches of slightly hotter material called faculae (Latin for "torches"). Because they are not much brighter than the rest of the surface, they are often hard to see, except near the limb.

With a standard H-alpha filter, you may see bright loops called prominences extending from the Sun's disk. They can also be seen with a narrowband H-alpha filter as dark filaments on the disk itself.

Another kind of activity—a flare—is always worth looking for, although it is rare to see

**ERUPTIONS** *on the Sun's surface (left) release an enormous amount of energy. A 1635 illustration (top) wrongly depicts the Sun as a burning mass of carbon.*

one. The first was discovered in 1859 by two English amateurs, Richard Carrington and Richard Hodgson, who were working independently. Flares are brilliant explosions of pent-up magnetic energy and radiation. They come and go, often within minutes, but most are invisible without an H-alpha filter. White-light flares, which can be seen with any solar filter, are rare but powerful events. They may release as much as 10 percent of the Sun's energy from a tiny area in a few minutes, and can disturb the electromagnetic environment of the whole inner Solar System. When the flare's particles arrive at Earth as a high-speed gust in the solar

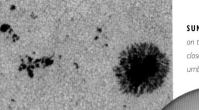

**SUNSPOTS** *can be observed on the Sun's disk (below). Up close (left), they show a dark umbra and a lighter penumbra.*

wind (see p. 125), they create geomagnetic disturbances such as auroras. In extreme cases, they can produce surges of current that cause power grids to fail, leading to widespread blackouts.

## SUNNY SIDE UP

The most famous solar features are the sunspots. Records from China show that, since ancient times, observers have noticed dark spots on the Sun. (These naked-eye observations were made when the Sun was low in the sky or partly obscured by fog, mist, or clouds. Do not try to copy their technique, however—many astronomers in antiquity probably went blind.)

In the early 1600s, Galileo used his telescopes to show that sunspots were not clouds above the photosphere as some claimed, but features on the photosphere itself. Eventually, he and others established that sunspots occur within about 35 degrees of the solar equator. In the nineteenth century, Richard Carrington noticed that sunspots near the Sun's equator took about two days less to circle the Sun than did spots in higher latitudes. He concluded that different parts of the Sun rotate at different rates, and that this differential rotation occurs because the Sun is a gaseous body. It was an important new finding.

Sunspots are places where the photosphere is about 3,600 degrees Fahrenheit

(2,000° C) cooler than the surrounding sunscape. This temperature difference makes the spots look darker. Sunspots occur where the Sun's magnetic field has emerged through the photosphere and stopped some of the rising energy from reaching the surface.

Sunspots have a dark central region, called the umbra, surrounded by a striated, lighter-colored halo, known as the penumbra. They vary greatly in size, shape, and complexity, and can appear as single, small, black dots or as complicated groups with so many discrete features that it is difficult to tally them. The largest groups can reach a size of 60,000 miles (100,000 km) across, many times the diameter of Earth.

In 1769, the Scottish solar astronomer Alexander Wilson noted that the largest sunspots appeared saucer-like when seen near the limb. He proposed that these spots were depressions in the solar surface. Today's astronomers still call this the Wilson Effect,

**HOT AND COOL** *This infrared image shows the temperature of a sunspot's umbra and penumbra. The hottest areas are white and the coolest areas are black.*

but ascribe it to changes in the transparency of the solar gas over a spot when it is seen at an extreme angle.

## OBSERVING SUNSPOTS

As the Sun rotates on its axis, you can watch a group of sunspots travel from one edge of the disk to the other over a period of about 10 days. A few long-lived spots will come around for a second period of visibility.

Sunspot observations are often best around mid-morning. At this time of day, you can see the Sun without a lot of atmosphere in the way and before it has heated your surroundings too much. To count sunspots visually through your telescope, use a full-aperture solar filter and a magnification that shows the whole disk. Note the number and positions of the sunspot groups, then switch to an eyepiece giving 65 power, the agreed-upon standard. Revisit each group and count the individual spots. The total estimated number of sunspots, $R$, is given by the formula: $R = 10g + s$, where $g$ is the number of groups, and $s$ is the total number of spots seen. Summaries of these numbers are regularly published in astronomy magazines.

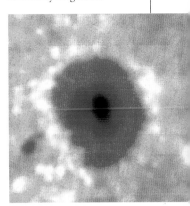

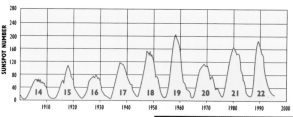

## SKETCHING SUNSPOTS

Sketching sunspots is easiest using the projection method. Decide on a standard size of sketch—6 inches (150 mm) is common—and prepare blank forms ahead of time with a circle for the Sun's diameter and places to note date and time, weather, seeing con - ditions, and the equipment you used. Tape the blank onto the projection screen, adjust it so the Sun just fills the circle, and with an HB or softer pencil, lightly trace the locations of the sunspots.

By making regular sketches, you can produce a record of the changing face of the Sun.

**SUNSPOT ACTIVITY** *follows a clear 11-year cycle (above). An infrared image taken in 1981 (right) shows the Sun near the peak of a sunspot cycle, with the spots positioned in two bands around 10 degrees north and south latitude.*

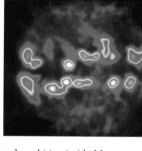

## THE SUNSPOT CYCLE

In the nineteenth century, German amateur astronomer Heinrich Schwabe was sys - tematically scanning the Sun's surface, hoping to catch in transit a planet then thought to be orbiting inside Mercury. To avoid mistakes, he tabu - lated all the sunspots he saw. Schwabe never found the planet—which does not, in fact, exist—but he did make the discovery that the number of sunspots rose and fell in a pattern lasting about a decade.

Today, after observations over many years, scientists have determined that sunspots follow an 11-year cycle, during which the number of spots goes from a minimum to a maximum. Solar astronomers reckon Cycle 0 as being the cycle that peaked in 1750. On this reckoning, Cycle 23 ended in late 2008, and Cycle 24 is expected to peak around 2013. Each cycle has an asymmetric shape, rising to maximum activity in about five years and declining over about six years.

At the start of a new cycle, the first spots appear around latitudes 35 degrees north and south. As the cycle pro - gresses, sunspots become more numerous and appear at latitudes

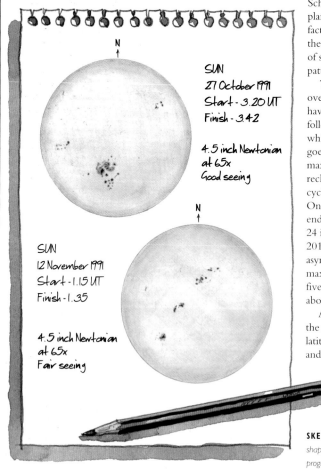

SUN
27 October 1991
Start - 3.20 UT
Finish - 3.42

4.5 inch Newtonian
at 65x
Good seeing

SUN
12 November 1991
Start - 1.15 UT
Finish - 1.35

4.5 inch Newtonian
at 65x
Fair seeing

**SKETCHES** *showing the location and shape of sunspots can record the progression of solar activity over time.*

## SUNNY RADIO SIGNALS

Because most cosmic radio signals are very weak, radio astronomy usually demands large, sophisticated equipment. The Sun, however, is by far the strongest radio source in the sky and this presents opportunities to the amateur with an interest in electronics.

Solar flares induce changes in the Earth's ionosphere—the layer of ionized particles in the upper atmosphere. The ionosphere usually reflects very low frequency (VLF) signals (5 to 50 kilohertz), permitting long-distance radio communication. Disruption of the ionosphere by a solar flare, however, can cause communications fade-out. You can monitor this effect with a radio receiver, antenna, and recorder that cost less than a medium-size reflector telescope.

The Sun itself can be detected with a short-wave receiver for 1 to 30 megahertz frequencies and a simple antenna. A radio-frequency preamplifier will improve the sensitivity of the system. You can even collect the Sun's signals using a backyard satellite dish, with minor modifications. The Small Radio Telescope (right) at Haystack Observatory, Massachusetts, uses a 9 foot (2.8 m) satellite dish to receive signals of 1 to 4 gigahertz frequencies.

To find out more, see John Potter Shield's *Amateur Radio Astronomers Handbook.*

*The adventure of the sun is the great natural drama by which we live, and not to have joy in it … is to close a dull door on nature's sustaining and poetic spirit.*

The Outermost House,
HENRY BESTON (1888–1968),
American naturalist and writer

Astronomers label this period the Maunder Minimum, and note that it coincides with a period that climatologists now call the Little Ice Age, when winters were more severe than normal, and summers were shorter and cooler.

Spurred by interest in global warming, scientists are actively exploring connections between activity on the Sun and the climate on Earth. Our climate system is highly complex. The Sun is clearly important, but solar variations are too small to explain recent rapid climate change.

nearer the Sun's equator. By the end of the cycle, sunspots appear around 7 degrees north and south latitude. As one cycle is tailing off, the first spots of the next cycle are already appearing at high latitudes.

Sunspots are highly mag - netized, and astronomers have discovered that there is actually a 22-year magnetic cycle superimposed on the ordinary 11-year cycle. When they measure the magnetic fields of sunspots during any 11-year cycle, they find that the spots leading each group across the disk show opposite magnetic polarities in the two hemispheres. Then, in the next 11-year cycle, all leading spots switch polarity.

**THE LITTLE ICE AGE,** *a period of very cold weather, coincided with the Maunder Minimum (right). London's River Thames froze over and was the site of frequent frost fairs (above right).*

DOES THE SUN VARY?
Compared with some stars, which can vary in apparent brightness by factors of a hundred or more, the Sun seems like constancy itself. On the other hand, there are those 11-year sunspot cycles. Furthermore, astronomical history shows that sunspots can virtually disappear for decades on end, with notable climatic consequences.

For example, the British amateur astronomer E. Walter Maunder drew attention to the fact that from 1645 until 1715 sunspots virtually disappeared for reasons still unknown.

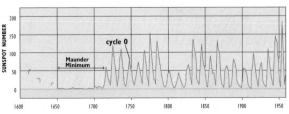

# SOLAR ECLIPSES

*Watching the Moon creep across the Sun,*

*blot out its brilliance, and then slowly reveal it*

*once more is an experience that fills us with awe.*

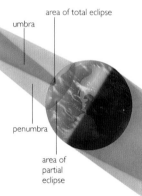

area of total eclipse

umbra

penumbra

area of partial eclipse

**THE LUNAR UMBRA** *traces a narrow path on Earth. Within the penumbra, people see only part of the Sun eclipsed.*

A total solar eclipse is one of Nature's grandest spectacles, and is something everyone should try to see at least once in their life.

In the folklore of many cultures, tales speak of a monster attacking the Sun, only to let it free again when the right prayers are said. The ancient Chinese described eclipses as a dragon devouring the Sun—in fact, their word for eclipse came from their word meaning "eat."

Eclipses have even stopped wars. Herodotus writes of an eclipse on 28 May 585 BC. The Medes and Lydians of Asia Minor had been fighting for more than five years and were in battle that day, when the Sun grew dark unexpectedly. At the frightening sight, both sides immediately sought peace. Though neither side knew it, the eclipse had been predicted by Thales of Miletus, the first such prediction on record.

## A GREAT COINCIDENCE

A solar eclipse occurs whenever a New Moon passes directly between the Sun and Earth. We do not see an eclipse at every New Moon because the Moon's orbit inclines about 5 degrees to the ecliptic. In most months, the Moon passes above or below the Sun instead of across it.

Solar eclipses can be total, partial, or annular, depending on how much Sun is covered by the Moon. Total eclipses occur when the Moon covers the entire disk of the Sun. A total eclipse may last just a second or two; the longest possible total eclipse is only about 7½ minutes.

It is pure accident that the apparent sizes of the Moon and the Sun are roughly the same. The Moon is about 400 times smaller than the Sun—but it also happens to be about 400 times closer. In fact, since the Moon formed, its orbit has been moving very slowly outward. From our point of view, it is steadily shrinking. Eventually, it will orbit too far from Earth to ever completely cover the Sun, and total solar eclipses will be a thing of the past.

**TWO GIRLS AND A LLAMA** *watch a solar eclipse in Chile in 1994 (above). This multiple-exposure shot captured the stages of an annular eclipse (right).*

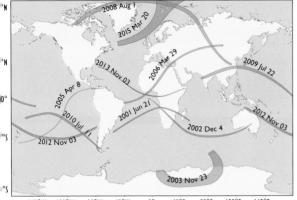

**ECLIPSE TOURS** *The world map (left) shows the paths of total solar eclipses until the year 2015. Use it to plan your eclipse travels. This group (below) visited India in 1995 to observe a solar eclipse.*

Because Earth is rotating and the Moon is moving, the lunar shadow, or umbra, traces a curving path on Earth. The umbra barely reaches Earth, so the path is relatively narrow—where the Moon is directly overhead, the path will be at most about 170 miles (270 km) wide. This means that total eclipses are selective, occurring only rarely at any particular place, although in most years there are one or two eclipses somewhere on Earth.

In recent decades, a new branch of the travel industry has developed, offering solar-eclipse expeditions to remote destinations. Tour-company advertisements start to appear in astronomy magazines about a year before the event.

## RING OF SOLAR FIRE

Total eclipses are rare, but partial eclipses occur more frequently. While enjoyable to view with the right equipment, partial eclipses lack the overwhelming drama of a total eclipse and few people will travel very far to see one.

A possible exception is the annular eclipse, a special type of partial eclipse. Annular eclipses happen because the Moon's orbit around Earth, and Earth's orbit around the Sun, are not circular but elliptical. Thus the apparent

sizes of the Sun and Moon vary by a small amount. On occasion, the Moon may appear too small (or the Sun too large) for a total eclipse to occur, even though other circumstances are right for it.

The result is that at maximum eclipse the Moon does not cover the Sun completely, and a ring (or annulus) of the Sun's uneclipsed surface

surrounds the black circle of the Moon. Annular eclipses will not show you the Sun's corona (unless the eclipse is about 99.99 percent total), but they can be quite spectacular all the same, especially if the Sun is sitting near the horizon.

## AN ECLIPSE FOR EINSTEIN

Total solar eclipses are dramatic, but few have been awaited more eagerly than the one in Brazil on 29 May 1919. At this eclipse, a British expedition made the first attempt to confirm (or refute) Albert Einstein's general theory of relativity, published in 1916.

A key prediction of the theory was that a ray of light grazing the edge of the Sun would be bent by the Sun's gravity by a small, but detectable, amount. A total eclipse offered the only chance of "turning off" the Sun's light so that astronomers could measure the positions of background stars and find the deflection—if it existed.

The 1919 eclipse was the first for which major expeditions could be arranged following World War I (1914–18). Teams were sent to two separate sites in case of clouds, with one expedition headed by Britain's foremost astrophysicist, Arthur Stanley Eddington.

After the eclipse, when the astronomers measured their photos, the deflection of starlight came out exactly as Einstein had predicted. What made the eclipse notable (and even poignant) at the time was that it was British astronomers confirming a theory developed in the depths of wartime by a German physicist. Einstein (right) suddenly found his name blazoned across every newspaper. Literally overnight, he became the most famous scientist in the world.

## TURNING OFF THE SUN

As with any observation of the Sun, you must take special precautions before viewing a solar eclipse. However, it is not dangerous to be outdoors, as some over-cautious warnings may lead you to believe. You can watch the progress of the eclipse through a telescope rigged for solar observation (see p. 128) or you can hand-hold a solar filter. When the Moon has fully covered the Sun, you can look directly with the naked eye, but you must turn away as soon as the Sun begins to reappear.

There are many ways to record an eclipse. To photograph the partial stages, use the same techniques as you would for the uneclipsed Sun (see p. 123). During totality, other kinds of photographs become possible (see Box).

Making a video of an eclipse provides a record of the whole event—including all the enthusiastic (and often funny) comments from bystanders. You will need a camcorder, a solar filter, a tripod or a

**AFTER FIRST CONTACT**, *the Moon slowly creeps across the Sun (above). Moments before second contact, the last rays of sunlight create Baily's beads (right).*

telescope mount, and adequate power (plenty of batteries or a line-current adapter). You can remove the filter during totality, but be ready to replace it when the Sun starts to reappear at the end. The Earth's rotation will carry the Sun and Moon two solar diameters every four minutes. You will either have to track them with the camcorder on an equatorial mount (see p. 78), or zoom back enough that they stay within the frame.

## FOUR CONTACTS

A total solar eclipse has four stages: first contact, second contact (or start of totality), third contact (or end of totality), and fourth contact. In your eclipse-trip planning, check astronomy magazines or the Web for the exact time of each stage.

The eclipse begins at first contact, the moment when the lunar disk first touches the solar one. First contact can be de-tected only through a telescope prepared for solar viewing.

Over the next hour, the partial phase unfolds, as the

**A HAND-HELD FILTER** *protected this Parisian's eyes during the 1912 eclipse.*

Moon steadily creeps across the Sun. At first, there is little apparent change. But then you will notice a difference in the quality of the light. Shadows grow sharper, while the light becomes more like that on an overcast day. The air grows cooler, and insects, birds, and other animals begin to act as if it is nightfall. If you are standing near a tree that is in leaf, look down. The small gaps between the leaves act like pinhole lenses, throwing hundreds of images of overlapping crescent Suns onto the ground.

The diminished light may tempt you to look at the Sun with the naked eye, but the Sun is still bright enough to damage your vision.

A few minutes before second contact—the start of totality—changes begin to occur in rapid succession. The sky grows quite dark, and the air feels distinctly cool. A breeze usually picks

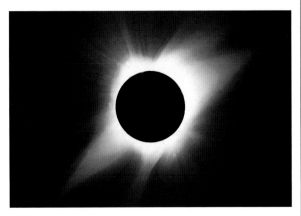

*TOTALITY reveals the Sun's corona—its outer atmosphere—with streamers that reach out a solar diameter or more.*

up, perhaps blowing dust across delicate equipment. In the gloom, observers may fumble over camera settings and mislay eyepieces and filters.

Then, the last rays of sunlight stream through valleys on the edge of the Moon, a phenomenon called Baily's beads, after an English solar observer, Francis Baily, who first described them, in 1836. The last of the beads may linger a moment, creating what is known as the diamond-ring effect. Then comes a flash of pinkish light from the chromosphere, the first layer above the Sun's surface—and totality begins.

## SOLAR FLOWER

The sky darkens to deep twilight, and planets and stars can be seen above the glow that circles the horizon. It is now safe to remove solar filters, but they should be kept handy. Around the black disk of the Moon is a soft fringe of pearl-colored light. This is the Sun's corona, its outer atmosphere.

The corona is much too dim to be seen except at totality.

Now it is time to turn a telescope on the edge of the Moon, where ruddy tendrils stand out. These are solar prominences—huge eruptions of gas suspended by magnetic force above the Sun's surface. They appear to extend up from behind the Moon. The prominences on one edge of the Moon slowly disappear while those on the other side emerge from behind the lunar limb. This is a reminder that the Moon is moving on and totality is near its end.

When only a few instants of the total phase are left, reach for your solar filter. Abruptly, there will be a bright burst of light from the Moon's eastern edge, then another, and daylight floods back. This is third contact, and the end of totality. The corona disappears into the rapidly brightening sky.

For many observers, the end of totality means the end of the eclipse, despite the hour or so that remains until fourth contact—the exit of the Moon from the Sun's disk. People pack up their equipment, and talk excitedly about the sights just seen, reliving the brief moments of totality. Precious images are reviewed and telescopes are packed away. The grand show has ended—that is, until the next eclipse!

## PHOTOGRAPHING A TOTAL ECLIPSE

To capture the feel of a total eclipse, some photographers use a camera with a wide-angle lens perched on a tripod and aimed where the Moon will be at mid-totality. The eclipsed Sun will be small in the frame, but the photo will catch the main ingredients of the scene: the Sun and Moon, the brighter planets and stars, the horizon glow, and the blurry silhouettes of observers.

Other photographers concentrate on the corona. With a camera piggybacked on a telescope (see p. 90) and a 1,000 mm or 2,000 mm lens, you could take a long exposure, showing its maximum extent. Or you could make a series of exposures starting with short ones and running progressively longer to capture the corona in stages, from the brightest parts near the lunar limb to the outer streamers. Another possibility is to photograph the eclipsed Sun through a telescope, perhaps using eyepiece projection for extra magnification (see p. 92). This will capture the pinkish prominences that reach above the lunar disk.

With any photo, the correct exposure time depends on your camera and f-ratio. Try the following maximum exposures.

- outer corona $1/2$ second at f/8
- inner corona $1/15$ second at f/8
- prominences $1/125$ second at f/8
- diamond ring $1/500$ second at f/8
- scenic shots 1 second at f/2.8

*Attaching a camera to a telescope is just one method of eclipse photography.*

# LUNAR ECLIPSES

*You need not travel far, nor wait for many years,
to behold the eerie sight of the Moon as it moves
into Earth's shadow and turns dusky red.*

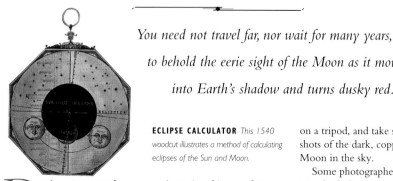

**ECLIPSE CALCULATOR** *This 1540
woodcut illustrates a method of calculating
eclipses of the Sun and Moon.*

During most months, the tilt of the Moon's orbit in relation to Earth's orbit around the Sun ensures that the Full Moon passes north or south of the Earth's shadow. But at roughly six-month intervals, the Full Moon lies near (or inside) the shadowy cone of darkness that extends far into space directly from Earth. It is during these "eclipse seasons" that lunar and solar eclipses take place. Lunar eclipses can be partial or total, depending on how deeply the Moon goes into the shadow.

The shadow of our planet points at the stars like a finger of night. It starts off as wide as Earth itself, but shrinks by the time it crosses the Moon's orbit. Although it is still almost three times as wide as the Moon, the shadow is a small area in astronomical terms.

*What is there in thee, moon,*

*that thou shouldst move*

*My heart so potently?*

*Endymion*, JOHN KEATS
(1795–1821), English poet

Assuming skies are clear, anyone who is on the night side of Earth while a lunar eclipse is happening can see it. Observers often find lunar eclipses a good deal more relaxing than solar ones because the event usually lasts hours instead of the minutes that measure solar eclipses.

## ECLIPSE PHOTOS

As a lunar eclipse progresses, viewers with binoculars and telescopes watch the Moon closely, timing when the shadow reaches individual craters and then estimating how dark the eclipse is. Other observers get a camera, put it

on a tripod, and take scenic shots of the dark, coppery Moon in the sky.

Some photographers piggyback the camera on an equatorially mounted telescope. With the telescope tracking the stars, they take a multiple exposure that captures the Moon in stages as it moves through Earth's shadow.

Another popular multiple-exposure photograph that is even easier to take is to frame the Moon in the sky next to a tree or building in the foreground. Then, without moving the camera (or changing its aim), you take a series of identical exposures, precisely 5 or 10 minutes apart. The result will be a striking scenic view with a number of Moons marching across the

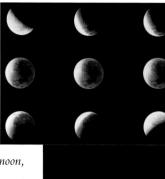

**THE PHASES** *of a lunar eclipse can
be photographed in close-up (left), or
they can be captured as they occur in
the sky in a single scenic image (below).*

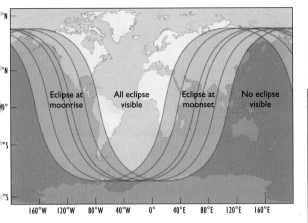

Eclipse at moonrise | All eclipse visible | Eclipse at moonset | No eclipse visible

160°W 120°W 80°W 40°W 0° 40°E 80°E 120°E 160°E

**ECLIPSE 2003** *This map for 9 November 2003 shows how much of the world can experience a total lunar eclipse.*

## DANJON SCALE

To compare lunar eclipses, an observer can note the degree of darkness using a five-step scale, named for the French astronomer André Danjon, who devised it.

**L = 4** Strikingly bright copper-red or orange eclipse; bluish tint where the inner shadow (umbra) meets the outer shadow (penumbra).

**L = 3** Brick-red eclipse; umbra has light gray or yellow rim.

**L = 2** Deep red or rust-colored eclipse; umbra usually has a very dark center and relatively bright outer edge.

**L = 1** Dark eclipse; gray or brownish coloration; surface features hard to make out.

**L = 0** Very dark eclipse; Moon nearly invisible.

Make your estimate at maximum eclipse. You can use decimal steps between the classes: 3.5 or 2.3, for example.

sky, changing color and brightness as the eclipse progresses.

You can capture the partial phases with very short exposures at f/8. During totality, use ¼ second at f/4. If the eclipse seems especially dark, you should double (or even quadruple) the exposure. Because it is so hard to guess the Moon's brightness ahead of time, you should plan to take a range of exposures. Make one shot at the "normal" exposure, followed by one at twice that exposure, and a third at four times the normal exposure. At least one of the images should come out just right.

As the Moon dwindles in brightness during the eclipse, beautiful star fields all over the sky become more prominent. This effect is strongest for observers at dark sites well away from light pollution.

## MOON COLORS

The spookiest thing about a lunar eclipse is the eerie reddish color that covers the Moon. If Earth had no atmosphere, lunar eclipses would simply turn the shadowed part of the Moon jet-black. But Earth's air acts like a simple lens or prism. It bends part of the Sun's light into the shadow and stains this light a deep, ruddy copper color. In essence,

what illuminates the Moon during a lunar eclipse is the light of every sunset and sunrise happening on Earth.

Experienced eclipse gazers try to guess how dark the shadow will be. With some eclipses, you see a pale amber Moon. But especially in years following big volcanic eruptions, the shadow can become so dark that the Moon virtually disappears, even when seen through a telescope. In 1883, the volcano Krakatau near Java exploded, sending several cubic miles of pulverized rock high into the atmosphere. At the next lunar eclipse several weeks later, the Moon all but vanished. Similarly, dark eclipses have followed other big eruptions, such as that of Mount Agung on Bali, which produced two "black Moon"

eclipses in 1963 and 1964. More recently, a very dark eclipse followed the eruption of the Mexican volcano El Chichón in 1982.

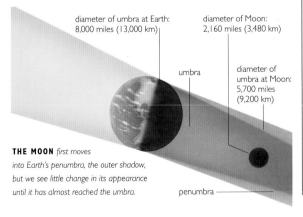

diameter of umbra at Earth: 8,000 miles (13,000 km)

diameter of Moon: 2,160 miles (3,480 km)

umbra

diameter of umbra at Moon: 5,700 miles (9,200 km)

**THE MOON** *first moves into Earth's penumbra, the outer shadow, but we see little change in its appearance until it has almost reached the umbra.*

penumbra

135

# OBSERVING *the* MOON

*For as long as people have thought about the world around them, they have looked up at the Moon and pondered its varying ways.*

The Moon has often been seen as an inferior companion to the spectacular Sun or as a fickle deity ruling the night sky. Its phases heralded many important cultural and religious events. In today's urban culture, lunar folklore holds a smaller place than it once did, but some still refer to people as being "moonstruck."

The light of the Moon aided hunters and farmers from the very beginning, and nearly every culture recognized the Moon's influence on the ebb and flow of the tides. The Moon also provided humankind with its first calendar—the word "month" even comes from the word "moon."

The Moon is the first celestial object most new telescope owners look at—and with good reason. No matter what size instrument you use, the sight is spellbinding. Craters,

**THE FULL MOON** *rising over water (above) and a crescent Moon bathed in earthshine (right) make fine photographic subjects. Many people claim they can see the Man in the Moon (top).*

rays, and mountains all parade in testimony to the Moon's violent geological past.

To the naked eye, the Moon shows two kinds of area—one dark and the other light. Ancient astronomers believed the dark areas were oceans like those on Earth and named them using the Latin word for sea: mare, pronounced MAH-ray (plural maria, MAH-ree-uh). These terms persist, despite our knowledge that the "seas" are really smooth sheets of congealed lava that oozed from beneath the Moon's surface after meteorite impacts. The lighter areas, known as highlands or terrae, are pieces of ancient lunar crust, battered by innumerable meteorites.

## MOON-ROVING

"Tides" raised in the rocky body of the Moon by Earth's gravity have caused the Moon to rotate on its axis in exactly the same time as it takes to orbit Earth. This is why the same side is always turned toward us.

As it orbits, the Moon shows a varying amount of its illuminated face because the angle between the Sun, Earth, and itself changes. These variations are called phases.

While Full Moon has its attractions, the best time to explore the Moon's cratered surface with a telescope is when sunlight strikes it at a shallow angle, creating long

| MOON FACTS | |
|---|---|
| **Distance from Earth** | |
| 238,856 miles (384,401 km) | |
| **Mass** 0.012 × Earth's mass | |
| **Radius** 0.272 × Earth's radius | |
| **Apparent size** 31 arcminutes | |
| **Apparent magnitude** | |
| −12.7 at Full Moon | |
| **Rotation** 27.3 Earth days | |
| **Orbit (relative to stars)** | |
| 27.3 Earth days | |
| **Orbit (seen from Earth)** | |
| 29.5 Earth days | |

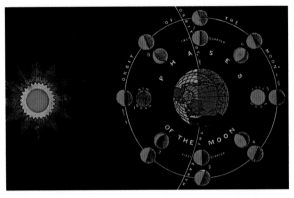

**THE LUNAR PHASES,** *as explained in this 1840 diagram, are: (1) New Moon; (2) Waxing Crescent; (3) First Quarter; (4) Waxing Gibbous; (5) Full Moon; (6) Waning Gibbous; (7) Last Quarter; and (8) Waning Crescent.*

shadows and throwing the features into sharp relief. This happens twice a month for about a week at a time, centered on the dates of the First and Last Quarter phases.

Start with the lowest magnification your telescope has and slowly increase the power as one feature or another catches your eye. If the view grows blurry, drop back in power. At first, the variety of features may be bewildering, but compare your view with one of the eight Moon maps on pages 142–149. Identify any large dark maria you can see, and use them as a guide to take you farther. You will see the most details near the terminator—the dividing line between the illuminated and dark portions of the Moon.

## A Two-week Day

Any feature of the Moon will change appearance during the two weeks that sunlight falls on it—the lunar day for that area. A fascinating exercise is to follow the play of sunlight across a big crater such as Copernicus (see Map 6, p. 147).

A day or two after First Quarter, as the Sun rises over Copernicus, the peaks of the rim turn into a ring of light. As dawn creeps inside, the light reveals terraces on the crater's walls and its central peaks—and finally a flat floor of lava.

At Full Moon, sunlight pours down on the crater,

flattening the view of the terrain but revealing rays—streaks of pulverized rock that shot from the newborn crater in the first moments after the meteorite impact.

As the Sun descends from lunar noon, shadows return and Copernicus takes on form and shape once more.

At lunar sunset, the crater is completely hidden by shadows. Then, one fine evening you see a young Moon low in the west and can start preparing to watch the show again.

## Sketchy Details

Anyone trying to sketch the Moon's face soon gives up hope of capturing all its details. Instead, try mapping just one small area throughout its phases. You will see how even small lighting changes can reveal crater terraces—or make a rille pop into view on a mare.

Pick a region or a feature, and settle down with your telescope. Use a magnification that captures the whole feature in the field of view. Lightly sketch the main parts of the feature using a sharp HB pencil, and then shade in details with a 2B pencil, smudging them with your finger or touching up with an eraser.

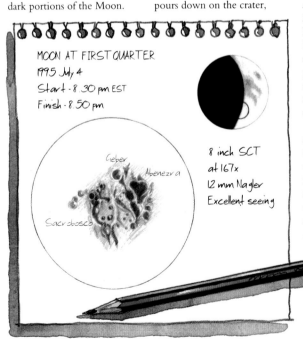

MOON AT FIRST QUARTER
1995 July 4
Start - 8.30 pm EST
Finish - 8.50 pm

Gieber
Abenezra
Sacrobosco

8 inch SCT
at 167x
12 mm Nagler
Excellent seeing

**WHEN YOU FINISH** *a sketch, note details such as the date, the telescope used, and the seeing conditions.*

# READING *the* MOON'S HISTORY

*One of the biggest lunar questions has been—*
*where did such a body come from?*

WALLACE AND GROMIT *visit the Moon in the animated film* A Grand Day Out. *They discover the Moon is made of cheese and slice off pieces for their picnic.*

Some scientists dub the current theory for the Moon's origin "The Big Whack." It emerged in 1984 after more than 10 years' study of rock and soil samples taken from the six Apollo landings (1969–72) and it still offers the most complete account of the birth of the Moon. According to this theory, the proto-Earth formed alone. By the time Earth had reached nearly its present size, it had melted throughout and separated into a core and mantle.

At this point, roughly 4.5 billion years ago, Earth was struck a glancing blow by another nearly formed planet about the size of Mars. This impact sped up Earth's rotation, probably tipped its axis, and heated its surface intensely. The force of the collision vaporized the impactor and sent a great spray of superhot vapor into orbit around Earth. This vapor quickly cooled into a ring of debris and in only a few tens of thousands of years, the fragments coalesced into the Moon.

But the newborn Moon looked very different from the one you might gaze at tonight. It loomed much larger in the sky because it orbited less than half as far away. A thin, brittle crust covered a global ocean of molten rock. Bright glowing wounds opened with every meteorite impact—and such impacts were constant at first, as the new Moon swept up the remaining swirling debris.

## A QUIET WORLD

Millions of years passed. The lunar crust cooled, thickened, and stiffened. Collisions continued at a great rate, but because the crust was tougher, impacts began to excavate basins and craters, and to pile up debris around them.

About a billion years after the Moon coalesced, the last few large impacts blasted a handful of basins on the Moon's Earth-facing side. Pools of dark lava flooded the basins, filling them and spilling over to make the face of the Man in the Moon. Although impacts continued to occur, the rate dropped off dramatically.

Finally, the bombardment all but ceased. For the last three billion years, the Moon

APOLLO ASTRONAUTS *left behind footprints (above) that will remain for millions of years because the Moon has virtually no erosion. During the Apollo 17 mission, Harrison Schmitt (left) spent 22 hours collecting rock samples.*

ROUNDED HILLS *flank the valley where Apollo 17 landed (left). This lunar rock (below) came back with Apollo 16.*

## LIQUID ROCK

Lunar geologists describe the maria as vast lava flows, many of which fill large shallow basins excavated by impacts. After a basin forms, millions of years may pass before lava flows into it. In the meantime, impacts can form craters on the basin floor. Look at Archimedes in Mare Imbrium—its walls are intact but its floor is flooded by lava.

When lava floods a basin, it flows easily and for long distances, being about as runny as motor oil at room temperature. Mare lavas are made up of several layers that may form a total thickness of less than 2 miles (3 km)— paper-thin in relation to their large extent.

Full Moon is a good time to look for differences in hue that reveal separate episodes of flooding. In places, the channels that fed the flows are visible as thin, winding grooves, known as rilles, that meander and then disappear. The Apollo 15 astronauts landed near the Hadley Rille, a sinuous rille at the foot of the Apennines (see Map 6, p. 147).

Where the basin floor buckles under the weight of the lava, the mare surface pushes up in what is known as a wrinkle ridge. Where mare lava bends over a rise, the brittle lava cracks in straightish lines to form long troughs, called grabens. Several sets of grabens are visible on the east-ern shore of Mare Humorum, from about 11 days after New Moon (see Map 7, p. 148).

has been a fairly quiet place. At present, it is struck only rarely. The crater Copernicus is 800 million years old, and the most recent major impact formed the crater Tycho about 109 million years ago. Compared with Earth, and its constant erosion, mountain building, and volcanic activity, the Moon is a dead world indeed, although unusual events are sometimes reported by skywatchers (see p. 140).

## IMPACTS, CRATERS, AND BASINS

A telescope shows an amazing array of lunar features. Most are the result of impacts by meteorites, asteroids, and comets. When a piece of rock, ice, or anything hits the Moon at several miles per second, it blows up in a cloud of vapor, blasting a scar on the Moon's surface. Small impacts make tidy, bowl-shaped craters, such as Aristarchus, which is only 25 miles (40 km) across (see Map 5, p. 146). These small craters are surrounded by aprons of ejected rock fragments, seen as bright areas around the younger craters.

Bigger impacts make larger craters with a more complex structure. A good example is Copernicus, 57 miles (92 km)

across, with central peaks formed from pieces of deep lunar crust thrust upward by the shock of the impact. Landslides have terraced the inside of the crater's walls, and hardened pools of impact-melted rock are littered across its floor (see Map 6, p. 147).

Craters that are larger still tend to resemble walled plains. They have flat or convex floors and their central peaks may form concentric rings. These big craters rarely have rays, however, because rays disappear in about a billion years through the action of tiny meteorites striking the surface—and all of the largest craters are older than that.

The biggest craters are called basins. Filled by Mare Imbrium (see Map 6, p. 147), the Imbrium basin spans 720 miles (1,160 km)—about the size of Texas—and was created by a small asteroid about 3.9 billion years ago. The Apennine Mountains form the southeastern part of the Imbrium basin's rim.

# MAPPING *the* MOON

*Since the invention of the telescope, many lunar observers have devoted their efforts to mapping the seemingly endless geological detail of the Moon.*

In 1610, Galileo Galilei was the first person to use a telescope to survey the Moon in detail through a full cycle of phases. His sketches were crude, but they were soon improved on by other astronomers. These early maps, with their plains, mountains, and craters, are still amazing to see, especially since they were done with unwieldy telescopes much inferior to today's amateur scopes.

Many observers hoped to record changes in the Moon's features, but despaired of ever capturing by hand all the detail visible in their eyepieces. About a century ago, photography stepped in to offer a solution, and Moon charts today are usually based on photos, even if the final map is drawn or painted.

The era of lunar mapping has not yet ended. The Space Age brought new maps of the unseen far side of the Moon,

long a source of speculation, as well as much more detailed maps of the near side. But it was only in 1996 that scientists were finally able to map the Moon's global topography precisely, thanks to the laser altimeter on the Clementine spacecraft. The altimeter fired a laser beam at the ground, timing its return very accurately. With both Clementine's

**MAPS OF THE MOON** *range from Galileo's 1616 sketches (left) to Clementine's 1994 topographic map (top), which shows highlands as pink and red, and lowlands as blue and purple.*

orbit and the speed of light being known, elevations of the landscape could be determined.

## IS THE MOON ACTIVE?
Searches for activity on the Moon began with Galileo almost 400 years ago and have continued ever since. Observers have claimed to see mists obscuring craterlets on Plato's floor, a glow on the central peak of Alphonsus, and the "disappearance" of the Linné crater, to name just a few examples. Are these transient lunar phenomena (TLPs) the result of meteorite strikes or gaseous escapes from the Moon's interior—or are they merely tricks of the light and an over-eager imagination?

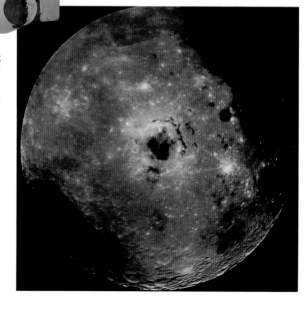

**THE LUNAR FAR SIDE**, *which is partly visible in the left half of this image, shows almost no areas of mare. This may be because its crust is too thick for lava to break through the surface easily.*

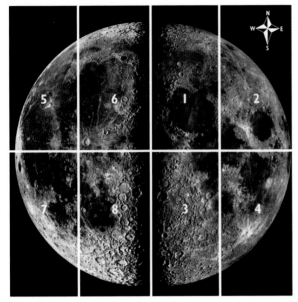

**THE NEAR SIDE** *is divided into eight Moon Maps on pages 142–149. Maps 1 to 4 show the Moon at First Quarter, and Maps 5 to 8 show it at Last Quarter.*

East and west on these maps match the directions used for Earth, rather than those usually used for the sky. This convention was chosen by the International Astronomical Union, which agreed with NASA's astronauts that for people on the Moon, the Sun should rise in the east and set in the west.

You may sometimes see slightly more or less of the Moon than our maps show because the tilted, elliptical orbit of the Moon causes "librations," which give us occasional glimpses of extra territory around its limb edges.

*It is a most beautiful and delightful sight to behold the body of the Moon.*

GALILEO GALILEI (1564–1642), Italian astronomer

Many TLP reports grew from the long-standing belief that the Moon's features were wholly or largely volcanic in origin. Since volcanoes are always erupting somewhere on Earth, it was natural to expect similar activity on the Moon.

But scientists now know that lunar features, especially craters, are overwhelmingly the result of impacts. The Moon's real volcanic activity mainly produced the maria, and research has shown that even the youngest mare has been cold, hard rock for at least a billion years. Instruments left by Apollo astronauts reported that the Moon's interior is seismically quiet, indicating that there is no volcanic activity.

While the Moon's geo - logical engine may have shut down ages ago, it is no less interesting a place. In fact, its surface records something of great value: a history of the Solar System's earliest days. The record of this period on Earth has all but vanished, thanks to erosion. But in the Moon, nature has conveniently placed a veritable geo-museum in our cosmic backyard.

## NAVIGATING THE MOON

The eight Moon maps on the following pages portray the Moon with north up. This is how it appears to naked-eye observers in the Northern Hemisphere. Depending on what type of telescope you use, your telescope image may be oriented differently (see p. 76). Also bear in mind that the Moon appears south up in the Southern Hemisphere.

## ICE ON THE MOON

If rocky asteroids and meteorites can hit the Moon, so can comets. But a comet would have a different effect from an asteroid because comets are extremely rich in water. The impact of a large comet would create a big crater, but would also give the Moon a tenuous and temporary atmosphere loaded with water vapor. Any water vapor that drifted near the lunar poles might condense into ice, especially in those parts that remain untouched by sunlight. The craters in these regions could retain ice for millions of years.

Recently, scientists using several spacecraft suggested they may have found evidence for ice just under the surface near the Moon's south pole (right). However, in 1999 the Lunar Prospector spacecraft deliberately crashed into a north pole crater, hoping to create a plume of water vapor. Nothing was seen in that experiment.

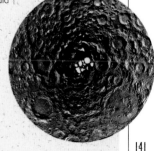

## FEATURES OF MOON MAP I

**M**are Serenitatis **(23)**, the Sea of Serenity, is a sheet of lava 380 miles (610 km) across that marks the Man in the Moon's left eye. It shows best toward First Quarter. The distinct tints in the lava suggest several flooding episodes. In the mare's east lies **Serpentine Ridge (31)**, a superb wrinkle ridge formed when the basin subsided and squeezed the lava. On the northeastern shore lies the flooded crater **Posidonius (21)**, with a curved ridge crossing its floor. The bright streak near **Bessel (29)** is a ray from Tycho, a young crater some 1,400 miles (2,200 km) away (see Map 8). The western rim of the mare, formed by the mountain ranges **Montes Apenninus (35)** and **Caucasus (24)**, is shared with with the Imbrium basin (see Map 6). North of Serenitatis lie several large craters and the eastern reaches of **Mare Frigoris (7)**, the Sea of Cold.

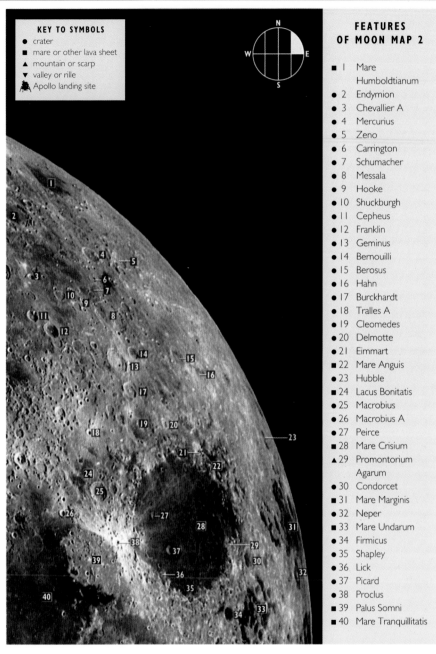

**KEY TO SYMBOLS**
- ● crater
- ■ mare or other lava sheet
- ▲ mountain or scarp
- ▼ valley or rille
- 🚀 Apollo landing site

N
W — E
S

**FEATURES
OF MOON MAP 2**

- ■ 1 Mare Humboldtianum
- ● 2 Endymion
- ● 3 Chevallier A
- ● 4 Mercurius
- ● 5 Zeno
- ● 6 Carrington
- ● 7 Schumacher
- ● 8 Messala
- ● 9 Hooke
- ● 10 Shuckburgh
- ● 11 Cepheus
- ● 12 Franklin
- ● 13 Geminus
- ● 14 Bernouilli
- ● 15 Berosus
- ● 16 Hahn
- ● 17 Burckhardt
- ● 18 Tralles A
- ● 19 Cleomedes
- ● 20 Delmotte
- ● 21 Eimmart
- ■ 22 Mare Anguis
- ● 23 Hubble
- ■ 24 Lacus Bonitatis
- ● 25 Macrobius
- ● 26 Macrobius A
- ● 27 Peirce
- ■ 28 Mare Crisium
- ▲ 29 Promontorium Agarum
- ● 30 Condorcet
- ■ 31 Mare Marginis
- ● 32 Neper
- ■ 33 Mare Undarum
- ● 34 Firmicus
- ● 35 Shapley
- ● 36 Lick
- ● 37 Picard
- ● 38 Proclus
- ■ 39 Palus Somni
- ■ 40 Mare Tranquillitatis

The northeastern limb of the Moon is best seen when the Moon is about five days old, or about two weeks later, just after Full Moon. The dominant feature is **Mare Crisium (28)**, the Sea of Crises, some 340 miles (550 km) across and visible to the naked eye. From Earth, it looks elongated north and south, but it is actually round, with an extension on the eastern side. Just to the west of Crisium lies the small, bright-rayed crater **Proclus (38)**, 18 miles (29 km) across. The irregular ray pattern indicates that the meteorite was traveling northeast at a shallow angle when it struck. North of Proclus, **Macrobius (25)** shows a dark floor, as does **Cleomedes (19)**, which sports a rille on its northern floor. Scientists think the lava that flooded the Crisium basin also underlies these two craters and other parts of the region where it shows up as dark patches.

While Mare Nectaris, the Sea of Nectar, is shared with Map 4, its western features lie here in Map 3. **Mare Nectaris (19)**, 220 miles (350 km) across, is a thin sheet of lava that fills just the inner part of the much larger Nectaris basin, 540 miles (860 km) in diameter. On the southern shore of the mare sits the crater **Fracastorius (23)**, with one wall melted away. **Theophilus (15)** is the youngest and best-preserved crater of a trio lying west of the mare. North from Theophilus, **Mare Tranquillitatis (5)**, the Sea of Tranquillity, contains the landing site of Apollo 11, which carried the first men to the Moon. **Rupes Altai (25)**, the Altai Scarp, is a curved, east-facing cliff that marks part of the Nectaris basin's rim. Running southwest from the scarp are the southern highlands, an ancient terrain where craters of all sizes interrupt each other in a saturated jumble.

## FEATURES OF MOON MAP 4

- ● 1 Taruntius
- ● 2 Secchi
- ● 3 Webb
- ■ 4 Mare Spumans
- ■ 5 Mare Smythii
- ● 6 Censorinus
- ● 7 Messier A
- ● 8 Messier
- ● 9 Lubbock
- ■ 10 Mare Fecunditatis
- ● 11 Langrenus
- ● 12 Isidorus
- ● 13 Capella
- ● 14 Gutenberg
- ● 15 Goclenius
- ● 16 Gaudibert
- ● 17 Magelhaens
- ● 18 Bellot
- ● 19 Crozier
- ● 20 Vendelinus
- ● 21 Colombo
- ▲ 22 Montes Pyrenaeus
- ● 23 Rosse
- ● 24 Cook
- ● 25 Holden
- ● 26 Balmer
- ● 27 Monge
- ● 28 Santbech
- ● 29 Biot
- ● 30 Wrottesley
- ● 31 Petavius
- ● 32 Humboldt
- ● 33 Snellius
- ● 34 Reichenbach
- ● 35 Furnerius
- ■ 36 Mare Australe
- ● 37 Rheita
- ▼ 38 Vallis Rheita
- ● 39 Metius
- ● 40 Fabricius
- ● 41 Steinheil
- ● 42 Watt

### KEY TO SYMBOLS
- ● crater
- ■ mare or other lava sheet
- ▲ mountain or scarp
- ▼ valley or rille
- 🛰 Apollo landing site

N
W — E
S

The region on the Moon's southeastern limb contains few "show-stopper" features, but it is easy to explore using **Mare Fecunditatis (10)**, the Sea of Fertility, as home base. The two best times to investigate this region are about five days after New Moon and about two days after Full Moon. Mare Fecunditatis itself is roughly circular and fills an ancient impact basin 430 miles (690 km) in diameter. Lying in its center are two tiny craters, **Messier (8)** and **Messier A (7)**, with two rays extending west from the latter. **Langrenus (11)** is a large, pale crater with a small central peak. To its south lies a similar crater, **Petavius (31)**, with terraced walls and rilles on its lava-filled floor. Around the medium-size crater **Furnerius (35)** lies a bright patch of rays coming from a diminutive but very fresh crater called Furnerius A.

## FEATURES OF MOON MAP 5

### KEY TO SYMBOLS

- ● crater
- ■ mare or other lava sheet
- ▲ mountain or scarp
- ▼ valley or rille
- ★ surface feature
- 🛰 Apollo landing site

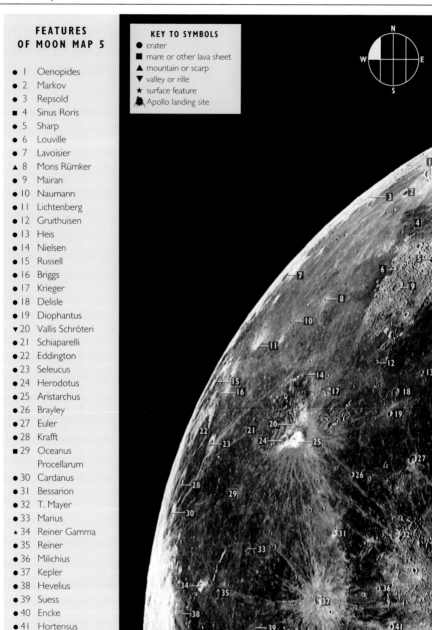

Compared with other parts of the Moon, this section looks almost empty. Most of the area is covered by a vast sheet of lava—**Oceanus Procellarum (29)**, the Ocean of Storms. Sitting atop a large plateau that saw some of the Moon's most violent activity is one of its most interesting craters—**Aristarchus (25)**, just 25 miles (40 km) across yet visible to the eye from Earth thanks to its dazzling brightness. Next to it is the crater **Herodotus (24)**, the source for Schröter's Valley, or **Vallis Schröteri (20)**, a winding sinuous rille carved by flowing lava.

Another crater to explore here is **Kepler (37)**, which looks like a miniature Copernicus (marked on Map 6). See also the mysterious feature **Reiner Gamma (34)**, a swirl of light markings on the mare surface. Compared with other material on the Moon, this feature has a strong magnetic field.

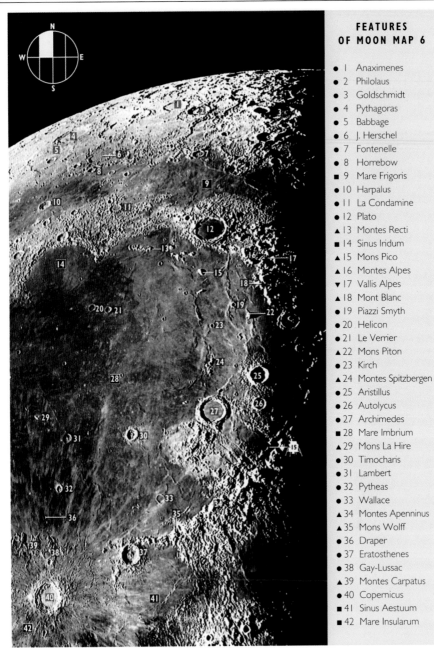

As the Moon enters its gibbous phase, Mare Imbrium and the Apennine Mountains emerge from their lunar night. **Mare Imbrium (28)**, the Sea of Rains, fills the Imbrium basin, a depression created in a gigantic impact that affected most of the Moon's near side.

Marking the basin's rim are **Montes Apenninus (34)**, the Apennines, plus **Montes Alpes (16)**, the Alps, and **Montes Carpatus (39)**, the Carpathian Mountains. Craters that formed in the time between the basin impact and the lava eruptions that filled it are **Archimedes (27)**, **Plato (12)**, and the Bay

of Rainbows, **Sinus Iridum (14)**. More recent features are **Eratosthenes (37)** and the splendid **Copernicus (40)**, which formed some 800 million years ago. Copernicus has a large ejecta apron—bright streaks of rock flung out by the impact—and a complex internal structure (see p. 139).

## FEATURES OF MOON MAP 7

- 1 Riccioli
- ■ 2 Grimaldi
- 3 Hermann
- ■ 4 Oceanus Procellarum
- 5 Flamsteed
- 6 Lansberg A
- 7 Wichmann
- 8 Lansberg
- 9 Reinhold
- 10 Euclides
- ▲ 11 Montes Riphaeus
- ■ 12 Mare Cognitum
- ■ 13 Mare Orientale
- ■ 14 Lacus Aestatis
- 15 Crüger
- 16 Sirsalis
- 17 Hansteen
- 18 Billy
- 19 Zupus
- 20 Gassendi
- 21 Lubiniezky
- 22 Mersenius
- 23 Liebig
- ■ 24 Mare Humorum
- 25 Hippalus
- 26 König
- 27 Palmieri
- 28 Doppelmayer
- 29 Campanus
- 30 Lagrange
- 31 Piazzi
- ■ 32 Lacus Excellentiae
- ■ 33 Palus Epidemiarum
- 34 Schickard
- 35 Phocylides
- 36 Hainzel
- 37 Schiller

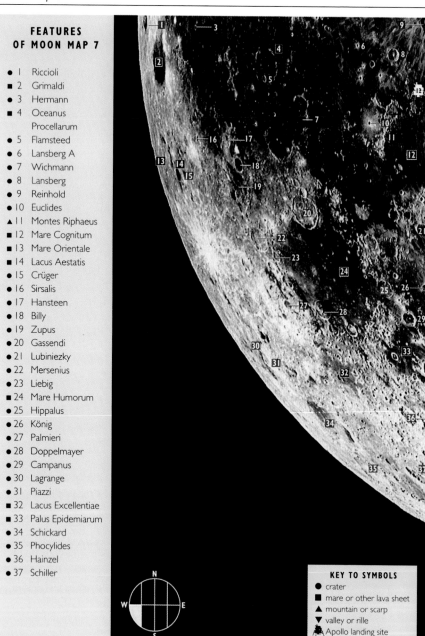

**KEY TO SYMBOLS**
- ● crater
- ■ mare or other lava sheet
- ▲ mountain or scarp
- ▼ valley or rille
- 🛰 Apollo landing site

In the last days before Full Moon, the Sun reaches the area of the southwestern limb. The main landmark here is the 220 mile (350 km) expanse of **Mare Humorum (24)**, the Sea of Moisture, which you can see with the naked eye. Two craters stand on its shores,

both partly breached by lava: **Gassendi (20)** in the north and **Doppelmayer (28)**, the more engulfed crater, in the south. Between Mare Humorum and the limb lies **Schickard (34)**, a large subdued crater with patches of dark lava covering about half its floor. Near Schickard,

scientists have found terrain they call cryptomare—dark patches of lava that seem to have oozed up from beneath craters and hills during a late phase of flooding. Perhaps also from this period is **Grimaldi (2)**, an isolated patch of mare that fills a 140 mile (225 km) basin near the western limb.

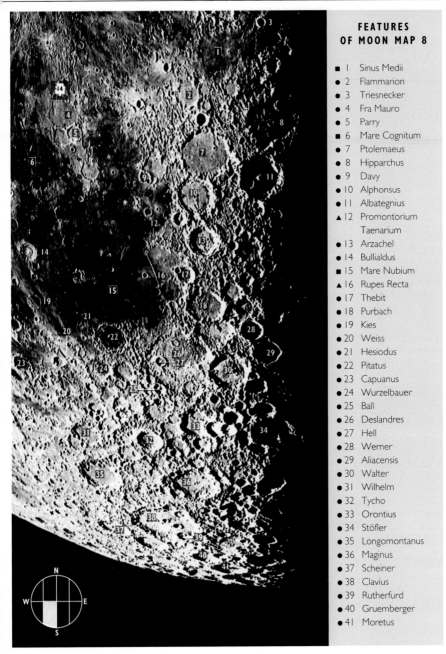

**FEATURES
OF MOON MAP 8**

- ■ 1 Sinus Medii
- ● 2 Flammarion
- ● 3 Triesnecker
- ● 4 Fra Mauro
- ● 5 Parry
- ■ 6 Mare Cognitum
- ● 7 Ptolemaeus
- ● 8 Hipparchus
- ● 9 Davy
- ● 10 Alphonsus
- ● 11 Albategnius
- ▲ 12 Promontorium
  Taenarium
- ● 13 Arzachel
- ● 14 Bullialdus
- ■ 15 Mare Nubium
- ▲ 16 Rupes Recta
- ● 17 Thebit
- ● 18 Purbach
- ● 19 Kies
- ● 20 Weiss
- ● 21 Hesiodus
- ● 22 Pitatus
- ● 23 Capuanus
- ● 24 Wurzelbauer
- ● 25 Ball
- ● 26 Deslandres
- ● 27 Hell
- ● 28 Werner
- ● 29 Aliacensis
- ● 30 Walter
- ● 31 Wilhelm
- ● 32 Tycho
- ● 33 Orontius
- ● 34 Stöfler
- ● 35 Longomontanus
- ● 36 Maginus
- ● 37 Scheiner
- ● 38 Clavius
- ● 39 Rutherfurd
- ● 40 Gruemberger
- ● 41 Moretus

In the evenings that follow First Quarter, the Sun rises over some of the Moon's finest sights. In the south lie two notable craters, giant **Clavius (38)**, which spans 140 miles (225 km), and smaller **Tycho (32)**. Catch Clavius when the Sun first strikes, revealing a convex floor with a curving chain of craterlets. Tycho, only 109 million years old, lies at the hub of a splash pattern of rays (best seen at Full Moon) that stretches for more than 1,000 miles (1,600 km). Look for a trio of craters—**Arzachel (13)**, **Alphonsus (10)**, and **Ptolemaeus (7)**—aligned north-south. **Rupes Recta (16)**, the Straight Wall, is a fault scarp that changes from a dark line to a bright one toward lunar noon. **Deslandres (26)**, to the south, is a ruined crater 146 miles (235 km) across. Nearby **Pitatus (22)** has several rilles on its floor and partly flooded wall.

The unquiet republic of the maze
Of Planets, struggling fierce towards heaven's free
wilderness.

*Prometheus Unbound,*
PERCY BYSSHE SHELLEY (1792–1822), English poet

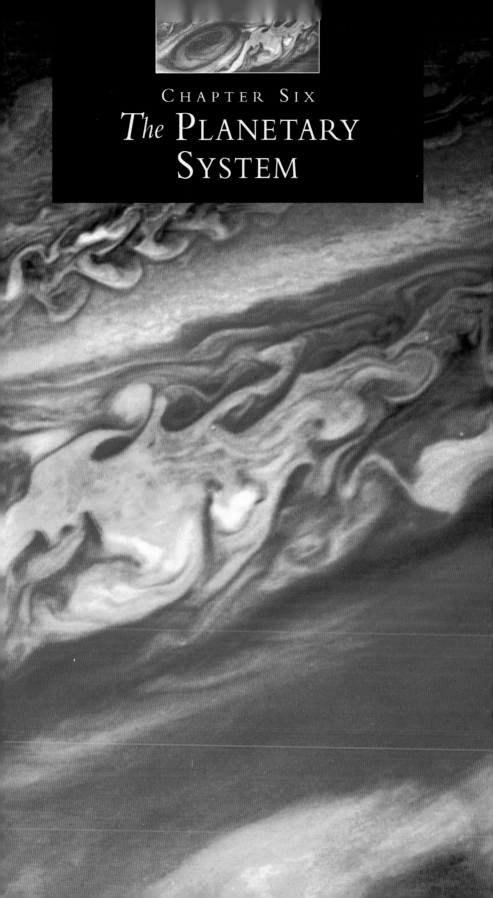

CHAPTER SIX
# *The* PLANETARY SYSTEM

# THE BIRTH *of* *the* SOLAR SYSTEM

*According to one theory, an ancient supernova began the process that resulted in the creation of our Solar System about 4.6 billion years ago.*

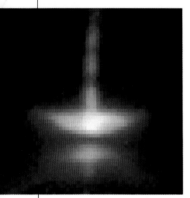

**THE CENTRAL STAR** *of HH30 is hidden from view but emits powerful "jets" and illuminates the upper and lower surfaces of a thin disk in this HST image.*

**HOW THE PLANETS FORMED** (right) *Triggered perhaps by an exploding star, the planets formed from a disk of gas and dust surrounding the Sun.*

Imagine a large, cold, dark cloud that has rested in space for a great length of time. Somewhere nearby, a star became unstable as it ran out of fuel, then blew itself apart, catapulting much of its contents into the cloud. As a result, the cloud was enriched with some heavier elements

*He, who through vast*

*immensity can pierce,*

*See worlds on worlds*

*comprise one universe,*

*Observe how system into*

*system runs, What other*

*planets circle other suns, . . .*

*May tell why Heaven has*

*made us as we are.*

*An Essay on Man,*
ALEXANDER POPE (1688–1744),
English poet

from the supernova, including carbon—the basis of life—and began to collapse. As it shrank, it began to rotate faster and its particles clumped together. Most of the material gravitated toward the center, where a mighty proto-sun began to grow and heat up.

Was this how the Sun began to form? For some reason, a gas cloud did begin to collapse to form the Sun, but what happened to the rest of the cloud as the Sun grew stronger? Scientists believe that grains of material from the disk consolidated into solid lumps of material. Because there were so many of these bodies, they kept colliding, growing into larger bodies called proto-planets, which became the planets we see today, while some of the leftover small pieces evolved into comets.

According to other theorists, there were many proto-planets that increased in size for a time, but they broke up as a result of repeated collisions with smaller objects and each other. Only the few largest ones were left to consolidate

and cool, becoming the nine planets we know today.

The disk of material, known as the accretion disk, continued to rotate, and temperatures soared at its center. As its core—the Sun—"turned on," it blew away the remnants of the cloud, leaving an infant Solar System consisting of a group of small, warm inner planets nestled close to the Sun, some larger, cold outer planets, and small, frozen comets at the outer edges—plus a good deal of rocky debris. And all this took place quite quickly, cosmically speaking. It is estimated that it may have been no more than 100 million years between the time the cloud began to collapse and the moment the Sun ignited. The discovery of planets around nearby stars seems to confirm that planetary systems are common, although perhaps the Sun's family is not typical.

**RELATIVE DISTANCES OF PLANETS FROM SUN** *(below) From left to right: Sun, Mercury, Venus, Earth, Mars, Asteroid Belt, Jupiter, Saturn, Uranus, Neptune, and the dwarf planet Pluto.*

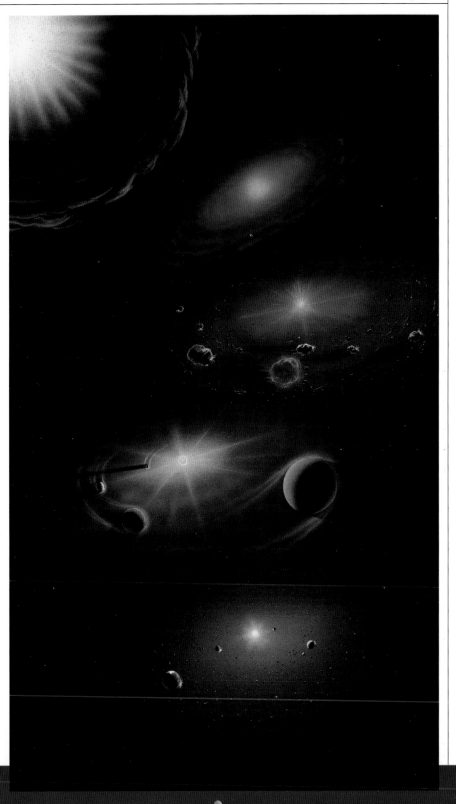

# OBSERVING PLANETS

*Telescopes may provide less spectacular views than spacecraft do, but they let you check out Mars, Saturn, or Jupiter on any clear night of the year.*

**AN ORRERY** *(left) is a mechanical model of the Solar System. The first such device was made for Charles Boyle, the Earl of Orrery (1676–1731).*

The Space Age has revolutionized our understanding of the planets, and future spacecraft missions will no doubt continue to dramatically alter the picture. Often as not, however, it is an amateur astronomer who alerts scientists that a new dust storm has begun on Mars, or that the placid equatorial zone of Saturn has been disturbed by a huge spot.

A telescope is also the closest thing to a spaceship most observers are ever likely to have. No one has yet stood on the sands of Mars, but you can peer into your eyepiece and see hazy clouds lit by afternoon sunlight over the Red Planet's volcanoes.

## ROCKS AND GAS

You can sort planets into two categories—two ways. The first pair of categories refers to the planets' physical nature. Earth is a terrestrial planet, so too are Mars, Venus, and Mercury. Jupiter, on the other hand, is a gas giant, along with Saturn, Uranus, and Neptune. The terrestrial planets are small and rocky with relatively thin atmospheres. The giants are at least a dozen times more massive and consist of deep, heavy atmospheres that surround small, rocky cores.

Pluto is an oddball that does not fit either category well. In 2006, the International Astronomical Union demoted it to the status of a dwarf planet. It is probably one of many in the Outer Solar System.

## INTERIOR PLANETS

The other way to sort planets is into interior and exterior ones. The two interior planets are Mercury and Venus; they circle the Sun inside Earth's orbit. Exterior planets are those from Mars outward; they stay outside of Earth's orbit.

This difference matters because it controls how and where you look for the planet. An interior planet always stays near the Sun in the sky; you will never see one at midnight, for example. As Mercury or Venus orbits the Sun, you will see it first in the evening sky, moving day by day out from the Sun's glare. After a few weeks, it reaches a point of maximum separation from the Sun called greatest eastern elongation, usually the time of best visibility.

The planet passes between Earth and the Sun at a point called inferior conjunction, when it briefly disappears into the solar glare. Greatest western elongation follows in the predawn, again bringing good visibility. The planet completes its orbit on the far side of the Sun, and disappears once more. This ends one apparition, or viewing season, and begins the next.

**INSIDE AND OUT** *Interior planets such as Venus reach best visibility at the points of greatest elongation, while exterior planets such as Mars are best seen at opposition. All planets disappear into the Sun's glare at conjunction.*

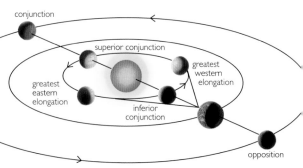

conjunction

superior conjunction

greatest western elongation

greatest eastern elongation

inferior conjunction

opposition

**THE EIGHT PLANETS** *of the Solar System orbit the Sun in ellipses, a fact discovered by Johannes Kepler in the seventeenth century.*

## EXTERIOR PLANETS

Exterior planets have a simpler routine. In all cases, Earth is orbiting faster than they are, so we overtake them. We first notice the planet rising just ahead of the Sun, low in the predawn sky. As weeks pass, it rises earlier each night.

Many weeks before op - position, the planet's motion against the background stars stops, and it then appears to move backward (westward). This retrograde motion is caused by Earth's greater speed—in effect, the outer planet behaves like a car traveling in the adjoining lane as you pass it on the highway.

At opposition, the planet reaches a point where it rises as the Sun sets. This is the time of best visibility. The planet then slips into the evening sky, where it sets after the Sun. After several weeks, it stops moving retrograde and recommences eastward motion. Eventually, the planet disappears into the Sun's glare and reaches conjunction. Conjunction is followed by the predawn start of a new apparition.

*The Planets in their*

*stations list'ning stood,*

*While the bright Pomp*

*ascended jubilant ...*

Paradise Lost,
JOHN MILTON (1608–74),
English poet

## FINDING PLANETS

Planets orbit close to the ecliptic, the plane of Earth's orbit. This makes them easier to find—a line marking the ecliptic is usually drawn on sky charts.

Unlike constellations, however, whose seasons of visibility can be predicted easily, planets are on the move. Astronomy magazines provide regular monthly guides and publish feature articles on upcoming events. Various annual handbooks and almanacs do likewise on a yearly basis. Planetarium-type computer programs can show you what is visible on any particular night.

## OBSERVING TIPS

Planets are generally bright, so a dark observing site is less critical than it is for deep-sky objects. And since planets seldom stray more than a few degrees from the ecliptic, the site does not need to provide access to the entire sky.

Planet watching demands no special type of telescope. Steadier and sharper views let you see more with less effort, so the most desirable items are a sturdy equatorial mounting and high-quality optics.

Because atmospheric seeing is rarely steady, most planet observing is done using eye - pieces that yield about 200x magnification or less. But planet watchers should have on hand at least one eyepiece that provides 300x or more.

You may not be able to use it very often, but as all planets have apparent sizes smaller than many of the craters on the Moon, large magnifications are called for at times. Many amateurs also use a set of filters—either glass eyepiece filters (see p. 85) or the less expensive gelatin ones available from camera stores.

For sketching, a clipboard with a dim red light attached will be handy. Purchase a set of circle and ellipse templates from an art store, along with pencils in several grades (2H to 3B), a stub stick, and a white eraser. Photography with film or CCDs can be challenging, but is another satisfying way to record the planets.

Observing planets is a skill like any other—you get better with practice. Beginners who take their first look at Jupiter, for instance, are often disappointed. "Is that all there is to see?" they think, when their telescope shows a tiny oval disk with just two faint, dusky bands crossing it. Likewise, reports of colors seen in astronomical objects often strike newcomers as exaggerated.

When the sight in the telescope eyepiece is unfamiliar, we need to educate our brain to recognize and understand the image it is receiving. The best way to see more detail is to spend lots of time with your telescope exploring the sky.

# MERCURY

*Few skywatchers can boast of spotting Mercury on more than isolated occasions. How can a planet that becomes as bright as the brightest stars be so elusive?*

Of all the planets that are visible to the naked eye, Mercury is the least often observed. The reason lies in its orbit.

Because Mercury is the closest planet to the Sun, its year lasts just 88 days, and it never strays more than 28 degrees from the blaze of light surrounding the Sun. Observers usually glimpse it just before sunrise or soon after sunset, and they can never see it in a fully dark sky.

For all this, antiquity's astronomers knew of Mercury and carefully logged its appearances. In Mesopotamia, where clear desert twilights and low horizons made it easier to see, this come-and-go object was named Nabu; he was a scribe and a messenger of the gods. The planet's swift apparitions also prompted the Greeks to name it after a heavenly messenger—Hermes. The Romans translated the name into Latin as Mercurius.

## FIRE AND ICE

In 1974 and 1975, the Mariner 10 spacecraft photographed about half of Mercury. The other half was mapped by the Messenger mission beginning in 2008. The spacecraft revealed an ancient Moon-like face with craters and scarps, but without

**MERCURIUS,** *sculpted here by l'Antico (1460–1528), was the messenger of the gods in Roman mythology.*

the Moon's dark lava sheets. Mercury has an unusually large iron core, perhaps the result of a catastrophic collision early in its history that removed part of the planet's mantle. Mercury also has a magnetic field, but with only 1 percent the strength of Earth's. Although the planet's gravity is too weak to retain any significant atmosphere, there is a tenuous layer of helium and hydrogen (probably captured from the solar wind), as well as sodium (possibly from the surface rocks).

With a diameter of only 3,024 miles (4,878 km), Mercury is the smallest planet. It also has the most eccentric orbit. Mercury's temperatures range more than any other planet. In daytime, the equatorial regions can reach about 800 degrees Fahrenheit (430° C), while at night, the surface temperature can fall to −300 degrees Fahrenheit (−185° C). Mercury may have subsurface ice in permanently shadowed craters at the poles.

## OBSERVING MERCURY

From the Northern Hemisphere, Mercury is best seen at an evening apparition during March or April, and

**VISITING MERCURY** *An artist's impression (above) shows Mariner 10 flying by Mercury. The spacecraft's cameras sent digital images to Earth, including this close view of grabens— depressions caused by fracturing (right).*

that looks like a tiny gibbous Moon devoid of features. At greatest eastern elongation, it appears half lit. As it approaches inferior conjunction, it grows in size while its illuminated portion shrinks to a crescent. After passing between Earth and the Sun, it repeats the phases in reverse order, finally reaching superior conjunction and starting the cycle anew.

Mercury typically presents a small disk that shimmers from the poor seeing near the horizon. Occasionally, with good seeing, you can use 200x magnification or more, yielding a view like that of the naked-eye Moon.

**A SMALL TELESCOPE** will show you Mercury's phases (right), but the planet's cratered surface can be seen only in Mariner 10 photographs (above).

at a morning one in September or October. (In the Southern Hemisphere, the best evening apparitions come in September or October, and morning ones in March or April.) At these seasons, Mercury's orbit tilts most steeply to the horizon, so it reaches maximum separation from the Sun and best visibility.

At a good evening apparition, you can follow Mercury for about two weeks before the date of greatest elongation, and a week after; for morning apparitions, it is a week before and two weeks after.

Mercury's apparent size varies according to its phases. When Mercury first appears in the evening sky, it is coming around the far side of its orbit toward us. Seen in a telescope, it presents a warm-gray disk

For better seeing, try viewing in twilight, when Mercury is higher in the sky, and there is less contrast between sky and planet. This strategy works best for a morning apparition—you locate the planet low in a relatively dark sky and track it as it rises and the sky brightens.

If your telescope has an equatorial mounting, you can see Mercury in full daylight. Find the coordinates of the Sun and Mercury in a magazine,

**A CRESCENT MOON** is about to hide, or occult, Mercury (right). Occultations are often closely followed by amateurs.

*You must not expect to see at sight … Seeing is in some respects an art which must be learned.*

William Herschel (1738–1822), English astronomer

computer program, or on the Web. Work out the difference in their positions, then center the Sun (with proper precautions; see p. 123), and step off the distance to Mercury. Scan with a low-power eyepiece to locate the planet. A computerized "Go To" telescope makes it easy, but still beware of the nearby Sun.

### Traversing the Sun

At rare intervals, observers can see Mercury cross in front of the Sun. The next two such transits of Mercury will occur on 9 May 2016 and 11 November 2019, and will last about five hours. Look for Mercury as a tiny dot crawling across the Sun's face. It will resemble a small sunspot, but will move detectably after several minutes.

**THE FACTS BOXES** (such as Mercury Facts, below) list statistics for each planet, mostly in relation to Earth. The distance between Earth and the Sun is known as an astronomical unit—or 1 AU.

### MERCURY FACTS

**Distance from Sun** 0.39 AU
**Mass** 0.06 × Earth's mass
**Radius** 0.38 × Earth's radius
**Apparent size** 5 to 13 arcsecs
**Apparent magnitude** −2 to +3
**Rotation (relative to stars)** 58.7 Earth days
**Orbit** 88.0 Earth days

# VENUS

*Venus is a planet of paradox. No other planet comes closer to us—or appears larger or brighter—but your telescope will show you nothing of its surface, thanks to a veil of globe-girdling clouds.*

Venus was well known to the ancients. The Greeks believed it was two distinct objects: Phosphoros in the morning sky and Hesperos in the evening.

The Mayans saw Venus as the god Kukulkán, a figure who symbolized the death and rebirth of the universe. They based a complex 418-year calendar on the observation that in the space of eight years Venus makes five complete apparitions. To the Aztecs, several centuries later, Venus was Quetzalcoatl, the feathered serpent who symbolized the power of life emerging from earth, water, and sky.

## THE EVENING STAR

Venus circles the Sun once every 225 days. But because Earth is also moving, it takes 584 days before Venus reappears in the same part of our sky. This makes the typical Venus apparition a far more leisurely business than Mercury's frantic scurry. It begins when Venus emerges from the Sun's glow, low in the evening twilight. At this point, it looks small (about 10 arcseconds across) and round in phase.

Each night, the planet climbs higher, moving farther from the Sun. At greatest eastern elongation, it stands highest above the western horizon. It is in half-lit phase, but it has grown to about 25 arcseconds. Now at magnitude −4.5, the planet will be brighter than anything in the sky except the Sun or Moon. Although it does not move visibly, Venus resembles an airplane with landing lights on, and has caused many a UFO report.

After greatest elongation, Venus loses altitude but grows even brighter as it approaches Earth. The point of greatest

**VENUS AND THE MOON** *Venus is sometimes the brightest point of light in the sky apart from the Moon (below). An occultation of Venus by a crescent Moon (right). A mask of Quetzalcoatl (top).*

brilliancy occurs about 4 weeks before inferior conjunction.

Venus is then bright enough to cast a shadow. In a dark location, hold a sheet of white paper facing Venus. Let your eyes dark-adapt, then move your hand just in front of the paper. The shadow will be sharp-edged compared with shadows cast by the Sun or

**NEAR AND FAR** *From Earth, Venus appears largest in its crescent phase (below left). Up close in 1979, Pioneer saw cloud details (below right). Magellan used radar to map the surface through the clouds in 1993 (far left).*

Moon. This is because Venus is a point-source of light, rather than a broad disk.

A week before inferior conjunction, Venus is in crescent phase, but has grown to nearly 60 arcseconds across—just within the resolving power of human eyesight. Not everyone's vision is acute enough to see the crescent, however, and binoculars can help.

At inferior conjunction, Venus overtakes Earth, leaves the evening sky, and shifts into the morning. Thereafter, it rises before the Sun, repeating the stages of its evening performance, but in reverse order.

Superior conjunction, when Venus lies on the other side of the Sun from us, marks the close of one apparition—and the start of the next.

### TELESCOPIC VENUS

Ironically, the very brightness that makes Venus so easy to find works against the telescope user. Seen in your scope's eyepiece in a dark sky, Venus is too dazzling an object for

good or even comfortable viewing. Experienced Venus-watchers observe at twilight—or in full daylight by offsetting the telescope from the Sun (see p. 157).

Unlike the other terrestrial planets, Venus hides its surface features under a veil of clouds. The clouds, roughly 30 miles (50 km) above the surface, are made of sulfuric acid. They are almost as reflective as snow, which is why Venus looks so bright. The atmosphere below the clouds is generally clear, but Earth-bound observers usually see just the opaque and featureless shroud.

Although atmospheric features are hard to detect, for more than a century visual observers have noted faint, temporary streaks—probably real cloud features. Likewise, the terminator sometimes shows an irregular edge instead of a smooth curve—the jags into the dark part of the disk could well be clouds at higher elevations.

A famous atmospheric feature is the "ashen light," an apparent brightening of the unlit portion of Venus occurring a month or two before and after inferior conjunction. Most observers say the light has a warm color to it. The ashen light has not been verified by spacecraft measurements, but scientists think it could be a kind of airglow, similar to auroras on Earth.

To see cloud effects, consider using filters. The best gelatin filters to use are blue or violet ones (Kodak's Wratten 38A or 47). Step up to the pricier glass ones if the results seem worthwhile.

### PENCIL AND CAMERA

The toughest part of sketching Venus is to avoid exaggerating its features, which are all of low contrast. Prepare sketch blanks ahead of time, using a standard scale, such as 1 mm per arcsecond, and checking the size of Venus in an almanac, magazine, or software.

It is fairly easy to take photos of a crescent disk, but the cloud features of Venus call for advanced techniques (see p. 92). For best results, photograph in ultraviolet (UV) light by using a UV *transmitting* filter, such as a Wratten 18A. A reflector telescope with a long focal length will give a suitably large image, but avoid Schmidt-Cassegrains, which have glass corrector plates that absorb UV light. Use eyepiece projection to enlarge the image, but make sure the lens elements are made of quartz or fluorite.

### VENUS FACTS

**Distance from Sun** 0.72 AU
**Mass** 0.81 × Earth's mass
**Radius** 0.95 × Earth's radius
**Apparent size** 10 to 64 arcsecs
**Apparent magnitude** −4.0 to −4.6
**Rotation (relative to stars)** 243 Earth days, retrograde
**Orbit** 225 Earth days

## THE LAVA PLANET

To peel away the clouds and map Venus's geological features, scientists used an imaging radar on a spacecraft named Magellan. Building on work by US and Soviet probes and Earth-based radar, Magellan inventoried a museum of volcanic features, completing its global survey in 1994.

Sixty percent of Venus's surface is a lava plain. The rest falls mainly into two continents: an equatorial one named Aphrodite Terra, and a northern one called Ishtar Terra. These stand a few miles higher than the plains. The remainder of the surface consists of individual mountains and broad-based volcanic peaks.

On the plain, Magellan revealed long channels that mark past lava flows, and found fissures and vents from which

**MAGELLAN DATA** was used to create a global mosaic of Venus (right), and a computerized image of the intensely deformed highland of Ovda Regio (above).

vast sheets of lava once poured, often in repeated eruptions. Scientists identified pancake-like lava domes, and found strange oval features called coronae that have no counterpart on Mercury, Earth, or Mars. The coronae may have formed when a rising blob of molten rock pushed up the surface, cooled, and subsided, leaving a distorted ring around a central depression.

### TRANSITS OF VENUS

Every 100 years or so, observers on Earth can watch Venus at inferior conjunction pass across the face of the Sun. These rare events are called transits of Venus, and they occur in a pair, 8 years apart. The last two transits were in 1874 and 1882; the next pairs 8 June 2004 and 6 June 2012. In 2004, Venus crossed the southern part of the Sun, and in 2012, it will cross the northern part.

In earlier times, transits of Venus let astronomers measure the distance from Earth to Venus, and, by extension, the scale of the Solar System. After the telescope was invented, a handful of astronomers made individual efforts for the 1639 transit. But for the 1761 and 1769 events, the British and French sent out expeditions all over the globe, among them the famous exploring voyage of James Cook, which went to Tahiti for the 1769 transit.

Better distances for the Solar System did emerge from these efforts, but the most notable finding was about Venus itself. Observing the 1761 transit, the Russian scientist Mikhail Lomonosov discovered that Venus has an atmosphere. He noted the halo it produced around the black dot of the planet as it slipped onto the solar disk and off again.

*The portable observatory used by James Cook.*

### CONTINENTAL DIVIDE

The continents preserve some of Venus's oldest terrain—the tesserae. These regions, which include Ovda Regio (shown above), have undergone a phenomenal amount of folding and faulting. They look like islands of an older surface that escaped the volcanic flooding that covered most of Venus.

Venus may have had a plate-tectonics cycle. Portions of both Aphrodite and Ishtar resemble the regions on Earth where plate tectonics have been active, building new crust and destroying the old.

Maxwell Montes, a rugged range in Ishtar, contains the highest point on Venus and stands more than 7 miles (11 km) above the global lava plain. It is much too massive to be supported as a single block of crust. Only plumes of molten rock rising from the mantle could sustain it.

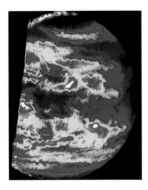

## HAVING AN IMPACT

Like every planet, Venus has been struck repeatedly by meteorites. The craters uncovered in Magellan's images show a spectrum of types, many familiar from other moons and planets. The largest is a flooded basin called Mead, which is 175 miles (280 km) wide. Most craters show signs of post-impact lava floods, almost as if they tapped a source of molten material below the surface.

Venus's atmosphere is dense enough to filter out smaller meteorites, and all its craters appear to be bigger than 3 miles (5 km) in diameter. Magellan did, however, find dark splotches where the surface seems to have been pulverized. The culprit may have been a shock wave—an aerial hammerblow from a meteorite that broke up before striking the ground.

## REPAVE THE PLANET

About 600 million years ago, something reset the geo-clock on Venus's surface, and nobody knows why or how. Impact craters are by far the most common landform in the Solar System, but Venus has less than a thousand of them. Given the rate at which new impacts occur, this means that the planet has somehow erased all traces of the first 85 percent of its history. There are various theories to explain this.

**THIS INFRARED IMAGE** *of Venus's lower cloud deck shows areas of thinnest clouds as white and red. High-speed winds cause the streaky appearance.*

Venus may have a thin, soft crust that cannot preserve ancient terrain. Or episodes of plate tectonics might have erased old features, but these periods alternate with times (like now) when it is quiescent. Or maybe Venus has spasms of volcanism. Or perhaps we are seeing Venus soon after its geo-engine shut down. Or some combination of the above. The fact is, scientists do not yet know.

Some scientists think that Venus is still volcanically active. The amount of sulfur dioxide—a common volcanic gas—in the atmosphere has changed over the last 15 years, perhaps indicating a recent eruption followed by a settling-out of the gas. Also, the tallest mountains are too high to be held up by Venus's crust alone, so either they are recently built or they are supported by hot rock still rising from below.

**VOLCANIC VENUS** *The giant volcano Maat Mons in a computer-generated view (below), and a lava dome, with the informal name of "Tick Mons," seen from above (right).*

## GREENHOUSE EFFECT

Spacecraft have also studied conditions above the ground. The atmosphere of Venus is 90 times heavier than Earth's. And with temperatures of about 890 degrees Fahrenheit (480° C), Venus has the Solar System's hottest surface. Because Venus is closer to the Sun than Earth is, the sunlight it receives is twice as strong. After filtering through the clouds, the light is colored orange and is about as bright as on an overcast day on Earth.

When sunlight warms the surface rocks, they try to radiate the heat back into space at infrared wavelengths. But the heavy carbon-dioxide atmosphere is largely opaque to infrared light and acts like a thick blanket to keep the heat in. As a result, it is as hot at the poles of Venus as at the equator, and the night side is no cooler than the day side.

# EARTH

*Compared with the other planets in our Solar System,
Earth is astonishingly active. Its surface is continually recycled
and its atmosphere sustains a stunning variety of life.*

Earth is the largest of the four rocky, or terrestrial, planets. While radically different from one another, Mercury, Venus, Earth, and Mars are easily distinguished from the giant planets—Jupiter, Saturn, Uranus, and Neptune—by their size, density, and chemical composition.

Terrestrial planets are composed mainly of dense, rocky materials, such as basalt, and metals, such as iron. The giant planets are more like the Sun in composition: they have enormous atmospheres, mainly of hydrogen and helium, wrapped around rocky centers. The terrestrial planets have far greater densities than the gas planets, but are low in mass—Earth has only one-fifteenth the mass of the smallest giant, Uranus.

### EARTH'S BIRTH
Some 4.6 billion years ago, the Solar System formed out of a cold cloud of interstellar gas and dust. The densest part of the cloud coalesced into the proto-Sun, while the remainder flattened into a far-flung disk called the solar nebula. Particles in the nebula

**A FEMALE FIGURE** *has often symbolized Earth, as in this 2000 BC bas-relief of the Sumerian earth goddess.*

collided and stuck, gradually accreting to form the planets.

Earth's characteristic properties were instilled at birth. It formed hot, and was heated further by the decay of radioactive elements in its interior. The heat turned Earth molten and caused the heaviest materials, such as nickel and iron, to sink to the center, while lighter elements, such as silicon, rose to the surface. Even the impact of the small planet that probably led to the birth of the Moon (see p. 138) did not disrupt this arrangement for very long.

### PEELING THE LAYERS
Although Earth's interior is nearby in astronomical terms—its center lies less than 4,000 miles (6,400 km) below our feet—it remains frustratingly out of reach. To look inside our planet, scientists rely on earthquakes, studying how seismic waves rebound and refract as they travel through the interior. In simple terms, the picture they give of Earth resembles an onion made of rock.

At the center is the core, which extends upward to a point 1,800 miles (2,900 km) below the surface. The core is about 90 percent nickel and iron, and has two parts. The inner core is more than hot enough to be liquid—about

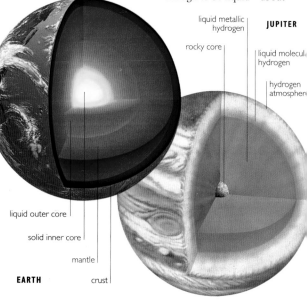

*CROSS SECTIONS of Earth and Jupiter (not to scale) show the contrasting structures of terrestrial and gas-giant planets.*

liquid metallic hydrogen

**JUPITER**

rocky core

liquid molecular hydrogen

hydrogen atmosphere

liquid outer core

solid inner core

mantle

**EARTH**

crust

**IGNEOUS ROCKS** *are forged by heat from the Earth's interior. The hot, molten rock (right) is often forced to the surface, resulting in volcanic eruptions (above).*

9,000 degrees Fahrenheit (5,000° C)—but it is kept solid by the enormous pressure, which is nearly four million times that at the surface.

In the outer core, the pressure is lower, temperatures are about 900 degrees Fahrenheit (500° C) cooler, and the iron remains liquid. This liquid nickel-iron flows in turbulent currents that give rise to Earth's magnetic field. The field reaches up through the planet, past the surface where it controls our magnetic compasses, and out into space where it affects the flow of charged solar particles blowing past Earth (see p. 125).

Above the core is a layer called the mantle, composed of high-density iron and magnesium-silicate minerals. While not liquid like the outer core, the rocks of the mantle are hot enough to flow, and the energy that they carry powers all the tectonic activity we see at the surface.

The rocks that make up Earth's surface form a thin but distinct layer called the crust. We live on the continental type of crust. It is made of lightweight granitic and carbonate rocks and is be - tween 20 and 40 miles (30 and 65 km) thick. A second kind

of crust, oceanic, is made of heavier and denser basalt. It is only 3 miles (5 km) thick.

Most oceanic crust is covered by miles of sea water, so, until recently, geologists mainly studied the continental crust. However, there are a few places, such as Iceland, where oceanic crust is exposed.

## TYPES OF ROCK

Crustal rocks and minerals appear in a bewildering array of types, since chemical elements can combine in myriad ways. But in broad terms, geologists recognize three kinds of rock.

Rocks that form in a molten state are called igneous. These include basalt, which erupts as molten lava, and granite, which forms deep within Earth and is exposed by erosion or geological activity.

Sedimentary rocks are formed by the slow accumulation of layers of particles, such as sand, mud, or organic debris. Shale, sandstone, and limestone are common sedimentary rocks.

Metamorphic rocks are those that have been altered by heat, pressure, or both. Metamorphism makes a rock harder and denser, but traces of the rock's original nature are usually preserved. The process can turn shale into slate, sandstone into quartzite, limestone into marble, and so on.

Earth is a geologically active planet, and these three types of rock are involved in a continual process of recycling, driven by weathering, heat, and pressure. Although Earth formed some 4.6 billion years ago, the oldest known rocks are about 3.9 billion years old, and these are rare exceptions; most of the surface is only about 100 million years old. This explains why we do not find the number of craters here that we do on the Moon or Mars. While Earth experienced at least as many impacts, most craters have been eroded by the weather or erased by geological activity.

**SEDIMENTARY ROCKS,** *such as sandstone (below left and right), often form in distinct layers and may be eroded into dramatic shapes.*

## CONTINENTS ADRIFT

Earth differs from other planets in that its crust is made up of slabs of thin rock called plates. Driven by currents in the upper mantle, the plates, which number more than a dozen, carry the continents on their backs like rafts.

Meteorologist Alfred Wegener argued for continental drift (as he called it) in the early 1900s. However, the idea was not generally accepted until the 1960s, when geophysicists found that new oceanic crust was constantly being formed in mid-ocean ridges on the seabed. The theory of how new crust moves away from the mid-ocean ridges and disappears at the edges of continents is called plate tectonics, and it has revolutionized geology.

As the edges of crustal plates collide, they rift, overlap, and fold, recycling their rocks and remaking the surface of the planet. The Himalaya Mountains, for example, mark where the Indian plate is being forced under the edge of the Asian plate. The Rift Valley of East Africa, on the other hand, marks where a continent is rifting apart, to eventually form a new ocean.

**LAND** *covers only 30 percent of Earth's surface (below). The collision of tectonic plates created landforms such as the Himalayas, seen here from space (left).*

## A WATERY PLANET

Conditions on Earth's surface depend on the interaction of two spheres: the atmosphere and the hydrosphere.

The hydrosphere is the water on, or near, the surface. This water is found in oceans, rivers, and lakes; it is found below the surface as groundwater; it is locked up in frozen polar caps and glaciers; and it exists as vapor carried in the atmosphere. Earth is the only planet where temperatures allow surface water to exist in solid, liquid, and gaseous states.

Oceans make up 97 percent of the hydrosphere and contain enough water to cover the planet 2 miles (3 km) deep. The only reason we are not underwater is that the continental crust (some 30 percent of Earth's surface) has an average elevation high enough to keep it above sea level.

The sea appears so eternal that few people think about how it began. Its origin is, in fact, a scientific puzzle. Some water must have formed with the planet and then erupted from volcanoes. But Earth's birthplace in the solar nebula was too hot to allow much water to exist. Comets, however, are loaded with ice and organic elements, and many scientists think that most of the hydrosphere was deposited by catastrophic comet impacts early in Earth's history. Much smaller impacts occur to this day, continually adding water to the hydrosphere.

## AN OCEAN OF AIR

Above Earth's surface lies the atmosphere, an ocean of air that sustains life, drives our weather, and aids the breakdown of rocks into their constituent minerals.

The composition of the atmosphere is constantly evolving. If you went back three billion years in time,

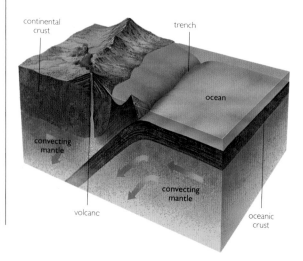

*continental crust*

*trench*

*ocean*

*convecting mantle*

*convecting mantle*

*volcano*

*oceanic crust*

**A DYNAMIC EARTH** *When an oceanic plate collides with a continental plate, the dense ocean crust is drawn under and melted in a process known as subduction.*

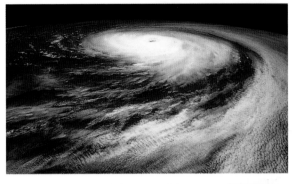

**EARTH'S ATMOSPHERE** *creates dramatic weather, such as this hurricane viewed from space (right), and forms a cocoon in which life thrives, as in this teeming tropical rain forest (below right).*

you would be unable to breathe. At that time, the atmosphere contained almost no oxygen—it was mostly carbon dioxide and nitrogen. Today, Earth's air is mostly oxygen and nitrogen.

Oxygen began to accumulate in the atmosphere only when primitive life forms (first bacteria and then plants) developed photosynthesis, excreting oxygen as a waste product of the process. Not only does oxygen let us breathe, it also reacts with sunlight to form ozone, which protects us from harmful ultraviolet rays.

Carbon dioxide acts as a "greenhouse gas," allowing sunlight to reach the surface, but preventing heat from

escaping. In this way, it helps warm our atmosphere, just as an abundance of it turned Venus into an inferno. Today, the burning of fossil fuels is releasing carbon dioxide into the air, subtly changing the constitution of the atmosphere and contributing to a global increase in temperatures.

## EARTH'S WEATHER
The interaction between the Sun's heat and Earth's atmosphere creates our weather. Other vital ingredients are the 23.5 degree tilt of Earth's axis, and the combination of land and water on Earth's surface. Because of its tilt, Earth is heated unevenly, and this is accentuated by the different heat-absorbing characteristics of land and water. In its quest for equilibrium, the atmosphere directs warm air to cold places, and cold air to warm, creating our basic system of weather patterns.

Today, Earth has a benign, stable climate with a narrow range of temperatures. In the past, however, there have been periods of rapid and dramatic climate change. Small variations in Earth's orbit around the Sun or in the tilt of its axis could be enough to drastically alter our climate, creating conditions far less conducive to life than those that currently exist.

## THE ORIGIN OF LIFE ON EARTH

No one has yet figured out exactly how Earth passed from a state of having all the chemical building blocks necessary for life to a state where life actually existed. But the change, while enormously important, may not have been all that much of a leap.

The oldest rocks in the world are about 3.9 billion years old, and ones showing fossil traces of life are barely younger. This suggests that either life conveniently arose when these rocks were formed (rather unlikely), or it arose even earlier. In any case, the date is important because at that time the Solar System was still a pretty hostile place. Large, devastating impacts by asteroids and comets were frequent, and Earth's surface looked more like the Moon's southern highlands than like any landscape on Earth today.

From what we know about today's Earth, life is outstandingly tenacious, adaptable, and resourceful. Given this tenacity, it is possible that life actually started on Earth several times, but was wiped out by impacts on each attempt but the last.

What does this say for the chances of finding life elsewhere, either in the Solar System or outside it? Actually, the chances look fairly good. Astrophysics shows that chemistry is chemistry the universe over, so nature ought to be able to duplicate in another place the pathway that led to life on Earth. But whether life can persist elsewhere, and for how long, we can only guess.

### EARTH FACTS

**Distance from Sun** 92,752,000 miles (149,600,000 km) = 1 AU
**Mass** $6 \times 10^{21}$ tons
**Radius** 3,954 miles (6,378 km)
**Rotation (relative to stars)** 23 hours 56 minutes
**Orbit** 365.24 days

165

# MARS

*The Red Planet is Earth's closest cousin,
and recent discoveries have raised the tantalizing
possibility that life once existed on Mars.*

Mars, with its ocher color, relatively swift motion among the stars, and changeable brightness, has long been associated with war, disaster, death, and pestilence. To many ancient people, who feared that changes in the sky portended changes for the worse on Earth, such an association was natural. Our name for the planet is that of the Roman war god, who took on the same attributes as his Greek forerunner, Ares.

## THE MARTIAN CYCLE

Depending on when you look at it, Mars's apparent size can vary greatly—from 4 to 25 arcseconds. A typical apparition begins with Mars low in the east in the glow of morning twilight. As weeks pass, the planet grows. It moves eastward against the stars and does not rise out of the dawn quite as quickly as they do. Earth, orbiting faster than Mars, begins to overtake it, and Mars's apparent east-ward motion slows and stops. Mars then spends about 11 weeks moving westward against the background stars

before halting once more and resuming eastward motion. In the middle of the westward (or retrograde) part, Mars reaches opposition and best visibility around midnight, and is at its maximum apparent size for the apparition.

Finally, over the following months, Mars slips into the early evening sky and eventually disappears in the twilight glow. After remaining hidden in the Sun's glare for several weeks while it passes conjunction, Mars reemerges at dawn to begin its next apparition.

Mars comes to opposition about every 780 days, or 2 years 7 weeks (see Table, p. 168). Because it has an eccentric orbit, Mars's distance from Earth and the Sun at opposition varies, so its apparent size and brightness also change. The most favorable opposition is when Mars lies closest to both Earth and the

Sun. Such oppositions are called perihelic, and they occur when opposition is in July, August, and early September—about once every 17 years. At these times, Mars is just 35 million miles (56 million km) from Earth, with an apparent size of about 25 arcseconds and a magnitude of –2.6. At aphelic oppositions (in January, February, and early March), Mars is 63 million miles (101 million km) from Earth. Its disk is only about 14 arcseconds wide and it shines at magnitude –1.0, more than four times dimmer.

## OBSERVING MARS

On first viewing Mars through a telescope, beginners are often struck by how small it appears. The planet is only half the size of Earth and, even at favorable oppositions, it appears no larger than a lunar crater. First views typically show an ocher disk, a few faint markings, and perhaps a whitish polar cap.

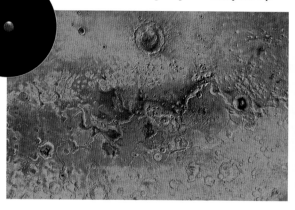

**MARS, GOD OF WAR** *(top), in a Diego Velázquez (1599–1660) painting. A telescope view of the real Mars (above) reveals only a small disk, but Viking images show a cratered surface (right).*

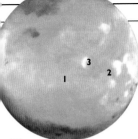

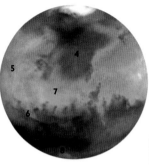

The

**THREE HUBBLE VIEWS** *of Mars in 1995. At 160 degrees longitude, features include: (1) the Amazonis region; (2) the Tharsis region; and (3) Olympus Mons.*

**AT 270 DEGREES LONGITUDE,** *look for: (9) Syrtis Major; (10) the Hellas basin; and (11) the Elysium basin.*

**AT 60 DEGREES LONGITUDE,** *the main features are: (4) Mare Acidalium; (5) Ascraeus Mons; (6) Valles Marineris; (7) Chryse basin; and (8) Argyre basin.*

powerful telescopes on Earth give views of Mars only about as detailed as naked-eye views of the Moon. This helps explain why Martian studies progressed little before the Space Age.

Early observers of Mars saw dark markings and thought they were either old seabeds or areas of vegetation. Today, these are known to be lava flows and boulder fields.

In 1877, the Italian astronomer Giovanni Schiaparelli reported seeing a network of straight lines he called *canali*, meaning channels. This was widely translated as canals, suggesting these features were artificial and fueling popular speculations about life on Mars. As it turned out, the canals were merely optical illusions.

In 1964, the first close-up views of Mars were taken by NASA's Mariner 4 spacecraft. The features in these photos that most shocked scientists and the public alike were the craters—very few people had expected to see such Moon-like features. The fact that a Moon-like Mars was such a shock reveals the depth of most people's assumptions that Mars was just a "little Earth."

GETTING EQUIPPED

Small telescopes can show you something of Mars, but to see details you will need at least a 5 inch (125 mm) refractor or an 8 inch (200 mm) reflector,

and long focal ratios (f/8 to f/10 or more) are preferable. Good seeing is also essential.

With Mars, unlike most planets, colored filters will really help your observations. High-quality glass filters that screw into your eyepiece are available from telescope-accessory dealers. Many ob-servers, however, prefer to start out with less-expensive

gelatin filters, such as Kodak's Wratten filters, which can be bought from camera stores.

A useful set would include a blue (Wratten 44a) or blue-violet (W47); a green (W58) or a yellow (W12 or 15); an orange (W21 or 23a); and a red (W25). To use these filters, simply hold them up to the eyepiece and look for features that appear enhanced.

## THE WAYWARD RED PLANET

The Red Planet and its motions challenged every astronomer from antiquity onward. By the late 1500s, the Danish astronomer Tycho Brahe (1546–1601) had amassed decades of observations of Mars and the other planets. His observations were by far the most accurate to date, yet he failed to weld them into a coherent system of the universe. Then Tycho hired a young Austrian mathematician-astronomer named Johannes Kepler (1571–1630) and put him to work on the orbit of Mars. When Tycho died shortly afterward, Kepler inherited his position and, more importantly, his observations.

Kepler (right) battled with Mars for much of a decade, finally emerging with the first two of his three laws of planetary motion. These include the conclusion that planets orbit in ellipses, with the Sun at one focus. (Every prior astronomer had insisted on combinations of circles.)

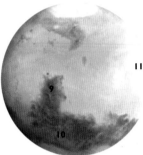

Kepler's work paved the way for Isaac Newton's *Principia* (1687), the cornerstone of all modern physical science.

## A VARIABLE PLANET

With the Moon, a look through any telescope tends to instantly impress. Mars, on the other hand, is a planet with delicate, elusive features, and the experienced eye will see much more than the beginner's. Mars has a great many features, but you will not catch them all on your first night's viewing. Experience is crucial, and the key to getting the most out of your observations of Mars is to view the planet as many times as you can.

Backyard observers who are patient and continue to watch Mars over many hours will see darkish markings move across the disk, gradually slipping out of sight as the Martian day passes by. Over the months leading up to opposition, you can see Mars's features at ever-better resolution as its globe grows and your eye becomes more practiced.

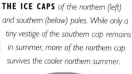

**THE ICE CAPS** *of the northern (left) and southern (below) poles. While only a tiny vestige of the southern cap remains in summer, more of the northern cap survives the cooler northern summer.*

## MARTIAN SEASONS

If you observe Mars from one opposition to another, you can see the Martian weather and seasons change. Mars has a tilt similar to Earth's and experiences four seasons, each lasting about twice as long as ours because Mars's orbit is that much larger.

Martian seasons do not correspond exactly with Earth's, because Mars's polar axis points in the direction of the star Deneb, whereas Earth's points to Polaris. A handy rule of thumb to remember is that the season on Mars at each opposition is one season in advance of Earth's at that time.

For Mars's northern hemisphere, spring and summer generally feature a clear atmosphere with little dust in the air. Whitish clouds,

**LOCAL DUST STORM** *In this Viking 2 picture of Mars's surface, an arrow indicates a bright, turbulent dust cloud 190 miles (300 km) across.*

however, can appear near the sunrise line and over high elevations. Frost deposits turn the large impact basins of Argyre, Hellas, and Elysium into bright patches. You will find that filters excel at revealing details such as these, especially when alternated with unfiltered views.

The northern polar cap vanishes under a hood of cloud during its fall and winter. The plummeting temperatures renew the ice cap by condensing carbon dioxide from the atmosphere. The carbon-dioxide ice comes and goes with the seasons, but a permanent water-ice cap in the north remains intact throughout the summer.

| UPCOMING OPPOSITIONS OF MARS | | | |
|---|---|---|---|
| **Date** | **Apparent size in arcseconds** | **Distance (x 1,000,000)** | |
| | | **in miles** | **in km** |
| 3 March 2012 | 13.9 | 62.6 | 100.8 |
| 8 April 2014 | 15.2 | 57.4 | 92.4 |
| 22 May 2016 | 18.6 | 46.8 | 75.3 |
| 27 July 2018 | 24.3 | 35.8 | 57.6 |
| 13 October 2020 | 22.6 | 38.6 | 62.1 |
| 8 December 2022 | 17.2 | 50.6 | 81.5 |
| 16 January 2025 | 14.6 | 59.7 | 96.1 |
| 19 February 2027 | 13.8 | 63 | 101.4 |
| 25 March 2029 | 14.4 | 60.2 | 96.8 |

*If you are bored with these weary calculations, take pity on me who had to go through at least 70 repetitions of them.*

JOHANNES KEPLER (1571–1630), Austrian mathematician-astronomer, writing about his observations of Mars

Because the southern-hemisphere summer occurs at perihelion—when Mars lies closest to the Sun—sunlight is 44 percent stronger at this time. This makes southern summers hotter, and winters colder, than northern ones.

The southern polar cap is mostly carbon-dioxide ice. During summer, it shrinks to a tiny remnant (with a core of water ice), but does not vanish completely. About every 17 years, when opposition coincides with perihelion, the southern polar cap is tilted toward Earth, showing clearly as it shrinks under the Sun's strong rays.

## DUST STORMS

Vast lava flows and boulder fields, such as Syrtis Major, usually appear as dark markings on the surface. However, their visibility varies because winds carry dust across them, obscuring them in whole or in part. Blowing dust can also hide smaller markings or make them difficult to identify.

Sometimes you will see no markings at all for days or weeks on end, thanks to one of the global dust storms to which Mars is prone. These storms occur most often during the southern hemisphere's spring and summer. The last global dust storm was in 2007. It posed a serious problem to the Mars Exploration rovers Spirit and Opportunity by covering their solar panels with dust.

## CAPTURING MARS

Sketching Mars is relatively easy. Many observers use standard-size blank forms—for example, 2 inches (50 mm) to the Martian diameter—but these are too large for times when the Martian disk is small. They will tempt you to put down details you cannot really see. A better course is to use an image scale of 2 mm per arcsecond. Circles work well for much of the time, but three to four months before and after opposition, Mars's disk looks distinctly gibbous—much like the Moon three or four days before Full—and you should modify your sketch blank to reflect this shape.

Photographing Mars, on the other hand, is a lot more difficult. The small planet demands a long focal length, and this means you need exposures of up to a second. "Windows" of good seeing seldom last this long, so capturing a sharp image on film depends mostly on luck. Many Mars photographers now use CCD cameras. These are just as subject to seeing requirements, but they let you quickly take many short exposures, at least some of which will catch Mars during moments of crisp seeing.

| MARS FACTS | |
| --- | --- |
| **Distance from Sun** | 1.52 AU |
| **Mass** | 0.11 × Earth's mass |
| **Radius** | 0.53 × Earth's radius |
| **Apparent size** | 4 to 25 arcsecs |
| **Apparent magnitude** | +1.8 to −2.6 |
| **Rotation (relative to stars)** | 24.6 hours |
| **Orbit** | 687 Earth days |

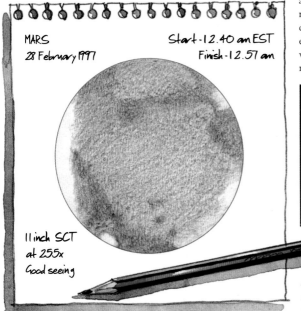

MARS
28 February 1997

Start - 12.40 am EST
Finish - 12.57 am

11 inch SCT
at 255x
Good seeing

### RECORDING MARS
*Observers wanting to keep a permanent record of Mars can try sketching the planet (left). Or they might photograph it—CCD cameras can produce images such as this (above).*

## RED-ROCK GEOLOGY

Of all the Sun's other planets, Mars is the most Earth-like. Its polar axis tilts 25 degrees to its orbit (Earth's tilts 23½ degrees), the Martian day lasts just 41 minutes longer than ours, and it has clouds, seasons, and polar ice caps. Despite these similarities, Mars is a frigid desert world, geologically inactive and hostile to life as we know it. Its atmosphere is too thin to allow liquid water at the planet's surface and offers no barrier to ultraviolet light from the Sun.

**SURFACE FEATURES** *of Mars include Candor Chasma (left)—part of the giant Valles Marineris— and Olympus Mons (above).*

Mars has two distinct types of terrain, each occupying about half of the planet: in the south are the older highlands with many craters, while in the north, a relatively un-cratered plain lies a few miles lower. What causes the difference and where did all that crustal rock go?

The southern highlands have geological puzzles of their own. They contain the largest volcanic region on Mars—Tharsis—with its four immense volcanoes. Tharsis is about the size of North America and appears to have been built by episodes of crustal uplift followed by intense volcanism, which deposited lava on top of the uplift. Tharsis is topped by Olympus Mons, which is 16 miles (25 km) high, and 300 miles (500 km) wide.

From the number of craters on the highlands, scientists have calculated that this terrain is at least three billion years old. However, some portions, such as the slopes of Olympus Mons, appear quite young in geological terms. They have no craters and may be only a few million years old.

Radiating east from Tharsis is a huge crack called Valles Marineris—2½ miles (4 km) deep and long enough to reach across the United States from coast to coast. It seems to have started as a tectonic fault, then, as the fault tapped sources of groundwater, the walls collapsed and eroded, opening up the valley. Parts of it are now 60 miles (100 km) wide.

Other regions flanking Tharsis show evidence of actual rivers. The largest channels drain from the highlands across the crustal boundary and onto the northern lowlands in the region of Chryse, where the Viking 1 spacecraft landed. Scientists believe that these flows were catastrophic and that they occurred when internal heat or meteorite impacts released groundwater in sudden floods. Although the volume of water in these flows would have been enough to fill Earth's Amazon River 100 times, scientists think the flows were brief, because Mars probably did not have enough water to sustain them continuously. Evidence of a watery past is accumulating in recent high resolution images coming from the Mars Global Surveyor spacecraft. Even more astounding is possible evidence of relatively recent water flows.

**THE NORTHERN PLAINS** *as recorded by the Viking 1 lander. The photo was taken at about noon, local Mars time.*

## A WETTER PAST

Running water, lakes, possibly rain, a warmer climate, and a thicker atmosphere—ancient Mars might well have been remarkably Earth-like. So what happened? Two things may be the keys to understanding present-day Mars.

First, being a small planet, Mars's geological engine seems to have run down. If it has not stopped entirely, it is nowhere near as active as Earth's. It also seems that Mars has never had a plate-tectonics cycle—as one scientist put it, Mars appears to be a one-plate planet. A rigid crust could explain why Tharsis is such a large region. On a planet with active plate tectonics, such as Earth, the crust above an erupting hot spot would keep moving and a volcano would never grow as big as Olympus Mons.

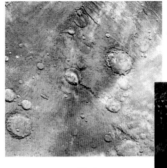

**A THIN LAYER OF FROST** *on the surface of Mars (above), composed of water and carbon-dioxide ice.*

The second key to present-day Mars is its lack of a large satellite like our Moon. Mars has only two small moons, Phobos and Deimos, which are probably asteroids captured from the nearby main belt. Both are mere pebbles compared with Mars. Lacking the stabilizing influence of a large moon, Mars's axial tilt could change abruptly (in

**MINERAL** *globules (below) in a Martian meteorite may have been formed by primitive organisms.*

geological terms), throwing its climate from warm, wet conditions into a global ice age and back again, perhaps several times over in the course of a billion years.

## LIFE ON MARS?

Possible similarities between Earth and ancient Mars beg the question: if life emerged here, why not on Mars?

Maybe it did. A meteorite from Mars that was discovered in Antarctica may contain traces of life. This meteorite, designated ALH84001, is believed to have been blasted off Mars by a comet or asteroid impact, after which it wandered through space before crashing to Earth. ALH84001 contains gas bubbles whose chemistry matches the Martian atmosphere that was sampled by the Viking landers.

What excited scientists was the presence of curious structures, thought to be the fossilized traces of microscopic, bacteria-like organisms. This is now thought to be unlikely. But even if the features in this particular meteorite fail the test, the increasing evidence of water in Mars' past ensures the question of ancient life on Mars remains a tantalizing possibility.

---

## WAR OF THE WORLDS

On the evening of 31 October 1938, radio producer Orson Welles broadcast a version of the H. G. Wells novel, *The War of the Worlds*. The novel tells of a Martian attack on London. It was published in 1897, at a time of enormous public interest in Mars, occasioned by recent close oppositions.

In preparing his radio play for an American audience 40 years later, Orson Welles had a stroke of genius and changed the locale and timing of the Martian attack. Instead of London, the Martians landed near Princeton, New Jersey. And instead of 1897, the show led listeners to believe that Martians were invading that very night.

The result was a broadcast production so realistic that many who heard it took it for an actual news flash from the scene of an extraterrestrial invasion. The radio show tapped directly into a mother lode of latent paranoia, brought to hypersensitivity by the gathering clouds of war over Europe. Panic ensued in some places, and America's pre-war jitters edged up a notch. Sixty years later, it is easy to smile at those who fell for the ploy. But would we respond any more skeptically today if some TV network staged a modern version of the hoax?

*An illustration from H. G. Wells's science-fiction novel, The War of the Worlds.*

# JUPITER

*Rightfully called the "king of worlds," the planet
Jupiter is more massive than all the rest of
the planetary system put together.*

To the naked eye, Jupiter displays a brilliant white gleam that is unmistakable, especially in dark skies. In its slow course around the ecliptic, Jupiter paces out a long zodiacal "year"—it takes 12 years to orbit the Sun, spending about a year in each constellation of the zodiac. Modern skywatchers can often recall the constellation that Jupiter was passing through the first time they saw it. For many ancient cultures, the numerical coincidence seems to have indelibly marked this planet as the celestial symbol for the leader of the gods.

## BABYLONIAN BELIEFS
Jupiter was the chief of the Roman pantheon, while to the ancient Greeks he was Zeus. More than a thousand years earlier, this planet was associated with Marduk, the most important figure in Mesopotamian cosmology and the patron god of the city-state of Babylon.

As the story goes, Marduk took on Tiamat, the goddess of chaos, and her

**ZEUS** *was the Greek version of Jupiter, ruler of the gods. He took the form of an eagle to carry Ganymede (the namesake of a Jovian moon) to Mount Olympus.*

11 monsters. With scheming and titanic effort, Marduk defeated them one by one, and split Tiamat's body in two, thus dividing heaven from Earth. Marduk came to symbolize the rule of heavenly order over the universe. In this role, he placed the wandering star Jupiter specifically in charge of the night sky.

## BORDER PLANET
At Jupiter, the Solar System changes character. All the inner worlds are small, rocky, and in the case of asteroids, fragmentary. From Jupiter out to Neptune, the worlds are large with thick, gaseous atmospheres, and each commands a miniature solar system of moons and rings.

Jupiter is almost entirely made up of hydrogen and helium, so what you see is the cloudy top of a deep atmosphere. (See diagram of Jupiter, p. 162.) Traces of methane, ammonia, water, phosphine, and other hydrocarbons combine with solar ultraviolet light to produce the color and banding seen in a telescope.

The cloud tops lie at a level where the pressure is almost the same as at Earth's surface, while above them an atmosphere of hydrogen thins out into space. Descending below the clouds, the pressure increases until the hydrogen starts to behave like a liquid. About 12,000 miles (20,000 km) down, the hydrogen becomes a liquid metal. At Jupiter's center lies a rocky-iron core, perhaps 15 times the mass of Earth and having a temperature of 34,000 degrees Fahrenheit (19,000° C).

System II

System I

System II

**THE BRIGHT ZONES** *and
dark belts of Jupiter divide into
two rotation "systems."*

- north polar region
- north temperate belt
- north tropical zone
- north equatorial belt
- equatorial zone
- south equatorial belt
- south tropical zone
- south temperate zone
- Great Red Spot
- south polar region

## ROUND AND ROUND

Jupiter is still cooling off from its formation, and the heat fuels phenomena ranging from the planet's immense magnetic field to convection currents in its clouds. The light-colored zones tend to be high, cool clouds in regions of updrafts, while the darker belts mark warmer areas of downdrafts. Jupiter's rapid rotation—less than 10 hours—smears the cloud features into east-west stripes paralleling the equator. This rapid rotation also causes Jupiter's marked oval shape—the planet is 7 percent wider at its equator than at its poles.

Lacking a solid surface, Jupiter displays differential rotation, like the Sun does. The period for System I (the equatorial zone) is 9 hours 51 minutes, while System II (the higher latitudes) rotates a bit more slowly and has a period of 9 hours 56 minutes.

## OBSERVING JUPITER

Jupiter is one of the planet watcher's greatest delights. It offers a wealth of ever-changing detail that can be seen in almost any telescope. It also rewards patience, since observers who take the time to become familiar with this planet benefit the most.

As with any planet, taking advantage of nights of good seeing is the key to enjoyable

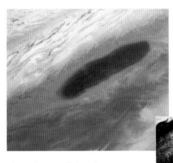

**A TELESCOPE VIEW** *of Jupiter (bottom) is impressive, but spacecraft images have revealed details in the Great Red Spot (below) and a brown oval (left).*

observing, and the view gets better the longer you look. If your telescope has a motorized equatorial mounting, it is worth taking the trouble to align it so that the drive tracks correctly (see p. 79).

What can you expect to see? With a 2.4 inch (60 mm) telescope and 50x to 100x magnification, you should be able to see two broad dusky belts paralleling the equator. Through a 4 inch (100 mm) scope at 100x or more, you can detect several more bands extending into higher latitudes. And with 6 inch (150 mm) or larger scopes, use 150x to 300x and the view will be overwhelming—details will cover most of Jupiter's disk and quickly pass in review as the planet rotates.

The bright zones and dark belts display subtle shades of brown, tan, yellow, orange, and blue-gray. The north and south equatorial belts are the most constant, but all the bands vary in strength and change position slightly. The edges of the bands become irregular, developing projections and indentations as storms stir up the clouds, and winds reach speeds of up to 250 miles per hour (400 km/h).

## STORM SYSTEMS

The most famous storm system on Jupiter is the Great Red Spot, which appears to be a vast high-pressure system, about twice the size of Earth. Its size has varied over the

decades, and it tends to become redder as solar activity increases (see p. 126). When the Great Red Spot is faded, look for the Red Spot Hollow, an indentation that surrounds it in the south tropical zone.

Jupiter has other storms, too: three white ovals reside in the south temperate belt, where they were first noticed in the late 1930s. A lasting characteristic of the white ovals is that they slowly drift eastward relative to the Red Spot. Smaller white spots appear now and then as well.

Additional cloud features to look for are festoons—dark, linear features linking two belts across an intervening zone—and condensations—dark spots or short lines that usually appear within the belts.

To enhance atmospheric details, experienced Jovian observers view through gelatin or glass filters. The most helpful is light blue (Kodak Wratten 80A or 82A), which can enhance the edges between belts and zones. A yellow or orange (W12 or 21) filter brings out festoons and polar details that are bluish in color.

---

### JUPITER FACTS

**Distance from Sun** 5.20 AU

**Mass** 318 × Earth's mass

**Radius** 11.2 × Earth's radius

**Apparent size** 33 to 50 arcsecs

**Apparent magnitude** −1.2 to −2.5

**Rotation (relative to stars)**
9.84 hours at System I;
9.93 hours at System II

**Orbit** 11.86 Earth years

173

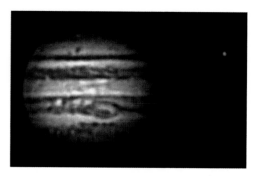

## SKETCHING TOOLS AND TECHNIQUES

Sketching Jupiter is a rewarding and useful way to train your eye to see more, and captur-ing the details on paper will give you an outstanding familiarity with the planet's features. You have to work quickly, however, spending no more than 10 minutes on each sketch because Jupiter's rotation is fast enough to alter the view on timescales longer han this.

Two approaches to sketching have been popular with Jupiter observers, one showing the whole Jovian disk, the other being the "strip sketch." The full-disk drawing uses a prepared blank outline, usually on a scale of 2 inches (50 mm) the planet's diameter. This is about 1 mm per arcsecond, and some observers standardize on that, varying the diameter of the blank as Jupiter's apparent size changes. (The planet's size ranges from 33 arcseconds near conjunction to 50 arcseconds at perihelic oppositions such as those of 2010 and 2011.) Whether you vary the blank's size or not, be sure that it

**THESE CCD IMAGES** *of Jupiter and Ganymede were taken 90 minutes apart. They clearly show the rapid rotation of the planet and motion of the moon.*

properly shows the oval disk. A 70 degree ellipse template from an art-supply store fits the shape perfectly.

The strip sketch focuses on a range of latitude, such as the zone containing the Great Red Spot and its two flanking belts. Because they do not attempt to capture all of Jupiter at once, strip sketches are often easier and considerably less nerve-wracking for beginners. By focusing on an area of interest, you can spend your 10 minutes per sketch more carefully.

Besides the Jupiter drawing, the page should include space to note the date and time, the instrument and magnification used, the weather and seeing conditions, and any other factors that seem relevant.

## JOVIAN PHOTOS

For a planet whose brightness averages magnitude −2.5 (exceeded only by Venus), Jupiter can be surprisingly difficult to photograph well. Its small size demands the

JUPITER
20 July 1994
Start - 7.35 pm EST
Finish - 7.45 pm

8 inch SCT at 167x
Fair seeing

\* Comet Shoemaker-Levy 9 impacts

**THIS SKETCH OF JUPITER** *captures details of the bands and belts, and the scars left by Comet Shoemaker-Levy 9 in 1994 (see also p. 196).*

use of eyepiece projection (see p. 93) and exposures of up to a second, during which seeing conditions may blur fine details.

Jupiter's limb is significantly darker than the center of its disk. While this is not obvious to the eye, it is readily detected in an image. In images processed for high contrast (to reveal belt and zone details), the limb regions can darken and merge with the sky background.

Almost all astrophotographers have now switched to CCDs, despite the high initial costs (see p. 94). The main reason is that electronic imaging affords greater flexibility in processing.

Astrophotographers resort to digital tricks to enhance their images. One way is to borrow a trick discovered by film photographers. This combines a negative image with a slightly out-of-focus positive image. This "unsharp mask" can be performed with a single key stroke in many imaging programs and can reveal wonderful details.

*The fairest thing*

*we can experience*

*is the mysterious.*

*It … stands at the*

*cradle of true art*

*and true science.*

The World As I See It,
ALBERT EINSTEIN (1879–1955),
German-born US physicist

## PARTICLES AND FIELDS

The Jovian world is more than what we see. If our eyes could see Jupiter's magnetosphere, it would be a highly impressive sight, with an apparent diameter equal to four times that of the Moon. The magnetosphere is a vast, complex structure full of fleeting charged particles. In fact, the density of radiation near Jupiter is high enough to kill an unprotected human

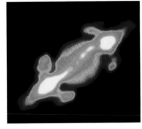

in just a few minutes. This radiation also causes spacecraft designers major headaches, since delicate electronics must be shielded from it.

Jupiter is one of the most intense radio sources in the sky. It produces emissions at millimetric, decimetric, decametric, and kilometric wavelengths. The decametric emissions, at frequencies between 0.6 and 39 megahertz, are the best known and first discovered. Within this band, Jupiter emits bursts of radiation triggered by interactions with its moon Io. These can last from a few seconds to several minutes.

**THIS RADIO IMAGE** *of Jupiter shows the planet in the center, with bright radiation belts on either side of it.*

## OCCULTATIONS OF JUPITER

The Moon occasionally covers up, or occults, a planet or star, producing a sight not to be missed. Occultations of Jupiter (below) are particularly captivating because the planet is relatively large in the telescope's field of view.

Consult astronomy magazines, the Web, or sky-charting programs to find out when occultations are going to happen, and whether they will be visible from your location. Amateur groups will often arrange parties to watch these events—check with your local club.

Set up your telescope at least 30 minutes before the scheduled disappearance. If you have an equatorial mounting, observing will be easier if it is aligned on the pole (see p. 79). Use a low-power eyepiece to center the telescope on the planet.

As the Moon approaches, get ready. It always seems to take forever to draw close, then the actual event runs its course quickly. After disappearing, the planet will emerge from behind the Moon's limb—it may be an hour later or it may be only a few minutes. The occultation predictions will tell you where on the limb to look for it.

**A VOYAGER I VIEW** *of Jupiter with its moons Io (far left) and Europa (left).*

## MINIATURE SOLAR SYSTEM

When Galileo discovered Jupiter's four large moons— Io, Europa, Ganymede, and Callisto—he was excited to see that Jupiter mimicked the Copernican model of the Solar System.

Being 5th to 6th magnitude in brightness, the four Galilean moons are conspicuous in any telescope. You can even duplicate Galileo's discovery with a pair of binoculars. Observe Jupiter at the same time each night, and plot the positions of the planet and the minute points of light lying beside it. After a week or two, a pattern will emerge, and you may experience some of the emotion that gripped Galileo.

From time to time, one or another of the Galilean moons will transit Jupiter's disk, where the dark shadow it casts will be more conspicuous than the moon itself. For observers with scopes of 10 inches (250 mm) or more, transits are a good time to look for the color of the moon because Jupiter provides a light background. Io has a yellowish cast, Europa looks grayish white, Ganymede tan-gray, and Callisto bluish gray.

These moons orbit in the plane of Jupiter's equator, which is presented nearly edge-on to us. Every six years, when Earth passes through this plane, we witness a series of mutual eclipses and occultations among the moons. Consult astronomy magazines or the Web, for dates of these events.

Jupiter has at least 63 other moons! Unfortunately, all of them are small, with the brightest, Himalia, reaching only 15th magnitude.

## FOUR WORLDS

Each of the Galilean moons has a unique character, but none is more bizarre than Io. This world, larger than our own Moon, is in a state of

## THE GALILEAN MOONS AND THE SPEED OF LIGHT

On 7 January 1610, Galileo Galilei (1564–1642), professor of mathematics at the University of Padua, aimed a new device of some glass lenses mounted in a piece of organ pipe—he called it a "perspicillum"—toward the southeastern sky. There, to the upper right of the gibbous Moon, was Jupiter, a bright dot of light.

With his telescope yielding about 20x magnification, Galileo (below) noted three little stars near Jupiter, one to the west and two to the east of it. The next night all three lay to the west of Jupiter. Two nights later, he saw only two stars, both to the east. This went on for two weeks before Galileo realized he was seeing a total of four worlds orbiting Jupiter. No one had ever seen these bodies before. Today, we know them as the Galilean moons.

By the 1670s, astronomers had determined the periods of revolution for the Galilean moons to within seconds of their modern values. But the predictions for when a moon would enter or leave Jupiter's shadow were often up to several minutes in error. These events seemed to occur earlier when Earth was nearer the Jovian system and later when it was farther away from it. In 1675, Danish astronomer Ole Römer realized these errors arose from the fact that the speed of light was not infinitely quick. As a result of this study involving the Galilean moons, the first scientific determination for the speed of light was made.

**PELE,** *the heart-shaped feature just left of center in this Voyager I image, was the first active volcano discovered on Io.*

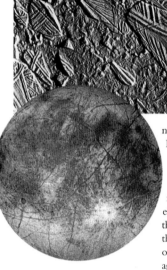

**ICE** *Ganymede's icy surface (below) is old and cratered, while Europa (below right), viewed close up by the Galileo spacecraft (right), shows young "rafts" of ice and few craters.*

continual volcanic eruption, thanks to gravitational tugs from Jupiter, Europa, and Ganymede. Io's volcanoes have covered it with a multicolored tapestry of sulfur deposits.

Europa, the next moon out, is a little smaller. Its surface is also young, but instead of lava flows, it is covered with ice floes floating on a deep ocean of liquid water.

The third Galilean moon, Ganymede, is bigger than Mercury. Its surface tells of a history with tectonic activity and impact cratering.

The fourth large moon is Callisto, about the size of Ganymede. It appears to be a sphere of ice and rock that has evolved little since it formed. The surface is pitted by countless impact craters, testimony to its long geological dormancy.

## JOVE'S RINGLET

Many scientists were surprised when Voyager 1 revealed a ring around Jupiter. The most tenuous of the gas-giants' ring systems, it is about 3,700 miles (6,000 km) wide and less than 20 miles (30 km) thick, and lies some 35,000 miles (56,000 km) above the cloud tops. It can be imaged from Earth only at near-infrared wavelengths.

## SOLAR SYSTEM POT-STIRRER

With its great mass, Jupiter exerts a major influence on the orbital mechanics of the asteroids and comets that pass near it. In the early Solar System, it prevented any planet from forming in the region that is now the asteroid belt by winnowing the population of large planetesimals in that vicinity. (Planetesimals are the small bodies that coalesced to form the planets.)

Today's asteroid belt has gaps in it where few objects orbit. Any that wander into these areas are soon removed by perturbations—changes in their orbit caused by the gravity of Jupiter.

Scientists are also starting to hold Jupiter largely responsible for the creation of the Oort Cloud of comets (see p. 192). The theory is that as comet

**THIS INFRARED IMAGE** *shows the plume (at bottom left) from the collision of fragment G of Comet Shoemaker-Levy 9 with Jupiter on 18 July 1994.*

nuclei move inward from the Kuiper Belt through perturbations with Neptune, they can end up near Jupiter. At that point, Jupiter's gravity may bend their orbits into elongated paths that approach the Sun and Earth. More often, though, it will fling them outward, to the Kuiper Belt again, into the Oort Cloud, or perhaps even out of the Solar System altogether.

Jupiter can also destroy comets, as the world saw in July 1994 with Comet Shoemaker-Levy 9. This hapless object passed close enough to Jupiter that tidal effects broke the comet apart. Jupiter then sent the comet on one last looping orbit around itself. When the comet returned to Jupiter, its 21 pieces crashed into the planet's southern hemisphere, leaving Earth-size dark splotches that could be seen even in small telescopes.

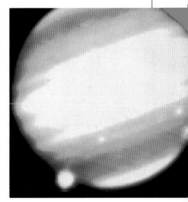

# SATURN

*While Saturn and its rings have captivated observers for hundreds of years, the full wonder of this planet and its satellites is only just being realized.*

**THE EARLY FRUITS** *of the Earth offered to Saturn, god of the harvest, from a fresco by Giorgio Vasari (1511–74).*

S aturn is the one object, apart from the Moon, that always gets a gasp of delight from first-time telescope viewers. Once known as *the* Ringed Planet, we now know that Jupiter, Uranus, and Neptune also have rings. However, Saturn's rings are the only ones you can see in any backyard telescope, and once seen, that tiny image is truly unforgettable.

## DISCOVERING SATURN

Saturn held few attractions for the ancients, who knew it as the slowest moving planet in the sky. Our name for the planet comes from Roman mythology, but the Greeks called it Kronos after Zeus's father, the overthrown ruler of the universe and a weary old man. Earlier still, Mesopo-tamian astronomers had called it "the old sheep" or even "the eldest old sheep."

The modern era for Saturn began in 1610, when Galileo Galilei turned his telescope on the planet. He thought it was a triple-bodied object. Other observers reported that it had "handles" or "ears," and fierce debate ensued.

Finally, in 1659, Christiaan Huygens (who discovered Titan, Saturn's largest satellite) announced that Saturn was circled by a broad, flat ring inclined to the ecliptic. This explained why the rings seem to disappear at about 15-year intervals. Twice each Saturnian year, Earth passes through the plane of the rings and they become too thin to be seen. The last such passage occurred in 2009, and the next will take place in 2024.

## THE SPACECRAFT ERA

In 1980 and 1981, the space-probes Voyagers 1 and 2 flew past the planet. Their close-up images of Saturn, its multi-tudinous rings, and the diverse family of moons revolutionized our view of the planet. The spacecraft revealed new details in Saturn's cloud belts, showed the rings were ribbons of particles moving in complex mathematical resonances, and mapped the intricate geologies of most of the larger satellites.

In 2004, the Cassini spacecraft entered orbit around Saturn to begin an extended mission observing Saturn. In January 2005, the Huygen probe was dropped into the atmosphere of the moon Titan.

## THE VIEW FROM HERE

Although Saturn orbits nearly 10 times farther from the Sun than Earth does, it is relatively bright, shining at about 1st magnitude. By 2015 it will be competing with the bright stars of Scorpius.

**SATURN AND ITS MOONS** *This image was composed from photographs taken by the Voyager spacecraft.*

**A STORM SYSTEM** *near Saturn's equator, revealed in a Hubble image (far left). The planet's rings are obvious even in a small telescope (left).*

Being a gas-giant planet, Saturn has a structure much like Jupiter's—mostly gas and liquid with a small, dense core. The visible surface is the top of an atmosphere containing hydrogen, helium, and other compounds such as methane, and the features we see are constantly shifting structures in the planet's cloud tops. Since it orbits farther from the Sun than Jupiter does, Saturn's environment is colder. This means it has less "weather" and displays fewer features.

To viewers with 3 inch (75 mm) scopes or smaller, Saturn's disk will appear featureless and creamy white. But in larger scopes, especially during good seeing, a pair of dusky bands paralleling the planet's equator becomes visible at about 20 degrees north and south latitude. These are called the north and south Equatorial Belts.

Viewing in twilight helps reveal these and other features. In full darkness, use a yellow filter to improve visibility. It helps to remember that belts are dark while zones are light.

The features contained between the two Equatorial Belts form what astronomers call System I, and they rotate once every 10 hours 14 minutes. System II includes everything else and has a rotation period of 10 hours 38 minutes.

At rare intervals, a white cloud breaks out in Saturn's Equatorial Zone. This occurred in 1990, when a small, bright patch appeared. Scientists think it was produced by a gigantic bubble of ammonia gas that rose from the depths and, upon reaching the chilly cloud tops, froze into white crystals and spread around the planet. Such eruptions seem to occur about every 30 years, but the cause remains unknown.

Another event occurs when Saturn lies 90 degrees away from the Sun in the sky—about 90 days before and after opposition. At this time, we can peer a little past one limb of Saturn and detect its shadow falling on the rings. The effect, though small, provides a visual hint of the planet's three-dimensional reality.

## CASSINI AND HIS DISCOVERIES

Giovanni Domenico Cassini (1625–1712) is famous for discovering the gap between the A and B rings of Saturn. He also proposed that the rings were made up of individual particles rather than being a solid body as many thought at the time.

Born in Italy, Cassini (below) taught at Bologna and determined the rotation period of Mars to within two minutes of its correct value. He later did the same with Jupiter. In 1669, Louis XIV of France invited him to take over the Paris Observatory. Cassini stayed in France for the rest of his life and became a French citizen.

His most important work came in 1672 when he used observations of Mars to determine the size of the planets' orbits. He derived a value for the Earth–Sun distance that was only about 7 percent too small—the most accurate estimate at that time.

In 1671, Cassini discovered the first of four new moons of Saturn. This moon now bears the name Iapetus. The following year he found another, Rhea, and on one night in 1684 he discovered two more moons, Dione and Tethys.

In 1675, he observed that the "ring" of Saturn contained a dark gap. To this day, it is known as the Cassini Division.

**THE UNDERSIDE** *of Saturn's rings, as seen by Voyager 1. Thousands of "ringlets" make up each of the rings.*

## LORD OF THE RINGS

If the Voyager spacecraft showed too many rings to number, the view from Earth is much simpler—we see only three. The outermost is called the A ring and it is 9,000 miles (14,500 km) wide. It is separated from the B ring by the dark gap of the Cassini Division, 2,600 miles (4,200 km) across. The B ring is both the brightest and, at 16,000 miles (26,000 km) across, the widest. On its inner edge is the gauzy C (or crepe) ring, 10,500 miles (17,500 km) wide.

While the B ring looks solid enough to walk on, in reality it, like the other rings, is a loose collection of particles orbiting in complex ways. This was dramatically illustrated in 1989, when Saturn passed in front of the star 28 Sagittarii. All over the world, professional and amateur astronomers watched the star flicker as it appeared to traverse the rings. The occultation revealed even more structure in the rings than Voyager had shown.

The Cassini Division is hard to see when the rings are nearly edge-on, as in 2009, but it becomes easier to see as the tilt increases. Voyager showed that the division is not empty, but simply contains less material than the rings. The division occurs because the gravity of the moon Mimas perturbs ring particles that orbit there, selectively removing them.

The C ring is also hard to view when the rings are relatively closed. When the rings are open, it is easy to spot the C ring in front of Saturn's disk.

Saturn's rings are probably transitory. They originated when one or more moons strayed too close to Saturn to withstand its tidal force and broke apart. Many scientists now think the rings will last only a few tens of millions of years before mutual collisions rob the particles of energy and they spiral into Saturn.

## WORLDS OF ICE

Saturn has over 60 known moons, many being mixtures of ice and rock. Most, however, lie beyond the reach of amateur-size telescopes. The largest is 8th magnitude Titan, visible even in a 3 inch (75 mm) scope. Titan circles Saturn every 16 days, and

**SATELLITES** *A Voyager image of Enceladus (top). A near-infrared Hubble image of Titan shows light and dark surface features (above). Titan is named after the giants of Greek mythology (right).*

*Oh Titan, warmed by a*

*hydrogen blanket,*

*ice ribbed volcanoes*

*jet ammonia*

*dredged out of a glacial heart.*

The Planets,
DIANE ACKERMAN (b. 1948),
American poet

when lying due east or west of Saturn, is about 4½ ring-diameters away from the center of the planet. This moon is a strange world indeed. At almost 3,200 miles (5,100 km) across, it is bigger than the planet Mercury. An opaque, smoggy atmosphere of nitrogen and methane blankets its surface, which may feature lakes of hydrocarbons.

Inside Titan's orbit, other moons to look for are magnitude 9.9 Rhea, magnitude 10.6 Dione, and magnitude 10.4 Tethys. Careful observation may also catch the magnitude 11.9 Enceladus.

Be sure not to overlook oddball Iapetus. Because its leading side is coated with dark dust, Iapetus is more than twice as bright when it lies west of Saturn than when it lies east of it—ranging from 10th down to 12th magnitude. Iapetus circles outside the orbit of Titan every 79 days at a distance of 13 ring-diameters.

CAPTURING SATURN
Saturn's low contrast and complex aspect make drawing it a major challenge. Many observers concentrate on portions—the Equatorial Zone or one side of the rings.

**CCD TECHNOLOGY** *allows amateurs to capture splendid color images of distant objects such as Saturn (right).*

There is no time to dawdle—Saturn's quick rotation means that 15 minutes is long enough to work on a sketch.

Photography of Saturn requires fairly advanced techniques. Saturn's small apparent size means you must use eyepiece projection on an equatorially mounted telescope and track the sky during the exposure (see p. 93). You also need the good luck to capture a few seconds of good seeing.

CCDs allow you to take many images in succession, which improves your chances of catching moments of good seeing. Software allows you to choose the best images and "stack" them into a single final image. The result, however, can be an outstanding record of a beautiful object.

## SATURN FACTS

**Distance from Sun** 9.54 AU
**Mass** 95.2 × Earth's mass
**Radius** 9.4 × Earth's radius
**Apparent size** 15 to 21 arcsecs
**Apparent magnitude** 0.6 to 1.5
**Rotation (relative to stars)** 10.2 hours
**Orbit** 29.5 Earth years

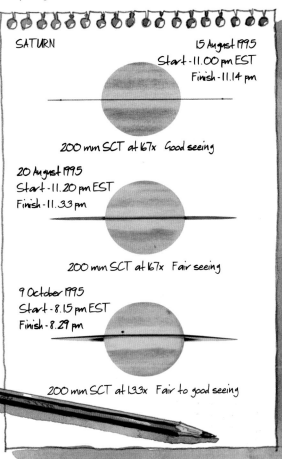

SATURN

15 August 1995
Start - 11.00 pm EST
Finish - 11.14 pm

200 mm SCT at 167x   Good seeing

20 August 1995
Start - 11.20 pm EST
Finish - 11.33 pm

200 mm SCT at 167x   Fair seeing

9 October 1995
Start - 8.15 pm EST
Finish - 8.29 pm

200 mm SCT at 133x   Fair to good seeing

**SKETCHES** *made at different times show the varying tilt of Saturn's rings. Sketching can often record details that would be blurred in photographs.*

# URANUS

*This distant, blue–green planet is something of a mystery. Even after Voyager 2's flyby in 1986, there remains a great deal for future astronomers to discover.*

**ARIEL** *(far left) with Prospero in a scene from Shakespeare's play* The Tempest. *All of Uranus's moons are named for Shakespearean characters.*

Uranus's brightness comes just within reach of the naked eye, and it was mistaken for a star many times before it was finally discovered.

The first recorded sighting was made in 1690 by England's Astronomer Royal, John Flamsteed, who cataloged it as 34 Tauri. Another observer, Pierre Lemonnier, logged Uranus as a star a total of 12 times—6 of those over one 9-day period in 1769. These early records later helped to establish the planet's orbit.

When William Herschel discovered Uranus in 1781, he initially thought it was "either a nebulous star or perhaps a comet." He soon confirmed, however, that the 6th magnitude object was a planet, circling the Sun every 84 years. It was named Uranus after the Greco-Roman god who

personified the universe and was the father of Saturn.

Being so distant, little was learned about Uranus until well into the twentieth century. Much of our current understanding came from the Voyager 2 flyby in 1986.

## A BIG BLUE WORLD

The planet's density, 1.3 times that of water, groups Uranus with the other giant planets, although it is often called an ice, rather than gas, giant. Its atmosphere is mostly hydrogen and helium, with some methane.

It is believed that Uranus consists of two, or possibly three, layers. At the surface, the mix is gaseous, but increasing temperature and pressure make it behave as a liquid about a third of the way to the center. The remainder of the planet is a hot, slushy mixture of water, methane, and ammonia, together with rocky components. It may also have a nickel-iron core. Uranus does not seem to have any internal heat source.

In a telescope, Uranus looks pale blue-green because of the methane gas in its atmosphere. Its disk almost always appears featureless, although past

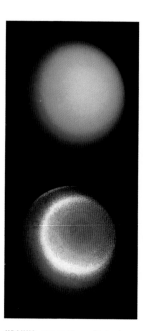

**URANUS** *appears blue and featureless in true color (top), and reveals hardly any more detail even with strong enhancement in false colors (above).*

observers have noted faint dark bands paralleling its equator.

## ORBITING SIDEWAYS

Uranus orbits almost "on its side" at a tilt of 98 degrees, possibly as a result of a collision with an Earth-size object early in its history. Its moons and rings may be leftover fragments from this impacting body. The tilt gives the planet peculiar seasons, with one pole or the other pointing toward the Sun for several decades. Such long cycles of

## URANUS FACTS

**Distance from Sun** 19.2 AU

**Mass** 14.5 × Earth's mass

**Radius** 4.0 × Earth's radius

**Apparent size** 3 to 4 arcsecs

**Apparent magnitude** 5.5 to 5.9

**Rotation (relative to stars)** 17.2 hours

**Orbit** 84.0 Earth years

sunlight and darkness heat the polar regions more than the equator. The resulting energy imbalance would explain the 450 mile per hour (720 km/h) winds detected during the Voyager 2 flyby.

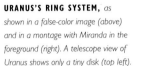

**URANUS'S RING SYSTEM,** *as shown in a false-color image (above) and in a montage with Miranda in the foreground (right). A telescope view of Uranus shows only a tiny disk (top left).*

## SMASHING MOONS

The first two of Uranus's five large moons, Titania and Oberon, were found by William Herschel in 1787. The third and fourth moons, Ariel and Umbriel, were not discovered until 1851. Miranda, the fifth, held out until 1948. These major moons certainly pose a challenge to observers: the brightest is Titania, at magnitude 13.7, and the dimmest, Miranda, is 16th magnitude.

Little was known of these moons until Voyager 2's visit. Its cameras revealed Oberon and Umbriel to be heavily cratered worlds of ice and rock, with no sign of geological activity. Ariel and Titania, on the other hand, appear to have undergone some kind of icy volcanism that has erased craters and produced long fault valleys.

The surprise was Miranda. It displays a geology so complex that, more than a decade after the Voyager visit, sci-

entists are still arguing about what causes its dominant feature—the grooved terrain known as coronae. One theory is that Miranda is a second-generation satellite, reassembled out of a shattered moon or moons. Its surface shows three enormous coronae, whose roughly oval patterns may mark places where large fragments sank into Miranda's interior.

Voyager 2 also discovered 10 more moons, all of them small, dark mixtures of ice and rock. Like the larger moons, these were named after Shakespearean characters, including Ophelia, Portia, Juliet, Desdemona, and Puck. There are now 27 known moons of Uranus.

## SLIM RINGS

In March 1977, astronomers watched Uranus slide in front of a 9th magnitude star and discovered that Uranus has rings. The Voyager 2 flyby confirmed that Uranus has 11 rings, as well as several ring-arcs, or partial rings.

Unlike Saturn's grand ring system, Uranus's rings are

**MIRANDA,** *seen in this computer reconstruction, displays a surface unlike any other in the Solar System.*

gauzily thin and difficult to detect from Earth. Most of the rings are not circular, being tugged out of shape by "shepherd" moonlets that confine their widths. The ring particles are as dark as coal and range from boulder- to house-sized fragments.

## WATCH THIS SPACE

At the time of the Voyager 2 flyby, Uranus displayed an all-but-blank appearance. This disappointed many scientists, who had hoped to find atmospheric features like those of Jupiter. Any features the planet had were hidden deep in the hydrogen-methane haze.

However, the Uranus that Voyager saw in 1986 may not tell us much about the planet we will see in the future. In 1985, the planet's south pole was pointing at the Sun. Past telescopic observations suggest that cloud markings appear on Uranus in the years around its equinoxes, when the Sun is over the planet's equator.

More recent observations with the Hubble Space Telescope have revealed more complex seasonal variations.

183

# NEPTUNE

*Although it is a dim sight from Earth, this deep-blue planet with its bright cloud bands is one of the most varied and interesting in the Sun's family.*

**NEPTUNE,** *the Roman god of water (left), is an appropriate namesake for the planet. A polar view (right).*

Neptune is four times the size of Earth, but because it orbits 30 times farther from the Sun, observers see only a small, 8th magnitude disk.

## DISCOVERY TRIUMPH

The discovery of Neptune was a great triumph of mathematical astronomy. By the 1840s, astronomers had known for years that Uranus was not moving as predictions said it should, and that there must be another planet affecting its orbit. Acting independently, two mathematician-astronomers, Urbain Le Verrier in France and John Couch Adams in England, took the discrepancies and tried to determine where this planet might be, assuming likely values for distance, mass, and so on.

Their research produced nearly the same result, but the British were slow to follow up Adams's prediction. Le Verrier had more luck, and on 23 September 1846, following Le Verrier's directions, Johann Galle and Heinrich d'Arrest at the Berlin Observatory found the planet on the first night of searching—in fact, within half an hour. In the end, after much embittered finger-pointing among British astronomers, Adams received joint credit with Le Verrier for the discovery.

## VOYAGING TO NEPTUNE

Because of the planet's distance from Earth, information about Neptune accumulated slowly. William Lassell saw Triton, the largest moon, within a month of the planet's discovery, but the second moon, Nereid, was discovered only in 1949, by Gerard Kuiper. In the 1970s and 1980s, astrono - mers received strong hints of a Neptune ring system, but could not make sense of the data.

In August 1989, Voyager 2 flew by the planet, and the modern era for Neptune began. Voyager found a big, blue ball, with many markings and cloud bands—a pleasant change from the bland Uranus the spacecraft had visited three years earlier. Voyager's instruments collected data about Neptune's composition. There may be

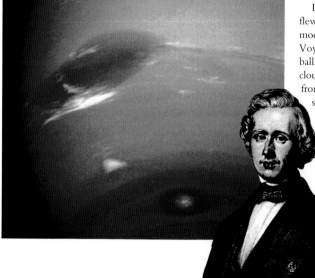

**A CHANGING FACE** *Neptune's Great Dark Spot (far left) seen by Voyager 2 disappeared. Another formed later. Urbain Le Verrier (left).*

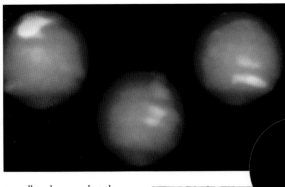

**NEW CLOUDS** *on Neptune, seen by the Hubble Space Telescope in 1995 (left), are invisible in the amateur telescope view (below). A false-color Voyager 2 image of Triton's surface shows possible nitrogen geysers (below left).*

## VIEWING THE BLUE WORLD

Seeing Neptune is not particularly difficult—almost any telescope can reach 8th magnitude. Knowing exactly where to look, however, can certainly be a challenge for beginners. Astronomy magazines regularly publish finder charts for Neptune, and any sky-charting software will give its current position. Until about 2012, the planet can be seen in Capricornus, a relatively star-poor region, making it easier to identify.

When you locate Neptune, chances are you will recognize it. At moderate power (about 70x), the planet shows a tiny disk and it appears distinctly blue-gray, because of the methane in its atmosphere. At higher power (150x or more), you can see the 2.3 arcsecond disk more clearly.

If you are hunting for moons, Triton is the only possibility. At 13th magnitude, it looks just like a star and is difficult to find. It calls for a telescope of 8 inches (200 mm) aperture or larger.

a small rocky core, but the bulk of the planet is probably a mantle of icy compounds. This merges into an atmosphere of hydrogen and helium. Methane in the top of the atmosphere gives the planet its pronounced blue color. The interior bubbles up more than twice as much heat as the surface receives from the Sun.

Neptune spins on its axis every 16 hours 7 minutes. In its equatorial zone, winds scream westward at 900 miles (1,500 km) per hour, powering huge storm systems.

Voyager froze Neptune in time with a snapshot, but the Hubble Space Telescope has since tracked its changing features. The spacecraft saw several storms, notably the Great Dark Spot (the size of Earth), which drifted toward the equator and vanished sometime after the flyby. Other noted markings, such as the Scooter, the second Dark Spot, and various white streaks, have also disappeared or changed greatly, while new features have emerged. A longer view is needed to fully understand the planet, so observations continue.

## GEYSERS AND ICEFIELDS

Voyager 2 confirmed the existence of five thin rings around Neptune and added six satellites to the two already known. The total is now 13. Voyager also took a close look at the largest satellite, Triton.

Two-thirds the size of our Moon, Triton displayed a varied face to the cameras— plains, impact craters, a strange cantaloupe terrain of pits and depressions crossed by ridges, and a southern hemisphere with a thin, pinkish ice cap of nitrogen. Triton also has plumes—geysers of nitrogen gas shooting 5 miles (8 km) straight up into a thin atmosphere of nitrogen and some methane. Winds then blow the gas plumes about 60 miles (100 km) downrange.

Triton is the only major satellite with a retrograde orbit. Scientists think it was captured by Neptune, maybe after drifting in from the Kuiper Belt (see p. 186). However, it has a much rockier composition than the comet-like objects orbiting in the belt. Perhaps it changed during capture.

Planetary scientists believe Triton may be providing us with a preview of Pluto, which is another Kuiper Belt refugee that we will get a first close-up of in 2015.

| NEPTUNE FACTS | |
|---|---|
| **Distance from Sun** | 30.1 AU |
| **Mass** | 17.1 × Earth's mass |
| **Radius** | 3.88 × Earth's radius |
| **Apparent size** | 2.3 arcsecs |
| **Apparent magnitude** | 7.9 |
| **Rotation (relative to stars)** | 16.1 hours |
| **Orbit** | 165 Earth years |

# PLUTO *and* BEYOND

*There is an undeniable thrill in seeing this distant member of the Sun's family, and Solar System buffs cannot call their observations complete without it.*

Percival Lowell, best known for his theories about Martian life, began the search for a distant planet in 1905. An accomplished mathematician, he believed that residual irregularities in the motion of Uranus could be explained by the gravitational pull of a planet orbiting beyond Neptune. Lowell's approach had good precedent: discrepancies in Uranus's motion had led directly to the discovery of Neptune in 1846.

However, in this case, Lowell was off track—we now know that the dwarf planet Pluto is much too small to have any such effect. But Lowell's obsession with "Planet X," as he called it, did eventually pay off. After Lowell's death in 1916, the observatory he founded in Flagstaff, Arizona, dropped the search for more than a decade. Then in 1929, Lowell

*Pluto, the grisly god,*

*who never spares,*

*Who feels no mercy, and*

*who hears no prayers.*

The Iliad (Book IX),
HOMER (c. 8th century BC),
Greek epic poet

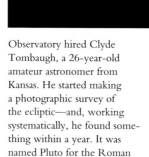

Observatory hired Clyde Tombaugh, a 26-year-old amateur astronomer from Kansas. He started making a photographic survey of the ecliptic—and, working systematically, he found something within a year. It was named Pluto for the Roman god of the underworld.

## VISITOR FROM THE KUIPER BELT

Pluto is an odd object. Its 248-year orbit is more eccentric and inclined than any of the planets, and, for part of it, Pluto travels closer to the Sun than Neptune does.

Astronomers now recognize Pluto as an object that wandered in from the Kuiper Belt, a region of the Solar System beyond the zone of the planets. The belt begins at roughly 35 astronomical units and extends to about 55 astronomical units. The bodies in the Kuiper Belt are icy planetesimals that never accreted into larger objects like Neptune.

Pluto was regarded as a planet for over 75 years. Then,

**THE BEST VIEW** *so far of Pluto and Charon (left), from the Hubble in 1994. Pluto as ruler of the underworld (far left). Clyde Tombaugh (below).*

in 2006, the International Astronomical Union refined the definition of a planet to require that it dominate its region of space. Pluto fails this test because it is only one of several similar objects discovered so far. The dwarf planet Eris, discovered in 2005, is now the largest known dwarf planet.

## A MOON FOR PLUTO

In 1978, James Christy discovered that Pluto has a moon. He named it Charon, after the mythological boatman who ferried the souls of the dead across the River Styx to Hades. When astronomers determined Charon's orbit,

**THE FAINT OBJECT CAMERA** *on the Hubble Space Telescope produced a map of Pluto (right). From the ground, Pluto looks like a faint star (below).*

they were startled to note that between 1985 and 1990, observers on Earth could see the Pluto-Charon system edge-on, and watch the two bodies eclipse and occult one another every 3.2 days.

By taking measurements during these eclipses, scientists worked out that Pluto has a diameter of 1,425 miles (2,300 km), and Charon is fully half its size: 760 miles (1,220 km). They are separated by 12,100 miles (19,500 km). Scientists also estimate the masses of Pluto and Charon together amount to less than 1/400 Earth's mass.

## A LAYER OF GAS

A second stroke of luck occurred in June 1988, when Pluto passed in front of a 12th magnitude star. As they watched the star blink out, astronomers discovered that Pluto has a thin, cold atmosphere consisting mainly of nitrogen and methane. The data also hinted that the atmosphere might have hazes. But the strangest thing is that Pluto's atmosphere is both dynamic and transient.

Pluto's weak gravity cannot hold the atmosphere forever, and it is escaping into space much like a comet's gases do. The atmosphere is also episodic: it is gaseous only when Pluto is around perihelion, the peak of its orbital "summer."

## SEEING PLUTO

Pluto is a tough challenge for any planet watcher. At about 14th magnitude, it is a very faint object. You need a telescope of at least 8 inches (200 mm) to find it—but a 10 inch (250 mm) or larger scope makes the job a lot easier. With an apparent diameter of less than 0.1 arcsecond, Pluto displays

| PLUTO FACTS | |
|---|---|
| **Distance from Sun** 39.5 AU | |
| **Mass** 0.002 × Earth's mass | |
| **Radius** 0.18 × Earth's radius | |
| **Apparent size** 0.08 arcsec | |
| **Apparent magnitude** 13.7 | |
| **Rotation (relative to stars)** 6.39 Earth days | |
| **Orbit** 248 Earth years | |

no visible disk and looks no different than a star.

So how do you know when you have found it? Basically, you locate the correct field of view, map (by hand or with a camera) every star you see down to 15th magnitude, and recheck the field a night or two later. The star that has moved in the interval is Pluto.

Astronomy magazines publish finder charts for Pluto, which are a great help for getting your telescope into the right area of the sky. Computer-driven telescopes (see p. 80) can make the job even easier. But actual identification still calls for observations over several nights to spot the moving dot.

Pluto lies north of Scorpius and is approaching the Milky Way for the first time since its discovery over 80 years ago. This makes it much more of a challenge to distinguish from the mass of background stars.

**SPACECRAFT** *are yet to visit Pluto and its moon Charon, but the NASA New Horizons mission is en route and will reach Pluto and Charon in 2015.*

# SMALL SOLAR SYSTEM BODIES: ASTEROIDS

*Rocky objects, smaller than planets,*

*that orbit the Sun, asteroids promise to tell us much*

*about the development of our Solar System.*

In 1766, Johann Titius of Wittenberg, Germany, divided the distance between the Sun and Saturn, then the farthest known planet, into 100 units or "parts." Mercury is thus 4 parts from the Sun, Venus is 7 (or 4 + 3), Earth 10 (or 4 + 6), and Mars 16 (or 4 + 12). Titius discovered that if you keep doubling the second number, you reach a planet, right out to Saturn—except that there is no planet at 28.

Johann Bode, of the Berlin Observatory, was intrigued by Titius's theory and started a campaign to search for the missing planet beyond Mars. So enthusiastic was he that the theory is now often called Bode's Law rather than the Titius–Bode Law.

## THE CELESTIAL POLICE

In 1800, a group calling themselves the Celestial Police came together in Germany to organize a systematic search for the missing planet. They divided the plane of the planets, the ecliptic, into several regions and assigned an astronomer to search each region.

On 1 January 1801, they were pre-empted by Giuseppe Piazzi of the Palermo Observatory in Sicily, who discovered a star-like object that behaved like a planet. Now called Ceres, this dwarf planet is in

the correct place to satisfy Bode's Law, but, at less than 600 miles (1,000 km) across, it is rather small. On 28 March 1802, the Celestial Police found a second asteroid, now known as Pallas, and later they found Juno and Vesta. Pallas and Vesta are the next largest asteroids in size after Ceres, but are only half its size.

In recent times, the pace of discovery has increased. We now know the orbits of more than 120,000 asteroids well enough to give them official numbers, and some 13,000 have official names. Only Ceres can be labeled a dwarf planet. The rest are simply "small solar system bodies".

No one knows how many asteroids there are in all, although we have probably found every object larger than

**ASTEROID 951 GASPRA** *has its color variations exaggerated in this view from the Galileo spacecraft.*

60 miles (100 km) across and about half of the 6 mile (10 km) objects. Estimates suggest there may be a million asteroids larger than ½ mile (1 km) in diameter.

Planetary scientists studying the main belt are beginning to see its current inhabitants as the remains of a much larger population of small bodies that formed and evolved under the gravitational control of Jupiter (see p. 177). They believe that collisions—with each other and with the planets—have steadily reduced larger asteroids to smaller ones, and smaller ones to dust and fragments.

Collisions may also hurl asteroid fragments into orbits that cross Earth's orbit, and these fragments are the main source of meteorites.

## TYPES OF ASTEROID

Scientists classify asteroids into two main kinds, based on their reflectivity. The darker kind reflect 5 percent or less of the sunlight falling on them, and resemble dark stony meteorites (see p. 198).

The brighter kind reflect about 20 percent of sunlight and look like light-colored stony meteorites. A rare third type resemble iron meteorites and may be the shattered metallic cores of ancient proto-asteroids.

## ELEANOR HELIN

One of the most successful asteroid discoverers, Eleanor Helin began her work in 1969 by researching a dozen or so Apollo asteroids. These asteroids approach the Earth from time to time as they orbit the Sun, and are capable of hitting us.

Using the 18 inch (450 mm) diameter Schmidt camera at Mount Palomar, Helin began a search for more of these asteroids. She recalls those difficult early days as the first woman observer working there regularly. "I'm sure Palomar didn't think I would last a year," she says.

Helin's Palomar program has ended, having discovered many asteroids and comets, but she now leads a new program. At the time of the Camp David Agreement, in 1978, Helin discovered a near-Earth asteroid. She later named it Ra-Shalom, after the Egyptian Sun god Ra and the Hebrew word for peace, to honor the peace accord between Israel and Egypt.

## WHERE ARE THEY?

**Main Belt Asteroids** The majority of asteroids with known orbits lie within the main belt of asteroids between the orbits of Mars and Jupiter. Despite the large number of asteroids, their total mass is only about 2 percent of the mass of the Moon. Fully a third of that lies in Ceres alone, orbiting near the middle of the belt.

Scientists have been trying to piece their histories together by defining different families. Working back in time from their orbits, it is possible to deduce that, for each family, a single large asteroid broke up after a collision to produce the fragments we see today. The belt is actually divided by Kirkwood gaps. These result

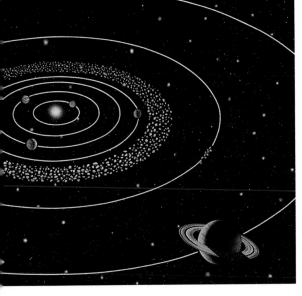

**THE ASTEROID BELT**, *and the Trojan asteroids which share the orbit of Jupiter. In reality, their positions are not this well localized.*

from the gravitational tug of the planets, especially Jupiter. If an asteroid lies in a gap its orbital period is a simple fraction of Jupiter's period (½, ⅓, ⅔, etc.) and so it periodically finds Jupiter pulling it in the same direction—and out of the gap.

**Trojans** There are two regions in Jupiter's orbit where asteroids can become trapped. Called Lagrangian Points, one is about one-sixth of an orbit in front of Jupiter and the other is one-sixth of the way behind it. Asteroids in these regions are called Trojans, and are traditionally named after figures from the legendary Trojan War. The Greeks (with 624 Hektor, a Trojan spy) precede Jupiter, and the Trojans follow it (along with Greek spy 617 Patroclus).

**Near-Earth Asteroids** In 1989, Henry Holt discovered an

**A CHART** *of the Solar System from 1857, showing, among other things, some of the known asteroids.*

asteroid that came within half a million miles (800,000 km) of Earth. In 2004, an asteroid 100 feet (30 m) across bypassed Earth by just 26,500 miles (42,600 km)—about one-tenth of the distance to the Moon. This was the closest miss ever seen of one of the Apollo asteroids—objects that cross Earth's orbit. There are over 2,000 of these "Earth grazers" known, and probably many thousands more of similar sizes. All are capable of hitting Earth some day.

## DISTANT ASTEROIDS

In 1977, the first example of a new group of asteroids was discovered. 2060 Chiron is around 125 miles (200 km) across but its orbit takes it from within Saturn's orbit to almost as far as Uranus. Dozens of similar objects, known as Centaurs, are now known. They may be former members of the Kuiper Belt, with sizes in between comets and the icy outer dwarf planet Pluto, or moons such as Triton.

## HUNTING ASTEROIDS

Their very name means "star-like," and it tells the telescope user not to expect details like those seen on the Moon or Jupiter. Asteroids always look just like points of light; only the Hubble Space Telescope or radar observations can image a few of their surfaces in any detail. Most asteroids are too faint for amateur instruments to detect. Still, there are several hundred within reach of a 3 inch (75 mm) telescope. For most observers, the question is: Where do I look?

Astronomical Web sites publish positions, called ephemerides, for the brightest asteroids, and more can be found in computer software, astronomy magazines, and annual astronomical hand-books (see p. 96).

When you have positions for an asteroid, plot them on a

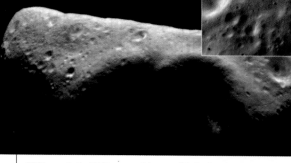

**EROS** *seen by the NEAR spacecraft in color (inset above). The landing site from 119 miles (192 km) out (above). The NEAR spacecraft took a series of shots*

*of the landing site as it approached the asteroid (left to right below): from 3773 feet (1150 m); from 820 feet (250 m); and from 394 feet (120 m).*

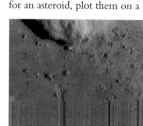

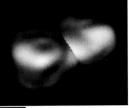

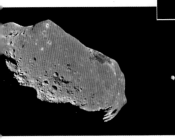

**ASTEROID** *243 Ida and its moon Dactyl (below) have been color-enhanced here to reveal mineralogical details. A computer model of near-Earth asteroid 4769 Castalia (right), "a contact-binary."*

star chart and observe the field. Start with one of the brighter asteroids—these just reach naked-eye visibility. When you locate the right area, compare your sky chart with the star field using your lowest-power eyepiece. If you are lucky, the asteroid will be easy to identify as the "star" not plotted on the chart.

If the asteroid is not obvious, you will have to identify the most likely area for it to be in and plot all the star-like objects you see, going down one magnitude fainter than that predicted for the asteroid. Wait several hours, or come back the next night, and re-observe the field. Your asteroid is the star that moved.

From time to time, an asteroid will pass in front of a star, briefly blocking its light. Such events, called occultations, are extremely valuable in determining an asteroid's physical size and shape. Often it is the only way to gain this information. Amateurs join professionals in expeditions to locations where they hope to see the event. As it is difficult to know precisely where the best view of the event will be visible, the more observers there are, the better. Anyone interested in helping with this kind of observing should contact their national amateur organization.

## FINDING NEW ONES

A handful of programs at professional observatories are devoted to the search for asteroids. One approach is to image a portion of sky through a telescope. As the chip records the field of view, it electronically subtracts a reference image for that area containing just the stars. Anything not in the reference image will stand out clearly. Asteroids are also discovered in surveys that regularly image the sky in search of other objects, such as Kuiper Belt objects. These surveys photograph a portion of sky, wait a day or two, and re-photograph it. By comparing the two images, any object that moved between the exposures—such as an asteroid candidate—can be identified.

Neither approach is perfect. The first produces results in real time but covers only a tiny portion of sky in each exposure. The second imposes delays, but covers a wider field of view. Patrol programs using imaging have become important as a way of uncovering Near-Earth Objects (NEOs). These are small asteroids or comets which come very close to, and may one day collide with, Earth.

### JOHN, PAUL, GEORGE, AND RINGO

Or as they are officially designated, minor planets 4147 Lennon, 4148 McCartney, 4149 Harrison, and 4150 Starr.

Unlike other astronomical objects, asteroids can carry the names of living people or commonplace objects. Whoever discovers an asteroid is allowed to suggest a name for it, and the name is usually accepted by the ruling body, a committee of the International Astronomical Union. The name is combined with a number reflecting the asteroid's order of discovery: 1 Ceres was the first asteroid found.

With more than 300,000 asteroids identified, fewer than half have names. The names of the first few hundred asteroids tend to be of females from classical mythology (16 Psyche, 34 Circe). Other religions and mythologies were explored (77 Freia, 1170 Siva), then wives and girlfriends followed (607 Jenny, 1434 Margot).

As the number of asteroids grew, the list broadened to include 2825 Crosby and 4305 Clapton, 3656 Hemingway and 4474 Proust, 4511 Rembrandt and 6677 Renoir, and even 6000 United Nations.

# COMETS

*Once interpreted as omens, comets remain a source of wonder and excitement. In recent years, our skies have played host to several spectacular examples.*

**THE BAYEUX TAPESTRY** *shows King Harold I being told of Halley's Comet before the Battle of Hastings in 1066.*

After a long period with few bright comets, the 1990s are remembered as the comet decade. Comets Levy in 1990, Shoemaker-Levy 9 in 1994, Hyakutake in 1996, and Hale-Bopp in 1997 captivated skywatchers around the world. Comet McNaught in 2007 was the brightest comet in over 40 years.

## ANCIENT COMETS

Comets are among the oldest objects in the Solar System, almost unchanged since they were formed billions of years ago. The primordial Solar System was a large, flattened cloud that spun slowly, its center building to become the Sun, its outskirts condensing to become the planets. Wandering through this cloud were the comets. As Jupiter and the other giant planets grew, their gravity slung many of these comets into a large sphere, now called the Oort Cloud and located well beyond Pluto's orbit. Others survive in the Kuiper Belt, just beyond Pluto and more closely confined to the plane of the Solar System.

As far as a light-year away from the Sun, the Oort Cloud contains billions of comets, their ices forever frozen unless disturbed by the gravitational tug of a passing star or by the Solar System's passage through the galaxy. A comet may then leave the cloud, either ejected from the Solar System al-together or following a new orbit that will head it toward the Sun. After 4½ billion years in deep space, the comet begins to boil off its gases, or sublimate, as it nears the Sun. A comet newly discovered may be in just this state, warming up for the first time and releasing ancient gases.

## WHAT IS A COMET?

Imagine a snowball several miles wide, loosely packed with rocky or muddy material. Far from the Sun, a comet consists of little else. Within the snowball are ices of water, cyanogen, and other materials, as well as what scientists call CHON particles—organic material containing carbon, hydrogen, oxygen, and nitrogen. As the snowball, or

**HALLEY'S COMET** *sketched in 1836 (above), and photographed from Earth in 1986 (above left). Also in 1986, the Giotto spacecraft revealed the comet's nucleus, which is surprisingly dark (left).*

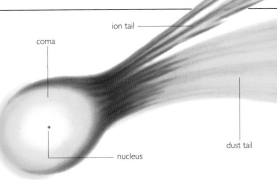

coma

ion tail

nucleus

dust tail

to Sun

nucleus, nears the Sun, its ices begin to sublimate, releasing gas and dust. This material forms a large head, or coma, around the nucleus, and then streams behind the coma to form a tail. The coma of a large comet might be thousands of miles across, its tail tens of millions of miles long. An active comet might bleed away only the topmost few feet of material during its months near the Sun. Even though the comet appears to be burning away, it has enough stored ice and dust to last for hundreds of revolutions about the Sun.

## COMET TAILS

There are two main types of comet tail. Ionized gas forms an ion tail, which appears bluish, while dust particles form a yellowish dust tail. The dust tail curves out from the comet under the pressure of sunlight. The ion tail reacts to the charged particles of the solar wind and is pushed directly away from the Sun. This means that both tails stream behind the comet when

approaching the Sun, but when the comet moves away from the Sun, the tails lead.

In addition to these tails, small jets of dust and gas may erupt. Such eruptions are triggered as the nucleus rotates and sensitive areas heat up quickly as the Sun rises over them. Seen through a telescope, jets are a striking sight.

## HOW COMETS TRAVEL

Like the planets and their moons, comets travel about the Sun in paths known as orbits. The orbits of the major planets are almost circular, but the orbit of a comet such as Halley's is a much more elongated ellipse. A comet coming in from the Oort Cloud has an even more extended orbit: it speeds round the Sun and then heads out for a trip that either never ends or will not bring it back for millions of years. On rare occasions, a comet travels in on an elliptical orbit, but then

encounters a large planet such as Jupiter, whose gravity puts it into a new, hyperbolic orbit that ejects it from the Solar System forever.

Comets that travel in elliptical orbits and return to the vicinity of the Sun within a period of 200 years are called short-period comets. The most famous of these is Halley's, whose return every 76 years allows each generation on Earth a chance to view it. Most short-period comets return much more often, with 6 years being the average. With a period of 3.3 years, Encke's Comet has the shortest period.

Comets Hyakutake and Hale-Bopp, the great comets of 1996 and 1997, are good examples of long-period comets. Returning for its first visit in 9,000 years, Comet Hyakutake passed close to Earth in March 1996. A year later, Comet Hale-Bopp rounded the Sun on its first visit in more than 4,000 years. An approach to Jupiter has shortened its orbit, so that its next visit will occur in only 2,400 years. It is possible that repeated encounters with Jupiter could shorten the period even more. In a hundred thousand years, Hale-Bopp might return as often as Halley's does now!

**COMET HYAKUTAKE** *The comet tail stretched across the sky during its closest approach to Earth, in March 1996.*

## WHAT IF YOU THINK YOU HAVE DISCOVERED A COMET?

**1** Check to see whether the object has a tail.

**2** Move the object around the field of view to make sure that it is not a reflection in your eyepiece from a nearby planet or bright star.

**3** If the object is faint, use high power to verify that the object is not a faint star, or group of stars, that might have appeared diffuse at low power.

**4** Check to see if the object is marked in a star atlas or catalog.

**5** Draw a simple sketch, showing the object's position in relation to nearby field stars.

**6** Look at it a half hour later to see whether it has moved.

**7** Make sure that your object is not one of the known comets in the sky. The positions of the brighter comets are published in major astronomy magazines and other sources.

**8** Have your sighting confirmed by an experienced observer. Tell that person the position, suspected nature, and direction of motion of the object you have found.

**9** If you still think that you have discovered a new comet, it is now time to notify the Central Bureau for Astronomical Telegrams in Cambridge, Massachusetts.

**10** If you have actually found a new comet, it may well be named after you.

---

*I think I've found a squashed comet!*

CAROLYN SHOEMAKER, American astronomer, on discovering Comet Shoemaker-Levy 9, 25 March 1993

## HOW TO SEARCH FOR COMETS

A comet can appear anywhere in the sky. In 1995, Alan Hale and Tom Bopp found Comet Hale-Bopp when they both happened to be looking at Messier 70, a globular star cluster in Sagittarius. Most comets, however, brighten only when they are near the Sun, and appear either in the western sky after dusk or in the eastern sky before dawn.

Through a telescope, most comets look like fuzzy stars or galaxies. The best way to search for them is to move your telescope slowly—about one field every few seconds. Many comet-like objects will be visible, but a star atlas will usually identify them as galaxies, star clusters, or nebulas.

## OBSERVING COMETS

Although some 25 comets are found each year, most of these are much too faint to appear in small telescopes. However, several comets of 11th magnitude or brighter appear annually, usually when they are close to the Sun. Until recently, they were almost always discovered by amateur astronomers. You can often see these comets in small scopes when the sky is dark and moonless. If the comet is brighter than 8th magnitude and the coma is condensed, it should be visible through 6 inch (150 mm) scopes. Dark skies are essential: a comet's ghostly light is easily swamped by a bright sky.

What makes comets so interesting to observe is their rapidly changing nature. A single comet can change radically in appearance as it swings past Earth, its geometry shifting so that we see different parts of the tail from night to night. It is also common to see the comet's coma change in shape and structure from week to week.

Occasionally, the ion tail of an active comet may appear to separate itself from the head. This "disconnection event" occurs when the comet's magnetic field reacts to changes in the solar wind. A new tail can be rebuilt in as little as a half hour, but the disconnection may recur every few days.

Less frequently, a comet will suddenly brighten by one or several magnitudes. This can be brought on by the onset of jet activity or by the ejection of a fragment from the nucleus. As Comet West rounded the Sun in 1976, the stresses of the Sun's gravity broke its nucleus into four pieces. This released great amounts of dust and gas,

**COMET WEST** *(left), rounding the Sun in 1976, and Comet Ikeya-Seki (below), photographed in 1965.*

**COMET HALE-BOPP** (below) *appeared in the northern skies in 1997. A false-color view of a comet (right).*

## CAPTURING A COMET ON FILM

Although photographing a comet is a challenging task, the results can be spectacular. Photographers have to be aware of two things: first, that the star field moves in response to the Earth's rotation; second, that the comet moves slowly in the field of stars. When taking long exposures, either through a telescope or with your camera piggybacked on a telescope, you will need to guide it (see pp. 90–93). You can guide your telescope on the comet if the coma is sharply defined. More often, however, the comet will be diffuse, without a solid point to focus on. In that case, center the guiding telescope on a star. If the exposure lasts for less than 10 minutes, it is unlikely that the comet will show perceptible motion among the stars.

causing the comet to brighten from magnitude 0 up to −2, which is as bright as Jupiter although more spread out.

### RECORDING YOUR OBSERVATIONS

Estimating brightness is one of the most common activities for a comet watcher, but it can also be one of the trickiest. Among experienced observers, a popular procedure is the Sidgwick, or "in-out," method.

First, fix in your mind the "average" brightness of the comet's coma. Unfortunately this "average" tends to vary among observers. Then choose a comparison star, and take your telescope out of focus until the star reaches the size of the in-focus coma. Compare the star's surface brightness with the memorized average brightness of the coma.

Repeat this procedure and try to find a star that matches the coma's brightness. You will usually find that the brightness is somewhere between the magnitudes of nearby stars.

Comets vary greatly in appearance. Observers often record how condensed the coma appears. A standard scale ranges from 0 (diffuse

image, no condensation) to 9 (a bright, star-like image).

If the comet has a visible tail, you may like to note its length. A tail equal in length to the Moon's diameter is 30 arc-minutes long. Also record the direction in which the tail is pointing. As the comet cruises past Earth, the orientation of the tail can change rapidly over a few days.

### CHARLES MESSIER: THE COMET FERRET

Parisian Charles Messier (1730–1817) found his first comet in 1759, and proceeded to locate new ones almost every year thereafter. He was so successful that King Louis XV called him the "Comet Ferret." Messier (below) worked at the Marine Observatory in Paris, supported by a pension from his friend, Jean Baptiste de Saron, President of the Paris Parliament and an expert in comet-orbit calculation.

The French Revolution forced Messier to leave Paris, but in September 1793 he found a comet in the constellation of Ophiuchus. By this time, de Saron had been accused as an enemy of reform and was in prison. Despite this, he used Messier's positions to calculate an orbit for the comet from his prison cell, predicting that it would move closer to the Sun, then swing away and reappear in the morning sky.

Messier confirmed de Saron's predictions on 29 December, and smuggled the news to the prisoner. On 20 April, just three months before the end of the Reign of Terror, de Saron was guillotined. Although Messier survived, he was left virtually penniless.

# COMET *and* ASTEROID IMPACTS

*Comets and asteroids are fascinating not just for what they are, but also for what they can do.*

**MIMAS,** *one of Saturn's moons, bears the scar of a devastating collision with a comet or asteroid billions of years ago.*

In July 1994, 21 fragments of Comet Shoemaker-Levy 9 collided with Jupiter, one about every six hours. The impacts produced energy greater than the combined force of every nuclear weapon on Earth, providing a graphic demonstration of what could happen if such an object were to strike our planet.

The Jupiter collisions brought comet impacts to the forefront of research by planetary astronomers and geologists, and offered a revised view of the evolution of life on Earth. Instead of a smooth, gradually changing process, it now appears that life on Earth progressed erratically. Long periods of relative stability were punctuated by periods of sudden change, possibly produced by comet and asteroid impacts.

We live in a peaceful time in Solar System history. Our sky is almost unchanging, with bright comets appearing only about once a decade. However, it was not always

this way—the youthful Solar System was a violent place and Earth was so hot that all its organic materials were vaporized away. As Earth cooled down, its sky was crammed with bright comets, and major impacts took place often.

## THE ORIGIN OF LIFE

It is possible that the combinnation of impacting comets and slowly falling cometary dust provided the building blocks for life on Earth—carbon, hydrogen, oxygen, and nitrogen. Observations of Halley's Comet in 1986 determined the presence of these materials in almost identical proportions as exist on Earth. It is possible that comets also brought our water supply. Astronomers have calculated that Comet Hale-Bopp, the great comet of 1997, carried

**FRAGMENTS OF COMET** *In February 1994, the Hubble Space Telescope captured Shoemaker-Levy 9 (below) on a collision course with Jupiter. It later recorded eight of the impact sites (right).*

with it one trillion tons of water, an amount 50 percent greater than all the water in North America's Great Lakes.

## MASS EXTINCTION

If comets can bring life, they can also take it away. Much evidence supports the theory that a comet or asteroid struck the Caribbean basin 65 million years ago. In this scenario, the first result was an incredible earthquake measuring 12 on the Richter scale. Millions of tons of dust surged upward to form a gigantic cloud. The excavated material rushed out with such force that it quickly circled Earth. For more than an hour, the surface of Earth was bombarded with this debris, sending temperatures soaring

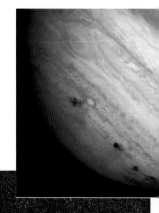

**IMPACTS, GREAT AND SMALL** *A 100,000x magnification of the crater left on a space satellite by a loose paint flake (left). Meteor Crater, Arizona (below).*

## SEARCHING FOR THE NEXT ONE

We do not know when Earth will be hit again. Statistically, a 6 mile (10 km) diameter comet—large enough to cause a mass extinction—should strike once every 100 million years. Smaller ½ mile (1 km) diameter objects could strike every 100,000 years, and would certainly still be devastating.

Astronomers have begun the process of systematically surveying the sky for impact hazards, plotting the orbit of every asteroid or comet that could inflict global damage. The numbers are daunting—there are millions of objects to be discovered and sifted through to identify the 2,000 or so most dangerous ones.

Next time a comet "stalks the sky," we should enjoy its beauty and mystery. And as we gaze, we should be mindful of the awesome potential for disaster should such an object strike Earth. But also remember: were it not for such an impact, 65 million years ago, we quite possibly would not be here today.

and setting off a global firestorm. Larger, slow-moving debris landed close to the main crater, in the Gulf of Mexico, creating miles-high tsunamis that devastated the coasts of Mexico and Florida.

Soon the whole planet was shrouded in a cloud of dust and soot, and for more than a month there was no sunlight anywhere on Earth. The huge amount of nitric oxide in the air created rain dense with sulfuric acid. As the air finally cleared, the temperatures rose again and Earth suffered a severe greenhouse effect lasting for centuries. It is widely believed that this impact, and the consequent climatic changes, caused the mass extinction that claimed 80 percent of all species, including the dinosaurs. This extinction, in turn, led to the rise of mammals.

## THE RISK OF IMPACT

Asteroids that remain in the main belt pose no threat to us. However, we know of some 200 asteroids with orbits that cross the orbit of Earth, and

**THE EXTINCTION** *of the dinosaurs may have been caused by a comet or large asteroid colliding with Earth.*

there are probably 10 times that number we do not know about, as well as an undetermined number of comets. Any of these objects could collide with our planet.

The threat of an impact is real. Earth already bears the scars of previous collisions. Meteor Crater is a 1 mile (1.6 km) diameter crater in northern Arizona, made 50,000 years ago by a piece of nickel-iron just 100 feet (30 m) wide. As recently as 1908, an object—possibly a 100,000 ton asteroid—collided with Earth, exploding at an altitude of about 6 miles (10 km) with the force of a 20 to 30 megaton nuclear bomb. The explosion occurred largely over uninhabited Siberian forest, but had the asteroid landed in a densely populated area, the impact would have been catastrophic.

# METEORS

*Comets and asteroids litter the Solar System with their debris,*

*some of which flashes across our skies*

*as "shooting stars."*

**GIANT ROCKS** *This enormous meteorite was found in Oregon, USA.*

The brief flash of light we call a meteor or, more popularly, a shooting or falling star, was once interpreted as a dead person's soul on its way to heaven, or even as a warning of death. The scientific explanation, however, is that a piece of space grit known as a meteoroid is burning up in Earth's upper atmosphere.

## SPEEDY METEORS

When meteoroids enter our atmosphere, they travel at enormous velocities. Meteoroids that overtake Earth travel at about 6 miles per second (10 km/s). Those that plunge head-on into our atmosphere, however, can reach speeds of 45 miles per second (75 km/s), more than 225 times the speed of sound.

The surface of the meteoroid, rapidly heated by friction to more than 2,000 degrees Fahrenheit (1,100° C), begins to vaporize, or ablate. The meteoroid's particles and the particles in the atmosphere collide violently and are stripped of electrons, creating the meteor's intense luminosity, which can be seen more than 90 miles (150 km) away.

Tiny meteoroids last less than a second before burning up completely. Gravel-size meteoroids produce brighter

and longer displays. Meteors as bright as the planet Venus are known as fireballs. These can cast shadows and last several seconds, leaving curling trails of glowing gas in their wake. Long-duration fireballs that explode or appear to fragment are called bolides.

The majority of meteoroids range in size from dust grains to small pebbles. Occasionally, larger meteoroids survive their fiery entry into Earth's atmosphere and reach the ground intact. When they do, they are called meteorites.

## SPORADICS, SHOWERS, AND STORMS

On any moonless night, you should be able to see four or five meteors per hour. These unexpected shooting stars—or sporadic meteors—can appear anywhere in the sky. Their rate tends to increase from local midnight until dawn, when the observer's location is carried toward Earth's leading orbital edge and placed in the path of more interplanetary particles.

You never really know how many sporadics you will see, but some meteors can be anticipated. On certain nights each year, the rate of meteors

### ROCKS FROM SPACE

When a meteoroid reaches the ground as a meteorite, it can help complete our picture of the universe. Many meteorites are pieces of matter unchanged since the earliest days of the Solar System, while others are bits of Mars or the Moon blasted into orbit by comet or asteroid impacts. Meteorites are divided into three main types according to their composition: stony meteorites; iron meteorites; and stony iron meteorites. Stony meteorites (above) are angular gray or brown stones, made up of stony minerals. They tend to be fairly smooth and often have a glassy black fusion crust, formed as the rock ablated in the atmosphere.

Iron meteorites (below) contain 90 percent iron and 10 percent nickel, making them heavy for their size. Their surfaces can be black or rusty brown and are often pitted with thumbprint-shaped depressions.

Stony iron meteorites bridge the stony and iron classifications. They have the coarsest surface, with crystals of olivine in a network of nickel-iron.

**A SHOOTING STAR,** *or meteor, provides a burst of movement in an otherwise still starry sky (left). Very bright meteors are known as fireballs (below).*

not only increases, but the meteors appear to radiate from a particular region of the sky. These predictable displays are called meteor showers.

Compared with the 4 or 5 meteors per hour you can expect to see on any dark, moonless night, the 20 to 50 meteors per hour of a meteor shower might be considered a downpour. Even a shower, though, appears like a sprinkle next to the meteor storm. A robust meteor storm can produce thousands of shooting stars per hour—sometimes 10 to 20 per second. Meteor storms, however, are rare events (see Box, p. 201).

## Meteor Streams

Meteor showers and storms occur when Earth passes through a meteoroid stream, which is a trail of dust and debris that orbits the Sun.

Most meteoroid streams are the detritus of comets, although some streams have been linked to asteroids. When a comet emerges from the deepfreeze of space and moves through the inner Solar System, its frozen surface is warmed by the Sun. As the comet's icy crust vaporizes, billions of tons of loose material spew

from its surface, and a concentrated path of debris forms in its wake. When Earth sweeps through this path, the meteoroids plunge into our atmosphere, where they burn up as meteors.

From the ground, it appears as if the meteors are streaming in from a particular region in the sky. That region is known as the radiant and its location gives the shower its name. For example, the meteor shower that appears to emanate from the constellation Gemini every 14 December is called the Geminids, which means "children of Gemini."

## How Many Meteors?

A meteor shower's activity is gauged by its zenithal hourly rate, or ZHR. This value, often quoted in the press and astronomy publications,

has sometimes been the source of misunderstanding and disappointment.

The ZHR is an ideal value. It is, by definition, the number of meteors a single observer could possibly see during a shower's peak with the radiant directly overhead in a dark, clear sky.

Most observers, however, will not see as many meteors as the ZHR suggests. Obviously, if peak activity occurs during the day or when the radiant is below the horizon, few or no meteors will be visible.

Also, the more ambient light—artificial or natural— at your location, the fewer meteors you will see. City observers will see only the brightest meteors, and a bright Moon will wash out fainter meteors, even in the country.

Luck is another factor. If you happen to be looking in the wrong direction of the sky, a fast, short-lived meteor can easily sneak past you.

**THE IMPACT** *of a large meteorite created Wolfe Creek Crater, in Western Australia. It is more than half a mile wide. Fortunately, most meteorites are small enough to fit into your pocket.*

## Preparing for a Meteor Shower

You do not need a telescope or binoculars to observe a meteor shower. The naked eye, in fact, makes the best instrument because of the large amount of sky it can take in at any one time. However, binoculars can help you locate fainter meteors. Scan in and around the radiant zone and be on the lookout for fast flashers zipping among the stars. You may even be able to see a meteor break up in flight and leave behind a "smoke" train.

Meteor watching often involves long sessions, so it is important to be comfortable. A deck chair or recliner is an easy way to keep your head inclined at the right angle. Position yourself so you face the direction of the meteor's radiant, but do not ignore other parts of the sky. Meteors in a particular shower will sometimes appear elsewhere in the sky, even on the opposite side of the sky from the radiant. This is why it can be especially worthwhile to monitor a meteor shower in a group, with each observer facing a different direction.

## When to Observe

While there is no need for expensive equipment, you do need to choose your observing time carefully. The table below gives the dates of peak activity for major showers. For the best observing times on those dates, however, you will need to check local newspapers, astronomy magazines, or the Internet. Because of Earth's rotation and orbital motion,

THE LEONID METEORS *show up as yellow streaks in this false-color photograph. The blue trails are stars.*

the exact time of a shower's peak varies slightly each year, usually by about six hours. Also, the shower's radiant may not be visible in your area at the time of maximum activity.

Say you live in Chicago and would like to observe the Perseid shower. The Perseids peak on 12 August. In 1997, the peak occurred at 18 hours Universal Time (UT), which converts to 1 pm in Chicago—obviously not a good time to look for meteors. You could observe either before dawn (some 9 to 10 hours before the peak), or in the evening (8 to 9 hours after the peak). An early morning session would probably be the better choice because the dawn side of Earth faces into its orbital path.

You also need to check when the shower's radiant will be above the horizon, and where in the sky it will be. (A simple plani-sphere or the sky charts in Chapter 8, on pages 248 to 271, can help you to determine this.) The radiant of the Perseids is in the

A BOLIDE *depicted in an 1870 painting. Bolides are rare, but you are most likely to see one during a meteor shower.*

| ANNUAL METEOR SHOWERS | | | | |
|---|---|---|---|---|
| Shower | Active period | Peak | Rate per hour | Parent comet |
| Quadrantids | 1–5 Jan | 3 Jan | 40–100 | |
| Lyrids | 16–25 Apr | 22 Apr | 15–20 | Comet Thatcher |
| Eta Aquarids | 19 Apr–28 May | 4 May | 20–50 | Comet Halley |
| Delta Aquarids | 8 July–20 Sept | 29 July | 20 | |
| Perseids | 17 July–24 Aug | 12 Aug | 50–100 | Comet Swift-Tuttle |
| Orionids | 10 Sept–26 Oct | 22 Oct | 25 | Comet Halley |
| Taurids | 15 Sept–26 Nov | 3 Nov | 12–15 | Comet Encke |
| Leonids | 14–21 Nov | 17 Nov | 10–15 | Comet Tempel-Tuttle |
| Geminids | 7–17 Dec | 14 Dec | 50–80 | Asteroid 3200 Phaeton |
| Ursids | 17–26 Dec | 22 Dec | 10–20 | Comet Tuttle |

constellation Perseus. For Chicago observers during August 1997, Perseus rose in the northeast at about 10 pm and stayed visible until dawn.

## THE METEOR LOG

Some meteor watchers go on to pursue serious meteor observing. Should you decide to do this, your observations could contribute important information about the meteoroids in a particular stream.

A log of each night's observations should note the following general details:
• the observing session's date, start and stop time, and time-out for breaks;
• the longitude and latitude of your observing site;
• sky conditions;
• the limiting naked-eye magnitude (the magnitude of the faintest star visible to your eye); and
• the direction of the area of sky you are observing, or, better yet, the right ascension and declination of the center of your viewing area.

Your observation records should include the number of meteors that you see per hour, and the magnitude and direction of each meteor. You can estimate magnitudes by comparing the meteor's brightness with that of the adjacent stars. To plot the direction a meteor takes, mark a track line on a star map showing where the meteor appeared and disappeared.

In addition, describe each meteor, noting its duration and the presence of smoke trains or fragmentation. If possible, include an estimate of the meteor's speed. The standard way to do this is to estimate how far it would have traveled had it persisted for one second (most meteors

last a fraction of a second). For example, if a meteor traveled 6 degrees in about half a second, its speed would be noted as 12 degrees per second.

This can be a daunting amount of detail to write down, especially when the meteors are frequent, so some observers use a tape recorder during the session and transcribe the information later.

If you are interested in contributing to meteor astronomy, you can approach the International Meteor Organization for additional information. Their web address is www.imo.net (see p. 99).

**COMETS** *shed tons of debris that can form meteor streams. Halley's Comet (right) has been linked to both the Orionid and the Eta Aquarid showers.*

*As I flit through you hastily,*

*soon to fall and be gone,*

*what is this chant,*

*What am I myself but one*

*of your meteors?*

Year of Meteors,
WALT WHITMAN (1819–92),
American poet

## THE LEONID METEOR STORM

One of the most famous meteor storms on record occurred on 17 November 1833, during the Leonid shower. All along North America's east coast—including Niagara Falls, New York (below)—stunned observers saw hundreds of meteors per minute. They described the meteors as falling like snowflakes or heavy rain. Estimates of the hourly rate range from 50,000 to 200,000 meteors.

Every 33 years, Earth passes through an especially dense portion of the Leonid stream, resulting in a brief storm. Following the 1833 event, thousands of meteors again rained down in 1866. The Leonid stream can be perturbed by Jupiter's gravity, which may explain the poor performances in 1899 and 1932. The storm returned in full force, however, on 16 November 1966, when an estimated 150,000 meteors were seen a few hours before dawn. Prediction techniques have improved remarkably recently, and the 1999 display was, as predicted, very modest compared with 1966.

# ARTIFICIAL SATELLITES

*When stargazing, you will occasionally notice a "star"*
*moving steadily across the sky—one of the hundreds*
*of artificial satellites orbiting our planet.*

Artificial satellites are the wallflowers of the sky. They are re-served, silent, and usually pass overhead without attracting a single glance. Yet, when they are seen, usually inadvertently, they can garner as much attention as a shooting star. At star parties, someone will often point up into the sky and shout, "Satellite!"

There are hundreds of satellites and spacecraft orbiting Earth today. They are used for everything from telecommunications, meteorology, and navigation, to Earth-resources monitoring, geophysics, and astronomy. Not all satellites are visible to the naked eye, but many are, and you can usually spot at least one on any night.

Satellites are placed into one of three types of orbit: equatorial, polar, or geostationary (see p. 38). Equatorial satellites travel from west to east. Because they are usually placed at low altitudes, they tend to look like bright stars and take just a few minutes to move across the sky. Polar-orbiting satellites are placed in high-altitude orbits and travel north to south or south to north. Because of their greater altitude, they look like faint stars and seem to move more slowly than equatorial satellites do. Geostationary satellites are placed at extremely high

**SKYLAB,** *the first US space station, could be seen orbiting Earth from 1973 until 1979, when it broke apart as it reentered the atmosphere over Australia.*

altitudes and remain fixed over the same point on Earth. They can be glimpsed only in telescopes, and even then they are no brighter than a 14th or 15th magnitude star.

When a satellite or a piece of space debris reenters Earth's atmosphere, it may flare up so brightly that it casts a shadow. Such objects are often mistaken for fireballs (see p. 198), which, in a sense, they are, except that they are artificial.

## SATELLITE SPOTTING

For people living from low to middle latitudes, satellites will appear anywhere in the sky an hour or two after sunset or before sunrise, when objects at great altitude are still bathed in sunlight. Observers living at high latitudes can expect to see satellites for longer periods.

A satellite's visibility depends on a number of factors:
• Altitude: Large satellites in low orbit can be brighter than 1st magnitude stars; those in high orbit are fainter.
• Elevation: The light from a satellite that appears low in the sky travels through more layers of air to reach the observer, and thus appears fainter than it would if it were overhead.

**SATELLITE TRACKS** *The dots among the star trails (left) are five geostationary communications satellites. A polar-orbiting satellite left a trail across this photo of the Vela supernova remnant (below).*

# Visual Satellite Observer's Home Page

**KEEP UP TO DATE** *by visiting the Satellite Observer's web site (left). It provides advance warning of events such as the space shuttle's reentry (below).*

From left to right the above images are: A Soyuz manned capsule. Satellite captured on long exposure photograph. The external tank caught just after separation during the STS shuttle mission.

A mirror of these pages with predictions for the UK, Australia, South Africa

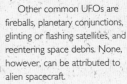

• Observer-Sun-satellite angle: A satellite is brightest when it is fully lit by the Sun; that is, when it is seen opposite the Sun, in the east after sunset or in the west at sunrise.
• Size: The larger the satellite, the easier it will be to see.
• Satellite surface: A polished surface reflects more light than a dull one does; a faceted surface makes a tumbling satellite appear to "twinkle."

## LUCKY SIGHTINGS AND PREDICTIONS

Satellite observing can be enjoyed on many levels. Most skywatchers simply enjoy keeping an eye out for satellites while they are doing other kinds of observing. For this, all you need is a lucky glance in the right direction at the right time. Satellites can even be rather intrusive, leaving their trails on long-exposure astrophotographs, or startling deep-sky observers by racing through their field of view.

Interested observers can obtain predictions for when artificial satellites will appear in their area. Favorite targets include the International Space Station, the space shuttle, and the Hubble Space Telescope. Predictions are supplied by groups such as the Belgian Working Group Satellites, or via Web sites such as the Visual Satellite Observer's Home Page (see pp. 98–99 for their Web address).

A few dedicated amateurs obtain software packages that enable them to make their own satellite predictions, and then diligently monitor satellite appearances, positions, and brightnesses. These observations can help scientists to accurately track satellite orbits and study how they are affected by changing conditions in the atmosphere and fluctuations in Earth's gravitational field.

To determine a satellite's position, you note exactly when and where the satellite crosses an imaginary line between two stars that are close to each other in the sky. This can be done by eye, but if you wish to submit your records, you will need to use a telescope or take photographs. For more information on contributing positional measurements of satellites, contact the Royal Greenwich Observatory, Satellite Laser Ranging Team.

## UFOS DO EXIST!

Do Unidentified Flying Objects exist? The answer is, yes—at least until they are identified.

A common UFO report is of an object that rapidly changes color, moves anomalously, and, sometimes, follows the observer. Disappointing as it may be to those who hope for close encounters with intelligent aliens, the two culprits usually responsible for such displays are the planet Venus and the brightest star in the night sky, Sirius.

At its most brilliant, Venus creates the illusion of being closer than it is, and may appear to trail after the observer. Sirius, being one of the brightest point sources in the sky, twinkles violently, especially when it is near the horizon. Twinkling, or scintillation, can involve rapid color changes and is caused by turbulence in the air.

Other common UFOs are fireballs, planetary conjunctions, glinting or flashing satellites, and reentering space debris. None, however, can be attributed to alien spacecraft.

*The popularity of the 1990s' television series* The X-Files *attests to the public's fascination with UFO sightings.*

# OTHER SOLAR SYSTEMS?

*The possibility of life elsewhere in the universe has always fascinated people. The discovery of extrasolar planets is a new clue in the mystery.*

We know for sure that there is one planet in our galaxy that supports life. Is Earth the only such planet, or is it part of some great web of life spanning the galaxy? Intelligent life might even exist on many worlds, distance being all that prevents us making contact.

The search for extraterrestrial intelligence began in 1960, when radio astronomer Frank Drake started Project Ozma, named after the Queen of Oz in a series of stories by L. Frank Baum. He hoped to find unusual signals from two nearby, Sun-like stars— Epsilon (ε) Eridani and Tau (τ) Ceti—but detected nothing. Attempts continue today, using the most sophisticated techniques. Project Phoenix, for example, is a targetted search which has used several large radio telescopes. The data for the SETI@home project

**USS ENTERPRISE** *from Star Trek: The Next Generation, presents a view of space travel in a future where contact with other races is routine.*

(see p. 99) comes from one of several programs where SETI observations are "piggy-backed" onto regular astronomical observations— taking data from whatever patch of sky is dictated by the main science observations.

Radio searches rely on finding signals, either deliberate or accidental. But there might be many life forms out there, ranging from

**ASSORTED ALIENS** *Blue men from Mars (above), a pointy-eared Vulcan (below left), and a wrinkled ET (below).*

## THE DRAKE EQUATION

Where would intelligent life be most likely to sprout? To provide a starting point, Frank Drake proposed the following formula to calculate N, the number of civilizations in our galaxy with the means of contacting us right now:

$$N = N_* f_p n_e f_l f_i f_c f_L$$

$N_*$ is the number of civilizations in our galaxy

$f_p$ is the fraction of stars with planets

$n_e$ is the number of planets with "suitable" environments

$f_l$ is the fraction of planets where life has arisen

$f_i$ is the fraction of times where life develops "intelligence"

$f_c$ is the fraction of intelligent civilizations capable of calling

between the stars

$f_L$ is the fraction of the star's lifetime the civilization survives.

Based on what little we know, we might guess that many of these factors are around 0.1. With around 200 billion stars ($N_*$) in our galaxy, we conclude that there are 200,000 possible contacts (N) out there.

Some scientists argue that the rise of intelligence might be very rare, so our estimate may be over-optimistic. After all, only one of Earth's uncountable species seems capable of technological communication. Maybe N really does equal 1—us!

microbes to civilizations, that do not use radio. Perhaps the only way to detect them is to range as far afield as we can to have a look. We have done this within our Solar System, but astronomers remain uncertain whether life like that on Earth could exist on any of the other planets in the Sun's domain. Tests by the Viking landers in 1976 at two locations on Mars, the most likely candidate, revealed ambiguous results that have recently excited new interest. The possibility of an ocean of water under the ice of Europa's surface offers a tantalizing new possibility.

## NEW PLANETS

The possibility of life as we know it beyond our Solar System depends on the existence of suitable planets or moons. For many years, measurements of the proper motion (see p. 208) of a few stars across the sky suggested the existence of massive planets pulling the parent stars back and forth. These observations are very difficult and most have now been discounted.

In 1995 astronomers announced the first positive results in programs to find planets orbiting Sun-like stars—other than the Sun itself. Many previous reports had proven to be false. It's hard to find a planet close to a much bigger, brighter star!

Just 11 years later, in 2006, the number of known "extra-solar" planets passed two hundred. Most of these were discovered because of their gravitational effect on their parent star. We know they are there but don't have good "pictures" of any of them.

**A NEW PLANET** *An artist's impression of a planet three times the mass of Jupiter, discovered orbiting the Sun-like star HD23079.*

## MEASURING VELOCITIES

When a planet orbits a star, it pulls the star back and forth. In fact, both the planet and the star are orbiting the "balance point" between them. The star is much more massive, so it doesn't move very far or very fast. These motions are revealed by the Doppler Effect that shifts light slightly in the spectrum of a star. Modern Doppler measurements allow astronomers to accurately measure stellar velocities as small as 3 feet per second (1 m/s).

## BIG PLANETS

The planets most easily found by this technique are massive planets close to their stars. They exert the strongest pull on the star. The planets are typically around the mass of Jupiter, but much closer to their star than Jupiter is to our Sun. In several systems there is more than one massive planet close to the star.

These planetary systems are not like our Solar System, but this is not surprising. As the Doppler technique improves and astronomers watch them for longer, it becomes possible to detect smaller or more distant planets. Planets the size of Uranus and Neptune are now being found, but Earth-like planets are currently beyond our reach.

Because of this, we don't yet know if there are planetary systems like ours—with Earth-sized planets and more distant Jupiter-sized giants. Obviously the planetary systems found so far are very different from ours. Did these giant planets form close to their stars or have they moved closer with time?

## OTHER METHODS

A few planets have been found by other means. For example, a fortunate alignment may cause a planet to periodically cross in front of its star, thus reducing the total light by a small amount. Others may cross in front of a much more distant star, causing a gravitational lensing effect (see p. 239) if the alignment is perfect.

Future satellites promise more discoveries, with the goal being to "see" the planets and study the information encoded in their light. The ultimate prize is to reach out and find an Earth-like planet in another solar system.

**DALEKS**, *from the long-running Dr Who TV series, are an example of the imagined threat from life elsewhere, a popular science fiction theme.*

Stars scribble in our eyes the frosty sagas,

The gleaming cantos of unvanquished space.

HART CRANE (1899–1932),
American poet

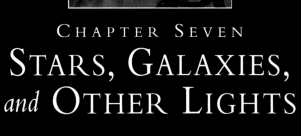

CHAPTER SEVEN

# STARS, GALAXIES,
## *and* OTHER LIGHTS

# STAR QUALITIES

*People have always sought to understand the stars. The journey begins when you go outside on a clear night and look up.*

**DIFFERENT COLORS** *of stars are captured in this photo of the Milky Way around the Southern Cross (left). A stained-glass window in France depicts the creation of the stars (far left).*

In ancient times, people saw stars as little more than ornamental lights affixed to the vault of the heavens. They arranged stars into sky pictures, or constellations, and gave names to the bright stars.

Ancient astronomers could hardly have guessed that stars generate light and energy deep in their cores by fusing hydrogen gas to form helium. Nor that these enormous spheres of hot gas are actually scattered throughout space.

## DISTANCES

The stars lie at mind-boggling distances. The nearest stars are a little more than 4 light-years away; those making up the dusty Milky Way around Auriga lie at an average distance of 2,500 light-years. These are still local in a galaxy spanning more than 100,000 light-years.

Distances to the nearest stars are obtained by measuring their annual parallax. This is the slight shift of a nearby star against the more distant background stars that occurs as Earth orbits the Sun. The nearer the star, the greater its displacement and the easier it is to measure. Distances measured by parallax have

improved dramatically in recent years with observations by the Hipparcos satellite. The *Hipparcos Catalog*, issued in 1997, gives distances of over 100,000 stars within 300 light-years of the Sun to better than 10 percent accuracy. A secondary list, the *Tycho Catalog*, has

less accurate distances to over one million stars. These are all our very near neighbors, but these distances form the first crucial steps in the cosmic distance ladder leading to remote galaxies. Future satellites, such as the Full-Sky Astrometric Explorer (FAME), promise to push the range of accurate parallax measurements out to 6,000 light-years— almost one-quarter of the way to the center of the galaxy.

### ANNIE JUMP CANNON

The monumental task of classifying the spectra of stars was undertaken in the early 1900s by Annie Jump Cannon (1863–1941) at the Harvard College Observatory, Massachusetts. Cannon (below) worked as a "computer," which meant someone who tirelessly examined hundreds of thousands of photographic plates of stellar spectra. The data was eventually compiled into the *Henry Draper Catalog*, which contains the spectral classification of 225,300 stars.

Cannon refined the patterns seen in the endless array of spectra into a single sequence. Her elegant classification system arranged the stars by color, rather than by the strength of hydrogen lines, as the first classification scheme did. Cannon's sequence went from the hottest, blue stars of type O, through the Sun-like G stars, and ended with the cool, red M stars. Almost a century later, her system is still in use.

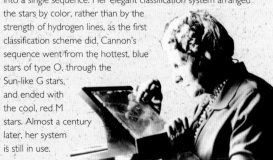

**THE PROPER-MOTION SHIFT** *of Kapteyn's Star is obvious in two photos taken in 1975 (far left) and 1990 (left) from the Palomar Sky Surveys.*

Beyond the range of accurate parallax measurement, astronomers estimate the star's distance by calculating the difference between its absolute and apparent magnitudes.

The star is first classified according to its spectrum (see p. 22 and below) and, given our knowledge of different types of stars, this reveals its true brightness—the absolute magnitude. Combined with its apparent brightness in our skies, this yields its distance. This technique works out to the nearest galaxies—which is as far as individual stars can be discerned.

## STAR MOTIONS

Compared with their distances from Earth, the space stars traverse is so minuscule that they appear immobile. Stars do move, though—the Sun, for example, has a space velocity of about 12 miles per second (20 km/s) in the direction of the star Vega.

A positional shift across the sky is called proper motion. For most stars, an appreciable difference can be discerned only after several thousand years. Hipparcos has contributed significantly here too, measuring proper motions for over two million stars.

Some stars, however, have unusually large proper motions. Kapteyn's Star, in the constellation of Pictor, moves 8.7 arcseconds per year. In 200 years, it will have moved the width of the Moon from its current position in the sky. At the star's distance of 12.8 light-years, this corresponds to a velocity of 100 miles per second (164 km/s).

However, this is only part of the story. The star is also moving along our line of sight—away from us at 150 miles per second (246 km/s). We know this by measuring its Doppler shift—a shift in the pattern of lines in its spectrum. With these velocities, Kapteyn's Star is a speedy interloper in the Sun's neighborhood.

## SPECTRAL CLASSES

Kapteyn's Star is a faint reddish star, telling us that it has a cool surface. "Reddish" and "cool" are vague descriptions to an astronomer. Using spectroscopy we can split the star's light into its component colors and reveal much more. In particular, the star's rainbow spectrum is crossed by dark spectral lines indicating the elements in the star's atmosphere.

Astronomers classify stars into seven main spectral classes according to the strength and pattern of the spectral lines. The hot, blue to white stars are type O, B, or A (in order of decreasing temperature); moderate, yellow to orange stars are F, G, or K; and cool red stars are type M.

The spectrum of Kapteyn's Star reveals a wealth of spectral lines, indicating a cool star of spectral class M. The spectrum of stars like the Sun is slightly less cluttered and has certain strong lines typical of a hotter surface and spectral class G. Sirius, the brightest star in our skies, is spectral class A, characterized by a relatively clear spectrum dominated by a few lines indicative of hydrogen gas. Other spectral classes are also used beyond the basic seven, and recently yet more have been suggested to encompass very cool stars, cooler than M-type stars, uncovered by recent infrared surveys.

**SPECTRA OF STARS** *show patterns of spectral lines which vary dramatically from hot O stars, through Sun-like G stars, to cool M stars.*

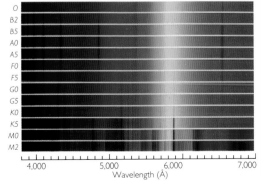

# DWARF, GIANT, *and* SUPERGIANT STARS

*With the exception of the Moon and the planets, every fixed point of light in the sky is a star—a nuclear powerhouse—and these stars range from dwarfs to supergiants.*

**SIRIUS A** *is a main sequence star some 10,000 times brighter than its white dwarf companion Sirius B, seen here as a dot 11 arcseconds away. Special methods to enhance the contrast produced this image.*

Our Sun is a star, and it appears large and bright only because we are close to it. Most stars are too far away to appear as more than points of light, even in the largest telescopes, but we still know quite a bit about them. We know, for example, that they differ in size, and that at least half of them consist of two or more stars locked in a gravitational embrace.

## WHAT IS A STAR?

Stars are large balls of hydrogen and helium, with a sprinkling of other elements, all in gaseous form. As gravity pulls the star's material inward, the pressure of its hot gas drives it outward, resulting in an equilibrium. A star's energy source lies at its core, where millions of tonnes of hydrogen are fused together every second to form helium. Although this process has been going on in our Sun for almost 5 billion years, it has used only a few percent of its hydrogen supplies. It is in the prime of its life—its main sequence phase.

## TYPES OF STARS

**Main Sequence Stars** Early this century, two astronomers tried to make sense of the wondrous variety of stars. Ejnar Hertzsprung, from Holland, and Henry Norris Russell, from the United States, came up with the Hertzsprung–Russell (HR) Diagram that plots the temperature at a star's surface

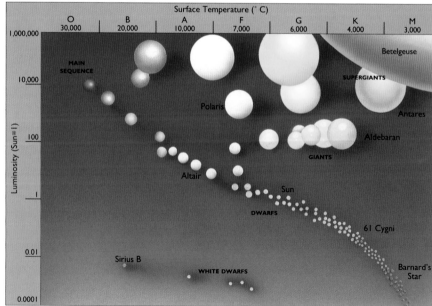

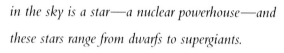

against its brightness, making allowance for its distance away from us.

Most stars, including the Sun, plotted nicely on a band across the diagram called the main sequence. These main sequence stars are often called dwarfs, although some are 20 times larger than the Sun and 20,000 times brighter.

**Red Dwarfs** At the cool, faint end of the main sequence are the red dwarfs, the most common stars of all. Smaller than the Sun, they are carefully doling out their fuel to extend their lives to tens of billions of years. If we could see all the red dwarfs, the sky would be thick with them, and the HR Diagram would be heavy with stars in its lower right corner. But red dwarfs are so faint that we can observe only the closest ones, like Proxima Centauri, the nearest star to Earth.

**White Dwarfs** Smaller than red dwarfs are the white dwarfs—typically the size of Earth but the mass of the Sun. A volume of white dwarfs the size of this book might have a mass of about 10,000 tonnes! Their position on the HR Diagram marks them as quite different from their dwarf cousins. They are "stars" whose nuclear fires have gone out.

**THE HERTZSPRUNG–RUSSELL DIAGRAM** (left) presents the basic groups of stars, together with an indication of their color and relative size. The relative numbers of stars in each portion of the diagram are not correctly presented. The majority of stars lie on the main sequence, stretching from upper left to lower right, with numbers increasing toward the faint, red end. Above the main sequence are many giants like Aldebaran and rare supergiants, like Betelguese. Tiny, faint white dwarfs lie across the bottom.

**Red Giants** After the main sequence stars, the most common stars are the red giants. They have the same surface temperature as red dwarfs, but are much larger and brighter and thus lie above the main sequence in the HR Diagram. These monsters typically have a mass similar to the Sun's, but, if they traded places with the Sun, their atmospheres would envelop the Solar System's inner planets. Most are, in fact, orange in color, but R Leporis, in Lepus, is so red that some have likened it to a drop of blood.

**Supergiants** Lying along the top of the HR Diagram are the largest stars of all—the rare supergiants. Betelgeuse, in Orion's shoulder, is close to 600 million miles (1,000 million km) across. Orion's other chief luminary is Rigel, a blue supergiant, which is one of the most luminous stars visible to the naked eye. Barely one-tenth the size of Betelgeuse, it is still almost 100 times the size of our Sun.

**SIZE COMPARISON** The Sun, a dwarf, is portrayed here as a yellow globe. The red giant star behind it (of which we see only a small portion) is 100 times larger! By contrast, a white dwarf star would be less than 1/50 the size illustrated. A neutron star might be 1,000 times smaller again, represented here by the grossly oversized black dot.

**BETELGEUSE** This is the only star, other than the Sun, for which we are currently able to produce an image of the star's surface, using techniques borrowed from radio astronomy.

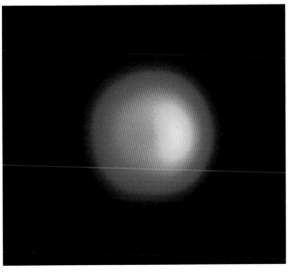

# STARS *and their* EVOLUTION

*Stars live a long time—way too long for us to observe a single star's life cycle. Instead, we build a picture of the life of a star by studying stars at different periods in their lives.*

Look at the Milky Way on a dark night and you will notice bands of darkness scattered along it—vast clouds of gas and dust that only become visible when they block out the light from more distant stars. Spread throughout the galaxy, these giant clouds provide the raw material for new stars.

## STAR NURSERIES

If you look through a pair of binoculars at Orion's sword, the middle star will appear misty. A telescope reveals a cloud of gas brightly lit by a group of bright blue stars—the latest in a series to be born from the gas permeating the sky in Orion. Their formation

was perhaps triggered by a star from an earlier generation exploding as a supernova within the last few million years.

Our Sun, like all other stars, is presumed to have been born in a similar association of clouds and stars, but its sibling stars have, over billions of years, dispersed in space.

## ALPHA CENTAURI

After birth, stars live the majority of their lives as well-behaved main sequence dwarfs, like the three stars making up the nearby Alpha ($\alpha$) Centauri system. The brightest star is much like the Sun and can expect to spend around 10 billion years on the main sequence, slowly growing warmer and brighter as it consumes the hydrogen fuel in its core. Its cooler, orange companion will do the same for perhaps 20 billion years, while the red dwarf Proxima Centauri may last for at least 60 billion years.

## A STAR'S LATER LIFE

What happens when a main sequence star like our Sun runs low on hydrogen to fuse into helium? As the fuel situation becomes critical, the star becomes brighter, larger, and redder and begins to fuse helium into carbon. No longer a main sequence dwarf, the star is now a red giant.

The bright star Capella, in Auriga, has two components

**THE LIFE OF THE SUN** *Prompted perhaps by a nearby supernova explosion, a gas cloud collapses to form the Sun and the planets. After a long life on the main sequence, the Sun will pass through a red giant phase before gently ejecting its outer layers, ending its days as a faint white dwarf.*

about three times more massive than the Sun, both evolving into red giants. Several other bright stars in the sky have already become red giants, like Arcturus in Boötes and Aldebaran in Taurus.

## MIRA THE WONDERFUL

In the constellation of Cetus, there is an old red giant called Mira that has roughly the same mass as our Sun. In the span of about a year this star becomes bright enough to be seen easily with the naked eye and then fades until you cannot see it without using a telescope. Several billion years from now, perhaps our Sun will regularly expand and contract in the same way Mira does.

*Stars are like animals in the wild. We may see the young but never the actual birth, which is a veiled and secret event.*

HEINZ R. PAGELS (1939–88),
American physicist

SUPERNOVA 1987A *photographed at the height of its outburst in early 1987. Overlaid in black are images made before the explosion, showing the outburst centered on a previously catalogued 12th magnitude blue supergiant known as Sanduleak -69°202. Pre-explosion images of supernova stars are extremely rare.*

# CELESTIAL GRAVEYARDS

*A star's life depends largely on how massive it is,*

*as does its ultimate fate as a white dwarf, neutron star, or black hole.*

A star like the Sun will live as a red giant for about a billion years. Then, as it exhausts its nuclear fuel, it will cast off its outer layers to form a planetary nebula (see p. 229) surrounding a blazing hot core. This core will slowly fade away over millennia as a cooling white dwarf. Something like this is the fate of most stars; however, very massive stars end their lives far more spectacularly in a supernova explosion.

## SUPERNOVAS

In fact, two types of explosion can create supernovas. The first occurs in a binary system when a white dwarf siphons off more matter from a companion star than it can support. The dwarf implodes and then rebounds violently from a rigid core in a massive explosion.

The second type of explosion arises when a massive star runs out of fuel. By the end of its life, a star with at least eight times the mass of the Sun has produced iron in its core via nuclear-fusion processes. Since nuclear reactions using iron consume energy instead of producing it, the star's internal furnace shuts down. Unable to keep the core hot enough, the gas in the core can no longer support the weighty outer layers, and the star caves in on itself. The material crushes in on the core and then rebounds, producing the explosion.

Both types of explosion destroy the old star. Most of the star's outer material is ejected, creating an expanding shell of gas called a supernova remnant.

**A PLANETARY NEBULA**, *such as the Dumbbell Nebula (M27) in the constellation of Vulpecula, represents the final moment of glory for most stars. The central core fades to obscurity as a white dwarf.*

The Crab Nebula (M1), in the constellation of Taurus, is all that is left of a mighty explosion that many people witnessed in 1054.

## NEUTRON STARS AND PULSARS

After an explosion, all that remains is the collapsed core, known as a neutron star, with a density far higher than even that of a white dwarf.

A white dwarf may pack a mass equal to the Sun's into a volume the size of Earth. In a neutron star, as much as two or three times that mass occupies a sphere just 12 miles (20 km) across. It is a star in name only as it is composed largely of neutrons and is not

**SUPERNOVA 1987A** *and the Tarantula Nebula (NGC 2070), as seen in 1984 (above), and in 1987 (right).*

**A BLACK HOLE** (right) is believed to emit energy from a surrounding accretion disk and ejected jets, as portrayed in this artist's impression.

**A PULSAR** in the Crab Nebula (M1) in Taurus is seen here in X-rays as it pulses on and off 30 times every second.

powered by nuclear reactions. How could we even hope to see something that small?

Apart from being small, yet massive, the other important properties of a neutron star are its intense magnetic field and rapid rotation. Both are consequences of taking part of a large normal star and shrinking it to a tiny neutron star. These features are what make them visible over interstellar distances.

Neutron stars emit beams of radio waves from the area around their magnetic poles. As the star rotates, the beams sweep past Earth, making it appear like light from a cosmic lighthouse. The resultant blinking has led astronomers to call these pulsars (from "pulsating stars"). The fastest pulsars rotate at almost one thousand times per second while others spin only once every few seconds. Some, like the pulsar in the Crab Nebula supernova remnant, also pulse in visible light and even X-rays. The Crab pulsar is young, less than 1,000 years old, and spins 30 times a second!

## BLACK HOLES

There is yet another fate that can befall a massive star—it can become a black hole.

The blue supergiant HDE 226868, in Cygnus, attracted astronomers' attention when they discovered that its position coincided with a powerful X-ray source, Cygnus X-1, that flickered once every few thousandths of a second. It also appeared to orbit an unseen companion every 5.6 days. The mass of this object was between 8 and 16 times the mass of the Sun—too massive to be stable as a neutron star. Instead, it would collapse indefinitely until it finally disappeared, leaving only a source of gravity so strong that even light could not escape from it. Most astronomers conclude that that is what had happened to the object alongside HDE 226868—it had collapsed to form a black hole. Gas from its companion star, HDE 226868, is raining down on the black hole, producing the X-rays we observe.

## ROBERT EVANS

Searching for supernovas, once the domain of amateur astronomers, is becoming dominated by automated patrol telescopes. However, the Reverend Bob Evans continues to search for them from his home in rural Australia—his 40 discoveries so far making him a legend in amateur ranks.

As he became familiar with galaxy after galaxy, by examining them whenever the weather was clear, he learned to spot intruding stars. In one year he made more than 15,000 galaxy checks, ending the year with a bounty of two supernovas for all his work.

Finding supernovas is clearly not easy. Most galaxies have bright starlit regions that look like exploding stars but on closer examination they turn out to be regions of gas. It takes persistence to put so much effort into a visual search, a quality that Bob Evans has in good supply.

# STELLAR COMPANIONS

*As a single star, our Sun is in a minority:*

*more than half the stars in the sky have at least one companion in space.*

Stars usually seem to occur as pairs, triples, or even clusters. Lone stars like the Sun are in a minority, although it generally takes the magnification of a telescope to reveal a star's multiple nature.

Sometimes nature tries to fool us by lining up two stars in our sky which are, in fact, widely separated in space. These optical doubles can only be distinguished from true binaries by many years of observation, which may eventually reveal their actual motion. The orbital motion of true binaries will also eventually be apparent, although this may take hundreds of years. Only 5 to 10 percent of stars are visual binaries—far enough apart to see the stars separately. A spectrograph can be used to closely examine the colors of the light received to read telltale binary signatures. A variation of this technique has recently led to the discovery of new planetary systems (see p. 204). An optical interferometer (see p. 35) can also reveal the presence of an unseen companion star.

**MASS EXCHANGE** *can occur in many close binaries. Here, a blue star leaks material slowly to the accretion disk around a black hole.*

## DOUBLE STARS

The closest example of a double star is Alpha (α) Centauri. Best seen from southern latitudes, this magnificent pair is the nearest stellar system to the Sun. A naked-eye view shows only one star—the third brightest star in the sky—but through a telescope you can easily see two. Strictly speaking, Alpha Centauri is a triple, since the system has a third member, Proxima Centauri—a small red dwarf that orbits the other two just over 2 degrees away.

Mizar, the middle star of the Big Dipper's handle, is one of the best-known double stars. Close by is a fainter star named Alcor that can be seen on a clear night. These two stars do not form a true binary system—in our line of sight, Alcor appears to be 12 arcminutes from Mizar. However, if you point a telescope at Mizar itself, you can see that it is two stars, forming a true double (with a separation of 14 arcseconds).

## MULTIPLE STARS

Castor is the second brightest star in the constellation of Gemini. Using a small telescope on a night with steady star images, you might be able to "split" the star into its two components (separation 3.9 arcseconds). You may also be able to see a third, fainter member farther afield. By analyzing the light from these

**OPTICAL VERSUS PHYSICAL DOUBLES** *A chance optical alignment is far different to a pair of stars physically in orbit about each other.*

**ALCOR AND MIZAR** *(in the lower left corner) are perhaps the most famous pair of stars for northern observers. They are separated by over one-third of the Moon's diameter.*

**RED GIANT LOSING MASS** *to a white dwarf companion (below). The gain in mass may cause the white dwarf to erupt periodically.*

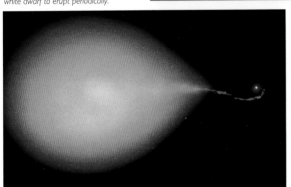

stars as seen through a spectroscope, we know that all three stars are double, so the system consists of six members locked in a gravitational embrace.

The bright star Alpha (α) Capricorni is a double so wide (separation over 6 arcminutes) you may be able to "split" it without any optical aid. It is a chance alignment, but a telescope makes all the difference since both the major stars are themselves true doubles (separations 46 and 7 arcseconds), and the fainter of one of these is yet another double.

## LIVING TOGETHER

The Sun–like stars of the Alpha (α) Centauri system orbit each other every 80 years. If placed in our Solar System with one star in place of the Sun, the other would be out near Neptune. The evolution of stars like this proceeds as if they were alone (see p. 212), but some stars are so close, orbiting in hours or days, that mass exchange can occur. The more massive of the pair evolves fastest, in time

becoming a red giant more than 100 times its original size. But, before it can reach full size, its companion may begin to steal its gaseous envelope. The loss

of mass by one star will slow its evolution but accelerate its companion's life. Eventually one star will complete its red giant phase and become a white dwarf, or perhaps explode as a supernova to leave a neutron star or even a black hole. When the second star expands, its gas will flow into the intense gravitational field of its tiny neighbor. This may cause the gas to pile up in an accretion disk around the companion, emitting intense radiation, including X-rays. Occasional outbursts from some of these objects lead them to be labeled as cataclysmic variables.

## A SAMPLER OF EASY-TO-OBSERVE DOUBLE STARS

This list comprises some of the prettiest doubles in the sky, in addition to those already mentioned. Each is plotted on the Constellation charts in Chapter 8.

Beta (β) Cephei: Both white; magnitudes 3.3 and 7.9 (separation 13 arcseconds).

Gamma (γ) Andromedae: Through a moderate-sized telescope this can be seen as a beautiful pair, the brighter member orange, the fainter bluish (separation 10 arcseconds).

Theta (θ) Orionis: The famous trapezium in the center of the Great Nebula in Orion. A small telescope will reveal four bright stars; a larger telescope will show two fainter ones as well.

Gamma (γ) Leonis: A beautiful pair that revolve about each other about once every four centuries (current separation 5 arcseconds).

Gamma (γ) Virginis: This pair revolves around each other in about 180 years. As the two stars circle each other the distance between them, from our point of view, appears to be shrinking (current separation 3 arcseconds).

Epsilon (ε) Lyrae: The famous "double double" in Lyra. Not far from bright Vega, each of these stars is itself a double (each with separation 2.5 arcseconds).

Alpha (α) Crucis: The brightest star in the Southern Cross (separation 5 arcseconds).

# OBSERVING DOUBLE STARS

*You can enjoy the different colors of a double, meet the challenge*

*of splitting a close pair, or precisely measure separations.*

You do not need a large telescope to observe double stars. A 3 inch (75 mm) refractor can produce excellent results, since its narrow field renders high-contrast, pinpoint star images. A 6 inch (150 mm) scope, though, is better for resolving close doubles.

You can observe double stars from the suburbs and even during a bright Moon. Because stars are point sources, ambient light generally does not overwhelm them.

## SEEING DOUBLE

Double-star observing is enjoyable on many levels. Some observers delight in aiming their telescopes at random stars and looking for companions, or they might try to "split" close pairs.

Others take a more serious bent—measuring, sketching, and photographing double stars, and then contributing their observations to a group such as the Double Star

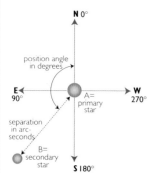

THE COLORS *and separations of three multiple-star systems (below). Double blue giants (top).*

Rigel (Beta Orionis)

Gamma Delphini

Ruchbah (Iota Cassiopeiae)

Section of the Webb Society. Such organizations particularly need help monitoring long-period binaries; with separations of at least 0.5 arcseconds, these are the binary type most suited to small telescopes. There also exists a great opportunity for amateurs with CCD cameras to make a significant contribution, as thousands of wide faint pairs have been unobserved for years.

## SPLITTING STARS

A double's separation is the apparent distance of the companion, or secondary star, from the brighter star, the primary. Albireo (Beta Cygni), a beautiful double with colors described as gold and sapphire, has a separation of 34 arcseconds. This is twice as wide as Saturn's apparent disk size, so

THE POSITION ANGLE *of a secondary star is measured eastward from a line extending north of the primary star. This system has a PA of 135 degrees.*

it is an easy object in a 3 inch (75 mm) telescope. Separations of less than about 5 arcseconds can be challenging in telescopes smaller than 6 inches (150 mm), especially when the seeing conditions are poor, the stars are of unequal brightness, or the stars are faint.

Splitting tough doubles is one area of astronomy where the optical quality of your telescope is paramount. High magnification, often avoided in other areas, is often essential for double-star observers.

The separation of double stars can be measured using a filar micrometer, a complex device that attaches to the telescope's eyepiece.

## POSITION ANGLE

The position of the secondary star with respect to the primary is called the position angle (PA). The primary star is designated component A, and the secondary, component B. Other components, if any, are listed as C, D, and so forth. You can estimate the position angle by eye, although a filar micrometer allows you to

make precise measurements. Position angles help astronomers keep track of the orbital motion of the secondary over the years. When its position angle is increasing, the motion is said to be direct. When it is decreasing, it is said to be in retrograde.

## MAGNITUDE AND COLOR DIFFERENCES

Double stars of different magnitudes and small separations can be both a joy and a challenge to observe. Tau⁴ Eridani's 5.7 arcsecond separation would not usually be a problem for, say, a 6 inch (150 mm) telescope to split. But in this case, the primary's magnitude of 4 can overwhelm the magnitude 10 secondary, particularly in poor seeing. Still, it is not impossible, and seeing a bright star next to a faint one with just a hairbreadth between them can be breathtaking.

Many doubles display dramatically different colors, which add to the overall splendor of the system. The contrast between the stars sometimes makes the colors appear unusual. One of the finest doubles in the sky, Eta Cassiopeiae, has yellow and red stars 12.7 arcseconds apart.

## OTHER TYPES

Some multiple-star systems contain more than two stars. One famous example is the Trapezium, a nexus of four prominent stars at the heart of the Great Nebula in Orion.

When the orbit of the fainter member of a binary system takes it in front of the brighter member, the system fluctuates in brightness. Such a system, called an eclipsing binary, is of interest to variable-star observers (see p. 221).

Some binary stars are so close together that they cannot be visually separated. However, a spectroscope (see p. 22) will detect tiny, periodic shifts in the motion of the brighter star. These shifts are caused by the pull of the invisible companion's gravity as it orbits. The A component of the visual double star Mizar in Ursa Major is an example of a spectroscopic binary.

**THE MOTION** *of the double star Capella is shown in these images taken two weeks apart with a special imaging technique. The stars are too close to separate in a conventional telescope.*

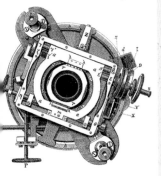

**A FILAR MICROMETER** *from the nineteenth century, used to precisely measure the separations of double stars.*

### DOUBLE DEDICATION

The first double star ever recorded—Mizar in the Big Dipper—was discovered accidentally in 1650 by the Italian astronomer Giovanni Battista Riccioli. Subsequent discoveries by other astronomers were also accidental. By 1779, enough observations had been compiled to inspire the indefatigable William Herschel (1738–1822) to begin a systematic search for these stellar curiosities. Two years later, he had discovered more than 800 new double stars, assessing each pair with a filar micrometer, a device that allowed him to precisely measure the separation and orientation of the components. Later measurements of these stars by Herschel and others revealed that some of them were, in fact, orbiting each other.

The American astronomer S. W. Burnham (1838–1921) kicked off a new age of double-star discoveries in 1873 when he published a list of 81 new pairs he had found with his 6 inch (150 mm) refractor. Over the next four decades, this tireless, sharp-eyed observer discovered an additional 1,340 double stars using telescopes of various sizes. In 1906, his observations were collected in the *General Catalogue of Double Stars.*

*The amateur astronomer S. W. Burnham.*

# OBSERVING
# VARIABLE STARS

*Not all stars shine with a steady, invariable light. Many fluctuate in brightness over periods ranging from minutes to years.*

Astronomers have listed some 30,000 variable stars of many different types and brightnesses. By observing these stars, amateurs can significantly contribute to the science of astronomy.

To begin observing variable stars, you need a telescope or binoculars, a list of variables noting their magnitude ranges and periods, and star charts showing their locations.

## HOW BRIGHT?

After locating a variable star, many observers estimate its apparent magnitude. There are various methods, but most use an interpolative method, comparing a variable's brightness with that of nearby stars.

The American Association of Variable Star Observers (AAVSO) provides finder charts for individual variables that show the magnitudes of surrounding stars. Say the variable is not as bright as comparison star A, but is brighter than comparison star B. From your AAVSO chart,

you know that star A is magnitude 8.5 and star B is 9, so you might estimate the variable's magnitude as 8.7.

The British Astronomical Society's method also uses comparison stars, but you do not need to know the magnitudes of the comparison stars until after the observing session.

Observing a star that is too bright for your telescope may result in an inaccurate magnitude estimate. A 4 inch (100 mm) telescope should not be used on stars brighter than 7th magnitude. The limit for 6 to 8 inch (150 to 200 mm) telescopes is magnitude 8.5. You can cut down the aperture of your telescope by making an aperture mask.

**MIRA,** *the brightest long-period pulsating variable, may go from magnitude 9 at minimum (above left) to magnitude 3 at maximum (above). The AAVSO logo (top).*

Some observers photograph variable stars. This provides a permanent record and allows them to estimate the magnitude away from the telescope.

For precise measurements of a variable's brightness, amateurs can use a photoelectric photometer, a device that attaches to the telescope's focus. CCD technology can also be used for photometry.

Groups such as the AAVSO or the Variable Star Section of the British Astronomical Association welcome visual, photographic, and photometric observations from amateurs, and can provide further advice about variable-star observing.

## VARIABLE TERMS

The brightest magnitude a variable star reaches is called its maximum, while the faintest magnitude is its minimum. The difference between the star's maximum and minimum is its amplitude, and the time

## A VARIABLE-STAR SAMPLER

| Name | Type | Magnitude range | Period (in days) |
|------|------|-----------------|------------------|
| Delta (δ) Cephei | Cepheid | 3.5–4.4 | 5.4 |
| Omicron (o) Ceti | Mira | 3.4–9.3 | 332.0 |
| Delta (δ) Librae | Eclipsing binary | 4.9–5.9 | 2.3 |
| Beta (β) Lyrae | Beta Lyrae type | 3.3–4.3 | 12.9 |
| Beta (β) Persei | Eclipsing binary | 2.1–3.4 | 2.9 |
| R Scuti | RV Tauri type | 4.5–8.2 | 140 |

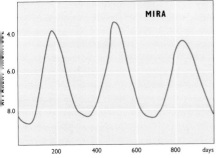

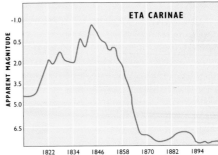

between successive maxima or minima is called the period. A light curve is a plot of the star's changing brightness over time.

Many variables have common names or Bayer Greek letters, but others are known by their official variable-star designation—for example, R Leonis or X Aquarii. The letters R to Z are used for the first variables discovered in a constellation. Additional variables are assigned RR to RZ, SS to SZ, and so on, up to ZZ. The sequence then progresses from AA to AZ, BB to BZ, and concludes with QZ. Since there are only 334 available letter combinations, the 335th variable in a constellation is labeled V335, and so on.

## CLASSES OF VARIABLES

There are three broad classes of variable stars: pulsating, eruptive, and eclipsing.
• **Pulsating** variables brighten and fade as their outer layers rhythmically contract and expand. They are further classified according to the pattern of their periods.

A well-known pulsating variable, Delta Cephei, is the prototype for the Cepheid variables. These have regular periods ranging from one to several days. The longer a

Cepheid's period, the greater its absolute magnitude. This relationship is so reliable that astronomers use Cepheids as "standard candles" to calculate the distances to nearby galaxies.

Long-period stars, such as Mira, are red giants that pulsate like Cepheids do, but their cycles are less regular and their periods are longer, ranging from 80 days to 5 years.

Other pulsating variables forming their own subtypes include RR Lyrae and RV Tauri. Semi-regular and irregular variables are also kinds of pulsating variables.
• **Cataclysmic,** or eruptive, variables exhibit sudden and large brightness outbursts. Many are novas, binary systems that erupt once in a cycle that can last thousands of years (see p. 222). Dwarf novas, such as U Geminorum, erupt more

**THE LIGHT CURVE** *of a pulsating variable such as Mira follows a regular pattern (above left). The cataclysmic star Eta Carinae has an erratic curve (above).*

frequently, with cycles lasting weeks. Supernovas are often classed as cataclysmic, but their outbursts happen just once and destroy the star.

Sometimes included in the cataclysmic category are the peculiar R Coronae Borealis stars. At irregular intervals, these rapidly fade by as much as 6 to 8 magnitudes, possibly because of an outbreak of carbon soot in their atmosphere.
• **Eclipsing** variables, such as Beta Lyrae, are binary systems in which one star eclipses the other during each orbital period. From Earth, we see this as a periodic decrease in light output, followed by a return to normal brightness.

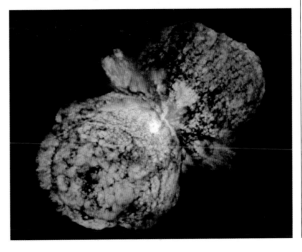

**ETA CARINAE,** *a strange cataclysmic variable, suffered a violent outburst 150 years ago that produced two polar lobes on either side of an equatorial disk.*

# NOVAS *and* SUPERNOVAS

*Novas and supernovas are produced by different processes with dramatically different finales: The violent eruption of a nova may recur, while a supernova signals complete stellar destruction.*

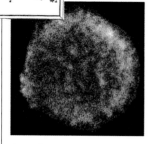

T he search for novas and supernovas is a fascinating and worthwhile amateur pursuit. These extraordinary exploding stars are of great interest to astronomers studying stellar evolution, nucleosynthesis (the creation of elements via nuclear reactions in stars), and cosmology.

## NOVAS

Novas are a class of cataclysmic variable star. They occur in binary systems made up of a white dwarf and a red giant. Hydrogen gas from the giant is drawn by gravity onto the surface of the white dwarf, and after hundreds or thousands of years, enough material accu - mulates to trigger a thermo - nuclear detonation. Over a period of a few days, the star brightens by perhaps 10 mag- nitudes and remains bright for several days before slowly

**THEN AND NOW** *In 1572, Tycho Brahe sketched a supernova in Cassiopeia (top left). All that remains of it today is Tycho's supernova remnant, seen here at X-ray wavelengths (left).*

fading again. The explosion may be repeated when suffic- ient new material builds up.

Famous novas include Nova Cygni 1975 and Nova Herculis 1934. Both of these stars reached 1st magnitude.

## SUPERNOVAS

More than 100 times more luminous than a nova, and capable of outshining all the stars in a typical galaxy, a supernova is the violent, con- vulsive explosion of a massive star. Naked-eye supernovas are very rare events—SN 1987A, in the Large Magellanic Cloud, occurred almost 400 years after the previous one!

## SUPERNOVA 1987A

On 24 February 1987, Ian Shelton, a Canadian astron- omer at the University of Toronto's telescope in Las

Campanas, Chile, had just begun looking for novas and variable stars. His targets were the two galaxies closest to the Milky Way—the Large and Small Magellanic Clouds.

Shelton was only two days into his program when he noticed an extra star near the Tarantula Nebula in the Large Magellanic Cloud. It was so bright that he realized it must be a supernova—the explosion of a dying star. Since the last bright supernova had been seen in 1604, Shelton could hardly believe what he saw. Astronomers monitored Supernova 1987A over the next three months as it became brighter and brighter until—at magnitude 3—it was almost as luminous as all the other stars in the Large Magellanic Cloud put together! The supernova dropped below naked-eye visibility within a year. It can still be seen faintly, but the Hubble Space Telescope reveals a remarkable view. Several rings surround the explosion site. The innermost

**NOVA CYGNI 1992** *in Cygnus erupted on 19 February 1992. A Hubble image taken 467 days later (left) shows a ring of gas around the star. Seven months later (right), the ring had expanded.*

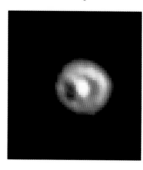

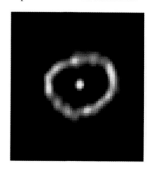

**TWO RINGS** *(right) appeared around Supernova 1987A a year after the explosion. These "light echoes" are sheets of dust lit up by the supernova. Nearby stars look black in this processed image.*

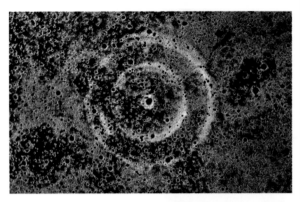

represents material gently blown off the star long before the supernova exploded. Now the debris from the explosion is reaching the ring, causing it to brighten. At the same time, radio emission from the remnant is picking up again. Astronomers are waiting to see what more it has in store.

## SEARCHING FOR EXPLOSIONS

Before a nova or supernova can be studied it must be found. The search for "transients" has become important in astronomy and automated telescopes now find many supernovas and novas. Visual hunters still have some successes. Searches should be methodical and thorough, with overlapping sweeps so that no area is missed. Dark, clear skies are especially important.

Nova hunters can use an instrument as basic as a pair of 7 x 50 binoculars or a wide-field telescope. Nearly all novas occur in a 20 degree strip centered on the galactic equator, which runs along the band of the Milky Way, so this is the best place to search. The Magellanic Clouds have also yielded a few nova discoveries, but novas in other galaxies are generally too faint to see.

Star charts showing all stars down to at least magnitude 8 are a must. Over time, you will come to know the star patterns within your chosen viewing field. If you see a star that "should not be there," it may be a nova, although you might be looking at a known asteroid or variable star at its peak. Check against a good star catalog or sky-charting program. At their maxima, novas look distinctly reddish or yellow.

Binoculars are of little use if your goal is to find supernovas. A number of supernovas can be seen each year in other galaxies, but since they usually peak below magnitude 14, you will need at least an 8 inch (200 mm) telescope.

To conduct a simple search, you survey the sky, galaxy by galaxy, looking for "new" stars

**SUPERNOVA 1987A** *began as one of many faint stars in the Large Magellanic Cloud (above left), but briefly became 10,000 times brighter in 1987 (left).*

**THE CRAB NEBULA** *in Taurus is a remnant of a supernova seen in 1054.*

with the help of photographs and charts. A faint star in or near a galaxy that does not appear on the chart could be a supernova. Accurate charts showing the magnitudes of the field stars in and around the galaxy can be obtained from the AAVSO or *The Supernova Search Charts and Handbook.*

If you think you have seen a nova or supernova, note its position and estimate its magnitude (see p. 220). Always double-check your observation, and have it confirmed by another competent observer, an observing organization, or an observatory. Give them the date and time of your observation, the instrument used, the position and brightness of the object, and the observing conditions. If your observation is confirmed, you can report it to the International Astronomical Union's Central Bureau for Astronomical Telegrams in Massachusetts.

# DEEP-SKY OBSERVING

*Locating and observing deep-sky objects*
*exposes you to the most remarkable and*
*exotic treasures the universe has to offer.*

When we look past the brightest stars in the sky, some of the specks of light we see are star clusters, nebulas, and galaxies. Astronomers have cataloged thousands of these deep-sky objects, many of which can be seen using binoculars and small to medium telescopes.

## OBSERVING TIPS

No matter what instrument you use, dark, clear skies are essential for deep-sky astronomy. Some galaxies are only slightly brighter than the normal background skyglow, so you need all the contrast you can muster from your instrument and your eyes. Before you begin observing, give your eyes 15 or 20 minutes to become dark-adapted. Then use a red-filtered flashlight to study star charts without ruining your night vision.

If you live in or around the city, you can effectively boost the contrast by using light pollution reduction filters on your telescope (see p. 84). These are especially useful for nebulas, many of which are normally invisible from cities.

You can also gain a little contrast with the technique of averted vision. By looking slightly to

one side of a deep-sky object, you avoid a "low sensitivity" patch on your retina.

For especially low-contrast objects, use averted vision and the "jiggle" method. Simply give the telescope tube a light tap when you think a nebula is in the field. The eye can detect contrast differences more easily if the image moves slightly.

Another hint is to keep both eyes open when looking through the eyepiece. Squinting shut the unused eye causes strain that can affect the active eye's vision. If you are distracted by the passive eye's close-up view of the telescope tube, simply cover the eye

with your hand or an eyepatch. To completely eliminate visual distractions, you can drape your head in a dark hood. Your breath may fog the eyepiece in cold weather, so keep the back end of the hood open.

Finally, concrete, asphalt, and other hard surfaces absorb heat during the day and slowly reradiate it at night, creating local air turbulence. Always try to observe from a grassy area.

## HOW BRIGHT?

Point sources such as stars have apparent magnitudes that indicate their visual brightness. The light of an extended object such as a cluster, nebula,

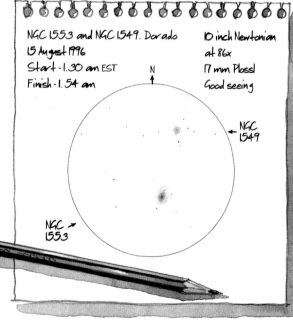

NGC 1553 and NGC 1549. Dorado
15 August 1996
Start - 1.30 am EST
Finish - 1.54 am

10 inch Newtonian
at 86x
17 mm Plossl
Good seeing

N

← NGC 1549

NGC → 1553

**A LARGE TELESCOPE** shows
*the spiral arms of NGC 2997 (top). Even*
*with a small or medium scope, you can*
*make satisfying galaxy sketches (right).*

**THE VEIL NEBULA** *(left) can hardly be seen through an unfiltered scope, but a light pollution reduction filter reveals its structure. A globular cluster (below).*

or galaxy, however, is spread out over its entire area, so you also need to consider the size of the object, indicated by its apparent diameter. The Pinwheel Galaxy (M33) in Triangulum has an apparent magnitude of 5.5, which would suggest a fairly bright object, but it has a large apparent diameter of 62 arc-minutes, twice that of the Full Moon. This means the galaxy has low "surface brightness" and is surprisingly hard to see.

Objects that have large di-ameters and faint magnitudes—as many galaxies do—require low magnification and atten-tion to contrast distinctions, while objects with small diam-eters and bright magnitudes—such as planetary nebulas—demand greater magnification and accurate positioning of the telescope's field of view.

Another factor that will affect the brightness of a deep-sky object is its altitude. If a galaxy, nebula, or star cluster is low over the horizon, its light must traverse more layers of our atmosphere than it would if it were overhead. Your best views will be when the object is high in the sky.

## DEEP-SKY CATALOGS

All of the brighter star clusters, nebulas, and galaxies are assigned numbers in deep-sky catalogs (see p. 97). Be aware that many objects have two designations: the Great Nebula in Orion, for example, is known as M42 and also as NGC 1976. Catalog numbers can help avoid the confusion that arises with common names. There are three objects sometimes known as the Pinwheel Galaxy. If you wish to specify the one in Ursa Major, you can call it M101 (or NGC 5457), and astrono-mers around the world will know which one you mean.

## RECORDING METHODS

Deep-sky observers are often so inspired by what they see that they wish to record it in sketches or photographs.

To sketch a deep-sky object, draw a circle representing the field of view. With a soft pen-cil, mark bright field stars and roughly sketch the object while at the eyepiece. You can refine the sketch later. Note details such as the date, the equipment used, and the seeing conditions.

To photograph most deep-sky objects, you need to attach a camera, usually now a CCD camera (see p. 94), to your telescope.

### THE WEBB SOCIETY

One of the most respected deep-sky observing organizations is the Webb Society. Founded in 1967, the society is named after the Reverend Thomas William Webb (1807–85), an eminent British amateur astronomer and author of *Celestial Objects for Common Telescopes*, first published in 1859. The book was a landmark work for amateurs wanting to explore the universe through their own telescopes. Webb (below) encouraged them to venture into deep space, writing that "in leaving our Sun and his attendants in the background, we are only approaching more amazing regions, and fresh scenes will open upon us of inexpressible and awful grandeur."

In the same spirit, the Webb Society today encourages the serious study of deep-sky objects. Its worldwide membership includes both professional and amateur astronomers. Through its various sections—Double Stars, Nebulas and Clusters, Galaxies, and the Southern Sky—the society provides a forum where amateurs can present their observations. Results of the society's work are published in their periodicals.

# STAR CLUSTERS

*Stars are grouped not just in pairs or triplets,*

*but also in clusters ranging from tens to*

*hundreds of thousands of members.*

**NGC 3293** *in Carina lies 7,600 light-years away, and is made up of several dozen relatively bright stars.*

A t the lower end of the clustering scale lie the open or galactic clusters. They are so called because they lie relatively close to us in the disk of the galaxy. They generally have only tens or hundreds of members, so we can see their individual stars clearly. These stars are younger than our Sun, some being among the youngest stars we can see.

The Pleiades (also known as the Seven Sisters or M45), in the constellation of Taurus, is a famous open cluster. You can see six or seven stars with the naked eye, but through a telescope you can see many fainter stars. Not far from the Pleiades is Aldebaran, the bloodshot eye of Taurus, the Bull; and the V-shaped grouping of stars near Aldebaran is the Hyades cluster, one of the closest of the open clusters.

It is often difficult to separate foreground or background stars from true members of a cluster. If, like the Hyades, the cluster is near enough to us, careful study allows the cluster stars to be picked out because they are moving with a common speed and direction across the sky. Our Sun is actually moving through an association of stars sometimes called the Ursa Major moving cluster, which is comprised of most of the Big Dipper stars and some others scattered across our sky, as revealed by their common motion.

**THE PLEIADES** *(M45) is probably the best-known open cluster, plainly visible to the naked eye in Taurus. The surrounding dust is only revealed clearly in photo-graphs such as this.*

## GLOBULAR CLUSTERS

Scattered throughout the sky lie more than 150 globular star clusters. These gigantic aggregations of stars are up to 13 billion years old—as old as the Milky Way itself. Through a small telescope they look like small fuzzy balls, but larger instruments (with 8 inch [200 mm] apertures or more) resolve the balls into untold thousands of stars.

Globular clusters have been studied almost since the invention of the telescope: Abraham Ihle found the large cluster in Sagittarius we now call M22 as early as 1665. In 1786 William Herschel noted that these clusters were mottled, "which denotes their being resolvable into stars."

Most of the globular clusters we see are of our galaxy, but not in it—they are in its halo, or outskirts. They contain very old stars and formed before the disk of our galaxy took shape. As a result they have a noticeably different composition to younger stars like the Sun or stars in open clusters. All stars are mostly hydrogen and helium, but the younger stars have significant traces of heavier elements such as calcium or iron. Stars in globular clusters are ancient, having formed long before the heavier elements were produced inside later generations of stars. The great age of the globular clusters also means that short-lived, massive blue stars that we see in open clusters

*. . . the perceptible Universe*

*exists as a cluster of clusters,*

*irregularly disposed.*

*Eureka,*
EDGAR ALLAN POE (1809–49),
American writer

**GLOBULAR CLUSTER 47 TUCANAE**
*(NGC 104) is one of the glories of the southern sky and shows a pronounced central core of stars.*

have long since died in the globular clusters. Their light is dominated by smaller yellowish stars like the Sun.

What would it be like to live inside a globular cluster? The sky would be awash with hundreds of stars as bright as Vega, and when night came the sky would be a twilight one. Perhaps the best part of the view from many of the globulars, though, would be seeing the glorious Milky Way galaxy spread across the sky.

| SELECTED OPEN CLUSTERS | | | | |
|---|---|---|---|---|
| Name | Constellation | Magnitude (total) | Approx. size (arcmin.) | Approx. distance (light-yrs) |
| M44 (Beehive) | Cancer | 3.1 | 95 | 600 |
| NGC 4755 (Jewel Box) | Crux | 4.2 | 10 | 6,400 |
| M39 | Cygnus | 4.6 | 30 | 1,060 |
| M35 | Gemini | 5.5 | 30 | 3,000 |
| h (The Double Cluster) | Perseus | 4.4 | 35 | 6,800 |
| Chi (χ) | | 4.7 | 35 | 7,600 |
| M6 (Butterfly) | Scorpius | 4.6 | 26 | 1,600 |
| M7 | Scorpius | 3.3 | 50 | 1,000 |
| M11 (Wild Duck) | Scutum | 5.8 | 14 | 6,100 |
| M45 (Pleiades) | Taurus | 1.2 | 110 | 430 |
| (Hyades) | Taurus | 0.5 | 330 | 150 |

| SELECTED GLOBULAR CLUSTERS | | | | |
|---|---|---|---|---|
| Name | Constellation | Magnitude (total) | Approx. size (arcmin.) | Approx. distance (light-yrs) |
| M3 | Canes Venatici | 6.0 | 18 | 34,000 |
| NGC 5139 (Omega (ω) Centauri) | Centaurus | 3.7 | 36 | 17,000 |
| M13 | Hercúles | 5.9 | 17 | 25,000 |
| M92 | Hercules | 6.5 | 11 | 27,000 |
| M15 | Pegasus | 6.4 | 12 | 34,000 |
| M22 | Sagittarius | 6.0 | 18 | 10,000 |
| M5 | Serpens | 5.8 | 17 | 25,000 |
| NGC 104 (47 Tucanae) | Tucana | 4.5 | 31 | 15,000 |

# CLOUDS AMID *the* STARS: *the* NEBULAS

*Of all the sights in the sky, the delicate clouds of gas and dust known as nebulas—the birthplaces and graveyards of the stars— are among the most stunning.*

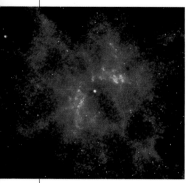

The Latin word for cloud, nebula, is the term used for the cosmic gas and dust that lie among the stars. Most nebulas are lit up by light from stars within them. The biggest nebulas do not shine at all— they can only be seen when they block the light of more distant stars.

## BRIGHT NEBULAS

Bright nebulas are glowing clouds of dust and gas that would be invisible if they were not associated with stars.

A bright nebula can be simply a reflection nebula, where starlight reflects off the nebula's dust, like a lighthouse beam illuminating a fog bank. When a nebula is near a very

**THE ROSETTE NEBULA** (NGC 2237–9) in Monoceros is a large nebula surrounding the open cluster NGC 2244. It is spectacular in photographs but quite faint to the eye at the telescope.

**PLANETARY NEBULAS** *are not always ring shaped. This Hubble Space Telescope image shows the faint eleventh magnitude NGC 2440 and its 16th magnitude central star.*

hot star, however, the intense ultraviolet radiation from that star excites, or ionizes, atoms within the gas, causing the cloud itself to emit light and become visible as an emission nebula. Emission nebulas are apparent in star-forming regions—good examples are the Great Nebula in Orion (M42) and the Lagoon Nebula (M8). Emission/reflection nebulas are a mixture of the two nebula types.

In photographs, reflection nebulas are a cool blue color and emission nebulas glow red, but our eyes are not sensitive enough to detect these colors through a telescope.

## DARK NEBULAS

"There's a hole in the sky!" More than two centuries ago, William Herschel, discoverer of Uranus and one of the greatest observational astronomers, discovered a new kind of object. Herschel thought that these objects might be "old

**A DARK NEBULA** (below) dimming and blocking light from the distant curtain of stars of the Milky Way in one dark patch, with more dark dust scattered elsewhere in the image.

regions that had sustained greater ravages of time" than the surrounding stars. His son, John Herschel, thought that they looked like gateways to something beyond. Deep in the southern sky, right beside the Southern Cross, he studied a dark region that looks so black we now call it the Coal Sack.

The Herschels had stumbled on dark nebulas—gas and dust with no nearby stars to light them, hiding the stars behind. They are usually seen against either the background of the Milky Way's stars or the brightly lit gas of a nebula.

## PLANETARY NEBULAS

Many, but not all nebulas are the birth-places of stars. After a star like our Sun evolves into a red giant, it enters a brief phase in which it blows off its outer layers. Eventually these layers become visible as a thin shell of gas around it. Nineteenth-century astronomers noticed that some were the shape and color of the planets Uranus and Neptune, so they called them planetary nebulas. We now know that these nebulas have nothing to do with planets. The Ring Nebula (M57) and the Dumbbell

**THE TRIFID NEBULA** (M20) in Sagittarius features red emission and blue reflection regions, plus the dark lanes which give the object its name.

Nebula (M27) are the most famous examples.

## SUPERNOVA REMNANTS

A considerably more massive star than the Sun dies much more violently—in a supernova explosion—and the gas ejected from the star sweeps up other gas from the interstellar medium to form a supernova remnant. The most famous of these is the Crab Nebula, in the constellation of Taurus, which looks like an oval glow in smaller telescopes. It exploded almost 1,000 years ago, and photos taken over a period of a few years show that it is still expanding.

**THE CONE NEBULA** in Monoceros is a spectacular dark nebula silhouetted against bright emissions but, like some other famous nebulas, it is difficult to see with small telescopes.

229

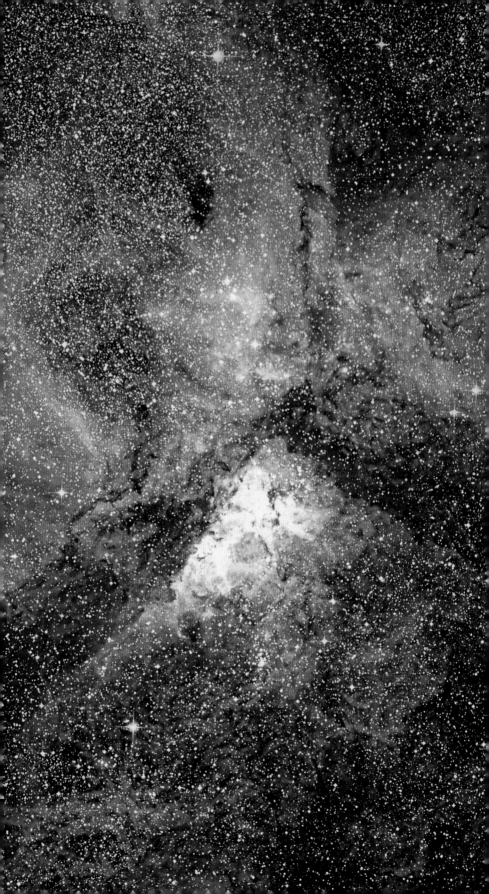

## SOME NEBULAS TO LOOK FOR

| Name | Constellation | Type | Mag. (total) | Approx. size (arcmin.) | Approx. dist. (lt-yrs) |
| --- | --- | --- | --- | --- | --- |
| NGC 3372 (Eta (η) Carinae) | Carina | Emission | 6 | 120 × 120 | 10,000 |
| NGC 2070 (Tarantula) | Dorado | Emission | 5 | 40 × 25 | 180,000 |
| M57 (Ring) | Lyra | Planetary | 8.8 | 2 × 1 | 2,300 |
| NGC 2237 (Rosette) | Monoceros | Emission | 6 | 80 × 60 | 5,500 |
| M42 (Great Nebula) | Orion | Emission | 3.5 | 85 × 60 | 1,600 |
| M8 (Lagoon) | Sagittarius | Emission | 5 | 80 × 35 | 5,200 |
| M17 (Omega) | Sagittarius | Emission | 6.9 | 45 × 35 | 5,000 |
| M20 (Trifid) | Sagittarius | Emission, reflection, and dark | 7 | 29 × 27 | 5,200 |
| M1 (Crab) | Taurus | Supernova remnant | 8.5 | 8 × 6 | 6,300 |
| M27 (Dumbbell) | Vulpecula | Planetary | 7.3 | 8 × 4 | 1,250 |

## BETWEEN THE STARS

The bright and dark nebulas are only the most obvious parts of the gas and dust pervading the space between the stars. The Great Nebula in Orion, for example, is only a small portion of the Orion Molecular Cloud, a cold dark cloud with a central density of perhaps one million particles per cubic centimeter— still billions of times less than the air around you on Earth. By contrast, the thin interstellar medium between the stars averages only one particle per cubic centimeter!

**THE VELA SUPERNOVA REMNANT** *(above) is what remains of an explosion that occurred 10,000 years ago. It is likely that the supernova that produced these luminous shards of gas was bright enough to have been seen in the daytime.*

**THE ETA CARINAE NEBULA** *(left) (NGC 3372) is one of the glories of the southern sky. It is an enormous star-forming region 20 times the size of the Orion Nebula, and appears about 2 degrees across in our sky.*

**BOK GLOBULES** *(right) are dense clumps of dust, seen against background emission, which may eventually collapse to form stars.*

# OBSERVING CLUSTERS
## *and* NEBULAS

*Our galaxy is a stellar crucible of coalescing clouds*
*of dust and gas—the building-blocks of stars.*

The raw material of stars can be found in nebulas—tenuous clouds of dust and gas. Stars that condense from the same nebula may become bound by each other's gravity into an open cluster. For a few hundred million years, they move together through the Milky Way like a school of fish. Eventually, tidal forces and interactions with other interstellar clouds disperse the stars.

### OPEN OBSERVING
Open star clusters come in many shapes and sizes. A rich cluster may comprise several thousand stars, while a sparse one may contain only a dozen or so. Some clusters, such as the Pleiades (M45) in Taurus, are large and bright enough to see with the naked eye. Most, however, require a telescope.

Large open clusters, such as the Beehive (M44) in Cancer, are best viewed using the low magnification of a wide-angle eyepiece. Small, compact clusters, such as the Jewel Box (NGC 4755) in Crux, can stand much higher magnifications. By increasing the magnification, you gain contrast, but you should avoid increasing it so much that the cluster becomes large and sparse. Aim for a balance between image scale, contrast, and detail.

A star cluster's overall visual presence is determined by its concentration—how compact or loose it is—as well as the distribution and individual magnitudes of its stars. Compact clusters made up of faint stars near the resolution limit of your telescope will appear nebulous, while large rich clusters can be hard to distinguish from background stars.

### CITIES OF STARS
In contrast to open clusters, globular clusters are tight, spherical aggregations of tens of thousands to hundreds of thousands of stars. Globular clusters are actually distributed in a spherical halo around the Milky Way, but from our perspective in the disk of the galaxy, they seem concentrated around the galactic center, in the constellations of Sagittarius and Ophiuchus (see p. 446).

Of the 150 globular clusters that have been cataloged, only a few can be glimpsed with the naked eye. In binoculars, most appear as mere smudges of light. Through a 4 to 6 inch (100 to 150 mm) telescope, they take on more definition, looking like cottony globes with a sprinkling of outlying stars; with high magnification, you may even be able to resolve individual stars within the core. A 10 inch (250 mm) scope aimed at a dark sky will reveal a spectacular granular sphere with strands of stars radiating out from the center.

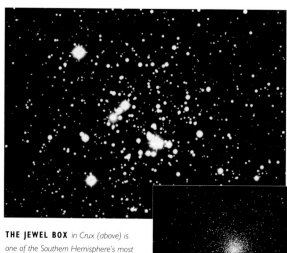

**THE JEWEL BOX** *in Crux (above) is one of the Southern Hemisphere's most beautiful open clusters, while the stunning Pleiades Cluster (top) is a favorite in both hemispheres. The globular M13 (right) is also known as the Hercules Cluster.*

**THE LAGOON NEBULA** *(left) is visible to the naked eye on a dark night, but you need a telescope to see its dust lane. The Helix Nebula (below) is the closest planetary nebula to Earth.*

A broadband light pollution reduction filter can help you locate faint globulars, as will averted vision (see p. 224).

When you find a globular, notice its shape. Some will look irregular, like a freeze-frame of bees swarming chaotically around a hive. Rich, rapidly rotating clusters can exhibit a distinct flattening. Examine the globular's disk for subtleties such as star chains, voids (apparent holes), or threads of dark dust.

One of the brightest globulars is M13 in Hercules, which is high overhead in Northern Hemisphere skies in late July. It is easy to find on one side of the "Keystone" in Hercules. Perhaps the grandest globular cluster of them all, however, is Omega Centauri (NGC 5139), best seen from the Southern Hemisphere around mid-year.

*A thousand little wisps,*

*faint nebulae,*

*Luminous fans and milky*

*streaks of fire ...*

The *Torch-Bearers*,
ALFRED NOYES (1880–1958),
English poet

It contains about a million stars packed into a region that measures about 150 light-years across. 47 Tucanae, even further to the south, rivals Omega Centauri in brilliance and is best viewed late in the year.

OBSERVING NEBULAS

Any telescope in the 4 to 6 inch (100 to 150 mm) range can show prominent nebulas in dark, clear skies, but the light grasp of larger scopes is a definite advantage when looking for less substantial nebulas. Under a good sky, a large telescope (of 10 inch [250 mm] diameter or more), will reveal the glowing gas and enable you to see tinges of red and green in some nebulas. But only long-exposure photographs really unveil the range of colors— particularly the striking red characteristic of hydrogen gas.

The easiest nebula to see is the middle star of the sword in the constellation of Orion —the Great Nebula (M42). If the sky is dark you can see it as a misty spot with the naked eye, but even from a city binoculars will clearly reveal it as a fuzzy cloud.

Reflection nebulas, as their name suggests, are lit by reflected light from nearby stars. Merope, one of the stars in the Pleiades, is surrounded by the classic blue glow of a reflection nebula. You cannot see it without a telescope, but on a clear, dark night out in the country a small telescope should reveal the faint cloudiness near Merope.

The Ring Nebula (M57) in Lyra is the best known of these nebulas. Visible in small telescopes as an out-of-focus star, in a 3 inch (75 mm) or larger one it looks like a soft ring. The Clownface or Eskimo Nebula in Gemini, another famous planetary, really does look like the face of a clown in a larger telescope, complete with nose.

Once you have located a nebula, boost your magnification for greater contrast until the image begins to degrade. Narrowband light pollution reduction filters can be very helpful, even in dark skies.

Dark nebulas are clouds of dust and gas dense enough to obscure the light from background stars. They are most visible when silhouetted against bright nebulas—two examples are the famous Horsehead Nebula (IC 434) and the dark lanes in the Lagoon Nebula (M8). Infrared telescopes often reveal the presence of new stars within these dark clouds.

# THE MILKY WAY

*Almost all the stars, star clusters, and nebulas visible in a small telescope are part of our local system of stars—the Milky Way Galaxy.*

**THE ORIGIN OF THE MILKY WAY**
*by Tintoretto (1518–94) shows Juno breastfeeding the infant Hercules. According to this legend, droplets of milk spilt upwards and became the stars of our Milky Way galaxy (below).*

For many thousands of years, Earth was thought to be at the center of everything, until Copernicus and Galileo persuaded us that the Sun ruled the heavens. By the eighteenth century, astronomers were wondering whether the Sun itself was near the center of the system of stars known as the Galaxy or the Milky Way (galaxias is Greek for "milky"). And was this the only galaxy?

In 1917 Harlow Shapley, at Mount Wilson in California,

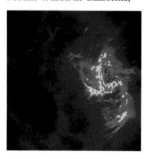

**THE GALACTIC CORE** *is seen here using radio waves. The spiral-like structure is a glowing cloud of gas some 10 light-years across, with a bright point of radio emission at its heart.*

used the distances to variable stars in the far-off globular clusters to show that the Sun is, in fact, some 50,000 light-years away from the center (now revised to about 28,000 light-years). In 1924, Edwin Hubble had shown that the Milky Way Galaxy was just one of many.

## SPIRAL STRUCTURE

Since then, scientists have been struggling to describe what the Milky Way looks like from the outside. This is a little like trying to paint a picture of the

exterior of a house when all you have seen is the inside of a room or two. But studying nearby buildings from your window may give you a clue as to how your house might look.

After Shapley established that Earth is far from the center of the Milky Way, astronomers wondered if our galaxy were pinwheel-shaped like many of our neighboring galaxies—M31 in Andromeda and M33 in Triangulum, for instance. We now know that the Milky Way is indeed a spiral galaxy that is flat, except at the center where there is a wide bulge. It contains some 200 billion suns, many of which we never see because so much dust and gas obscures our view.

The disk of the Milky Way is about 1,500 light-years thick, with spiral arms uncoiling to a distance of at least 100,000 light-years. Each star and nebula in this vast array orbits the galaxy's center more or less independently, our Sun completing an orbit in about 240 million years. Surrounding this galactic disk is a halo of older stars that stretches perhaps another 150,000 light-years.

**THE SPIRAL STRUCTURE** *of the Milky Way galaxy as it might appear from the outside—an artist's impression, of course!*

## WHAT IS AT THE CENTER?

Because the Milky Way consists of so much gas and dust, it hides its secrets well, one of the greatest being what lies at its center. For some time astronomers thought that a source of strong radio emissions called Sagittarius A was at the center. Now an even smaller source of intense radiation, known as Sagittarius A★, has been found in this complex region. It is thought be a vast black hole with a mass of 3.7 times the mass of the Sun. Material cascading into it would release the energy we can detect.

*. . . Torrent of light and river of the air, Along whose bed the glimmering stars are seen, Like gold and silver sands in some ravine. . .*

HENRY WADSWORTH LONGFELLOW (1807–82), American poet

## BART BOK AND THE BIGGER AND BETTER MILKY WAY

At the time that Harlow Shapley was redefining our place in the Milky Way, a Dutch youth named Bart Bok was longing to follow in his footsteps. By 1930, Bok had crossed the Atlantic to Harvard and was studying the Eta (η) Carinae Nebula in the constellation of Carina—one of the richest areas in the southern Milky Way.

In 1941, Bart and his wife Priscilla, also a well-known astronomer, published the first edition of their book *The Milky Way*, which became a classic. They continued their studies both of the Eta (η) Carinae region and the galaxy's spiral structure at Australia's Mount Stromlo Observatory and later at the University of Arizona. Happy to address any audience, Bok loved to talk about what he called "the bigger and better Milky Way"— a galaxy that becomes larger and more interesting the more we learn about it.

# GALAXIES

*Through a telescope, galaxies appear as
delicate whorls of soft light, some barely visible,
others distinct and surprisingly symmetrical.*

Galaxies are immense,
remote systems of gas,
dust, and billions of
stars held together by gravity.
They come in a wide variety
of shapes and sizes—from pinwheels, spheres, and footballs
to shapeless clouds.

Astronomers once referred
to some galaxies as "spiral
nebulas" because they were
spiral shaped and could not be
resolved into stars. By the mid-
1920s, however, powerful
telescopes had revealed that
the brightest of these fuzzy
objects contained stars
of their own. In 1926, the
American astronomer Edwin
Hubble sorted galaxies into
three broad categories: spirals,
barred spirals, and ellipticals.
With a few modifications,
his system is still used today.

## SPIRAL STRUCTURE
The most common galaxies
in the universe are dwarf
ellipticals and irregulars, but
by far the largest proportion
of bright galaxies—about
75 percent—are spirals. These
systems range from 15,000 to
150,000 light-years in diameter and may contain several
hundred billion stars in a flat -
tened disk. Within the disk,
spiral arms appear to emerge
from a bright central nucleus,
traced out by young, hot stars
and bright emission nebulas,
like lights around a Christmas

tree. Open star clusters and
interstellar dust and gas are
distributed throughout the disk.
The Andromeda (M31) and
the Whirlpool (M51) galaxies
are typical spirals.

In a barred spiral, the bright
stars and ionized gas of the
nucleus extend for thousands of
light-years from each side of
the center in a straight "bar."
From the end of each bar, the
arms wrap back around the
nucleus, as normal spiral arms

**COMPANIONS** *In 1850, Lord Rosse
sketched M51 and NGC 5195 (top). The
Andromeda Galaxy has two elliptical
companions: M32 and M110 (above).*

do. In obvious cases, each
bar looks something like a
scythe blade. Approximately
one-third of the spirals exhibit
a bar-like structure, and
some astronomers suspect that
all spirals contain at least a
weak bar running through the
disk. NGC 1365, in Fornax, is

### SOME GALAXIES TO LOOK FOR

| Name | Constellation | Type | Mag. (total) | Approx. size (arcmin.) | Approx. dist. (million light-yrs) |
|---|---|---|---|---|---|
| Large Magellanic Cloud | Dorado | Irregular | — | 600 | 0.18 |
| Small Magellanic Cloud | Tucana | Irregular | — | 250 | 0.21 |
| M31 | Andromeda | Spiral | 3.4 | 180 × 60 | 2.9 |
| M33 | Triangulum | Spiral | 5.5 | 60 × 40 | 3.0 |
| NGC 253 | Sculptor | Spiral | 7.1 | 25 × 7 | 10 |
| NGC 5128 | Centaurus | Elliptical | 7.0 | 18 × 15 | 11 |
| M81 | Ursa Major | Spiral | 7.9 | 18 × 10 | 12 |
| M104 | Virgo | Spiral | 8.3 | 9 × 4 | 50 |
| M87 | Virgo | Elliptical | 8.6 | 7 × 7 | 60 |

sometimes known as the Great Barred Spiral (see p. 387).

## ELLIPTICALS

Elliptical galaxies tend to be shaped like footballs or spheres. The largest, and rarest, ellipticals have a diameter of at least 100,000 light-years, and may contain more than 10,000 billion stars. Of the sky's brightest galaxies, large ellipticals—such as M84 and M86 in Virgo—make up about 20 percent. Much more common are the faint dwarf ellipticals, which contain just a few million stars and may be no more than 1,000 light-years across. The companions orbiting the Andromeda Galaxy, M32 and M110, are dwarf ellipticals.

## OTHER SHAPES AND TYPES

Galaxies with shapes that fall between the elliptical and spiral categories—such as NGC 1201 in Fornax—are called lenticular galaxies. Lenticulars have a central nucleus as spiral galaxies do, but, like the ellipticals, show little or no evidence of spiral structure. They contain mostly old stars and very little gas.

A fifth category contains those galaxies with no regular structure—irregular galaxies. Only 5 percent of bright

**A COLLISION** *with a spiral galaxy created the dark lane that crosses the elliptical NGC 5128 (Centaurus A).*

**THE BARRED SPIRAL** *nature of M95 in Leo is revealed only in telescopes of 10 inches (250 mm) or more.*

galaxies can be called irregular. These sprawling conglomerations of stars may have conspicuous bars, or even hints of spiral structure. Intense bursts of star formation dominate the appearance of some irregulars. Look for "hotspots" of bright young clusters and ionized gas. The Magellanic Clouds, easily seen with the naked eye from the Southern Hemisphere, are irregular galaxies that contain many star-forming regions.

A galaxy that appears to have suffered a severe disturbance is known as a peculiar galaxy. One member of this class is the elliptical NGC 5128. Seyfert galaxies, of which M77 in Cetus is the best known, are great spiral systems with bright centers. Named after Carl Seyfert, who first noticed them in 1942, these galaxies are part of a continuum of "active" galaxies with unusual, often violent, core activity.

A quasar or QSO (quasi-stellar object) is believed to be the highly energetic core of an active galaxy. They are thousands of times brighter than the rest of the galaxies they inhabit and thus are detectable over enormous distances.

## OBSERVING GALAXIES

Whether a galaxy is challenging or easy to observe depends on several factors, not the least of which is the galaxy's apparent magnitude. But magnitude means little or nothing unless you also consider its apparent diameter (see p. 225). Face-on galaxies have low surface brightness, which can make them difficult to locate against a bright sky background. Conversely, the light of edge-on spiral galaxies is concentrated in a narrow bar-shaped feature, so they have greater contrast.

With a 4 to 8 inch (100 to 200 mm) telescope, you can find many galaxies and see the prominent features of bright ones. An 8 inch (200 mm) will even reveal the star-like image of the brightest quasar—3C273 in Virgo. But to see a faint spiral's dusty arms or nebulas, you will need a larger telescope, not to mention very dark skies. A 12 to 16 inch (300 to 400 mm) telescope provides spectacular views of galaxies down to about magnitude 14.

Without a doubt, galaxies can be challenging to observe. But what other endeavor extends your vision across such great gulfs of space or allows you to look back millions of years into the past? This is deep-sky observing at its most profound.

# THE BIG PICTURE

*We live on the third planet out from a medium-sized sun,*

*about two-thirds of the way along one spiral arm of the Milky Way Galaxy.*

*Where does this put us in the universe as a whole?*

Around the turn of the century, Vesto M. Slipher was studying the sky from the Lowell Observatory in Flagstaff, Arizona. The director, Percival Lowell, was interested in investigating planets around other suns, and he thought that the spiral "nebulas" that were being discovered could be stars with new planetary systems forming around them.

To test this theory, Lowell asked Slipher to study the composition of the spiral nebulas using a spectrograph, which splits the light up into its component colors. Using a 24 inch (600 mm) refractor, Slipher had to spend as much as two full nights gathering enough light for a spectrum of a single nebula. The results baffled him: all the spectra showed large shifts of the features toward red light.

It was the work of Edwin Hubble, at Mount Wilson in California, that eventually resolved the mystery of these "redshifts." With the great 100 inch (2.5 m) reflector at Mount Wilson at their disposal, Edwin Hubble and Milton Humason took such clear photographs of nearby spiral nebulas that by 1924 they were able to resolve them into individual stars.

In 1929, Hubble demonstrated that redshifts were telling us that the galaxies are moving away from us at hundreds or thousands of miles per second. At these speeds, the light waves they leave behind are stretched out, noticeably reddening the light.

Hubble noted from his measurements that the fainter, and therefore, presumably, more distant galaxies, showed greater redshift. Thus, Hubble's Law states that the redshift of distant galaxies increases in proportion to their distance from us. Measuring redshift

**THE LOCAL GROUP** *is a confederation of some 40 or so galaxies, dominated by two giant spirals—the Milky Way and the Andromeda Galaxy (M31). Most members are faint dwarf ellipticals and irregulars that would be invisible from a greater distance.*

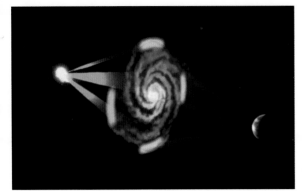

**THE EINSTEIN CROSS**, *as seen by the Hubble Space Telescope (below), is actually four images of one distant quasar produced by a galaxy (in the center) 20 times closer acting as a gravitational lens (as illustrated at left).*

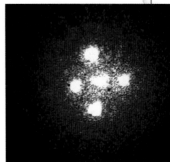

enables us to estimate distance in the universe.

## THE DISTRIBUTION OF THE GALAXIES

Shortly after Hubble proposed that the universe was expanding, he claimed that the galaxies were distributed evenly in space. To prove his point, he took a large number of pictures of small regions of the sky using the huge Mount Wilson reflector. With the exception of what he called a zone of avoidance around the Milky Way, where dust obscured the galaxies behind it, he found galaxies in roughly equal numbers everywhere.

Other cosmologists disagreed with Hubble's findings. In a survey of the Northern Hemisphere sky taking wider fields of view, Harlow Shapley and Adelaide Ames noted considerable discrepancies in the populations of galaxies throughout the sky. In some places they found large numbers and in others relatively few. Clyde Tombaugh, who discovered Pluto in 1930, confirmed Shapley and Ames's observations and went a step further by discovering, in 1937, a cluster of hundreds of galaxies in Andromeda and Perseus.

The greatest advance, however, came during the making of the Palomar Observatory Sky Survey with a new 48 inch (1.2 m) Schmidt telescope. Using its fine photographic plates, George Abell demonstrated that the galaxies are arranged unevenly into clusters and superclusters.

## THE LOCAL GROUP

The Milky Way and the Andromeda Galaxy (M31) are the largest members of a small group of about 40 galaxies called the Local Group. This cluster is part of a supercluster of galaxies whose other members can be seen in the constellations of Coma Berenices and Virgo. The Virgo Cluster contains 2,500 galaxies. Though it lies some 65 million light-years away, it covers more than 45 degrees of sky.

We now recognize other superclusters scattered across the cosmos. They appear to form enormous frothy structures in space, with vast "voids" between them. The Sloan Digital Sky Survey and the Anglo-Australian Telescope's 2dF galaxy and quasar surveys are mapping these patterns across slices of the sky to unprecedented distances. These immense structures are partaking in the general expansion of the universe. Galaxies within clusters are gravitationally bound together, but outside the clusters the expansion of space itself is inexorably pulling the clusters away from each other.

## GRAVITATIONAL LENSES

At the end of the 1970s, a pair of identical quasars were found on a Palomar Sky Survey photograph, with a faint but less distant massive galaxy lying between them. The galaxy and the quasar demonstrate part of Einstein's general theory of relativity—that gravitational sources can bend light. The gravity of the galaxy was acting as a lens, bending the light of the distant quasar so that it appeared double.

More exotic cases have since been discovered where galaxies are positioned in such a way that they cause more distant objects to break up into beauti-ful arcs and even rings, and one is so perfectly placed that the quasar far behind it is broken up into four images we call the Einstein Cross. Gravitational lenses help probe the distant universe by revealing galaxies and quasars much further away than could otherwise be seen.

**V. M. SLIPHER**
*(1875–1969)*
*is known for his observations of the planets, nebulas, and distant galaxies.*

CHAPTER EIGHT

# *The* CONSTELLATIONS

*Silently, one by one, in the infinite meadows of heaven*

*Blossomed the lovely stars, the forget-me-nots of the angels.*

*Evangeline,*
HENRY WADSWORTH LONGFELLOW (1807–82), American poet

# FINDING YOUR
# WAY AROUND

*Finding constellations can be quite a challenge for the beginner. The trick is to start by identifying some of the brighter stars, and then to "starhop."*

**ASTRONOMICAL KNOWLEDGE** *is used by navigators in this detail from a work by Cornelis de Baeilleur (1607–71).*

Finding your way around the sky can seem daunting, but it is really no harder than reading a road map, and a good deal more relaxing. When you drive along a highway, you have only a few minutes to match the road signs to the map before you risk missing an important turn. By comparison, the stars do not race by; instead they appear to move gracefully across the sky. And tomorrow night they will appear much the same.

But how do you find your way to a particular place in a sky that is so vast and so full of stars? By using a star at a time as a guide, you can "starhop" your way around.

## FINDING NORTH

For Northern Hemisphere dwellers, the sky provides a bright star close to the north celestial pole which is a good place to start. It also provides a convenient way to find it. Just find the Big Dipper (in Ursa Major), mentally draw a line joining the two stars at the end of the bowl, extend it five times, and you're at Polaris.

This is easy if the Big Dipper is in the sky, which it is from mid-northern latitudes every evening of the year. However, in fall and winter you may not be able to see it unless you have a clear northern horizon. From the southern United States, the Dipper lies below the horizon on winter evenings.

If the Dipper is low, then the W shape of Cassiopeia, on the other side of the pole, will be high. It does not point the way as clearly as the Dipper, but it will give you an idea of the direction of the Pole Star.

**CASSIOPEIA**

**CEPHEUS**

**Polaris**

**URSA MINOR**

**NORTH CELESTIAL POLE**

**URSA MAJOR**

**FINDING NORTH** *is easy for Northern Hemisphere skywatchers because the bright star Polaris lies very close to the north celestial pole.*

**FINDING SOUTH** *is made harder by the lack of a bright star near the pole, but other bright stars provide clues to the pole's position.*

## FINDING SOUTH

For Southern Hemisphere skywatchers, finding south is not quite so easy. But although the south celestial pole is not marked by a bright star, there are several well-known ways to find the pole. The easiest is to simply extend the long arm of Crux, the Southern Cross, four and a half times to reach close to the pole. The pole is very nearly marked by a star, Sigma (σ) Octantis, but this is too faint really to be useful.

## OTHER SIMPLE JOURNEYS

The Big Dipper can lead you to other stars and constellations in the sky. Joining the three stars in the handle forms a curved line or arc. If you extend the line away from the Dipper you can "arc to Arcturus," the brightest star in Boötes, the Herdsman, and then "speed to Spica," in the same direction—the bright star in Virgo. Judging these distances in the sky can be

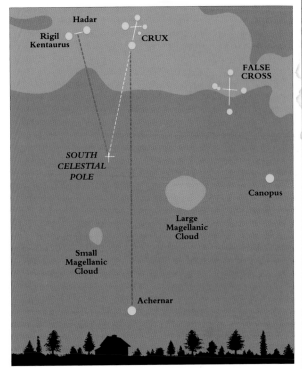

tricky to begin with, but it just takes a little practice.

You can devise your own starhopping journeys. Start at a bright star and move around the sky, one star at a time, until you reach your destination. Constellations containing a particularly bright star are marked with a special symbol in the constellation charts.

## BRIGHT CONSTELLATIONS

In finding your way around the sky, it rapidly becomes apparent that a few constellations are bright and easy to find. In the constellation charts that follow they are rated 1 on the visibility scale and serve as handy jumping-off points for your starhopping journeys across the sky.

For example, Orion is prominent in the sky from most locations in the first few months of the year. More or less opposite Orion in the sky is Scorpius, which is prominent in the northern summer (southern winter) sky. Other well-known landmarks include the Big Dipper in Ursa Major, the W shape of Cassiopeia, Leo with its sickle, the Great Square of Pegasus, and Crux, the Southern Cross.

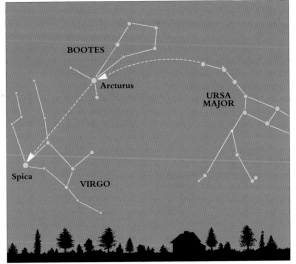

**ARC TO ARCTURUS** *and speed on to Spica is an expression well known to northern skywatchers learning their way between bright stars in the sky.*

# THE STARFINDER CHARTS

The Starfinder charts are your map to the night sky. Begin with the sky charts, and then zoom in on your destination with a constellation chart in this chapter or a starhopping map in the next.

The charts can be used anywhere in the world. Not all constellations will be visible from any given place; for example, constellations in the south polar region of the sky will always remain below the horizon for mid-northern latitude observers, and vice versa (see pp. 106–7).

On the charts, solid lines link some of the major stars of individual constellations. Today, 88 constellations are recognized, with boundaries defined by a commission of the International Astronomical Union in 1930. The lines form the configurations that are the basis of the stories associated with the constellations and the naming of them.

## THE SKY CHARTS

The bimonthly sky charts will help you find your way among the stars. By choosing a chart appropriate to the time and date of observation, you should be able to match a chart with your view of the sky.

There are twelve sky charts in all—six for Northern Hemisphere skywatchers and six for those in the Southern Hemisphere. Each is in two parts on facing pages—the left-hand chart looking north, the right-hand chart looking south—with considerable overlap in the middle. Each is designed to be held with the writing upright.

The size of the dots representing the stars indicate their relative brightness—the larger the dot, the brighter the star. A key to the symbols used appears with each chart. Most of the stars plotted are of magnitude 4.5 or brighter—bright enough to be easily seen with the naked eye in a rural setting.

## CHOOSING A SKY CHART

The constellations visible from any given part of the world appear to rotate one-sixth of the way around the sky every two months, so at the same time of night, every two months, we see an additional 60 degrees of sky in the east and lose sight of 60 degrees in the west (except at the poles).

Each chart is valid for specific times on specific days. For instance, Sky Chart 1 is valid for 12 am, 1 Jan; 11 pm, 15 Jan; 10 pm, 1 Feb; and so on, as indicated alongside the chart. Sky Chart 2 is valid four hours later on any of these dates, with the arrows showing the direction in which the sky rotates.

Each sky chart is also valid for other months, but at different times of the night (or day!). Choose a chart which is the closest match to the date and time you want, as indicated in the Sky Chart Tables. On these tables, sunset

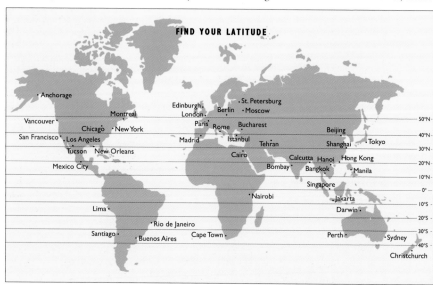

## SKY CHART TABLE: NORTHERN HEMISPHERE

| Local Time | 6pm | 7pm | 8pm | 9pm | 10pm | 11pm | 12am | 1am | 2am | 3am | 4am | 5am | 6am |
|---|---|---|---|---|---|---|---|---|---|---|---|---|---|
| DST | 7pm | 8pm | 9pm | 10pm | 11pm | 12am | 1am | 2am | 3am | 4am | 5am | 6am | 7am |
| 1 January | | | Chart 6 | | | | Chart 1 | | | | Chart 2 | | |
| 15 January | | Chart 6 | | | | Chart 1 | | | | Chart 2 | | | |
| 1 February | Chart 6 | | | | Chart 1 | | | | Chart 2 | | | | Chart 3 |
| 15 February | | | | Chart 1 | | | | Chart 2 | | | | Chart 3 | |
| 1 March | | | Chart 1 | | | | Chart 2 | | | | Chart 3 | | |
| 15 March | | Chart 1 | | | | Chart 2 | | | | Chart 3 | | | |
| 1 April | Chart 1 | | | | Chart 2 | | | | Chart 3 | | | | Chart 4 |
| 15 April | | | | Chart 2 | | | | Chart 3 | | | | Chart 4 | |
| 1 May | | | Chart 2 | | | | Chart 3 | | | | Chart 4 | | |
| 15 May | | Chart 2 | | | | Chart 3 | | | | Chart 4 | | | |
| 1 June | Chart 2 | | | | Chart 3 | | | | Chart 4 | | | | Chart 5 |
| 15 June | | | | Chart 3 | | | | Chart 4 | | | | Chart 5 | |
| 1 July | | | Chart 3 | | | | Chart 4 | | | | Chart 5 | | |
| 15 July | | Chart 3 | | | | Chart 4 | | | | Chart 5 | | | |
| 1 August | Chart 3 | | | | Chart 4 | | | | Chart 5 | | | | Chart 6 |
| 15 August | | | | Chart 4 | | | | Chart 5 | | | | Chart 6 | |
| 1 September | | | Chart 4 | | | | Chart 5 | | | | Chart 6 | | |
| 15 September | | Chart 4 | | | | Chart 5 | | | | Chart 6 | | | |
| 1 October | Chart 4 | | | | Chart 5 | | | | Chart 6 | | | | Chart 1 |
| 15 October | | | | Chart 5 | | | | Chart 6 | | | | Chart 1 | |
| 1 November | | | Chart 5 | | | | Chart 6 | | | | Chart 1 | | |
| 15 November | | Chart 5 | | | | Chart 6 | | | | Chart 1 | | | |
| 1 December | Chart 5 | | | | Chart 6 | | | | Chart 1 | | | | Chart 2 |
| 15 December | | | | Chart 6 | | | | Chart 1 | | | | Chart 2 | |

## SKY CHART TABLE: SOUTHERN HEMISPHERE

| Local Time | 6pm | 7pm | 8pm | 9pm | 10pm | 11pm | 12am | 1am | 2am | 3am | 4am | 5am | 6am |
|---|---|---|---|---|---|---|---|---|---|---|---|---|---|
| DST | 7pm | 8pm | 9pm | 10pm | 11pm | 12am | 1am | 2am | 3am | 4am | 5am | 6am | 7am |
| 1 January | | | Chart 12 | | | | Chart 7 | | | | Chart 8 | | |
| 15 January | | Chart 12 | | | | Chart 7 | | | | Chart 8 | | | |
| 1 February | Chart 12 | | | | Chart 7 | | | | Chart 8 | | | | Chart 9 |
| 15 February | | | | Chart 7 | | | | Chart 8 | | | | Chart 9 | |
| 1 March | | | Chart 7 | | | | Chart 8 | | | | Chart 9 | | |
| 15 March | | Chart 7 | | | | Chart 8 | | | | Chart 9 | | | |
| 1 April | Chart 7 | | | | Chart 8 | | | | Chart 9 | | | | Chart 10 |
| 15 April | | | | Chart 8 | | | | Chart 9 | | | | Chart 10 | |
| 1 May | | | Chart 8 | | | | Chart 9 | | | | Chart 10 | | |
| 15 May | | Chart 8 | | | | Chart 9 | | | | Chart 10 | | | |
| 1 June | Chart 8 | | | | Chart 9 | | | | Chart 10 | | | | Chart 11 |
| 15 June | | | | Chart 9 | | | | Chart 10 | | | | Chart 11 | |
| 1 July | | | Chart 9 | | | | Chart 10 | | | | Chart 11 | | |
| 15 July | | Chart 9 | | | | Chart 10 | | | | Chart 11 | | | |
| 1 August | Chart 9 | | | | Chart 10 | | | | Chart 11 | | | | Chart 12 |
| 15 August | | | | Chart 10 | | | | Chart 11 | | | | Chart 12 | |
| 1 September | | | Chart 10 | | | | Chart 11 | | | | Chart 12 | | |
| 15 September | | Chart 10 | | | | Chart 11 | | | | Chart 12 | | | |
| 1 October | Char 10 | | | | Char 11 | | | | Char 12 | | | | Chart 7 |
| 15 October | | | | Chart 11 | | | | Chart 12 | | | | Chart 7 | |
| 1 November | | | Chart 11 | | | | Chart 12 | | | | Chart 7 | | |
| 15 November | | Chart 11 | | | | Chart 12 | | | | Chart 7 | | | |
| 1 December | Chart 11 | | | | Chart 12 | | | | Chart 7 | | | | Chart 8 |
| 15 December | | | | Chart 12 | | | | Chart 7 | | | | Chart 8 | |

and sunrise are indicated by the edge of the lighter zones, but the sky is not really dark within about an hour and a half of these times. Also watch out for the difference between standard time and daylight saving (summer) time—DST in the charts and tables.

Across the bottom of each sky chart is a series of curved lines representing the horizon for observers at different latitudes. (The world map opposite will help you to determine your latitude.) Any stars lying beneath the horizon line applicable to your location will not be visible. In the upper-middle portion of each map, the zenith (the point directly overhead) for observers at each latitude is marked with a "plus" sign.

The ecliptic (the plane of the Solar System) is marked on each chart by a dotted line. The Moon and planets, if they are above the horizon, will be close to this line. A bright 'star' near the ecliptic, but not on the charts, will almost certainly be a planet. The wavy, pale-colored area that appears on the charts represents the band of the Milky Way.

## LONGITUDE OF YOUR OBSERVING SITE

Contrary to what you might think, this is not too important. Observers in Tucson and Tel Aviv (despite being in two distinctly separate longitude zones) will see much the same sky at the same *local* time— say, 9 pm at each location— because they share the same latitude zone. Small differences do arise because civil time is based on standard time zones rather than actual local time, but this effect is minor.

## STEP-BY-STEP GUIDE TO USING THE SKY CHARTS

1 First, determine the hemisphere and latitude of your location by referring to the world map on p. 244.
2 Use a Sky Chart Table on p. 245 to work out which chart is applicable for your date and time of viewing, then turn to the sky chart.
3 Decide whether you will be facing north or south, and refer to the relevant half of the chart.
4 Look at the horizon lines on the chart and work out which one is applicable to your latitude; also find the corresponding zenith point (the point directly overhead).
5 Choose two or three of the brightest stars featured (the biggest dots on the chart) and try to find them in the sky, noting where they are relative to the horizon and zenith.
6 Once you've identified one of these stars, try to trace out the constellation of which it is a part, then trace out the patterns of constellations nearby.

**Example** If you live in San Francisco and you plan to go skywatching at 10 pm (11 pm daylight saving time) on 1 April, here's what you should do.
1 Turn to the map on p. 244 to find the latitude of San Francisco, which is approximately 40° N (actually 37°48'). The horizon line and zenith relevant to you on all the Northern Hemisphere sky charts will therefore be those marked 40°N.
2 According to the Northern Hemisphere Sky Chart Table on p. 245, you should turn to Sky Chart 2 on pp. 250–1.
3 On the south-facing half of Sky Chart 2 you will see that the prominent constellation of Orion can be found low in the western sky with the brightest star Sirius near the south-western horizon.
4 If you want to find out more about Orion, turn to its constellation chart on p. 334. Together, the sky charts and constellation charts will guide you through the night skies.
5 For a deeper view of the sky near Orion, consult the starhopping key map on pp. 374–5. Starhop 7, starting on p. 402 probes the Orion area. Several other starhops cover adjacent areas of the sky.

## THE CONSTELLATION CHARTS

Once you're ready to zero in on a particular constellation, turn to the constellation charts. These are presented in alphabetical order, and with north at the top and east to the *left*— different from maps of Earth, but necessary to match our view of the sky.

*Within* the constellation, stars down to magnitude 6.5 are shown; those which fall *outside* the constellation are stars down to magnitude 5.5. All the stars marked are therefore visible to the naked eye under dark skies, but in towns or cities binoculars may be needed to pick out the fainter ones. (In general, the appropriate magnitude limits are: cities— 2 or 3, suburbs—4, distant suburbs—4.5 or 5, rural—5 to 6.5.) In addition to the naked-eye stars, the charts mark the locations of other important objects such as star clusters, nebulas, and galaxies; most of these require optical aids to be seen. Any deep-sky objects of particular interest are included, down to around magnitude 11.

A key to the symbols used on the charts appears below. Note also that many stars are named with letters of the Greek alphabet. These are shown clearly on the charts. Each chart is accompanied by descriptions of the main objects of interest to the amateur observer, with an indication of the type of instrument needed to see them. Many are within the range of binoculars or telescopes with front lens apertures of 2.4 inches (60 mm). Of course, most objects will reveal more detail when viewed with a larger telescope.

## CONSTELLATION CHART FEATURES

Each of the constellation charts feature the following inform-ation (see sample pages, above):
• a pronunciation guide and the constellation's common name

### KEY TO SYMBOLS

Magnitudes  ● –1  ● 0  ● 1  ● 2  ● 3  • 4  · 5  · 6 and under

Double stars  ●•  Variable stars  ⊘ ○  Open clusters  ◌ ○

Globular clusters  ⊕ ⊕  Planetary nebulas  ◇ ✧

Diffuse nebulas  ⌂□ □  Galaxies  ○ ○  Quasar  ⊗

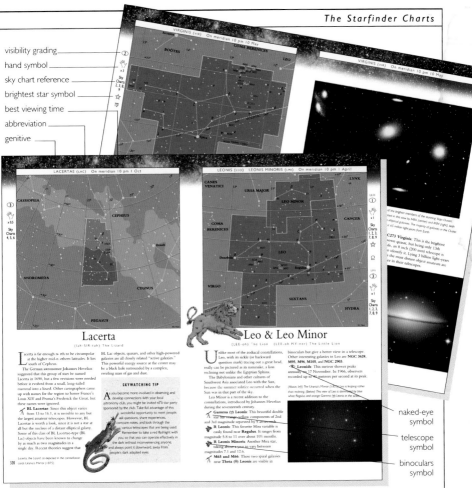

visibility grading
hand symbol
sky chart reference
brightest star symbol
best viewing time
abbreviation
genitive

naked-eye symbol
telescope symbol
binoculars symbol

• the genitive form of the constellation name, used when correctly naming objects within the constellation
• the standard three-letter abbreviation of the constellation's proper Latin name
• best viewing time, which is the approximate date the constellation is on meridian (that is, highest in the sky) at 10 pm at night—standard time, not daylight saving time
• a visibility grading, based on a scale from 1 to 4, representing the ease with which the constellation can be seen
• a hand symbol, which indicates the number of outstretched hand spans (each about 20 degrees across) that will cover the constellation east to west (left to right on the map). (See p. 69 for an explan-

ation of how to measure distances with your hand.)
• reference to specific sky chart(s), where the constellation is prominently featured
• where applicable, a symbol indicating that one of the 25 brightest stars is featured in the constellation
• picture symbols indicating whether an object is first easily viewed with the unaided eye, binoculars, or a small telescope.

## THE STARHOPPING CHARTS

More detail of the sky can be found on the starhopping charts. They cover smaller patches of sky to greater depth. Stars are marked down to magnitude 8.5—below naked-eye visibility but easily seen in binoculars. Deep-sky objects

reach down to magnitude 12.5, requiring an 8 inch (200 mm) telescope and a dark sky. Other aspects of the starhopping charts are similar to the constellation charts. See p. 370 for more clues on using them.

## A FINAL WORD

The bimonthly sky charts and the constellation charts are easy to use, but take your time—especially on your first night out—and make sure you start with the correct sky chart. When focusing on one constellation, begin with an "easy" one, like Orion or Pegasus, or one of those with a visibility grading of 1, so you can gain a bit of confidence before taking on the more elusive constellations. Good luck, and dark skies!

The northern sky is full of bright stars now. With Auriga and Gemini almost overhead, the sky seems filled with bright stars like Capella, Castor, and Pollux. The Big Dipper is standing on its handle as it rises in the northeast. Leo is comfortably high in the east, its sickle appearing at its side.

High in the west, Perseus and Andromeda are well placed, and the Pleiades are also conspicuous. Often confused with the Little Dipper because of their shape, the Pleiades in Taurus are a cluster of stars. The Little Dipper, hanging off Polaris, is far larger and harder to see.

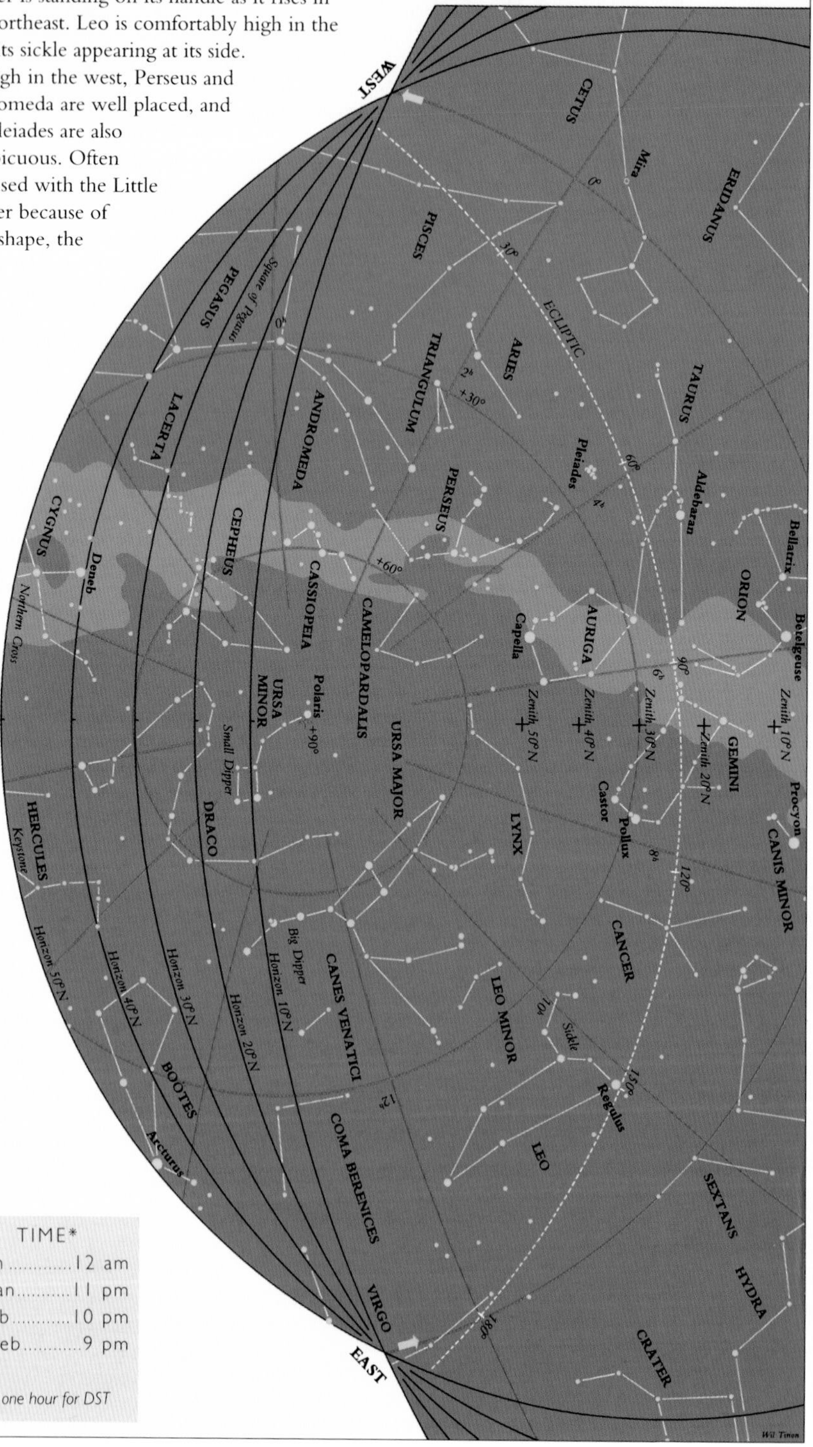

WEST
CETUS
Mira
ERIDANUS
0h
PISCES
30°
ECLIPTIC
PEGASUS
Square of Pegasus
ARIES
TRIANGULUM
2h
+30°
TAURUS
ANDROMEDA
LACERTA
Pleiades
60°
Aldebaran
PERSEUS
4h
CYGNUS
Deneb
CEPHEUS
+60°
CASSIOPEIA
Bellatrix
ORION
Northern Cross
CAMELOPARDALIS
Capella
AURIGA
Betelgeuse
NORTH
URSA MINOR
Polaris +90°
6h
90°
Zenith 50°N
Zenith 40°N
Zenith 30°N
Zenith 20°N
Zenith 10°N
GEMINI
Small Dipper
URSA MAJOR
Castor
Procyon
HERCULES
Keystone
DRACO
LYNX
Pollux
8h
120°
CANIS MINOR
CANCER
Big Dipper
Horizon 10°N
CANES VENATICI
Horizon 50°N
Horizon 40°N
Horizon 30°N
Horizon 20°N
LEO MINOR
10h
Sickle
150°
Regulus
BOOTES
12h
COMA BERENICES
LEO
SEXTANS
Arcturus
HYDRA
VIRGO
180°
EAST
CRATER
Wil Tirion

TIME*

1 Jan.............12 am
15 Jan...........11 pm
1 Feb............10 pm
15 Feb............9 pm
etc.

** Add one hour for DST*

Orion is the chief attraction of the winter sky. High in the heavens, it is characterized by its three belt stars, lined up neatly, between brilliant blue-white Rigel and red Betelgeuse. These nights, the sky has enough bright stars sprinkled liberally through it to form some interesting patterns. One of these is the "Heavenly G." Begin with Aldebaran, the bright red star in Taurus, and continue through the brightest stars of Gemini—Castor and Pollux—Procyon, in Canis Minor, and Canis Major's Sirius, the brightest star in the sky. The G's curve stops in Orion at Rigel, and then indents from Rigel to Betelgeuse.

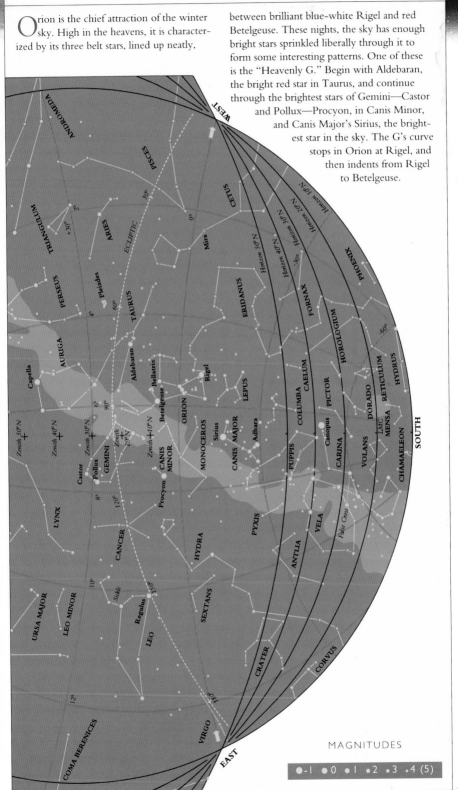

MAGNITUDES

●-1  ●0  ●1  ●2  ·3  ·4 (5)

The Big Dipper is the centerpiece of the northern sky right now, and you can use it as a key to finding other stars. The two stars at the end of the bowl point toward Polaris, and the stars in the handle "arc to Arcturus," the center star in Boötes, the Herdsman. Looking much more like a kite than a herdsman, Boötes is the primary figure in the eastern sky.

The kite's northern neighbor is Corona Borealis, the Northern Crown—a semicircle of stars. Hercules is low in the east. In the western sky lie Pollux and Castor, the bright twins of Gemini, with Capella in Auriga, the Charioteer, just to the north.

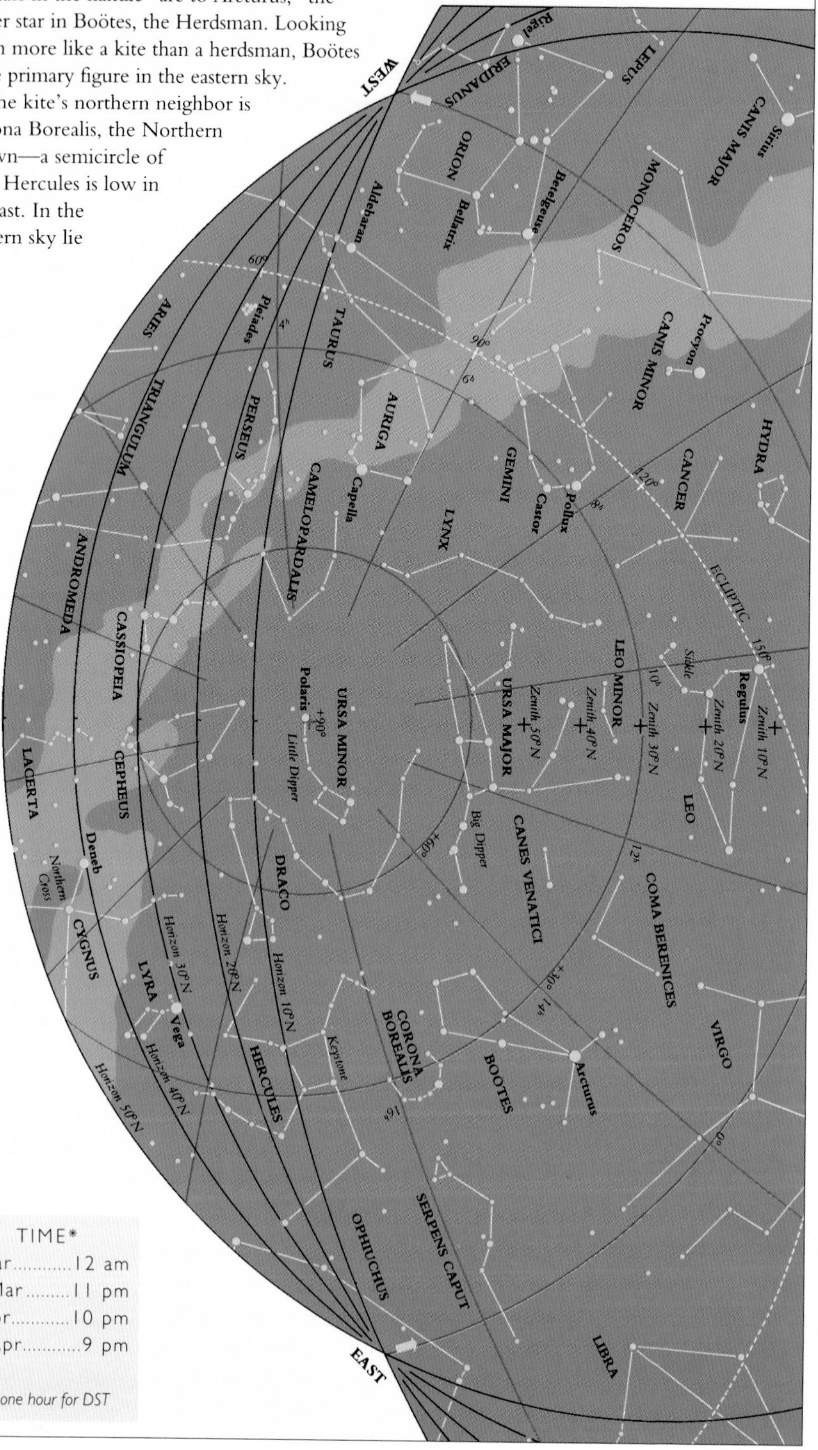

WEST

NORTH

EAST

TIME*
1 Mar............12 am
15 Mar.........11 pm
1 Apr............10 pm
15 Apr...........9 pm
etc.
* Add one hour for DST

250

Riding high in the sky in both hemispheres, Leo's sickle has a distinctive shape like a backwards question mark. To find it, fill the Dipper's bowl with water and poke some holes in the bottom. As the water spills out, it will be Leo taking a shower!

Starting to set, the bright stars of Orion still dominate the western sky. The three stars shining in a row form his belt. In the east is a widely-spaced group of four bright stars called Virgo's Diamond. Spica, Arcturus, Denebola (the star at the eastern end of Leo) and Cor Caroli (the bright star in Canes Venatici) mark the corners of this fanciful stone.

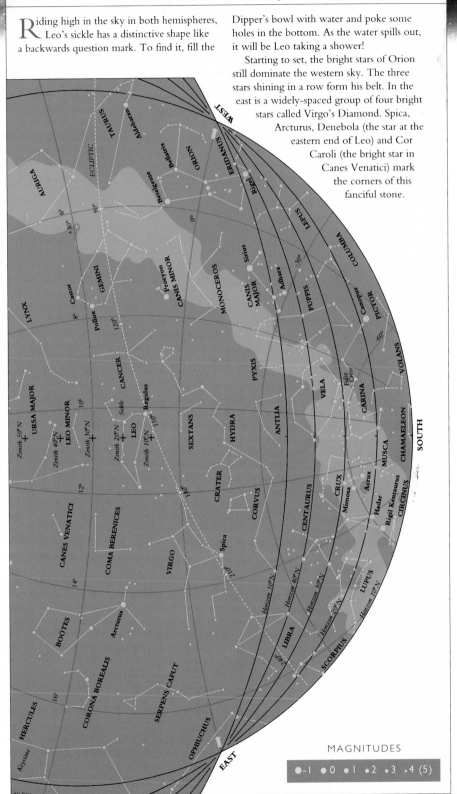

MAGNITUDES

-1  0  1  2  3  4 (5)

"Arc to Arcturus" is the chant for this chart. Although the Big Dipper is evident here, most other patterns are faint. Draco, the Dragon, takes up a good deal of room as it winds between the Dippers of Ursa Major and Ursa Minor, but its stars are difficult to see.

In the west, Leo's sickle is the salient feature. The eastern sky is much busier. In the northeast, the keystone of Hercules is easy to spot, as is the semicircle of stars that makes up Corona Borealis. The

Summer Triangle of Vega (in Lyra), Deneb (in Cygnus), and Altair (in Aquila) is rising, and the Milky Way is visible in a dark sky.

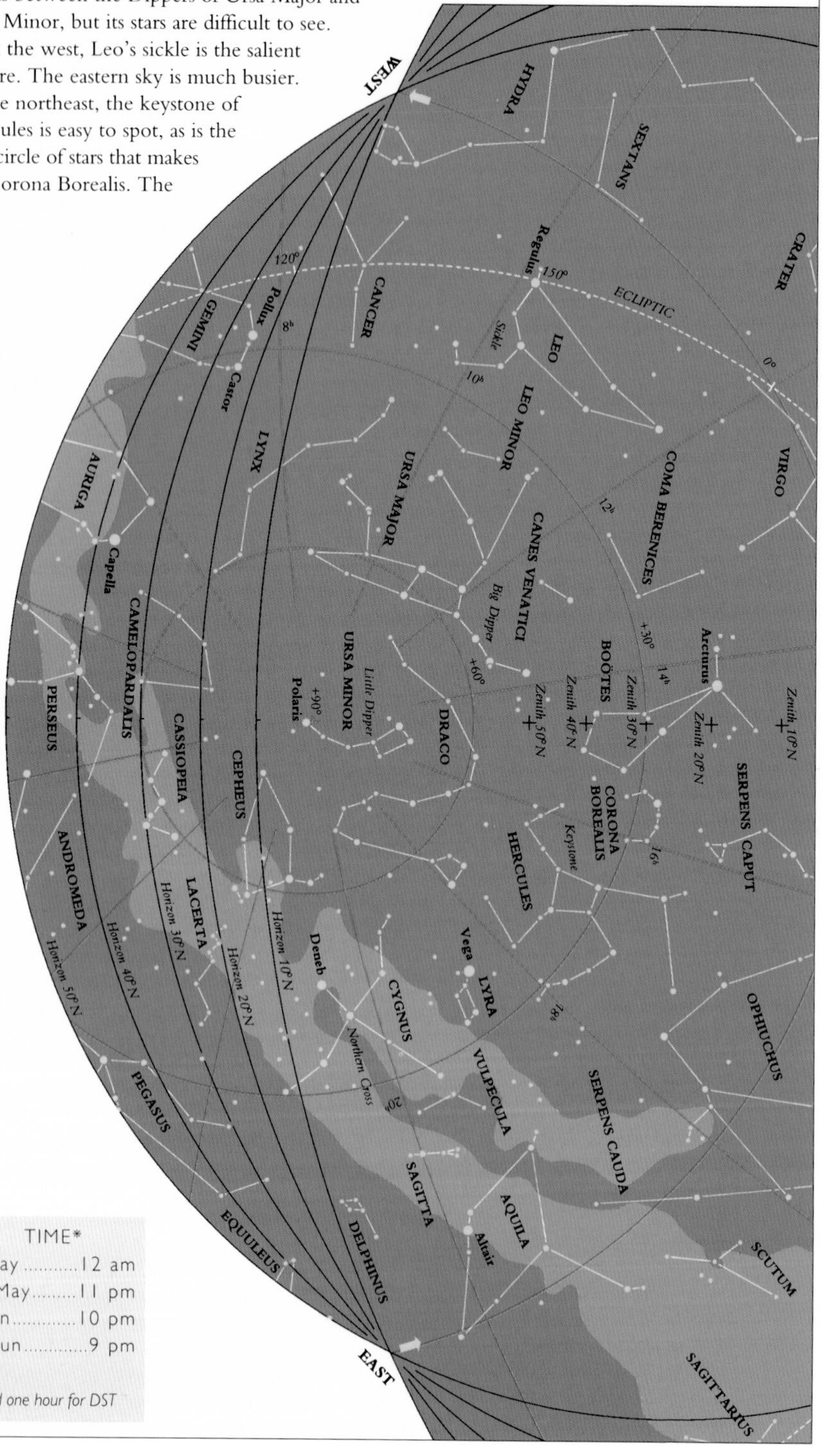

TIME*

| | |
|---|---|
| 1 May | 12 am |
| 15 May | 11 pm |
| 1 Jun | 10 pm |
| 15 Jun | 9 pm |
| etc. | |

* Add one hour for DST

252

Although the Big Dipper is off the top of this chart, we can still use its handle to "arc to Arcturus." Now we can continue the line and "speed to Spica," the brightest star in the otherwise faint constellation of Virgo. Just to the southwest is the squarish shape of Corvus, the Crow, an often forgotten but striking constellation. The sickle of Leo is setting in the west.

The rich constellations of Scorpius and Sagittarius are just rising in the southeast. You are looking straight into the center of our galaxy here, and if the sky is dark enough you should be able to see the Milky Way.

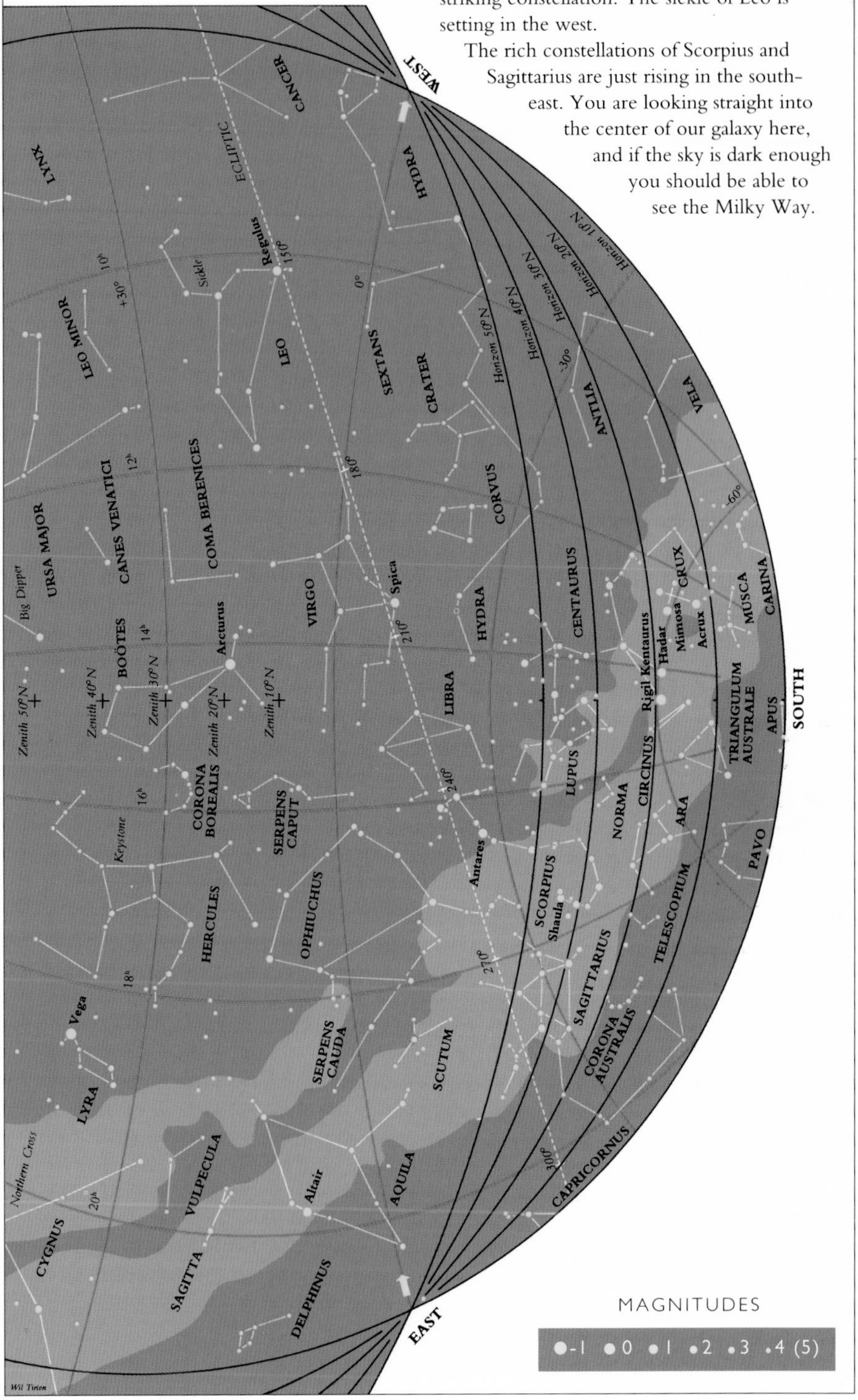

MAGNITUDES

●-1  ● 0  ● 1  ● 2  • 3  • 4 (5)

Wil Tirion

Bluish-white Vega commands the center of the northern sky, with the other members of the Summer Triangle—Deneb and Altair—off to the east. The Milky Way is high enough for you to be able to make it out, unless your sky is badly light polluted. Off to the east, the mighty Square of Pegasus makes an appearance. The four stars are easy to see since there is not much else there. Farther eastward is the circlet of Pisces.

The western sky is full of strange shapes: the keystone of

Hercules, the semicircle of Corona Borealis, and the kite of Boötes. The Big Dipper is low in the northwest.

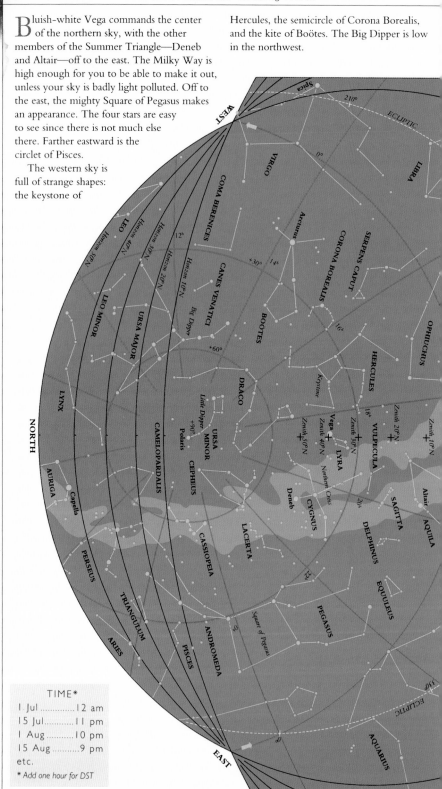

TIME*

| | |
|---|---|
| 1 Jul | 12 am |
| 15 Jul | 11 pm |
| 1 Aug | 10 pm |
| 15 Aug | 9 pm |
| etc. | |

* Add one hour for DST

254

The Summer Triangle of Vega, Deneb, and Altair command this view of the sky. Although Ophiuchus, The Serpent Bearer, takes up a good deal of room in the southwest, most of its stars are faint and the constellation is difficult to decipher. Crossing Ophiuchus, Serpens snakes along toward the semicircle of Corona Borealis, with Arcturus and the kite of Boötes nearby.

The galactic center, with its constellations of Sagittarius and Scorpius, dominates the southern sky. Libra is just to the west. The large but faint figures of Capricornus, Aquarius, and the Square of Pegasus, rule the east.

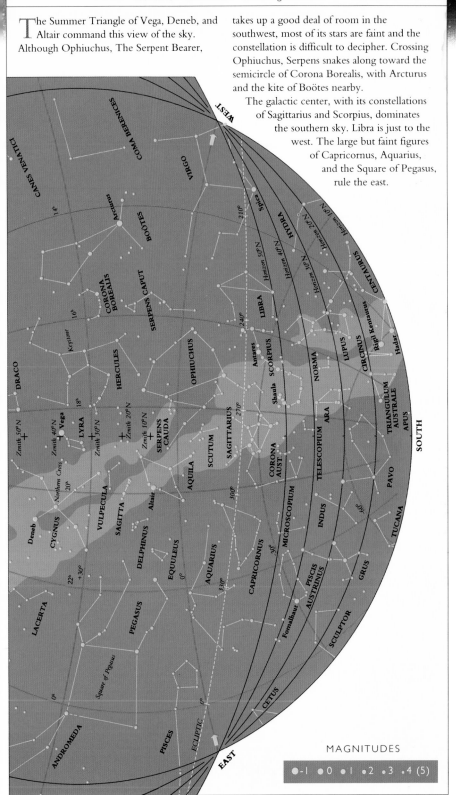

MAGNITUDES

●-1  ● 0  ● 1  ● 2  • 3  • 4 (5)

255

The misshapen W of Cassiopeia, the Queen, is now high in the northeastern sky. Her husband Cepheus, the King, looks like an upside-down house with its roof pointing roughly toward Polaris. To the east lies daughter Andromeda, still chained to a rock. Her savior, Perseus, is to the northeast, and Pegasus, his winged horse, is to Andromeda's south. There is also a giraffe (Camelopardalis) in the picture, but its stars are so faint that you can easily miss this constellation.

In ancient Mediterranean lore, whenever Capella rose in the evening, as on this chart, winter storms would not be far behind.

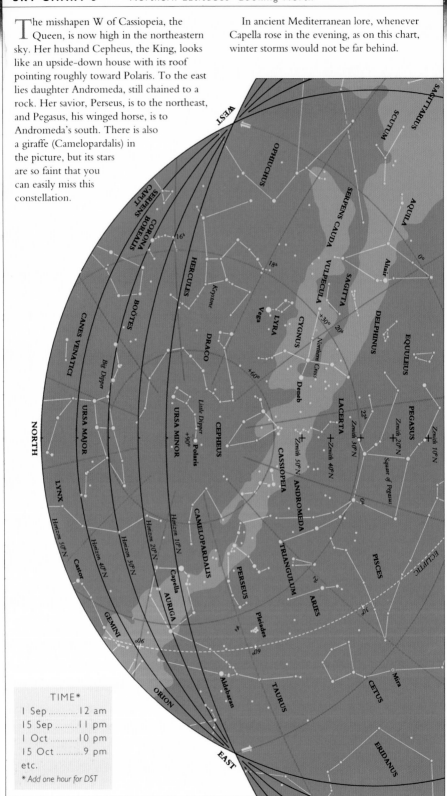

TIME*

| | |
|---|---|
| 1 Sep | 12 am |
| 15 Sep | 11 pm |
| 1 Oct | 10 pm |
| 15 Oct | 9 pm |
| etc. | |

* Add one hour for DST

With the Summer Triangle setting in the west, the Great Square of Pegasus now governs the sky. The distinctive circlet of faint stars forming the head of Pisces is immediately to the south. Far to the south of the Square of Pegasus, past Aquarius, is lonely Fomalhaut—the bright star in Piscis Austrinus.

The eastern sky, with few bright stars and few easy patterns, is dominated by the aqueous constellations of Pisces, the Fish; Cetus, the Whale; and Eridanus, the River. Aldebaran, the bloodshot red eye of Taurus, the Bull, is just rising, and the compact structure of the Pleiades lies above it in the northeast.

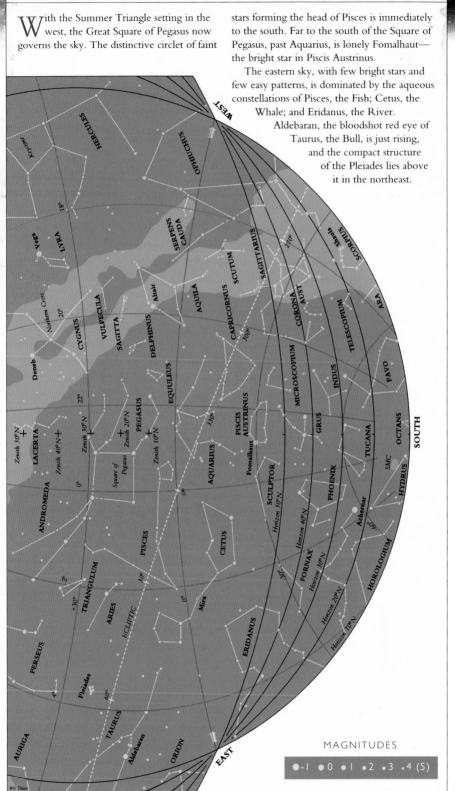

MAGNITUDES

●-1 ●0 ●I ●2 •3 .4 (5)

The eastern sky is now rich with bright stars, ranging from Capella to Castor and Pollux (the twins) and the dog stars Procyon and Sirius. If your sky is dark enough, you will be able to see the Milky Way arching overhead. It is rich in the west, as it goes through Cygnus, but thins out in the east. Far from the galactic center, this is the Milky Way's dimmest part.

Cassiopeia straddles the top of the sky, its W inverted into an M. Since the pointer stars of the Big Dipper may be too low at the moment to help find the pole, we can use Cepheus, whose pointed roof suggests the way toward Polaris.

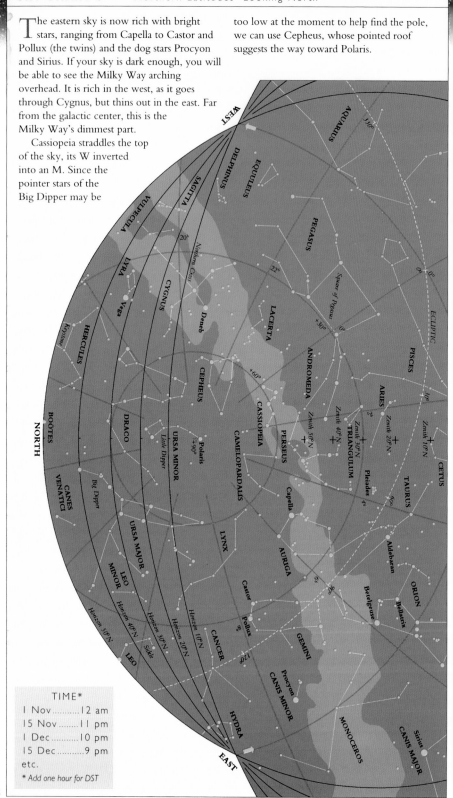

TIME*

| 1 Nov | 12 am |
| 15 Nov | 11 pm |
| 1 Dec | 10 pm |
| 15 Dec | 9 pm |
| etc. | |

* Add one hour for DST

The sky to the south seems divided into two camps. The Square of Pegasus overlooks the west. To the south is Aquarius, and Fomalhaut in Piscis Austrinus. Orion rules the eastern camp, its three belt stars pointing westward toward Aldebaran in Taurus, and eastward to Sirius.

A watery border between east and west is formed by Eridanus, the River, and Cetus, the Whale. Two small constellations continue this border. One is Aries, the Ram, and the other, consisting of three stars, is somewhat unimaginatively called Triangulum.

WEST

CYGNUS

Northern Cross

DELPHINUS

EQUULEUS

CAPRICORNUS

22ʰ

PEGASUS

Square of Pegasus

AQUARIUS

Fomalhaut

MICROSCOPIUM

INDUS

0ʰ

LACERTA

ANDROMEDA

PISCES

PISCIS AUSTRINUS

GRUS

TUCANA

TRIANGULUM

2ʰ

CETUS

SCULPTOR

PHOENIX

SMC

HYDRUS

Zenith 50°N
Zenith 40°N
Zenith 30°N
ARIES
Zenith 20°N
Zenith 10°N

Mira

Achernar

PERSEUS

4ʰ

Pleiades

6ʰ

TAURUS

ERIDANUS

FORNAX

RETICULUM

LMC

MENSA

SOUTH

Aldebaran

Bellatrix

Rigel

CAELUM

HOROLOGIUM

DORADO

PICTOR

60°

Capella

AURIGA

6ʰ

ORION

Betelgeuse

0°

LEPUS

COLUMBA

Canopus

CARINA

9ʰ

CANIS MAJOR

Adhara

PUPPIS

30°

GEMINI

Castor

Pollux

8ʰ

Procyon

MONOCEROS

Sirius

CANIS MINOR

Horizon 50°N
Horizon 40°N
Horizon 30°N
Horizon 20°N
Horizon 10°N

120°

ECLIPTIC

CANCER

HYDRA

EAST

Will Tirion

**MAGNITUDES**

● -1  ● 0  ● 1  • 2  • 3  • 4 (5)

259

Like the Dipper in the north, Orion is a useful key for pointing to other groups of stars in the sky. Draw a line from Rigel through Betelgeuse to find Gemini, the Twins. Then you can go southeast from the three belt stars to Sirius, the brightest star in both hemispheres, or northwest to Aldebaran in Taurus.

Just east of Orion is Monoceros, the Unicorn. The animal might be one of the most wonderful creatures ever created by human imagination, but this version is full of faint stars and is undistinguished.

Off to the east is the huge expanse of Hydra, the Serpent, and off to the west is Eridanus, the River, stretching almost as far.

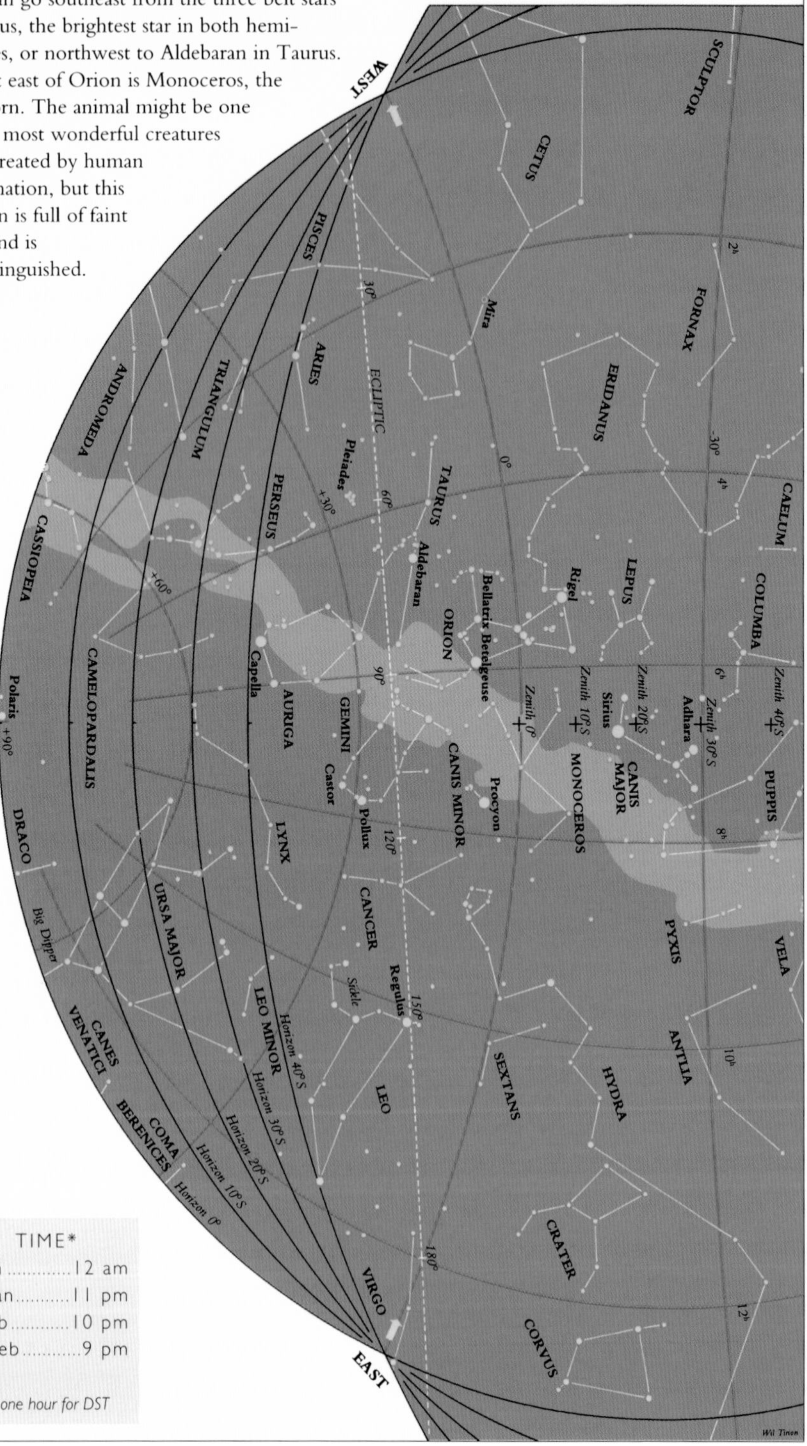

WEST

CETUS · PISCES · ARIES · TRIANGULUM · ANDROMEDA · CASSIOPEIA · PERSEUS · Pleiades · ECLIPTIC · Mira · SCULPTOR · FORNAX · CAELIUM · COLUMBA · PUPPIS · VELA · ERIDANUS · TAURUS · Aldebaran · Capella · AURIGA · ORION · Bellatrix · Betelgeuse · LEPUS · Rigel · CANIS MAJOR · Sirius · Adhara · Zenith · PYXIS · ANTLIA · GEMINI · Castor · Pollux · LYNX · CANCER · CANIS MINOR · Procyon · MONOCEROS · HYDRA · Regulus · Stäble · LEO MINOR · LEO · SEXTANS · CRATER · CORVUS · VIRGO

NORTH · Polaris +90° · DRACO · CAMELOPARDALIS · URSA MAJOR · Big Dipper · CANES VENATICI · COMA BERENICES

Horizon 40° S · Horizon 30° S · Horizon 20° S · Horizon 10° S · Horizon 0°

EAST

TIME*

| | |
|---|---|
| 1 Jan | 12 am |
| 15 Jan | 11 pm |
| 1 Feb | 10 pm |
| 15 Feb | 9 pm |
| etc. | |

* Add one hour for DST

Wil Tirion

Sirius and Canopus, the brightest and second brightest stars in the sky, command our view tonight, with Rigel and Achernar completing a large semicircle overhead. The sky is a narrow oasis of beauty, with the Milky Way straddling such lovely constellations as Puppis, Vela, Carina, and Crux. Just to the west lies the faint glow of the Large Magellanic Cloud (LMC), now at the crest of its tight circuit around the celestial pole.

Deep in the southeast lies the stunning constellation of Centaurus, surrounding on three sides the tiny constellation of Crux, the Southern Cross.

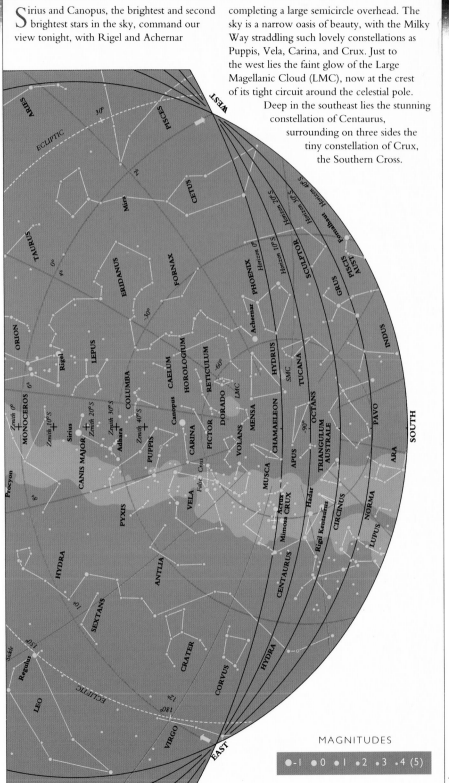

MAGNITUDES

●-1  ●0  ●1  •2  •3  •4 (5)

Regulus in Leo marks the sky's center at this hour, and a strange sight the sky is for visitors from the Northern Hemisphere. Leo's sickle, like everything else, appears upside down! For Southern Hemisphere dwellers, this is how Leo normally appears, just as Orion, in the west, has Rigel at the top and Betelgeuse at the bottom.

Spica and Arcturus outshine all other stars in the eastern sky, although their constellations of Virgo and Boötes are otherwise unremarkable.

Crater and Corvus are high in the east. With its brighter stars and rhomboid appearance, Corvus is the easier of the two to spot.

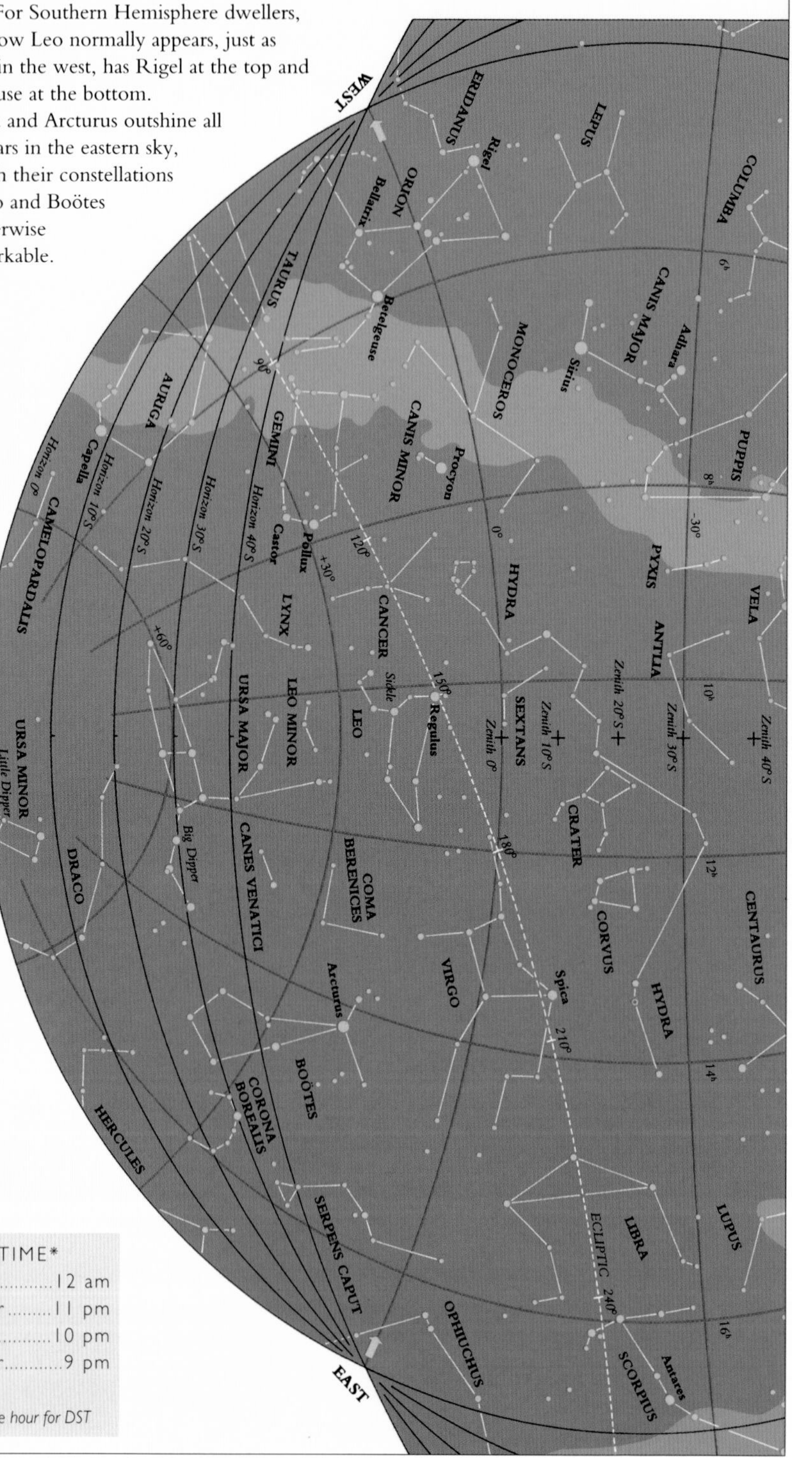

| TIME* | |
|---|---|
| I Mar | 12 am |
| 15 Mar | 11 pm |
| I Apr | 10 pm |
| 15 Apr | 9 pm |
| etc. | |

* Add one hour for DST

262

The southern Milky Way is here marching across the sky in all its glory. An elegant paradigm of stars begins with Alpha (α) Centauri (or Rigil Kentaurus) and Beta (β) Centauri, the "pointers" to the Southern Cross. The parade goes on through the rich star clouds of Carina and the bright star Canopus, ending with Sirius. There is little in the northern or southern sky to match this.

In the south are the faint constellations around the pole, to which the Cross is a pointer. Octans, Tucana, Reticulum, and Phoenix are something of a wasteland, but if your sky is dark you will see the two Magellanic Clouds.

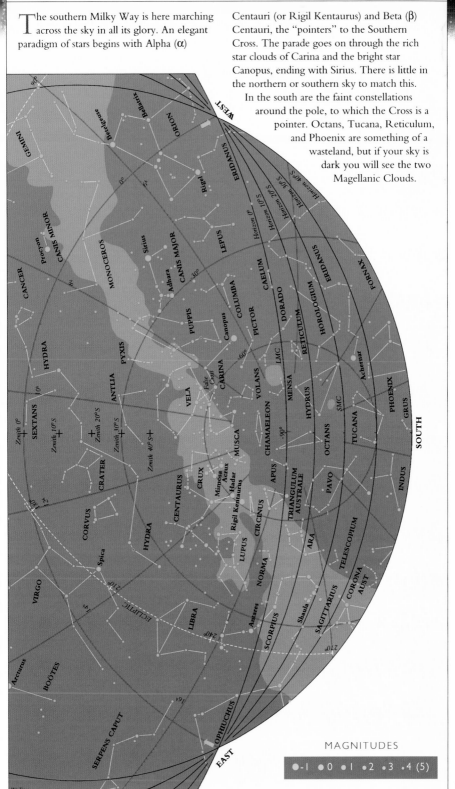

MAGNITUDES

● -1  ● 0  ● 1  ● 2  ● 3  ● 4 (5)

263

With Arcturus in the center, the sky at this point is filled with a variety of constellations. The galactic center is high in the southeastern heavens, its majestic forms of Sagittarius and Scorpius adding sparkle to a dark night. Rising in the northeast are Lyra, the Lyre, its bright star Vega leading the way, and Aquila, the Eagle, marked by Altair, its brightest star. Ophiuchus is a centerpiece of the eastern sky, its wide area making up for its lack of bright stars.

Leo, Virgo, Corvus, and Crater

dominate the west, this wide area being punctuated by only two bright stars—Regulus (in Leo) and Spica (in Virgo).

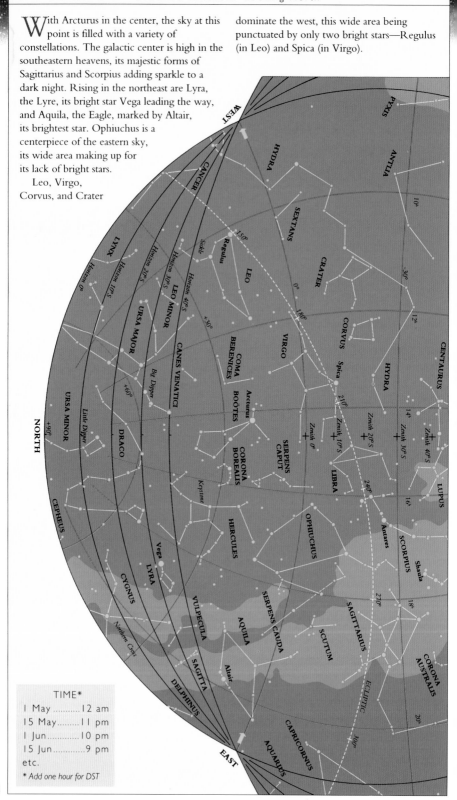

TIME*

| | |
|---|---|
| I May | 12 am |
| 15 May | 11 pm |
| I Jun | 10 pm |
| 15 Jun | 9 pm |
| etc. | |

* Add one hour for DST

264

From a dark sky, you can see the Milky Way in all its splendor stretch across the sky, through Aquila and Scutum to the galactic center in Sagittarius, and on to Scorpius. The Milky Way continues through Centaurus, with the "pointers" Alpha ($\alpha$) Centauri (or Rigil Kentaurus) and Beta ($\beta$) Centauri pinpointing Crux, now high in the south, and Vela, and Puppis. If our view of the center were not blocked by dark matter, the whole sky would appear much brighter. Farther south shine the Large and Small Magellanic Clouds, our two neighboring galaxies.

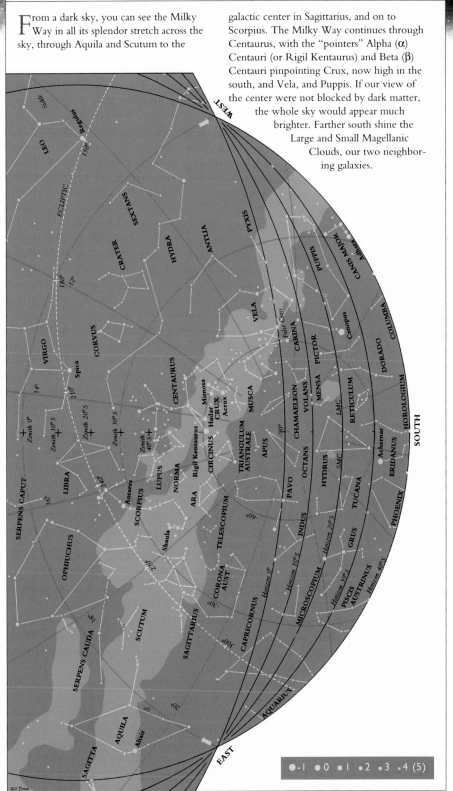

-1  0  1  2  3  4 (5)

Two birds fly through the northern sky here: Cygnus, the Swan, and Aquila, the Eagle, with its bright star Altair. Nearby, bright star Vega is crossing the meridian. Arching directly overhead, the Milky Way, with its bright heart in Sagittarius, should be visible, unless you are in the city. The constellations Delphinus, the Dolphin, and Sagitta, the Arrow, are small but distinct.

To the east, Capricornus and Aquarius are large but faint, while Piscis Austrinus, the Southern Fish, is easy to find only because of Fomalhaut, its one bright star.

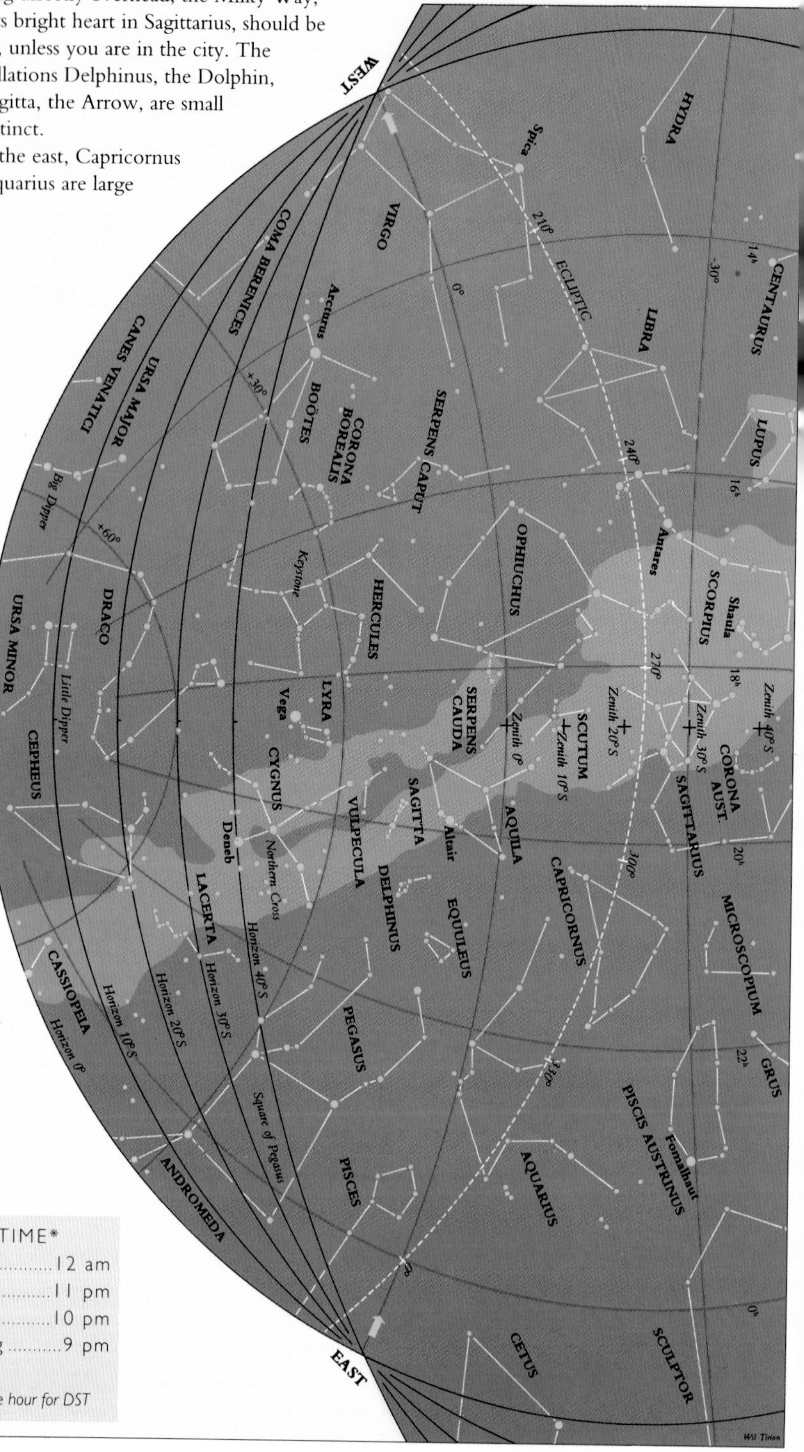

WEST

NORTH

EAST

HYDRA
CENTAURUS
LUPUS
VIRGO
Spica
COMA BERENICES
CANES VENATICI
URSA MAJOR
Big Dipper
Arcturus
BOÖTES
CORONA BOREALIS
SERPENS CAPUT
LIBRA
ECLIPTIC
Antares
SCORPIUS
Shaula
CORONA AUST.
SAGITTARIUS
MICROSCOPIUM
GRUS
OPHIUCHUS
HERCULES
Keystone
LYRA
Vega
SERPENS CAUDA
SCUTUM
Zenith 0°
Zenith 10°S
Zenith 20°S
Zenith 30°S
Zenith 40°S
DRACO
URSA MINOR
Little Dipper
CEPHEUS
CASSIOPEIA
CYGNUS
Denab
Northern Cross
SAGITTA
VULPECULA
DELPHINUS
AQUILA
Altair
EQUULEUS
CAPRICORNUS
PISCIS AUSTRINUS
Fomalhaut
PEGASUS
LACERTA
ANDROMEDA
Square of Pegasus
PISCES
AQUARIUS
CETUS
SCULPTOR

Horizon 0°
Horizon 10°S
Horizon 20°S
Horizon 30°S
Horizon 40°S

+30°
+60°
+90°
0°
210°
240°
270°
30°

TIME*
1 Jul..............12 am
15 Jul............11 pm
1 Aug...........10 pm
15 Aug...........9 pm
etc.
* Add one hour for DST

Wil Tirion

The sky is dominated by the glowing center of the Milky Way in Sagittarius, directly overhead. From there, the river of stars flows through the tail of Scorpius down to the Southern Cross and Carina.

Lying mainly between the bright stars Fomalhaut and Achernar in the southeast, and the Milky Way, you can find several birds, namely Grus, the Crane; Pavo, the Peacock; Tucana, the Toucan; and the fabulous Phoenix. Apus, the Bird of Paradise, is closer to the Southern Cross.

These birds compete with Musca, the Fly, and Volans, the Flying Fish.

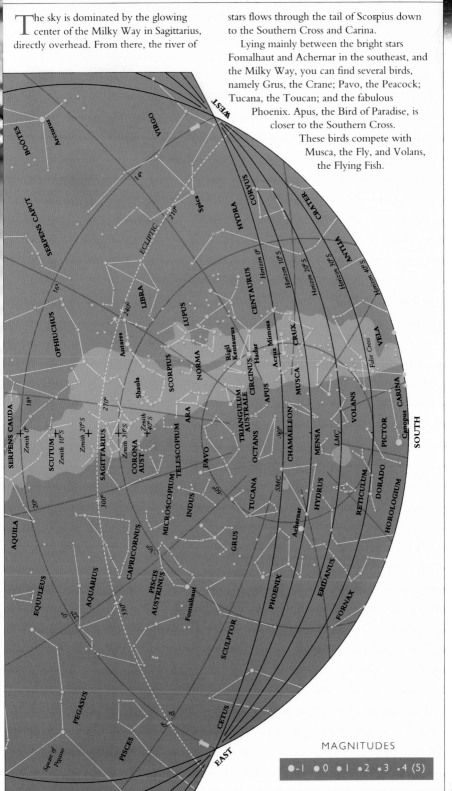

MAGNITUDES

●-1  ● 0  ● 1  ● 2  ● 3  ● 4 (5)

Dominated by Fomalhaut, the bright star now crossing the meridian, this view of the sky is unusual in that the entire eastern half lacks a single major star. The large but faint constellations of Eridanus, Cetus, and Pisces overlook this wintry portion of the sky, but it is the Square of Pegasus that really dominates.

The constellations of the Milky Way enrich the north and west. As you look farther west, the Milky Way widens as it approaches the

galactic center in Sagittarius. Capricornus, Microscopium, and Corona Australis complete this panorama.

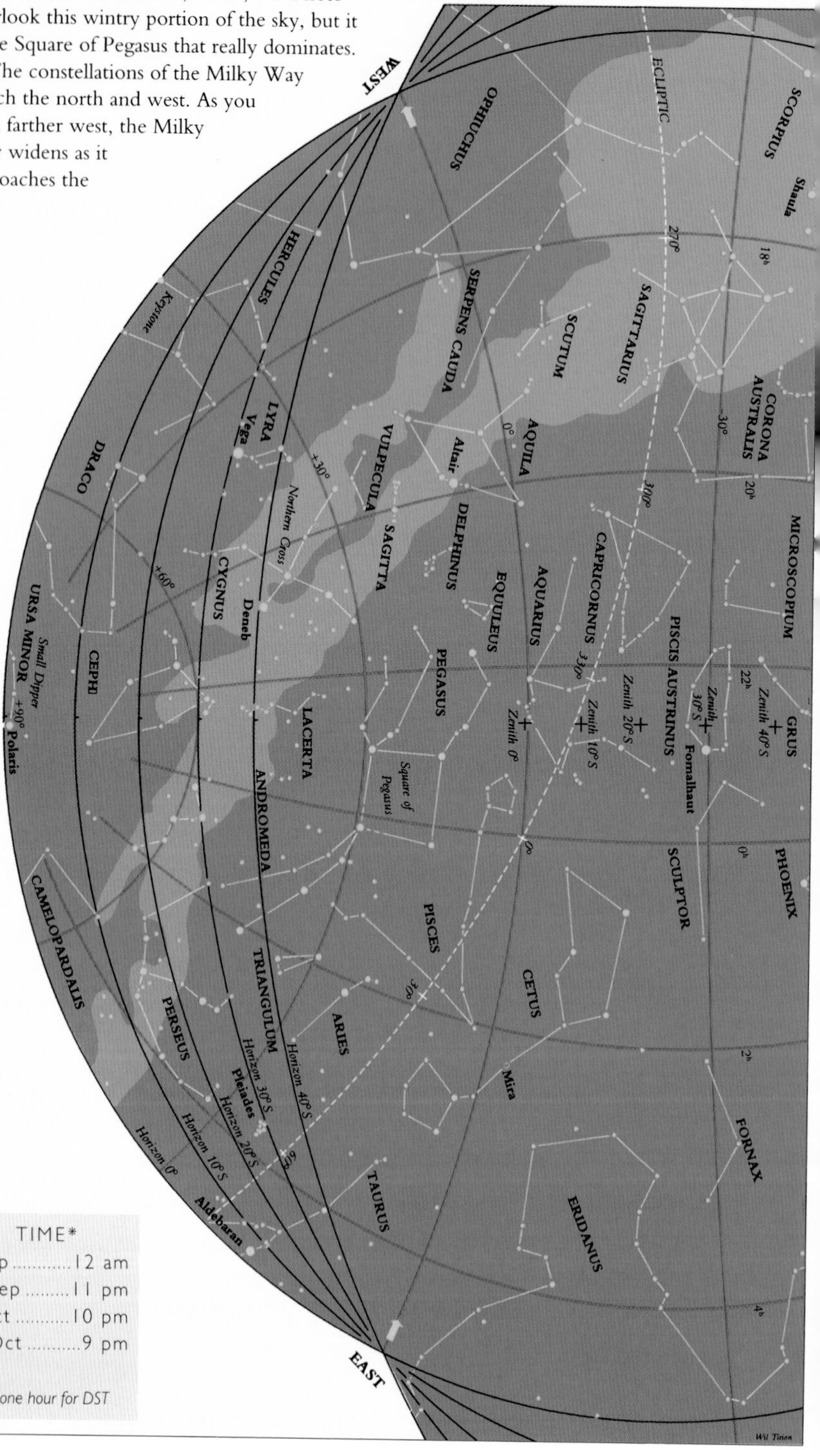

TIME*

| | | |
|---|---|---|
| 1 Sep | .......... | 12 am |
| 15 Sep | ........ | 11 pm |
| 1 Oct | .......... | 10 pm |
| 15 Oct | .......... | 9 pm |
| etc. | | |

* Add one hour for DST

A string of bright stars running southward livens up this view of the southern sky. Fomalhaut in Piscis Austrinus, the Southern Fish, is the northernmost star. The line runs through Phoenix to Achernar in Eridanus, the River, and ends in Canopus, the brightest star in Carina. The western sky is full of rich Milky Way constellations, with the center of the galaxy high in the west.

The eastern half of this sky is remarkably devoid of bright stars, with Cetus and Eridanus taking up most of the space. The Magellanic Clouds are again rising to prominence in the south.

WEST

SERPENS CAUDA

OPHIUCHUS

Altair
AQUILA

SCUTUM

LIBRA

240°

Antares
SCORPIUS
Shaula
SAGITTARIUS
CORONA AUST.
TELESCOPIUM
ARA
NORMA
TRIANGULUM AUST.
CIRCINUS
Rigil Kentaurus
LUPUS

270°

EQUULEUS

CAPRICORNUS

300°

MICROSCOPIUM

INDUS

PAVO

APUS

OCTANS

Hadar
Mimosa
Acrux
CRUX
Musca
CENTAURUS

22ʰ

330°

Zenith 0°
Zenith 10°S
Zenith 20°S
Zenith 30°S
Zenith 40°S

AQUARIUS

PISCIS AUSTRINUS
Fomalhaut

GRUS

TUCANA

SMC

HYDRUS

MENSA
CHAMAELEON

VOLANS

MUSCA

SOUTH

0ʰ

SCULPTOR

PHOENIX

Achernar

RETICULUM

LMC

DORADO

CARINA

False Cross

VELA

PISCES

CETUS

FORNAX

HOROLOGIUM

CAELUM

Horizon 0°

Horizon 10°S

PICTOR

Horizon 20°S

Canopus

PUPPIS

Horizon 30°S

Horizon 40°S

30°

Mira

ERIDANUS

LEPUS

COLUMBA

ECLIPTIC

TAURUS

EAST

18ʰ

20ʰ

60°

90°

MAGNITUDES

●-1  ● 0  ● 1  • 2  • 3  • 4 (5)

Back in the sky once more, Orion outshines all the other constellations in this view. To the east are Sirius and Canis Major. Lepus, the Hare, scampers to the south, under the watchful eye of Columba, the Dove. The bright star Aldebaran is in the north, with Capella closer to the horizon. Perseus is crossing the meridian, low in the north, with Cetus, the Whale; Eridanus, the River; and Fornax, the Furnace, higher in the sky.

With its large but faint constellations like

Pegasus, Pisces, and Cetus, the west forms quite a contrast, Fomalhaut being the only bright star in the area.

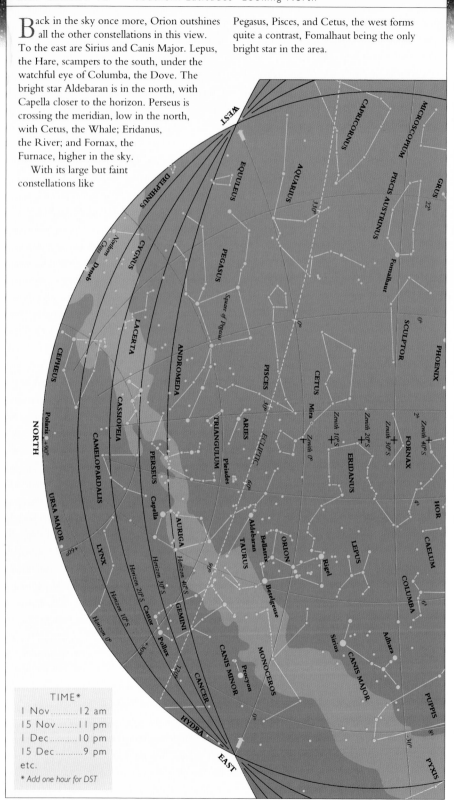

TIME*
| | |
|---|---|
| 1 Nov | 12 am |
| 15 Nov | 11 pm |
| 1 Dec | 10 pm |
| 15 Dec | 9 pm |

etc.

* Add one hour for DST

Four stars dominate this view: Betelgeuse, Rigel, Sirius, and Canopus. A fifth, Procyon, is just rising in the east. In the southeast are parts of a great dismembered ship—once a vast constellation known as Argo Navis—that must have given pause to south sea explorers: Puppis, the Stern; Carina, the Keel; Vela, the Sail; and Pyxis, the Compass.

There are also birds of all shades flying about: Columba, the Dove, and to the west, Tucana, Pavo, Apus, Grus, and Phoenix. Elsewhere are several small constellations, including Microscopium, Indus, Telescopium, Ara, and Triangulum Australe.

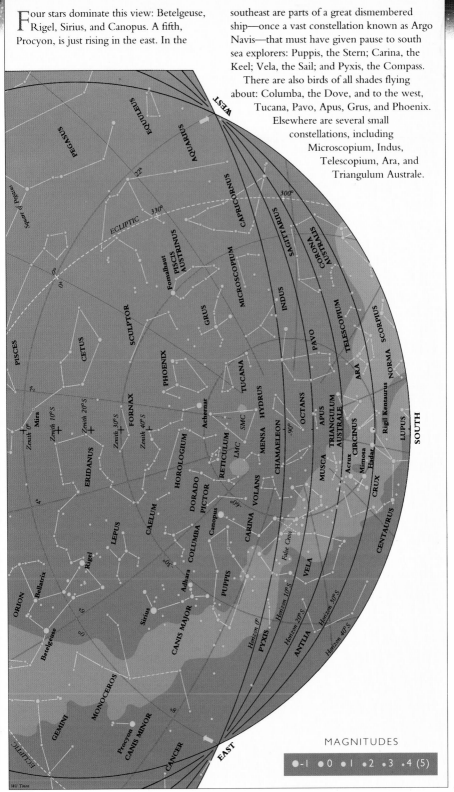

MAGNITUDES

●-1 ● 0 ● 1 • 2 • 3 • 4 (5)

Wil Tirion

# Andromeda

(an-DROH-me-duh) The Chained Princess

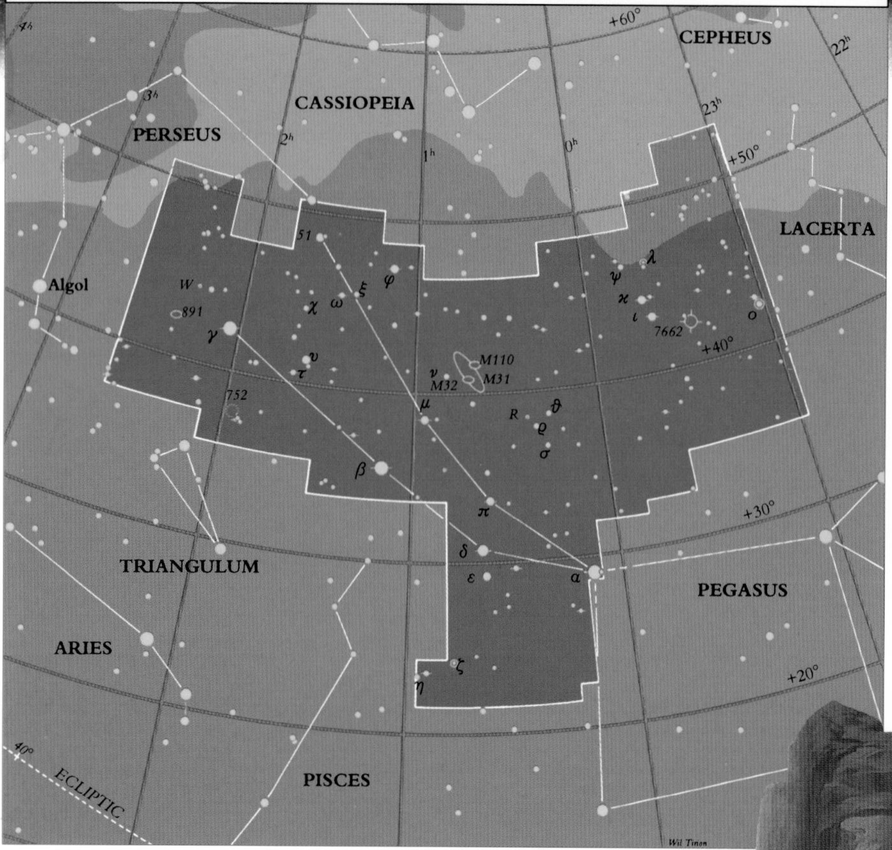

A ndromeda is one of the earliest
constellations to have been named and
the mythology surrounding it is a rich
and varied one, including the stories behind
several other star groups.

Andromeda was the daughter of Cassiopeia
and Cepheus, rulers of the ancient country of
Æthiopia. When Cassiopeia boasted that she was
more beautiful than the daughters of Nereus, the
angry sea god sent the monster Cetus to
ravage the kingdom. Advised by an oracle that
the sacrifice of their daughter to Cetus was the
only way to appease the god, the king and queen
duly chained Andromeda to a rock by the sea.
Perseus came to the rescue just in time, however,
swooping down astride the winged horse
Pegasus. Perseus was able to save Andromeda
from her cruel fate by revealing the hideous head
of Medusa to Cetus, which instantly turned the
great monster to stone.

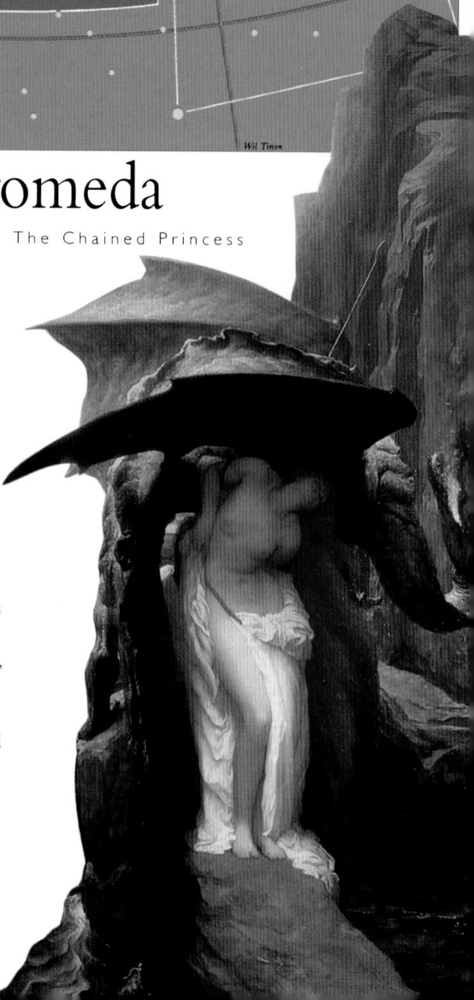

*A detail from a painting by Frederic Leighton (1830–96),
showing Cetus about to devour the helpless Andromeda.*

272

*The bright stars of Andromeda, featuring the Andromeda Galaxy—the nearest major galaxy to Earth.*

Although Andromeda is justly famous for the great and distant galaxy that resides within the constellation, its stars are not very bright. It is easy to find, however, located south of Cassiopeia's W, and just off one corner of the Great Square of Pegasus. In fact, Alpheratz, the star at the northeastern corner of the square of Pegasus, belongs to Andromeda.

👁 **The Andromeda Galaxy (M31)**: The closest major galaxy to us, the Andromeda Galaxy was first thought to be a nebula, and was listed in comet hunter Charles Messier's eighteenth-century catalogue of nebulas. A spiral galaxy much like our own Milky Way, it is a maelstrom comprising 200 billion suns and clouds of dust and gas. It is bright enough to be seen with binoculars from city sites and with the naked eye beneath a dark sky, being one of the most distant objects visible to the unaided eye. In the field of larger binoculars, or using a small telescope, you can see its two neighboring elliptical galaxies. **M32** is small and compact; **M110** is larger and more diffuse, and is therefore harder to see.

🔭 **Gamma (γ) Andromedae**: This is a beautiful double star. The brighter member of the pair is a golden yellow, and its companion is greenish blue.

**R Andromedae**: This Mira star has a range of 9 magnitudes.

*The Andromeda Galaxy, with M32 (left) and M110 (right). Only the central yellow regions appear in a small telescope.*

**NGC 752**: This open cluster lies about 5 degrees south of Gamma (γ) Andromedae and is easy to find because of its relatively bright stars. Because it is spread out over such a large area, it is actually easier to see through binoculars than through a telescope. If using a telescope, use it at its lowest power.

🔭 **NGC 7662**: A fairly bright planetary nebula, this blue-green object looks almost starlike through the smallest telescopes. But through a 6 inch (150 mm) telescope at moderate power, it becomes a graceful, glowing spot of gas about 30 arcseconds across.

🔭 **NGC 891**: This galaxy is a challenge even for 6 inch (150 mm) telescopes. However, with good eyes and a dark sky, you will see one of the best examples of a spiral galaxy, viewed edge-on.

*Andromeda! Sweet woman!*

*why delaying, So timidly*

*among the stars: come hither!*

*Join this bright throng, and*

*nimbly follow whither*

*They all are going.*

Endymion, JOHN KEATS (1795–1821), English poet

273

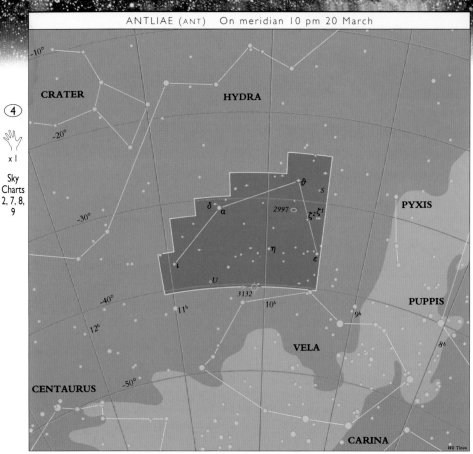

CRATER

HYDRA

-10°

-20°

CRATER

CENTAURUS

-30°

-40°

-50°

12ʰ

11ʰ

10ʰ

9ʰ

8ʰ

PYXIS

PUPPIS

VELA

CARINA

δ α

ϑ

s

2997

ζ² ζ¹

η

ε

ι

U

3132

Wil Tirion

4

x 1

Sky
Charts
2, 7, 8,
9

# Antlia

(ANT-lee-uh)  The  Air  Pump

Antlia Pneumatica, the Air Pump, named after seventeenth-century physicist Robert Boyle's invention, is a southern constellation. It was given this somewhat unpoetic name by Nicolas-Louis de Lacaille during the time he spent working at an observatory at the Cape of Good Hope, from 1750 to 1754. As a result of his observations of some 10,000 southern stars, de Lacaille divided the far southern sky into 14 new constellations, of which Antlia is one.

Antlia is a small, faint constellation just off the bright southern Milky Way, not far from Vela and Puppis. Its alpha (α) star is just barely the constellation's brightest star and has been given no proper name. It is quite red in color and possibly varies slightly in magnitude.

**NGC 2997**: This is a large, faint spiral galaxy, with a stellar nucleus. It is quite difficult to observe with a small telescope.

*NGC 2997: An impressive galaxy with its spiral arms traced out by blue stars, pink hydrogen clouds and dust.*

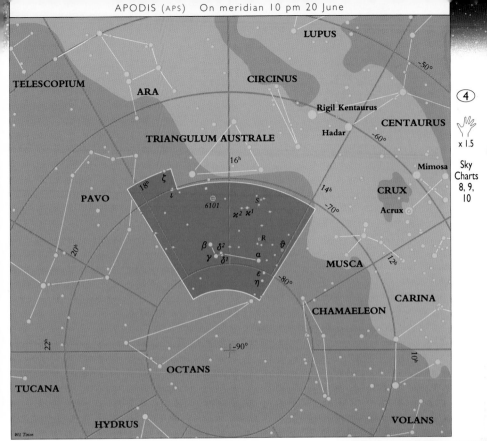

LUPUS

TELESCOPIUM

CIRCINUS

ARA

Rigil Kentaurus

-50°

CENTAURUS

Hadar

TRIANGULUM AUSTRALE

-60°

16ʰ

Mimosa

18ʰ  ζ

ι

PAVO

6101

κ²  κ¹

S

14ʰ

CRUX

-70°

Acrux

β  δ²

γ  δ¹

R

α

ϑ

MUSCA

12ʰ

ε

η

-80°

CARINA

CHAMAELEON

20ʰ

OCTANS

-90°

10ʰ

22ʰ

TUCANA

VOLANS

HYDRUS

④

x 1.5

Sky
Charts
8, 9,
10

Wil Tirion

# Apus

(ay-pus) The Bird of Paradise

This faint constellation is directly below Triangulum Australe, the Southern Triangle. Being close to the southern pole, it cannot be seen from most northern latitudes. *Apus* is an ancient Greek word that means "footless" and derived from *Apus Indica*, the name given to India's Bird of Paradise. This magnificent bird was offered as a gift to Europeans, but not before its unsightly legs were cut off.

*Apus, as seen in Johann Bode's Uranographia, complete with an early use of constellation boundaries.*

A 12 inch (300 mm) telescope captured this view of NGC 6101.

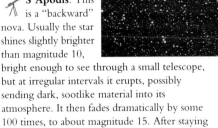

**S Apodis**: This is a "backward" nova. Usually the star shines slightly brighter than magnitude 10, bright enough to see through a small telescope, but at irregular intervals it erupts, possibly sending dark, sootlike material into its atmosphere. It then fades dramatically by some 100 times, to about magnitude 15. After staying faint for several weeks, it slowly returns to its original brightness.

**Theta (θ) Apodis**: This variable star ranges from magnitude 6.4 to below 8 in a semi-regular cycle over 100 days.

**NGC 6101**: A faint globular cluster, large and slightly irregular, NGC 6101 can be seen as a small, misty spot through a small telescope.

275

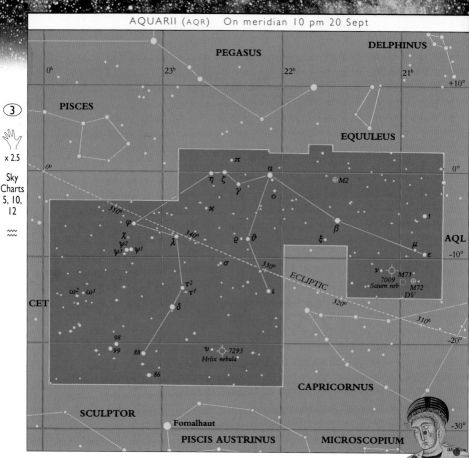

PEGASUS

DELPHINUS

0ʰ       23ʰ       22ʰ       21ʰ

+10°

PISCES

EQUULEUS

③

✋ x 2.5

Sky Charts 5, 10, 12

〰〰

π

⊕ M2

α

0°

η ζ
γ
ο

κ

β

3

AQL

350°

φ
χ
ψ² ψ³ ψ¹

λ

340°

ϱ θ

ξ

μ
ε

-10°

σ

330°

ECLIPTIC

ι

ν
7009
Saturn neb

M73
M72
DV

ω² ω¹

τ²
τ¹

δ

320°

310°

-20°

98
99   88

υ   7293
Helix nebula

86

CET

SCULPTOR

Fomalhaut

CAPRICORNUS

-30°

PISCIS AUSTRINUS       MICROSCOPIUM

# Aquarius

(ah-KWAIR-ee-us) The Water Bearer

The Water Bearer dates as far back as Babylonian times and is appropriately placed in the sky not far from a dolphin, a river, a sea serpent, and a fish. Of its many mythological associations, it was at times identified with Zeus pouring the waters of life down from the heavens.

*Aquarius, the Water Bearer, from a thirteenth-century Italian manuscript.*

**M2**: This fine globular cluster appears as a fuzzy spot of light through binoculars and small telescopes. It is possible, however, to see the cluster's mottled appearance through a 4 inch (100 mm) telescope, and to resolve it into stars through a 6 inch (150 mm) telescope.

*The Helix Nebula, at 450 light-years away, is the nearest planetary nebula to Earth.*

**The Saturn Nebula (NGC 7009)**: This small planetary nebula was named by Lord Rosse who, with his large reflecting telescope, first saw the protruding rays that made it look like a dim version of Saturn with its rings. It is visible through a telescope as a greenish point of light.

**The Helix Nebula (NGC 7293)**: The largest and closest of the planetary nebulas, this cloud takes up half the angular diameter of the Moon in the sky. Because its brightness is spread over a large area, it appears best with a low-power, wide-field telescope or binoculars under a dark sky.

**Delta (δ) Aquarids**: This strong meteor shower peaks every year on 28 July.

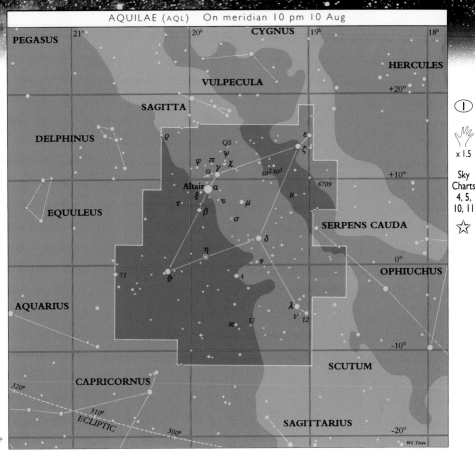

PEGASUS   21ʰ   20ʰ   CYGNUS   19ʰ   18ʰ

HERCULES   +20°

VULPECULA

SAGITTA

DELPHINUS   ϱ   QS   ε   ζ

φ π χ   +10°

o γ   ω² ω¹

Altair α   6709

EQUULEUS   τ ξ   ν μ

β   σ   SERPENS CAUDA

η   δ   0°

v   OPHIUCHUS

71   ϑ   ι

AQUARIUS   κ   U   λ   V 12   -10°

SCUTUM

CAPRICORNUS   SAGITTARIUS   -20°

320°   310°   ECLIPTIC   300°   Wil Tirion

Ⓘ   x 1.5   Sky Charts 4, 5, 10, 11   ☆

# Aquila

(uh-KWI-luh)  The Eagle

Identified as an eagle by astronomers of the Euphrates Basin, the constellation of Aquila takes its name from the bird that belonged to the Greek god Zeus. Aquila's main accomplishment was to bring the handsome mortal youth Ganymede to the sky to serve as his master's cup bearer.

Two major novas have appeared in Aquila. The first, in AD 389, was as bright as Venus, and the second, as recently as 1918, was brighter than **Altair**, Aquila's brightest star. One of the brightest stars in the whole sky, Altair is a prominent beacon on the Milky Way between Sagittarius and Cygnus.

👁 **Eta (η) Aquilae**: This supergiant star is a bright Cepheid variable that changes over a magnitude in brightness (3.5 to 4.4) in a period of little more than a week. At its brightest it rivals **Delta (δ) Aquilae**, and it fades to about the magnitude of **Iota (ι) Aquilae**.

**R Aquilae**: This Mira star varies in magnitude from 6 to 11.5 over a period of 284 days.

*Aquila, with brilliant Altair, astride the Milky Way's long ribbon of stars.*

**NGC 6709**: This pretty open cluster consists of a group of closely knit stars against an already rich background of stars. On 13 November 1984, Comet Levy-Rudenko was discovered by David Levy as a faint fuzzy object in the same field of view as this cluster.

277

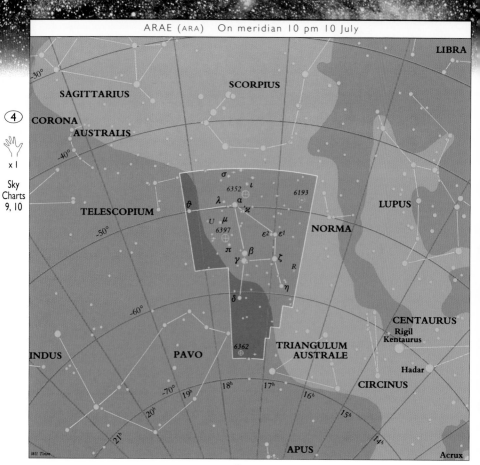

LIBRA

SAGITTARIUS

SCORPIUS

CORONA
AUSTRALIS

-30°

-40°

④

🖐
x 1

Sky
Charts
9, 10

TELESCOPIUM

σ

6352   ι

λ   α

θ   κ

U   μ

6397

π   γ   β   ζ

R

η

δ

LUPUS

NORMA

6193

ε²   ε¹

-50°

-60°

INDUS

PAVO

6362

TRIANGULUM
AUSTRALE

CENTAURUS
Rigil
Kentaurus

Hadar

CIRCINUS

-70°   19ʰ   18ʰ   17ʰ   16ʰ   15ʰ   14ʰ

20ʰ

21ʰ

APUS

Wil Tirion

Acrux

# Ara

(AR-uh)  The Altar

Located south of Scorpius, Ara's original Latin name was Ara Centauri—the altar of the centaur Chiron. Half man and half horse, Chiron was thought to be the wisest creature on Earth. Ara has also been referred to as the altar of Dionysus; the altar built by Noah after the flood; the altar built by Moses; and even the one from Solomon's Temple.

**U Arae**: This Mira-type variable is bright enough to be seen through a small telescope when it is at its maximum of magnitude 8. However, it then drops a full five magnitudes before rising again over a period of more than seven months.

**NGC 6397**: Possibly the closest globular star cluster to us, this bright cluster is placed between **Beta (β) Arae** and **Theta (θ) Arae**.

The cluster is relatively loose, so an observer with powerful binoculars should be able to detect it without difficulty and perhaps even resolve its faint stars. It may be as large as 50 light-years across.

*Ara, alongside the Milky Way, with NGC 6397 clearly seen between Beta (β) and Theta (θ) Arae.*

*(Above) Ara, the Altar, as depicted in the 1723 edition of Johann Bayer's Uranometria.*

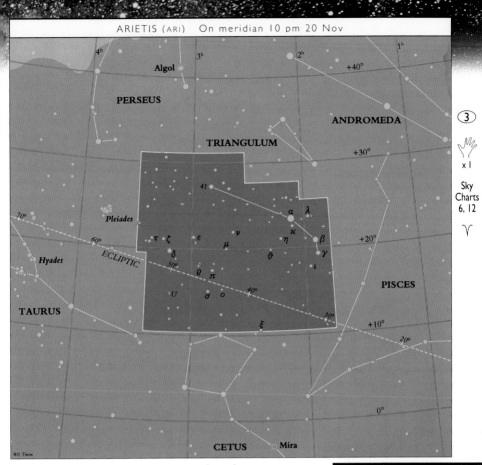

PERSEUS

Algol

ANDROMEDA

TRIANGULUM

+40°

+30°

③

🖐
x 1

Sky
Charts
6, 12

♈

41

Pleiades

α  λ
ν      κ
τ  ζ       η      β    +20°
ε   μ        θ      γ
δ                    ι
Hyades   ECLIPTIC   50°  π
ϱ
U           σ  o   40°

PISCES

TAURUS                         30°

+10°

20°

ξ

0°

CETUS    Mira

WS Tirion

# Aries

(AIR-eez) The Ram

The ancient Babylonians, Egyptians, Persians, and Greeks all called this group of stars the Ram. In one version of the Greek legend, the king of Thessaly had two children, Phrixus and Helle, who were abused by their stepmother. The god Hermes sent a ram with a golden fleece to carry them to safety on its back. Helle fell off the ram as it was flying across the strait that divides Europe from Asia, a body of water the Greeks called the Hellespont, the sea of Helle (now the Dardanelles). Phrixus was carried to safety on the shores of the Black Sea, where he sacrificed the ram and its fleece was placed in the care of a sleepless dragon. It was from here that Jason and the Argonauts stole it.

*The brighter stars of Aries, with the stars of Triangulum lying to the north.*

Aries is the zodiac's first constellation, since the Sun at one time was entering Aries on the day of the vernal equinox—the moment when it crosses from the southern to the northern half of the celestial sphere. However, because of Earth's precession, the Sun is now in Pisces at the vernal equinox.

Aries is well known and is not difficult to find, but it has few objects of interest.

**Gamma (γ) Arietis**: In 1664, Robert Hooke was following the motion of a comet when he chanced upon this beautiful double star. One of the earliest doubles to be found with a telescope, this star has a separation of 8 arcseconds, and is easy to find and observe.

*A thirteenth-century Italian view of Aries.*

279

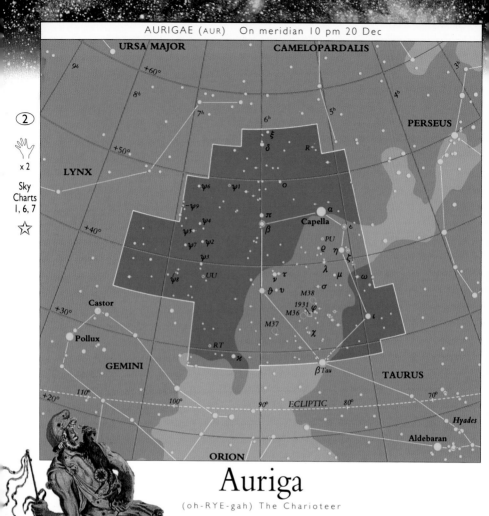

URSA MAJOR
CAMELOPARDALIS
+60°
PERSEUS
LYNX
+50°
+40°
Capella
+30°
Castor
Pollux
GEMINI
RT
TAURUS
+20°
110°
100°
90°    ECLIPTIC    80°
70°
Hyades
Aldebaran
ORION

② 
🖐 
x 2
Sky
Charts
1, 6, 7
☆

# Auriga

(oh-RYE-gah) The Charioteer

This lovely multi-sided figure is easy to find in the sky, largely because of bright Capella, the she-goat star, and her retinue of three little kids. Ancient legends portray Auriga as a charioteer carrying a goat on his shoulder and two or three kids on his arm. The charioteer is also seen as Erechtheus, the son of Hephaestus (the Roman god Vulcan), who invented a chariot to move his crippled body about.

Capella has been seen as the she-goat star since Roman times. Forty-two light-years away, Capella is similar to our Sun, only larger.

👁 **Epsilon (ε) Aurigae**: An extraordinary variable system, this supergiant star fades when its companion passes in front of it once every 27 years. During an eclipse, its brightness drops

by two-thirds of a magnitude. The deepest phase of the eclipse lasts a full year, which may indicate that the companion is surrounded by an enormous disk of gas and dust.

**M36**: This bright open star cluster is some 5 degrees southwest of **Theta (θ) Aurigae**, and contains about 60 stars of 8th magnitude and fainter.

**M37**: This is an exceptional open star cluster, almost the size of the Moon, and one of the finest in the northern sky. Binoculars will show this cluster as a misty spot. A small telescope will reveal its large number of stars.

**M38**: This small cluster of stars resembles the Greek letter π (pi) when seen in a small telescope.

*Lying 4,600 light-years away, M37 still presents a magnificent spectacle.*

*(Above) Auriga, with brilliant Capella on his back, as seen in a seventeenth-century atlas.*

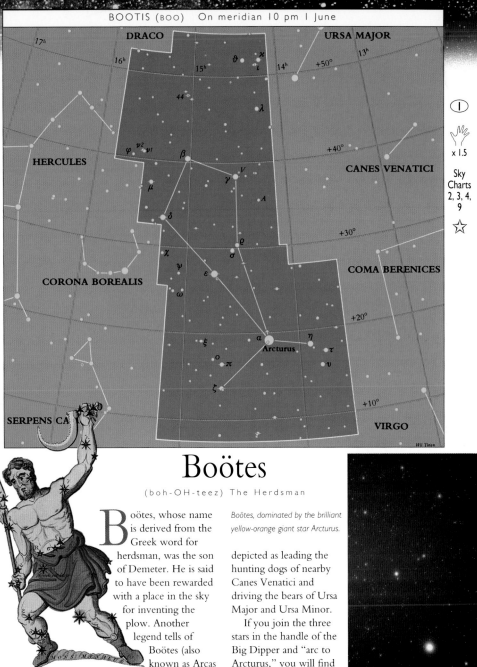

DRACO

URSA MAJOR

17ʰ

16ʰ

15ʰ

14ʰ

13ʰ

+50°

44

λ

HERCULES

φ   ν2  ν1

β

CANES VENATICI

μ

γ

+40°

δ

A

+30°

χ

ρ

σ

ψ

ε

COMA BERENICES

CORONA BOREALIS

ω

+20°

ξ

α

η

Arcturus

τ

o

π

υ

ζ

+10°

SERPENS CA

VIRGO

Wil Tirion

# Boötes

(boh-OH-teez) The Herdsman

Boötes, whose name is derived from the Greek word for herdsman, was the son of Demeter. He is said to have been rewarded with a place in the sky for inventing the plow. Another legend tells of Boötes (also known as Arcas and Arcturus), son of Zeus and Callisto. Callisto, changed into a bear by Zeus's jealous wife Hera, was almost killed by her son when he was out hunting. Zeus rescued her, taking her into the sky where she became Ursa Major, the Great Bear.

The name Arcturus (the constellation's brightest star) comes from the Greek meaning "guardian of the bear." Sometimes Arcturus is

*(Above) The Herdsman, with Arcturus at his knee, as depicted in the Urania's Mirror constellation cards (1825).*

*Boötes, dominated by the brilliant yellow-orange giant star Arcturus.*

depicted as leading the hunting dogs of nearby Canes Venatici and driving the bears of Ursa Major and Ursa Minor.

If you join the three stars in the handle of the Big Dipper and "arc to Arcturus," you will find this constellation.

**Arcturus: Alpha (α) Boötis**, this yellow-orange star is 37 light-years away from us, making it one of the closest of the bright stars. Arcturus's actual position in the sky has changed by over twice the Moon's apparent diameter in the last 2,000 years; astronomers say that Arcturus has a large *proper motion*. Comet Levy (1987y) was discovered by David Levy near Arcturus one September evening in 1987.

**281**

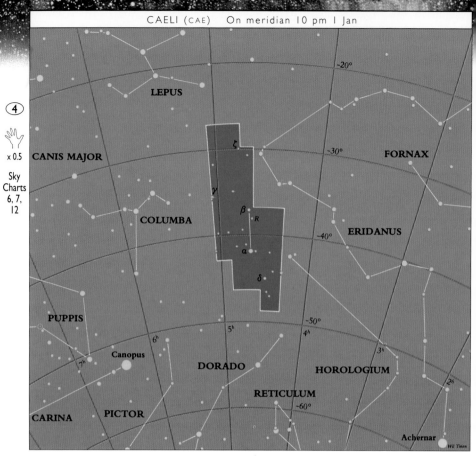

# Caelum

(SEE-lum) The Chisel

One of the least conspicuous of all the constellations, Caelum is one of the many regions in the Southern Hemisphere skies that was named by eighteenth-century astronomer Nicolas-Louis de Lacaille. It comprises a largely empty region of the heavens between the constellations of Columba, the Dove, and Eridanus, the River.

**R Caeli:** A bright Mira-type variable, this star changes from magnitude 6.7 to 13.7 over a period of about 13 months.

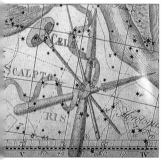

*A tiny patch of southern sky, Louis de Lacaille's Caela Sculptoris, the Sculptor's tool, is now known simply as Caelum, a single chisel. Its few faint stars, with a magnitude of 5 at best, could equally well have become part of nearby Columba or Eridanus.*

### SKYWATCHING TIP

Recording your observations is, in many ways, as important as making them. Whether the notes are in written, sketched, or taped form, the act of recording will enhance your powers of observation. Keep your notes simple and to the point. This is an example of what a Northern Hemisphere observer might record when attempting to locate Caelum:

• Tried to find Caelum, but could see no stars through the horizon haze to the south.

• Made out Alpha (α) and Beta (β) Caeli despite horizon haze.

• Saw bright meteor, magnitude 1. Began near Orion's belt, then went south through Lepus and vanished in Caelum with a bright burst. Greenish in color.

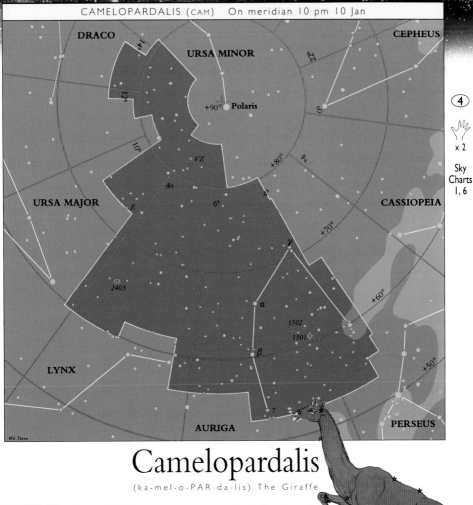

4

× 2

Sky
Charts
1, 6

# Camelopardalis

(ka-mel-o-PAR-da-lis) The Giraffe

What is a giraffe doing next to two bears and a dragon in the frigid sky near the North Star? Camelopardalis was dreamed up by Bartsch in 1624, who claimed that it represented the camel that brought Rebecca to Isaac. ("Camel-leopard" was the name the Greeks gave to the giraffe, as they thought it had the head of a camel and a

*The Giraffe, as featured in the constellation cards of Urania's Mirror (1825).*

leopard's spots.) The constellation lies in the large space between Auriga and the bears.

**Z Camelopardalis**: This cataclysmic variable star erupts every two or three weeks from its minimum of magnitude 13 to a maximum of 9.6, which is still quite faint. Its resemblance to other such variables ceases when, while fading, it stops changing and hovers at an intermediate magnitude. This "standstill" might last for months before the decline resumes. In the late 1970s, Z Cam stayed around magnitude 11.7 for several years.

**VZ Camelopardalis**: This bright star varies irregularly over the small range between magnitudes 4.8 and 5.2. Located close to Polaris, VZ Cam is visible every night of the year from most northern latitudes.

*At 12th magnitude, NGC 1501 is faint for small telescopes, but a 20 inch (500 mm) one shows it with its 14th magnitude star.*

283

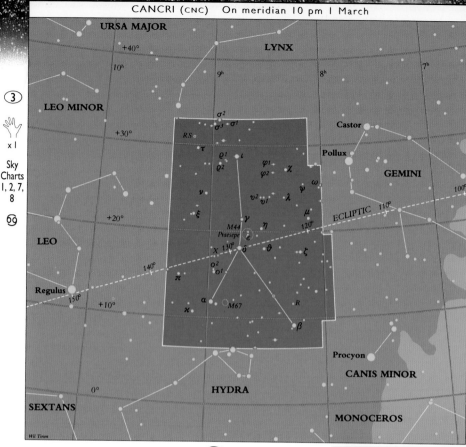

URSA MAJOR

LYNX

+40°

10ʰ                          9ʰ                    8ʰ                              7ʰ

③

🖐
x 1

Sky
Charts
1, 2, 7,
8

♋

LEO MINOR

+30°

σ²
σ³ σ¹
RS
τ
ϱ¹  ι
ϱ²      φ¹
φ²   χ
ν        ω
ξ        v²   λ   ψ
v¹        μ
γ              η            120°
M44
Praesepe   ε
χ  130°  δ        ϑ        ζ
ϱ²
ο¹
π
α   M67
κ                              R
β

Castor

Pollux

GEMINI

100°

110°

ECLIPTIC

+20°

LEO

Regulus   150°
+10°

140°

Procyon

CANIS MINOR

0°

SEXTANS

HYDRA

MONOCEROS

Wil Tirion

# Cancer
(CAN-ser) The Crab

I n Greek mythology, Cancer was sent to distract Hercules when he was fighting with the monster Hydra. The crab was crushed by Hercules's foot, but as a reward for its efforts Hera placed it among the stars. The zodiacal symbol represents the crab's claws.

Millennia ago, the Sun reached its summer solstice (its northernmost position in the sky—declination 23.5 degrees north) when it was in front of this constellation. It was then overhead at a northern latitude we call the Tropic of Cancer. As a result of precession, the Sun's most northerly position has now moved westward to the border of Gemini and Taurus.

Cancer lies between Gemini and Leo—two of the sky's showpieces. It has no star brighter than 4th magnitude and its only claims to fame

*A sixteenth-century Turkish representation of Cancer.*

*The faint stars of Cancer harbor the open cluster M44—a celestial showpiece.*

are its membership of the zodiac and beautiful M44.

👁 **The Praesepe or Beehive (M44)**: One of the sky's finest open clusters, this is easy to see through binoculars from the city and with the naked eye from a dark location. There are over 200 stars in the Praesepe. Spread over 1½ degrees, they are best seen with binoculars.

**M67**: This open cluster has 500 faint stars spread over ½ degree. Although you can find it with binoculars, your best view will be through a small telescope's low-power eyepiece.

**R Cancri**: This bright long-period variable is easily visible through binoculars when near its 6.2 magnitude maximum. It varies down to 11.2 and back in almost precisely a year.

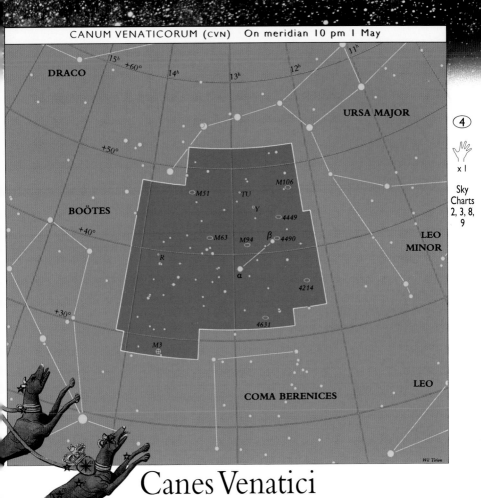

DRACO

URSA MAJOR

④

🖐
x 1

Sky
Charts
2, 3, 8,
9

+60°

15ʰ   14ʰ   13ʰ   12ʰ   11ʰ

+50°

M51   TU   M106

BOÖTES   Y   4449

+40°   M63   M94   β   4490

LEO
MINOR

R   α

4214

+30°

4631

M3

LEO

COMA BERENICES

Wil Tirion

# Canes Venatici

(KAH-nez ve-NAT-eh-see) The Hunting Dogs

This constellation, tucked away just south of the Big Dipper's handle, contains a wide variety of deep-sky objects. Conceived by Hevelius in about 1687, Canes Venatici are the hunting dogs, Asterion and Chara, held on a leash by Boötes as he hunts the skies of the north for the bears Ursa Major and Ursa Minor.

**Cor Caroli**: The heart of Charles, **Alpha (α) Canum Venaticorum** is believed to have been named by Edmond Halley after his patron, Charles II. It is a wide double (separation 20 arc-seconds), easily split by the smallest telescope.

**M3**: A rare gem of the northern sky, this globular cluster is midway between Cor Caroli and Arcturus. Some 34,000 light-years away and 200 light-years across, M3 begins to resolve into stars through a small telescope.

**Y Canum Venaticorum (E-B 364)**: Named La Superba by Secchi in the nineteenth century, this 5th magnitude star is splendidly red. It varies from magnitude 5.2 to 6.6 over 157 days.

**The Whirlpool Galaxy (M51)**: This famous galaxy appears as a round, 8th magnitude glow with a bright nucleus. A 12 inch (300 mm) telescope will show its spiral structure.

*(Left) The Hunting Dogs were held on a leash by Boötes, the Herdsman, in Urania's Mirror (1825). (Below) The spiral galaxy M51 and its companion NGC 5195 form one of the best-known images in astronomy.*

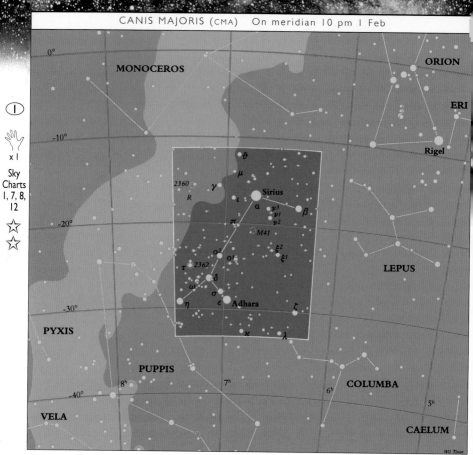

MONOCEROS

ORION

ERI

Rigel

2360
R
γ
ι
θ
μ
Sirius
α   ψ³
ψ1
β
π   ψ2
M41
ξ²
ξ1
LEPUS
τ   2362
ο²
ο1
δ
ω
σ
η   ε   Adhara   ζ
κ   λ

PYXIS

PUPPIS

COLUMBA

VELA

CAELUM

Wil Tirion

# Canis Major

(KAH-niss MAY-jer) The Great Dog

O ne of the most striking of all the constellations, the Great Dog is marked by the brilliant star Sirius, commonly known as the Dog Star—the brightest star in the entire sky. Sirius is said to be responsible for the Northern Hemisphere's hot, muggy "dog days" that occur in September. Legend has it that because Sirius rises at the same time as the Sun during

late summer, its brightness adds to the Sun's energy, producing additional warmth.

Canis Major and its neighboring constellation, Canis Minor, the Little Dog, appear in quite a number of myths. One legend has the two dogs sitting patiently under a table at which the Twins are dining. The faint stars that can be seen scattered in the sky between Canis Minor and Gemini are the crumbs the Twins have been feeding to the animals.

According to the ancient Greeks, Canis Major could run incredibly fast. Laelaps, as they called him, is said to have won a race against a fox that was the fastest creature in the world. Zeus placed the dog in the sky to celebrate the victory.

Another myth has the Great Dog and Little Dog assisting Orion while he is out hunting, his favorite sport. With his eye fixed on Lepus, the Hare, crouching just below Orion, Canis Major

*Canis Major, the Great Dog, as depicted in the Urania's Mirror constellation cards (1825).*

286

*The great star Sirius clearly dominates this view of Canis Major, although the large disk is not seen by eye—it is a photographic artifact due to the star's great brightness.*

seems ready to pounce. In other versions of the story, Sirius is Orion's hunting dog.

The ancient Egyptians had a great deal of respect for Sirius. After being close to the Sun for some months, the star would rise just before dawn in late summer, an event known as its heliacal rising. This would herald the annual flooding of the Nile Valley, the waters re-fertilizing the fields with silt. This event was of such importance to them that it marked the beginning of their year.

◉ **Sirius**: The sky's brightest star, Sirius is only 8.6 light-years from Earth. Its great brilliance is also due to its being some 40 times more luminous than the Sun.

In 1834, Friedrich Bessel noted that Sirius had a strange wobble to its position, indicating an unseen companion. In 1862, the famous telescope maker Alvan Clark, while testing a new 18½ inch (460 mm) refractor on Sirius, discovered the faint star we now know as the Pup. It is a white dwarf star, its density being so great that a piece of it the size of this book might weigh around 200 tons (181 tonnes). On its own, the Pup would be a respectable star visible through a telescope at magnitude 8.4, but its closeness to mighty Sirius makes it a difficult target, requiring a telescope of 10 inch (250 mm) aperture and very steady viewing conditions.

In ancient Greek and Roman astronomical records, Sirius is quite frequently described as being "ruddy" or "reddish" in color. Was Sirius red in the recent past? Current thinking sees this as unlikely, since other bright stars were sometimes also described as red. This might perhaps have been because of the colors that can be seen when bright stars twinkle.

◉ **M41**: A beautiful open cluster, M41 is surrounded by a rich field of background stars. If you look at it through a telescope, you will be able to see a distinctly red star near the cluster's center.

🔭 **NGC 2362**: This cluster of several dozen stars is tightly packed around **Tau (τ) Canis Majoris**. What is not clear is whether Tau (τ) is actually a member of the cluster or just a chance foreground star.

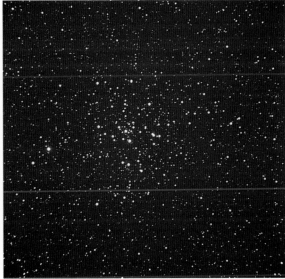

*M41 is a large, bright galactic cluster, almost the size of the Full Moon.*

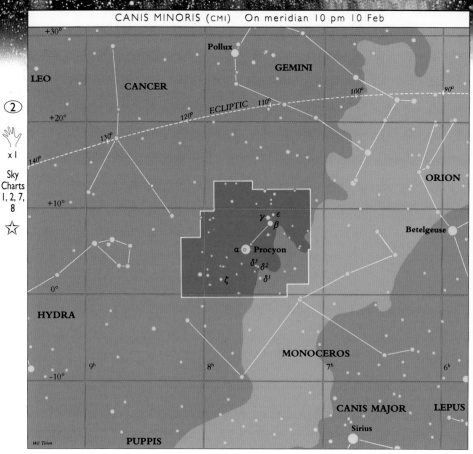

Sky
Charts
1, 2, 7,
8

# Canis Minor

(KAH-niss MY-ner) The Little Dog

anis Minor, Canis Major's playful smaller companion, has only two stars brighter than 5th magnitude—Procyon (Greek for "before the dog," as it rises before Sirius) and Gomeisa. Besides being one of Orion's hunting dogs, Canis Minor was also said to be one of Actaeon's hounds. One day Actaeon surprised Artemis, goddess of the chase and the forests, while she was bathing in a pond with her companions. Spellbound by her great beauty, he paused for a moment and she saw him. Furious that a mortal had seen her naked, Artemis turned him into a stag, set her pack of hounds upon him, and he was devoured.

**Procyon: Alpha (α) Canis Minoris.** This beautiful deep yellow star follows Orion across the sky. Only 11.4 light-years away, it is accompanied by a white dwarf that is much fainter than the Pup that accompanies Sirius.

**Beta (β) Canis Minoris:** This star is set in a beautiful field which includes one quite red star.

*Procyon at upper left, Sirius at the bottom, and Orion at right are beacons in the sky early each year.*

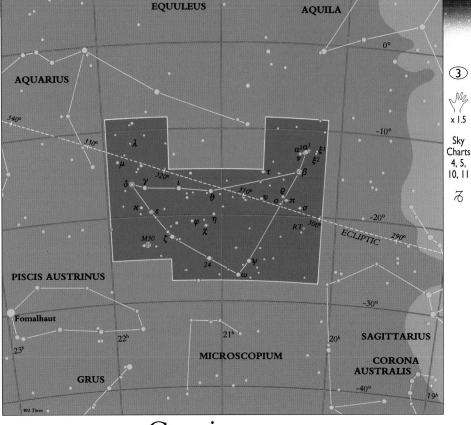

EQUULEUS

AQUILA

0°

AQUARIUS

-10°

340°

330°

λ

320°

μ

310°

δ

γ

ι

ϑ

α²α¹ ε ξ¹

ξ²

β

ν ο

ρ

π

τ

-10°

κ

ε

φ

η

χ

σ

-20°

M30

ζ

RT

300°

ECLIPTIC

290°

PISCIS AUSTRINUS

24

ψ

ω

-30°

Fomalhaut

22ʰ

21ʰ

20ʰ

SAGITTARIUS

23ʰ

MICROSCOPIUM

CORONA
AUSTRALIS

GRUS

-40°

19ʰ

Wil Tiron

3

x 1.5

Sky
Charts
4, 5,
10, 11

♑

# Capricornus

(kap-reh-KOR-nuss)  The Sea Goat, Capricorn

Capricornus has been named for a goat since the time of the Chaldeans and Babylonians. Sometimes it is shown as a goat, but more commonly it is depicted as a goat with the tail of a fish. This might relate to a story about the god Pan, who, when fleeing the monster Typhon, leaped into the Nile. The part of him that was underwater turned into a fish tail, while his top half remained that of a goat.

Several thousand years ago, the Sun reached its southernmost position in the sky (its winter solstice—declination 23.5 degrees south) when it was in front of Capricornus. During this time it was overhead at a southerly latitude we call the Tropic of Capricorn. It still carries this name, although the Sun, as a result of precession, is now in Sagittarius at the time of the winter solstice.

*The triangle of stars forming Capricornus is easily recognized, although the stars are no brighter than 3rd magnitude.*

Capricornus, the zodiac's least visible constellation, can be found by joining Aquila's three brightest stars in a line southward.

👁 **Alpha (α) Capricorni**: This double has a separation of 6 arcminutes—a naked-eye test for a night's clarity and steadiness. The pair is a double by coincidence, but each star is itself a true binary.

**M30**: Perhaps 26,000 light-years away, this globular cluster has a fairly dense center. It is not well resolved in small telescopes.

*Capricornus, as depicted in a fresco at the Villa Farnese, Caprarola, Italy (1575).*

289

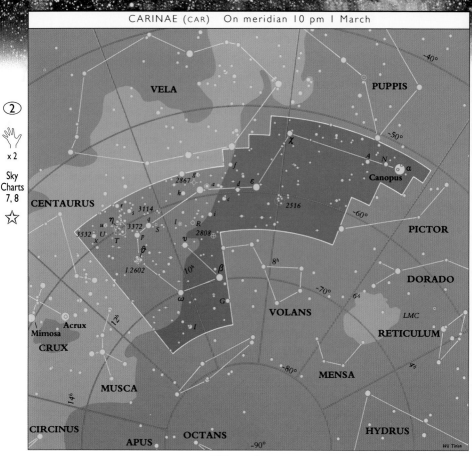

**2**

x 2

Sky
Charts
7, 8

☆

VELA

PUPPIS

CENTAURUS

2867

3114

3372

3532

2808

I.2602

2516

Canopus

PICTOR

DORADO

VOLANS

LMC

RETICULUM

Acrux
Mimosa

CRUX

MUSCA

MENSA

CIRCINUS

APUS

OCTANS

HYDRUS

-90°

Wil Tirion

# Carina

(ka-RYE-nah) The Keel

This Southern Hemisphere constellation is in the middle of one of the richest parts of the Milky Way and under a dark sky it is breathtaking. With binoculars, you can see at least half a dozen bright open clusters.

Carina is part of what was once a huge constellation known as Argo Navis, the Ship Argo—the vessel that Jason and his Argonauts sailed in on their search for the Golden Fleece. Argo Navis covered such a rich area of sky that it was divided into four separate constellations: Pyxis, Puppis, Vela, and Carina.

👁 **Canopus: Alpha (α) Carinae**, this yellow

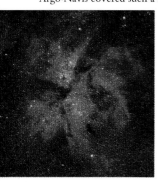

*The magnificent Eta (η) Carinae Nebula will reward any size of binoculars or telescope.*

supergiant is the second brightest star in the sky and is some 74 light-years away.

**Eta (η) Carinae**: In 1677, Edmond Halley noticed that this star had brightened. In 1827 it shot up to 1st magnitude, and for a few weeks in 1843 it tied with Sirius as the brightest star in the sky. In recent years, however, Eta (η) has been too faint to be seen without binoculars.

The star is famous primarily for the surrounding **Eta (η) Carinae Nebula (NGC 3372)**, the most exquisite nebula in the Milky Way. It is 2 degrees across, with dark rifts appearing to break it up. Superimposed on the brightest part of the nebula is the dark **Keyhole Nebula (NGC 3324)**.

👁 **NGC 3532**: A brilliant open cluster some 3 degrees from Eta Carinae, this is the finest of the clusters in Carina, with about 150 stars visible in a telescope at low magnification.

👁 **IC 2602**: This open cluster of scattered, bright stars around **Theta (θ) Carinae** is seen to best advantage in binoculars or in the eyepiece of a wide-field telescope.

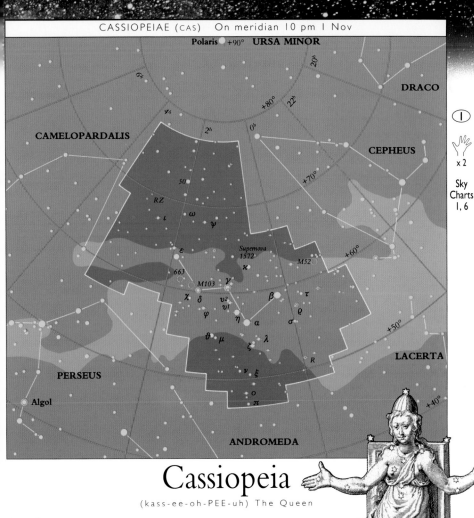

# Cassiopeia

(kass-ee-oh-PEE-uh) The Queen

This striking W-shaped figure is on the other side of Polaris from the Big Dipper. Most prominent in the Northern Hemisphere's winter sky, Cassiopeia is visible all year from mid-northern latitudes. In Greek mythology, she was queen of the ancient kingdom of Æthiopia—wife of Cepheus and mother of Andromeda.

The Romans saw Cassiopeia as having been chained to her throne, as a punishment for her boastfulness, and placed in the heavens to sometimes hang upside down. Arab cultures pictured the constellation as a kneeling camel.

*An engraving of Cassiopeia by Jacob de Gheyn (1621)*

👁 **Gamma (γ) Cassiopeiae**: This star lies at the center of Cassiopeia's W figure. Normally the constellation's third brightest star, it is an irregular variable. Over a few weeks in 1937, it was the brightest star in the constellation and almost as bright as Deneb in Cygnus. Known as a shell star, Gamma (γ) Cassiopeiae is slowly losing mass into a disk or shell that surrounds it, and alterations in the shell's thickness might be responsible for its variations in brightness.

🔭 **M52**: This group of about 200 stars is one of the richest in the northern half of the sky, but only one of several open clusters scattered throughout Cassiopeia.

**NGC 663**: A small open cluster of quite faint stars, NGC 663 is an attractive sight in a small telescope.

*The prominent "W" of Cassiopeia is an unmistakable feature in the northern sky.*

291

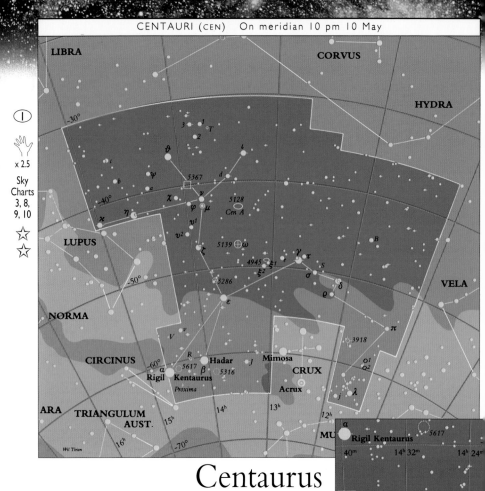

LIBRA

CORVUS

HYDRA

-30°

LUPUS

NORMA

CIRCINUS

LUPUS

VELA

-40°

-50°

3 1 T
2

ϑ
ψ
a
χ
φ µ
ν1
ν2
ζ
ε

5367
d
ν
5128
Cen A

5139 ⊕ ω
4945 ξ¹
ξ²
5286

γ τ
S
σ
ϱ δ

B

π

3918

ARA

TRIANGULUM
AUST.

NORMA

CIRCINUS

V ν
R
5617
α
Rigil Kentaurus
Proxima

-60°
Hadar J
β 5316

Mimosa

CRUX

Acrux

o¹
o²

j λ

15ʰ

16ʰ

14ʰ

13ʰ

12ʰ

MU

-70°

Wil Tirion

α  Rigil Kentaurus          5617

40ᵐ          14ʰ 32ᵐ      14ʰ 24ᵐ

Mag
5
6
7
8
9

2050
2000
1950

Proxima
Centauri

# Centaurus

(sen-TOR-us) The Centaur

These stars represent Chiron, who features widely in Greek myths. Chiron was one of the Centaurs—creatures that were half man, half horse. Unlike the other Centaurs, who were monstrous and brutal, Chiron was extremely wise, and tutored such humans as Jason and Hercules.

Hercules accidentally wounded him, and Chiron, in great pain but unable to die because he was immortal, pleaded with the gods to end his suffering. Zeus mercifully allowed Chiron to die, and placed him among the stars.

This huge constellation includes the bright Milky Way near the Southern Cross (Crux).

**Alpha (α) Centauri**: At the foot of the Centaur, this star is only 4.36 light-years away and is the Sun's nearest neighbor. One of the prettiest binary stars, its two components revolve around each other once every 80 years. The separation is currently just under 10 arcseconds, but this will close to 2 arcseconds by about the year 2035. Alpha (α) and **Beta (β) Centauri** are the bright "pointers" to the Southern Cross.

*The Centaur, from an engraving by Jacob de Gheyn (1621).*
*(Above) A chart of the region near Proxima Centauri.*

*The Southern Cross, in the lower right corner, is almost surrounded and dwarfed by Centaurus.*

**Proxima Centauri**: In 1915, R. T. Innes was measuring the proper motions of stars around Alpha (α) Centauri when he found a faint, 10.7 magnitude star moving at the same rate and in the same direction as the two stars of Alpha (α) Centauri but about 2 degrees away. A very small red dwarf star only 40,000 miles (25,000 km) across, this star is actually a little closer to us than the other two, but it is presumed to be their companion. It flares occasionally, jumping by half a magnitude or more, usually returning to its normal brightness within half an hour.

**Omega (ω) Centauri**: This globular cluster is, by most accounts, the finest example in the entire sky, with perhaps 1 million members. With the naked eye it is visible as a fuzzy star of 4th magnitude, so bright that Johann Bayer, in the early seventeenth century, gave it the designation Omega (ω). Only 17,000 light-years away, this is one of the closest clusters to us, second only to **NGC 6397** in Ara. Unlike most globulars, it is oval rather than round in shape.

*Omega (ω) Centauri is a spectacular sight in any telescope over about 6 inches (150 mm).*

A 3 inch (75 mm) telescope will show a large, fuzzy disk with mottled edges, while a 6 inch (150 mm) one will resolve it into stars. Viewed in an even larger telescope, under a dark sky, it looks magnificent, with the field of a low-power eyepiece overflowing with faint stars.

**NGC 5128**: Located only 4½ degrees north of Omega (ω) Centauri, this peculiar elliptical galaxy is distinguished by a strange dark band that crosses its center—probably the result of a collision with a spiral galaxy. A strong source of radio energy, known to radio astronomers as **Centaurus A**, this galaxy emits more than 1,000 times the radio energy of our own galaxy. The dark dust lane is apparent in dark skies with a 4 inch (100 mm) or larger telescope.

**NGC 3918**: This is a planetary nebula not far from the Southern Cross, presenting a classic blue-green disk about 12 arcminutes across, like a larger version of Uranus.

*The green star seen here in the dust lane of NGC 5128 is a supernova that was bright only when the green image of this red/green/blue composite was photographed.*

**293**

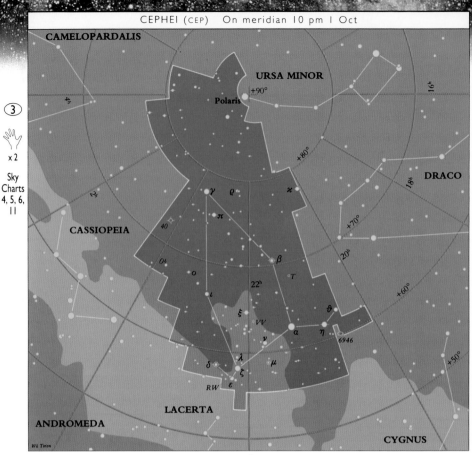

# Cepheus

(SEE-fee-us) The King

K ing of the ancient land of Æthiopia,
Cepheus was the husband of Cassiopeia
and father of Andromeda. His wife and
daughter are both represented by constellations,
and an account of the myth in which he is
involved is given under Andromeda (p. 272).

*(Above) NGC 6446 is a 9th magnitude spiral galaxy, but a*
*16 inch (400 mm) telescope will reveal the spiral arms.*

294

Cepheus is an inconspicuous constellation.
The constellation's five bright stars are easy to
find, only because they face the open side of the
W shape of Cassiopeia. It looks a little like a
house with a pointed roof. Although the top of
the roof does not really point to Polaris, it offers
the general direction to the pole at a time of year
when the pointer stars of the Big Dipper are not
readily accessible.

**Delta (δ) Cephei**: One of the most
famous of the variable stars, and the prototype
for the Cepheid variables, Delta (δ) Cephei's
variation was discovered by John Goodricke,
a deaf-mute teenager, in 1784. Its highest
magnitude is 3.5, as bright as neighboring **Zeta
(ζ) Cephei**, and it fades to 4.4, the brightness
of **Epsilon (ε) Cephei**. It completes a cycle
every 5.4 days.

**Mu (μ) Cephei**: This star is so strikingly
red that William Herschel called it the Garnet
Star. Using Zeta (ζ) and Epsilon (ε) as
comparison stars, you can watch it vary in
brightness irregularly over hundreds of days.

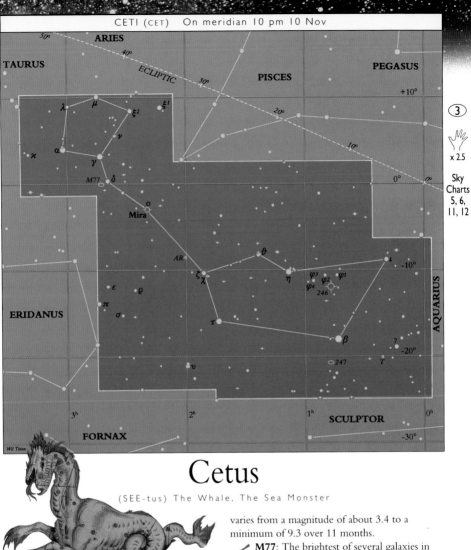

ARIES

TAURUS

ECLIPTIC

PISCES

PEGASUS

+10°

50°

40°

30°

20°

10°

0°

λ μ ξ²
ξ¹
ζ

ν

κ α γ

M77 δ

ο
Mira

AR
ϑ

ζ χ
η φ³
φ⁴ φ²
246
φ¹

ε ϱ

π

σ

τ

ι
-10°

ERIDANUS

β

7
-20°
T

υ
◯247

3ʰ 2ʰ 1ʰ SCULPTOR 0ʰ

FORNAX

-30°

AQUARIUS

Wil Tirion

③

🖐

x 2.5

Sky
Charts
5, 6,
11, 12

# Cetus

(SEE-tus) The Whale, The Sea Monster

K nown by the ancient Greeks as the
monster that was about to attack
Andromeda when Perseus destroyed it,
Cetus was later thought to represent the whale
that consumed Jonah. Cetus consists of faint
stars, but it occupies a large area of sky. His head
is a group of stars not far from Taurus and Aries,
and his body and tail lie towards Aquarius.

👁 **Mira: Omicron (o) Ceti**, known as Mira,
is the most famous long-period variable of all.
On 13 August 1596, David Fabricius, a Dutch
sky-watcher, noticed a new star in Cetus. Over
the following weeks, it faded, disappeared, then
reappeared in 1609. In 1662 Johannes Hevelius
named it Mira Stella, the Wonderful Star. Mira

varies from a magnitude of about 3.4 to a
minimum of 9.3 over 11 months.

🔭 **M77**: The brightest of several galaxies in
Cetus, M77 is a 9th magnitude spiral galaxy
with a bright core. A 4 inch (100 mm) telescope
shows a faint circular disk around the core.

*Mira appears near maximum brightness in this view of Cetus.*

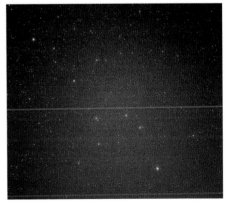

*(Above) Bayer's depiction of Cetus, from his Uranometria
(1603), in which Greek letters were introduced to label the stars.*

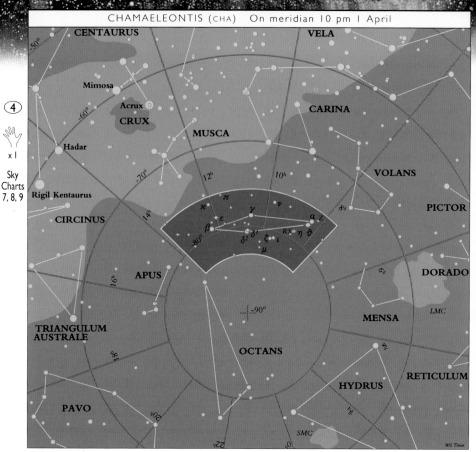

4

x 1

Sky
Charts
7, 8, 9

# Chamaeleon

(ka-MEE-lee-un) The Chameleon

Johann Bayer drew this constellation early in the seventeenth century, following descriptions that had been given by certain early south sea explorers.

The chameleon is a small lizard found in Africa that can change color to match its surroundings.

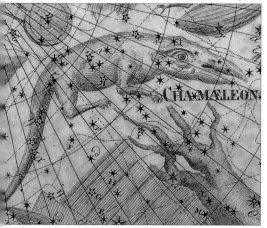

Modern researchers have found that the animal changes its color in response to sudden changes in light and temperature, and to emotional shock.

One of the smallest and least conspicuous of the constellations, Chamaeleon does a good job hiding in the sky too. Consisting of a few faint stars, it lies close to the south celestial pole, south of Carina and right beside the south polar constellation of Octans.

**Z Chamaeleontis**: This faint variable star erupts periodically. At its minimum it shines at magnitude 16.2, invisible except in 12 inch (300 mm) or larger telescopes. However, every three to four months it undergoes an outburst, rising within a few hours to about magnitude 11.5, and for a few days it is visible through a 6 inch (150 mm) telescope. Even so, this does not constitute an easy target in a corner of the sky with few stars.

*The Chameleon, as depicted by Johann Bode in his Uranographia (1801), extends beyond the modern boundaries of the constellation.*

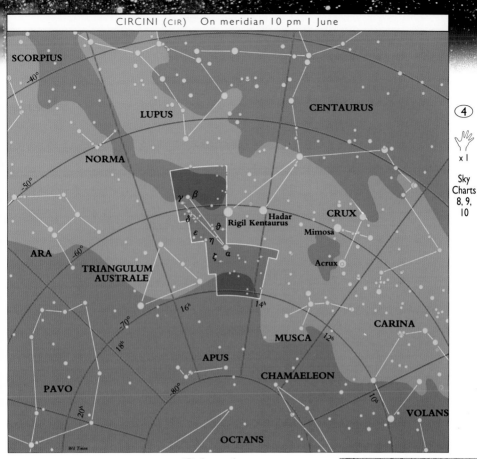

SCORPIUS
-40°

LUPUS

CENTAURUS

NORMA

-50°

γ β

δ
ε
θ    Hadar
Rigil Kentaurus

η
ζ  α

CRUX

Mimosa

ARA
-60°

TRIANGULUM
AUSTRALE

Acrux

16ʰ

14ʰ

-70°

18ʰ

CARINA

MUSCA  12ʰ

APUS

CHAMAELEON

PAVO
-80°

20ʰ

VOLANS

10ʰ

OCTANS

Wil Tirion

4

x 1

Sky
Charts
8, 9,
10

# Circinus

(SUR-seh-nus) The Drawing Compass

The early explorers in the south seas were less interested in mythology than in the modern instruments that they relied on to find their way around uncharted waters. The Drawing Compass is one of a number of obscure constellations that were designated by the

*Johann Bode's Uranographia (1801) was the first reasonably complete atlas of stars that can be seen with the naked eye, as can be seen by comparing his map of Circinus with the modern map.*

French astronomer Nicolas-Louis de Lacaille. He worked at an observatory at the Cape of Good Hope from 1750 to 1754, where he compiled a catalogue of more than 10,000 stars.

**Alpha (α) Circini**: This, at only 3rd magnitude, is the constellation's brightest star. It lies just near the much brighter Alpha (α) Centauri. It is 53 light-years away and has a faint 9th magnitude companion.

## SKYWATCHING TIP

Scale is one of the hardest things to become accustomed to when observing the night sky. If, in your early attempts, you are unable to find a single constellation it might be because you have no idea what size pattern you are looking for.

Firstly, look at the edge of the constellation chart to see how many hand spans will cover the constellation from east to west. Then compare the new constellation chart with some familiar ones.

When using the starfinder charts at the start of this chapter, bear in mind that star groups at the edges of charts will appear larger than they really are, because of distortion.

297

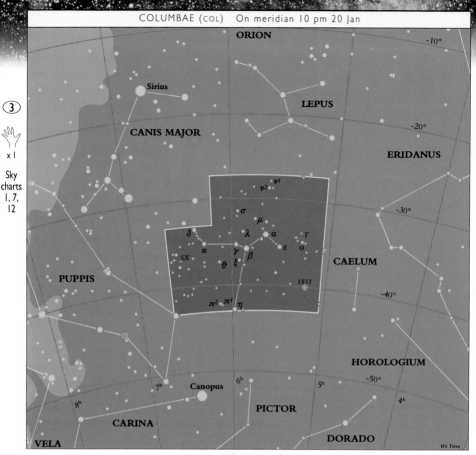

ORION

-10°

Sirius

CANIS MAJOR

LEPUS

-20°

ERIDANUS

ν2  ν1

σ

μ

δ        λ    α      τ

κ        γ      ε  ο

-30°

SX        θ  ξ  β

CAELUM

PUPPIS

1851

-40°

π2  π1  η

HOROLOGIUM

-50°

7ʰ       Canopus      6ʰ              5ʰ                    4ʰ

8ʰ

PICTOR

CARINA

DORADO

VELA

Wil Tirion

③

🖐 x 1

Sky charts 1, 7, 12

# Columba

(koh-LUM-bah) The Dove

Immediately south of Canis Major, Columba is a modern constellation named by Petrus Plancius, a sixteenth-century Dutch theologian and mapmaker. This inconspicuous group of stars honors the dove that Noah sent out from the ark after the rains had stopped, to see if it could find dry land.

**T Columbae**: A Mira variable, this star has a maximum magnitude of 6.7. It drops to magnitude 12.6 and then rises again over a period of seven and a half months.

**NGC 1851**: Bright and large, this 7th magnitude globular cluster appears as a misty spot through binoculars under a good sky. A 6 inch (150 mm) telescope will begin to resolve the cluster's brightest stars.

*Johann Bode's drawing of the Dove in Uranographia (1801) coincides quite closely with the official extent of Columba. (Left) About 11 arcminutes across, NGC 1851 is quite a large globular cluster.*

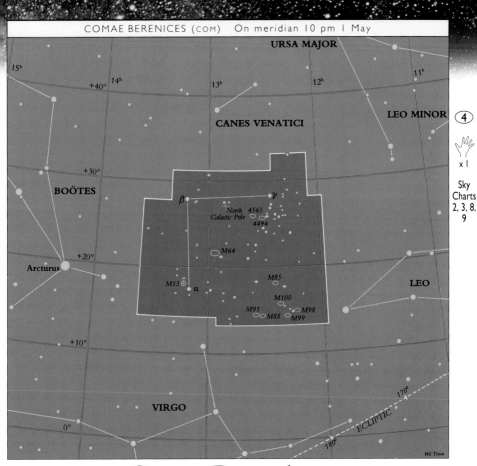

URSA MAJOR

15ʰ    14ʰ    13ʰ    12ʰ    11ʰ
+40°

CANES VENATICI

LEO MINOR

④

x 1

Sky
Charts
2, 3, 8,
9

+30°

BOÖTES

β    γ

North    4565
Galactic Pole    ○
4494

+20°

M64

Arcturus

M53 ⊕    α

M85

LEO

M100

M91    M98
M88    M99

+10°

VIRGO

170°

0°    ECLIPTIC

180°

Wil Tirion

# Coma Berenices

(KOH-mah bear-eh-NEE-seez) Berenice's Hair

Between Arcturus and Denebola (Beta [β] Leonis), Coma Berenices has no bright stars and is hard to distinguish, but it is a remarkable area of sky. It is a sprinkling of faint stars superimposed on a cloud of galaxies—the northern end of the Virgo cluster of galaxies. Fainter still, beyond the range of most amateur astronomers, is the Coma cluster of galaxies.

The story behind the constellation is, this time, about real people. Berenice, the beautiful wife of the ancient Egyptian king Ptolemy III, promised to sacrifice her long golden hair to Aphrodite if her husband returned safely from battle, which he did. Her hair was placed in the temple, but it disappeared. The king was about to put the temple guards to death when the court astronomer announced that Aphrodite, delighted with the gift, had placed it in the sky for all to admire.

Berenice's hair, committed to the stars, from Urania's Mirror (1825).

The Blackeye spiral galaxy with its dark dust lane.

**M53**: This fine globular cluster is about 3 arc-minutes in diameter and is located close to **Alpha (α) Comae Berenices.**

**The Blackeye Galaxy (M64)**: This is one of the most unusual galaxies in the sky. It looks like an ordinary spiral galaxy, with tightly wound arms, but when viewed with a 4 to 6 inch (100 to 150 mm) telescope or larger, a huge cloud of dust can be seen dominating its center, giving it the look of a black eye.

**NGC 4565**: Under a dark sky, a small telescope should show this faint object as a pencil-thin line of haze. It is a spiral galaxy seen edge-on, with a dust lane that becomes apparent in 8 inch (200 mm) telescopes.

299

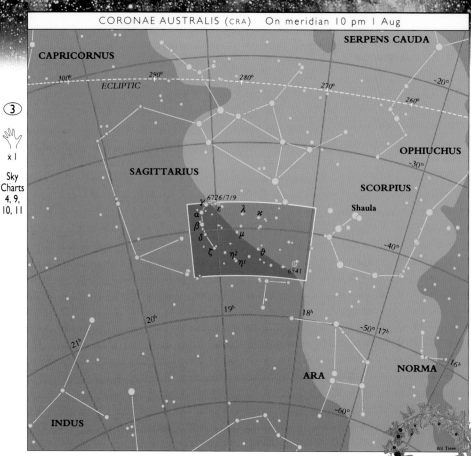

SERPENS CAUDA

CAPRICORNUS

ECLIPTIC

SAGITTARIUS

OPHIUCHUS

SCORPIUS

Shaula

6726/7/9

6541

ARA

NORMA

INDUS

Wil Tiron

# Corona Australis

(kor-OH-nah os-TRAH-lis) The Southern Crown

One of the 48 original constellations catalogued by Ptolemy in the second century AD, this small semicircular group of faint stars is inconspicuous, especially from the Northern Hemisphere. It lies just south of Sagittarius and is said to represent a crown

*In Uranometria (1603), Bayer's Southern Crown is a crown of laurels, as given to victorious athletes in ancient times.*

of laurel or olive leaves. One story has it that the crown belongs to Chiron.

Another story relating to the crown comes from Ovid's *Metamorphoses*. Juno discovered that her husband, Jupiter, was the lover of Semele, a human. Masquerading as Semele's maid, Juno suggested that Semele ask Jupiter to appear before her in all his glory. Jupiter was appalled at her request, but did not refuse it. When she saw him in his splendor she was consumed by fire. Her unborn child was saved, however, to become Bacchus, god of wine, who honored his mother by placing the crown in the sky.

**NGC 6541**: This globular cluster presents a small nebulous disk to smaller telescopes. An 8 inch (200 mm) telescope only begins to resolve the edge into stars.

*NGC 6726–7 at the top, and comet-like NGC 6729 just below it, lie almost on the border with Sagittarius.*

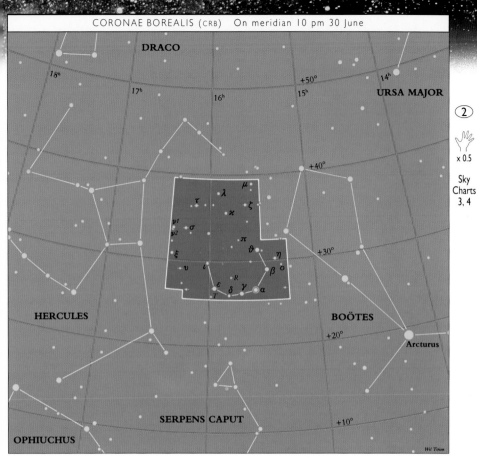

DRACO

URSA MAJOR

HERCULES

BOÖTES

Arcturus

SERPENS CAPUT

OPHIUCHUS

*Wil Tirion*

②

x 0.5

Sky
Charts
3, 4

# Corona Borealis

(kor-OH-nah bor-ee-AL-is) The Northern Crown

J ust 20 degrees northeast of Arcturus lies the
Northern Crown, a small semicircle of stars
that are faint but very distinct.

There are legends from many cultures to
explain its presence in the sky. The Greek myth
claims the crown belongs to Ariadne, daughter
of Minos, King of Crete. Ariadne was reluctant
to accept a marriage proposal from Dionysus
(who was in mortal form), since she did not wish
to marry a mortal after being deserted by
Theseus. To prove he was a god, Dionysus took
off his crown and threw it into the heavens as a
tribute to her. Satisfied, Ariadne married him and
became immortal herself.

**R Coronae Borealis**: One of the more
remarkable stars in the sky, R Cor Bor, as
it is generally known, is a nova in reverse.
Normally shining at magnitude 5.9,
at completely irregular
intervals the star will

*The Northern Crown appears as a
royal crown in Urania's Mirror (1825).*

*A small but clear semicircle of faint stars makes Corona Borealis
quite easy to recognize.*

suddenly fade, sometimes by as much as
8 magnitudes, as dark material erupts in its
atmosphere. It then slowly recovers as the
material dissipates.

**T Coronae Borealis**: Now shining at
magnitude 10.2, in 1866 this star suddenly
rose to magnitude 2. Known as a recurrent nova,
the star repeated the performance unexpectedly
in 1946, and will probably do so again.

301

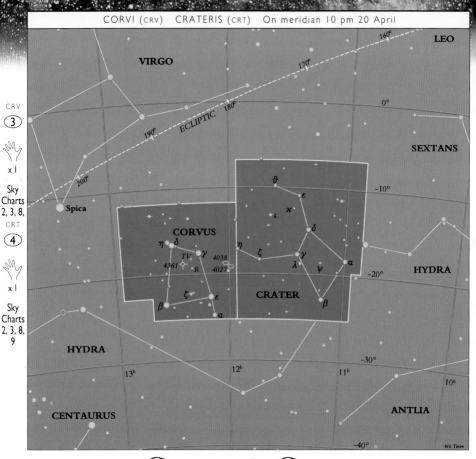

LEO

VIRGO

160°

170°

ECLIPTIC   180°

190°

200°

0°

SEXTANS

CRV
③
✋
x 1

Sky
Charts
2, 3, 8,
CRT
④
✋
x 1

Sky
Charts
2, 3, 8,
9

Spica

CORVUS

η δ
TV γ 4038
4361 R 4027

η ζ

ϑ
ε
ι ϰ
δ

-10°

γ
λ ψ α

CRATER

ζ
β ε
α

β

HYDRA

-20°

-30°

HYDRA

13ʰ

12ʰ

11ʰ

10ʰ

CENTAURUS

ANTLIA

-40°

Wil Tirion

# Corvus & Crater

(KOR-vus) The Crow     (KRAY-ter) The Cup

A rc to Arcturus, speed to Spica, then turn west and you will see a small foursome of stars that the ancients all called the Crow or the Raven. Crater is a fainter constellation alongside that looks like a cup.

Sent one day by Apollo for a cup of water, Corvus was slow in returning as he had been waiting for a fig near the spring to ripen. Bringing the cup (Crater) of spring water and a water serpent (Hydra) back in his claws, he told Apollo that he had been delayed because the serpent had attacked him. Apollo, knowing Corvus was lying, placed all three in the sky. The Cup is to the west of Corvus, within reach, but the serpent prevents him from drinking from it.

*The Crow and the Cup,*
*from the constellation cards*
*of Urania's Mirror (1825).*

**R Corvi**: This Mira-type variable star ranges from magnitude 6.7 to 14.4 over a period of about 10 months.

**Tombaugh's Star**: This very faint cataclysmic variable star, **TV Corvi**, was discovered as a nova by Clyde Tombaugh in 1931, while searching for planets. David Levy chanced upon a reference to it while researching a biography of Tombaugh. After examining 360 photographs, he found it had repeated its outburst 10 times. In 1990, he observed an explosion of the star with a 16 inch (400 mm) telescope.

**The Ring-tailed Galaxy**: Also called the Antennae or Rat-tailed Galaxy, **NGC 4038** and **NGC 4039** form a faint, 11th magnitude pair of galaxies that are interacting or colliding. Needing an 8 inch (200 mm) telescope to see, it is still one of the brightest pairs of connected galaxies.

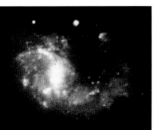

*NGC 4027 is a faint,*
*disturbed spiral galaxy*
*in Corvus.*

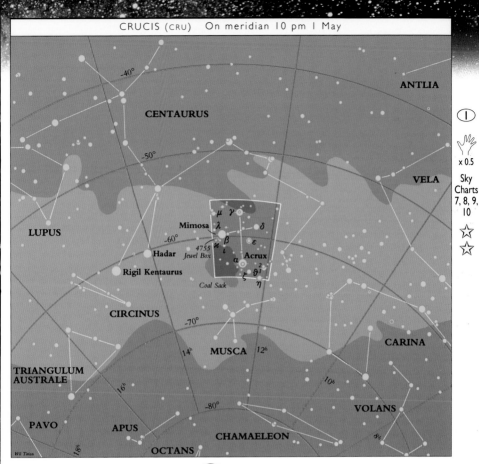

ANTLIA

CENTAURUS

VELA

LUPUS

Mimosa

Hadar   4755
Jewel Box

Rigil Kentaurus

Coal Sack

Acrux

CIRCINUS

MUSCA

CARINA

TRIANGULUM
AUSTRALE

PAVO   APUS

VOLANS

CHAMAELEON

OCTANS

Wil Tirion

# Crux

(KRUKS) The Southern Cross

The most famous southern constellation, the Southern Cross appears on the flags of several nations. Its distinctive pattern of stars helped guide sailors for centuries, the upright of the cross pointing the way to the south celestial pole. Because it lies so far south, Crux was not mapped as a separate entity until 1592. Until then, it formed part of Centaurus.

The cross contains the most striking pair of opposites—the Jewel Box and the Coal Sack—embedded within the southern Milky Way.

**Acrux**: This is the popular name for **Alpha (α) Crucis**, the bright double star at the foot of the cross, separated by about 4½ arcseconds. A third star, quite bright at magnitude 5, lies 90 arcseconds away.

**Gamma (γ) Crucis**: Also known as **Gacrux**, this wide double star marks the northern end of the cross. An optical double, it consists of a magnitude 6.4 star lying almost 2 arcminutes from a bright orange primary.

**The Jewel Box:** Superimposed on **Kappa (κ) Crucis**, this is one of the finest open

*One feature of the Jewel Box is the contrasting colors of the line of three stars at its heart.*

clusters. Although small, it sparkles in any instrument, and has several stars of contrasting color.

**The Coal Sack**: This is one of the largest and densest dark nebulas in the sky. It lies just east of Acrux and is clearly visible in a dark sky against the star clouds of the Milky Way.

303

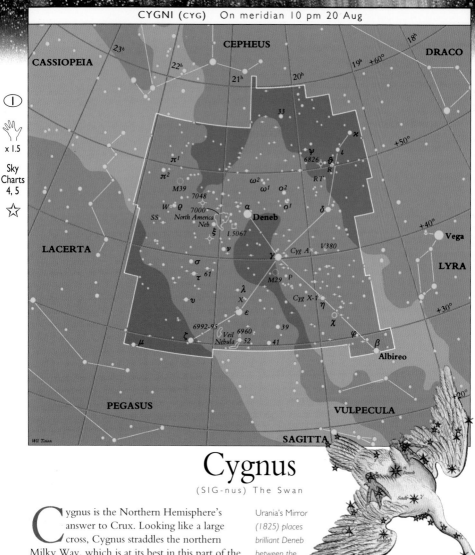

CEPHEUS

CASSIOPEIA

DRACO

23ʰ

22ʰ

21ʰ

20ʰ

19ʰ  +60°

18ʰ

+50°

33

π¹

π²

M39

7048

W   Ω   7000

North America
Neb

SS

ξ   L 5067

ψ
6826

RT

ι

ϑ
R

ω²

o²

ω¹

α

o¹

δ

Deneb

Vega

LACERTA

ν

γ

Cyg A

V380

LYRA

+40°

σ

τ   61

M29   P

+30°

υ

X

λ

Cyg X-1

η

ε

χ

φ

β

6992-95

ζ

Veil   6960
Nebula   52

39

41

Albireo

μ

PEGASUS

VULPECULA

+20°

Wil Tirion

SAGITTA

# Cygnus

(SIG-nus) The Swan

Cygnus is the Northern Hemisphere's answer to Crux. Looking like a large cross, Cygnus straddles the northern Milky Way, which is at its best in this part of the sky. If you are under a dark sky you may be able to see the Milky Way divide into two streams in Cygnus. A dark nebula between us and the more distant stars causes this apparent divergence.

Since the time of the Chaldeans, many civilizations have seen this constellation as a bird of some sort. One story claims that Cygnus is Orpheus, the great hero of Thrace, who sang and played his lyre so beautifully that wild animals and even the trees would come to hear him. It is said that Orpheus was transported to the sky as a swan, so that he could be near his cherished lyre. Another myth claims that Cygnus is Zeus in the disguise of a swan—the form he took to seduce Leda of Sparta.

Deneb (Alpha [α] Cygni): Deneb means "tail" in Arabic, which is where this star is positioned on the swan. On a par with Rigel in Orion, it is one of the mightiest stars known—

*Urania's Mirror (1825) places brilliant Deneb between the feet of the swan.*

25 times more massive and 60,000 times more luminous than the Sun. About 1,500 light-years away, Deneb is by far the most distant star of the famous Summer Triangle, which it forms with Vega and Altair. Vega is 25 light-years away and Altair only 17.

Albireo (Beta [β] Cygni): Whether you are observing this star from the dark of the country or from the middle of a city, Albireo, at the foot of the cross, is one of the prettiest sights in the sky. Without a telescope it is seen as a single star; a telescope transforms it into a spectacular double with a separation of 34 arc-seconds. One member is golden yellow with a magnitude of 3, and the other is bluish with a magnitude of 5.

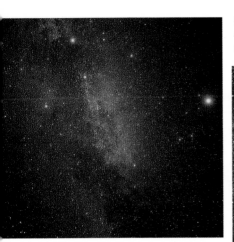

(Left) The Northern Cross in Cygnus lies astride the bright northern Milky Way. The North America Nebula (NGC 7000), visible just to the left of Deneb, is clearer in the image below.

**61 Cygni**: Dubbed the Flying Star because of its rapid motion relative to more distant stars, this double is easily separated in small telescopes. The two components revolve around each other over the course of about 650 years. 61 Cygni is the first star (apart from the Sun) to have its distance measured. In 1838, Friedrich Bessel used parallax measurements to derive a distance of 10.4 light years, close to the modern value of 11.4 light years.

**The North America Nebula (NGC 7000)**: One of the sky's best examples of a bright nebula, this giant cloud is illuminated by Deneb, which lies only 3 degrees to the west. Because of its size, the nebula is difficult to see in

a telescope: it is best seen with the naked eye on a dark night. Photographs show this nebula to look surprisingly like the shape of North America, but this resemblance is not readily apparent to the eye when observing.

**M39**: This loosely bound open star cluster is seen at its best through a pair of binoculars. On a clear night you might be able to see it with the naked eye.

**Chi (χ) Cygni**: At maximum brightness, typically magnitude 4 or 5, this long-period variable is bright enough to be seen with the naked eye. It fades to about magnitude 13 and then climbs back in a period of a little more than 13 months.

**SS Cygni**: One of numerous faint variables in Cygnus, SS Cygni is a dramatic cataclysmic variable star, erupting every two months to magnitude 8 but normally remaining faint at magnitude 12.

**The Veil Nebula (NGC 6960, 6992, 6995)**: The lacy remnants of an ancient supernova, this beautiful nebulosity requires at least a 6 inch (150 mm) telescope. NGC 6960, the nebula's western arc, passes through 52 Cygni, which makes it easier to find but harder to see.

**The Blinking Nebula (NGC 6826)**: This planetary nebula has a relatively bright central star. If you concentrate on the star, the surrounding cloud disappears.

NGC 6992 and 6995 form the eastern arc of the Veil Nebula.   **305**

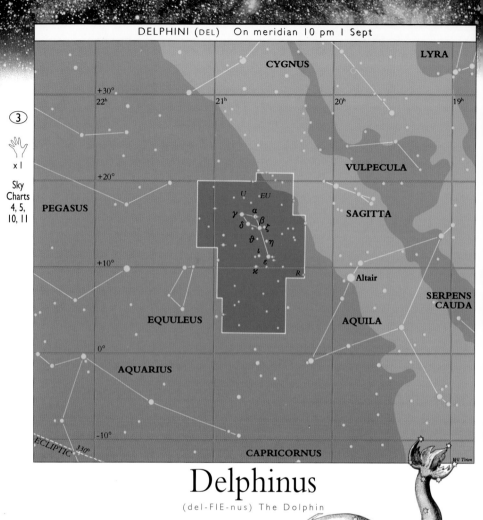

LYRA

CYGNUS

+30°
22ʰ          21ʰ          20ʰ          19ʰ

③

✋
x 1

Sky
Charts
4, 5,
10, 11

VULPECULA

+20°

U    EU

γ   α
δ   β ζ
ϑ   η
ι
ε
κ          R

SAGITTA

PEGASUS

+10°

Altair

SERPENS
CAUDA

EQUULEUS

AQUILA

0°

AQUARIUS

-10°

ECLIPTIC   330°

CAPRICORNUS

Wil Tirion

# Delphinus

(del-FIE-nus) The Dolphin

A small constellation with a distinctive shape, Delphinus has been thought of as a dolphin since ancient times. It is said that the mermaid Amphitrite agreed to marry Poseidon, from whom she had been trying to escape, on the advice of a dolphin. Poseidon was so pleased with the little dolphin that he placed him among the stars. The constellation is also sometimes known as Job's Coffin, the origin of which is obscure.

This small group of faint stars looks a little like a kite. Its alpha (α) star is named Sualocin and its beta (β) star is known as Rotanev. These names honor a relatively recent

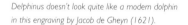

*Delphinus doesn't look quite like a modern dolphin in this engraving by Jacob de Gheyn (1621).*

observer, Niccolo Cacciatore, long-time associate of the famous nineteenth-century observer Giuseppe Piazzi. Star atlases at the time included these names without comment, but the Reverend Thomas Webb worked out that the names, spelled backward, are Nicolaus Venator—the Latinized version of Cacciatore's name.

**Gamma (γ) Delphini**: This is an optical double with a separation of 10 arcseconds. The brighter is magnitude 4.5 and the fainter, which is slightly green, is 5.5.

**R Delphini**: This Mira star has a magnitude range of 8.3 to 13.3 over a period of 285 days.

*NGC 6934 is a 9th magnitude globular, marked by the relatively bright star nearby.*

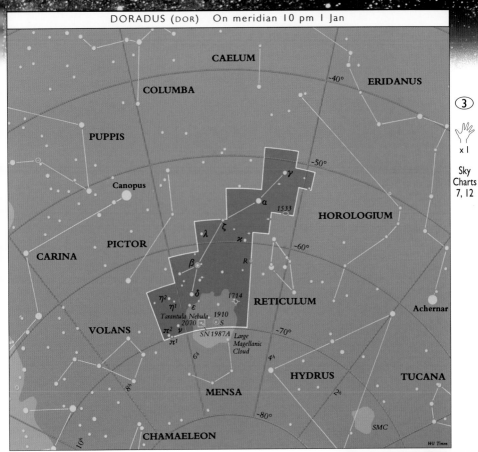

CAELUM

COLUMBA

ERIDANUS

~40°

PUPPIS

Canopus

~50°

γ

α

1533

HOROLOGIUM

PICTOR

λ ζ

κ

~60°

CARINA

β

R

η²

η¹ ε

1714

RETICULUM

Achernar

δ

Tarantula Nebula 1910

VOLANS

π²

ν

2070

S

SN 1987A Large

π¹

Magellanic

Cloud

6ʰ

4ʰ

~70°

HYDRUS

TUCANA

MENSA

2ʰ

8ʰ

~80°

CHAMAELEON

SMC

10ʰ

Wil Tirion

③

x 1

Sky
Charts
7, 12

# Dorado

(doh-RAH-doh) The Goldfish, The Dolphinfish

Far to the south, this constellation was first recorded by Bayer in his star atlas of 1603. Dorado does not honor the tiny fish in many a home aquarium, but the tropical dolphinfish, the mahi-mahi, member of the Coryphaenidae family, which can be more than 5 ft (1.75 m) long. Since they swim fast and often leap out of the water in play, sailors used to consider their appearance a good omen.

*A cluster of hot, young stars centered around 30 Doradus illuminates the massive Tarantula Nebula.*

*The Tarantula Nebula is the largest of the numerous pink nebulas in the Large Magellanic Cloud.*

◉ **The Large Magellanic Cloud (LMC)**: This is a companion galaxy to the Milky Way, lying 180,000 light-years away—less than one-tenth the distance to the Andromeda Galaxy (M31). As a result, it spans about 11 degrees of the sky, presenting its contents to the scrutiny of Southern Hemisphere observers. It was from this galaxy that supernova 1987A blazed forth. The LMC is plainly visible in a dark sky, but it is easily lost in the glare of city lights.

◉ **The Tarantula Nebula (NGC 2070)**: Also known as the 30 Doradus Nebula, this is one of the finest emission nebulas in the sky, despite its distance from us. It is perhaps 30 times the size of the more famous Great Nebula in Orion (M42).

✦ **S Doradus**: A single super-luminous star within the open cluster **NGC 1910**, S Doradus varies irregularly in brightness between magnitudes 9 and 11. It is one of the most luminous stars known.

307

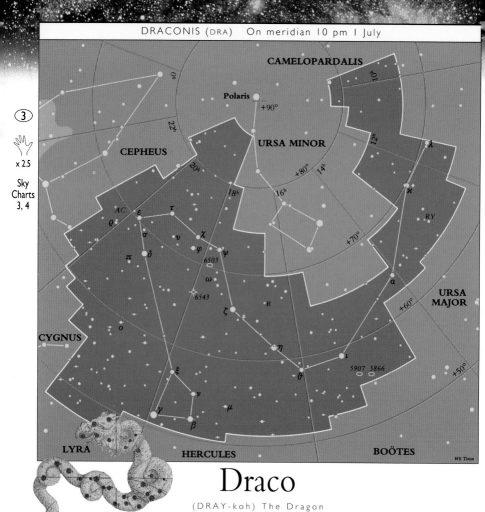

CAMELOPARDALIS

Polaris
+90°

URSA MINOR

CEPHEUS

+80°

AC

URSA MAJOR

CYGNUS

+70°

6503

6543

+60°

+50°

5907 5866

LYRA          HERCULES          BOÖTES

Wil Tirion

# Draco

(DRAY-koh) The Dragon

This constellation is circumpolar from much of the Northern Hemisphere and is best seen during the warmer months. A large, faint constellation, the Dragon is hard to trace as it winds about between Ursa Major, Boötes, Hercules, Lyra, Cygnus, and Cepheus.

The Chaldeans, Greeks, and Romans all saw a dragon here, while Hindu mythology claims the creature is an alligator. The Persians saw a man-eating serpent.

Draco has been identified with a number of ancient Greek stories. A dragon guarded the entrance to the Hesperides, where the golden apples grew, and was killed by Hercules. And Athena threw a dragon into the sky, after it attacked her while she was fighting the Titans.

Thuban, the brightest star in the constellation, was the pole star in ancient times, but Earth's precession has since moved the pole to Polaris.

**Quadrantids:** This is one of the strongest meteor showers. The time of maximum activity is around 3 January, and it lasts only a few hours.

**Draconids:** This meteor shower consists of particles from Periodic Comet Giacobini-Zinner. In 1933 and 1946, the shower's date of 9 October closely followed the comet's crossing of Earth's orbit, and the result was a storm of meteors.

**NGC 6543:** This 8th magnitude planetary nebula lies midway between the stars **Delta (δ)** and **Zeta (ζ) Draconis**. It is bright blue-green in color, but high power is needed in order to make out its small, hazy disk.

*(Above) Draco, the Dragon, as portrayed by Johann Bayer in his Uranometria (1603), curls its way through the heavens.*

*(Right) NGC 6543 is one of the brighter planetary nebulas.*

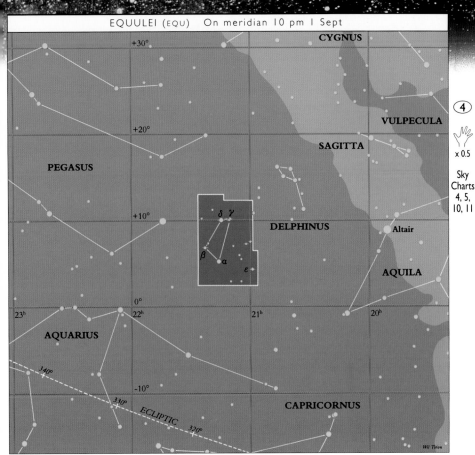

+30°
CYGNUS
+20°
VULPECULA
SAGITTA
PEGASUS
DELPHINUS
+10°
δ γ
β
α
ε
Altair
AQUILA
0°
23ʰ     22ʰ     21ʰ     20ʰ
AQUARIUS
340°
-10°
330°
ECLIPTIC
320°
CAPRICORNUS

Will Tirion

④

× 0.5

Sky Charts 4, 5, 10, 11

# Equuleus

(eh-KWOO-lee-us) The Little Horse

With the exception of Crux, Equuleus occupies a smaller space than any other constellation. It lies just to the southeast of Delphinus and because it has no bright stars it is of limited interest. **Alpha (α) Equulei**, its brightest star, is named Kitalpha—Arabic for "little horse."

The famous Greek astronomer Hipparchus is thought to have made up the constellation in the second century BC. It has been said to represent Celeris, brother of Pegasus (the Winged Horse), given to Castor (one of the twins represented by Gemini) by Mercury.

### SKYWATCHING TIP

When skywatching, it's a good idea to keep a record of what you see. Not only will this help you to remember your observations, but the effort to record details will stimulate careful and meaningful viewing. A notebook that's large enough for drawings as well as notes is ideal. The information you take down should include: date, time, and location of observation; instrument(s) used; seeing conditions; and sketches of your sightings.

Nebulas, star clusters, and galaxies present some of the most interesting targets for amateur astronomers. In general, low magnification and a wide field of view are necessary for deep-sky observing, but don't be afraid to experiment.

*Equuleus, as represented in Bayer's Uranometria (1603). It is likely that it came to be known as the Little Horse to distinguish it from nearby Pegasus, the Winged Horse.*

ORION
TAU
Mira
CETUS
32
0°
μ  ν  ξ
o¹
-10°
o²
ζ  ρ³  ρ¹
ρ²  η
δ  ε
β
ω
Rigel  ψ
λ
π
1535
γ
z
-20°
53
1300
τ¹
54
τ²
τ⁵  τ⁴  τ³
τ⁶
τ⁹  τ⁷
τ⁸
LEPUS
-30°
3ʰ  FORNAX  2ʰ
υ¹
υ²
4ʰ
υ⁴
5ʰ  υ³
-40°
g
f
θ
6ʰ
e
PHOENIX
COLUMBA
CAELUM
κ
-50°
HOROLOGIUM
Wil Tirion

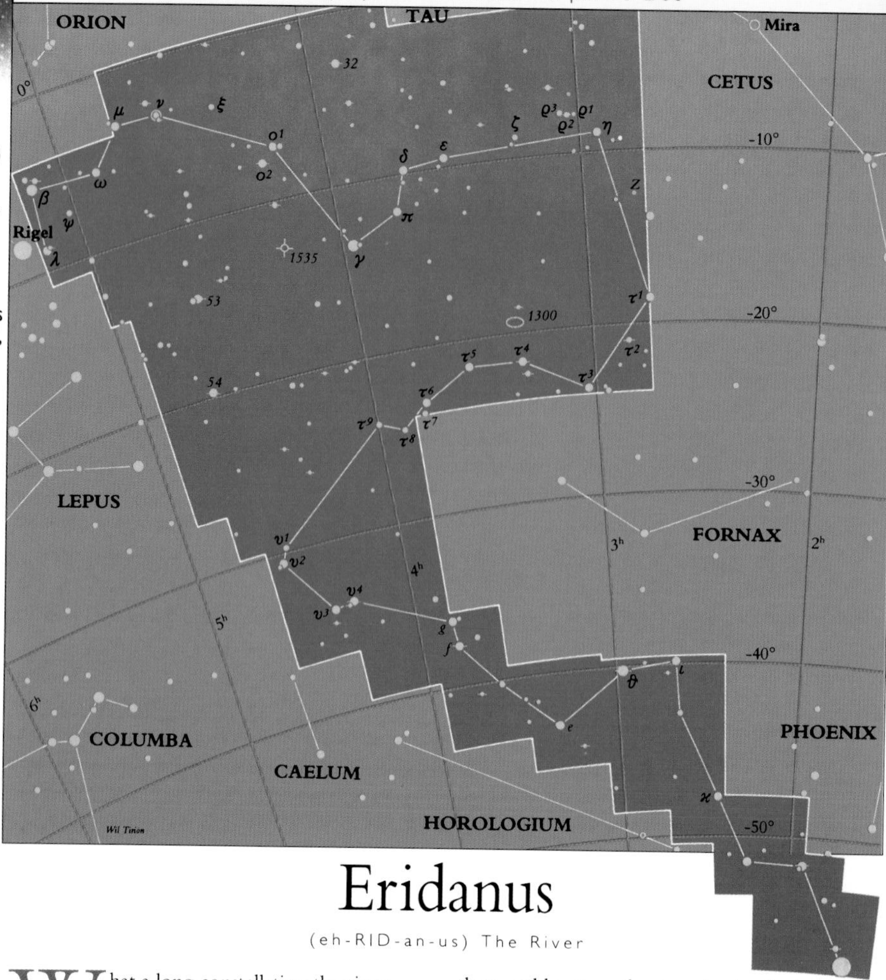

# Eridanus

(eh-RID-an-us) The River

What a long constellation the river Eridanus is! Its source lies immediately to the west of Rigel in Orion, with a star called Cursa or Beta (β) Eridani. It flows southward until it reaches its mouth in **Achernar (Alpha [α] Eridani)** near the south celestial pole—a very bright star few northern observers ever see. From the Southern Hemisphere, an observer can follow the full course of the river, even though the stars are faint.

This constellation has been seen as a river since ancient times—usually the Euphrates or the Nile. For early observers in Southwest Asia, the river extended only as far south as Acamar, or **Theta (θ) Eridani**, because

they could not see the stars that lay further south. In Book II of *Metamorphoses*, Ovid writes of Phaethon being tossed out of the chariot of the Sun to drown in Eridanus.

**Omicron 2 (o₂) Eridani**: This remarkable triple consists of a 4th magnitude orange dwarf, a 9th magnitude white dwarf, and an 11th magnitude red dwarf. The red and white dwarfs form a pair (separation 8 arcseconds), and are separated from the brighter star by over 80 arcseconds. The white dwarf is the only one of its class that is easy to see in a small telescope.

**Epsilon (ε) Eridani**: Only 10.5 light-years away, this star is a smaller version of our Sun. Radio telescopes have been pointed towards it in a so far unsuccessful search for signals indicating intelligent life.

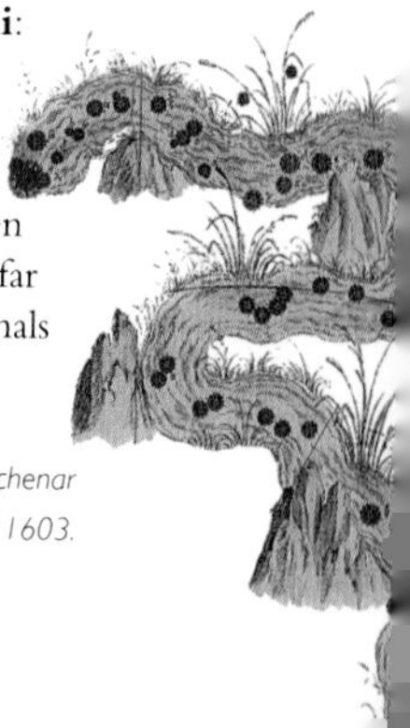

*Really a target only for larger telescopes, NGC 1300 is a classic barred spiral.*

*Eridanus, the River, with brilliant Achenar at its mouth, in Bayer's drawing of 1603.*

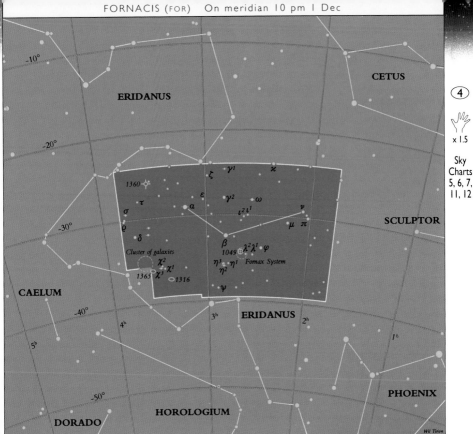

CETUS

ERIDANUS

④

⋈

x 1.5

Sky
Charts
5, 6, 7,
11, 12

-10°

-20°

-30°

1360

τ

σ

ρ

δ

Cluster of galaxies

1365    χ²    χ¹    ●1316

CAELUM

-40°

5ʰ    4ʰ

ζ    γ¹    χ

ε    γ²    ω

ι²ι¹

α

β
1049    λ²λ¹    φ

η³    η¹    Fornax System
η²

ψ

3ʰ    ERIDANUS    2ʰ

ν

μ    π

SCULPTOR

1ʰ

PHOENIX

-50°

DORADO    HOROLOGIUM

Wil Tirion

# Fornax

(FOR-nax)  The Furnace

NGC 1365, a barred spiral
galaxy, and one of the
brighter Fornax galaxies,
at 9th magnitude.

When Nicolas-Louis de Lacaille invented this constellation out of several faint stars in a bend of River Eridanus, he was honoring the famous French chemist Antoine Lavoisier, who was guillotined during the French Revolution in 1794.

**The Fornax Galaxy Cluster**: While there are no bright points of interest in Fornax, if you have a large telescope you will enjoy this challenging cluster of galaxies near the

Fornax–Eridanus border. With a wide-field eyepiece, you may see up to nine galaxies in a single field of view. The brightest galaxy, at 9th magnitude, **NGC 1316** is also the radio source Fornax A.

**The Fornax System**: Although this dwarf galaxy—a diminutive member of our Local Group of galaxies—appears to be unusual, galaxies like it may be common in the universe. Spherical in shape, it is a large group of very faint stars that includes a few globular clusters. It is too faint to see with an amateur telescope, but one globular cluster, **NGC 1049**, is, at magnitude 12.9, visible in a 10 inch (250 mm) telescope under a good sky.

A portion of the Fornax Galaxy Cluster, in a field about
1 degree across.

311

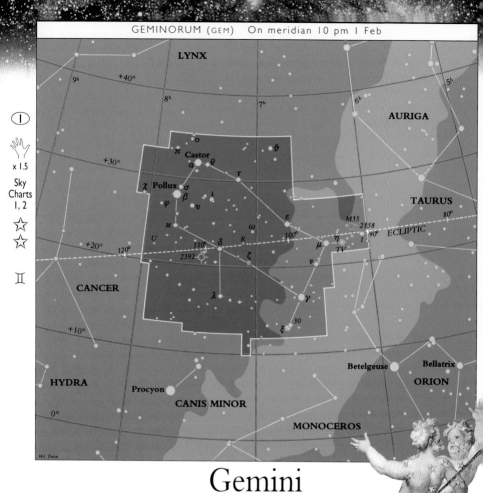

LYNX

9h    +40°

8h          7h        6h        5h

AURIGA

o

π   Castor

α   ρ        ϑ

+30°              τ

χ Pollux   σ

β   ι              TAURUS

φ   υ                          80°

κ          ε        M35

+20°   120°   U   110°   δ   R   ω   100°   2158   ECLIPTIC
                    2392              η   1   90°
                    ζ        μ        TV
                              ν

CANCER              λ                      γ

+10°                        ξ   30

Betelgeuse   Bellatrix

HYDRA              Procyon          ORION

0°              CANIS MINOR

MONOCEROS

Wil Tirion

# Gemini

(JEM-eh-nye) The Twins

A familiar pattern in the sky, Gemini is part of the zodiac. Various cultures have seen the stars as twins—either as gods, men, animals, or plants. The Greeks named the constellation's two brightest stars Castor and Pollux, after the twins who hatched from an egg from their mother Leda, following her seduction by Zeus. The twins were among the heroes who sailed with Jason in the quest for the Golden Fleece. They helped save the *Argo* from sinking during a storm, so the constellation was much valued by sailors.

William Herschel discovered Uranus near Eta (η) Geminorum in 1781, and Clyde Tombaugh discovered Pluto near Delta (δ) Geminorum in 1930.

*M35, with NGC 2158 just to the west.*

*Gemini, as depicted in a fresco from an Italian villa (1575).*

**Castor (Alpha [α] Geminorum):** This sextuple star can be seen only as a double through a small telescope. Its current separation is about 3 arcseconds.

**Eta (η) Geminorum:** This bright semi-regular variable varies from magnitude 3.2 to 3.9 and back over about eight months.

**M35:** This bright open cluster is beautiful through binoculars and spectacular in a small telescope. **NGC 2158** is a smaller, fainter open cluster on its southwest edge. It appears in small telescopes as a smudge, being about 16,000 light-years away—five times the distance to M35.

**The Clownface or Eskimo Nebula (NGC 2392):** This strange-looking, 8th magnitude planetary nebula has a bright central star. The blue-green tint of its 40 arcsecond disk gives it away.

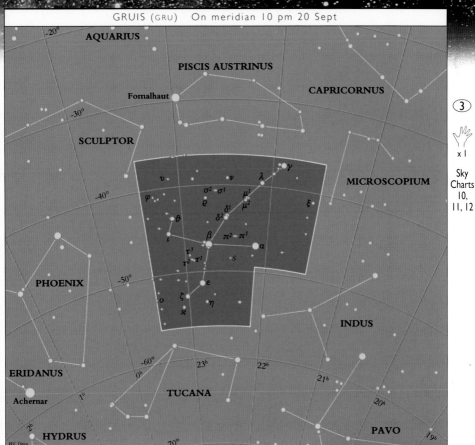

-20°
AQUARIUS
PISCIS AUSTRINUS
CAPRICORNUS
Fomalhaut
-30°
SCULPTOR
MICROSCOPIUM
-40°
$\nu$  $\nu$  $\lambda$  $\gamma$
$\sigma^2$ $\sigma^1$  $\mu^1$
$\varphi$  $\varrho$  $\delta^1$  $\mu^2$  $\xi$
$\vartheta$  $\delta^2$
$\iota$  $\beta$  $\pi^2$  $\pi^1$
$\tau^3$  $\alpha$
$\tau^2$ $\tau^1$  $S$
PHOENIX   -50°
$o$  $\zeta$  $\varepsilon$
$\kappa$  $\eta$
INDUS
-60°
23ʰ      22ʰ
ERIDANUS  0ʰ                     21ʰ
Achernar  1ʰ
TUCANA                     20ʰ
2ʰ                              19ʰ
HYDRUS        -70°             PAVO
Wil Tirion

③

🖐
x 1

Sky
Charts
10,
11, 12

# Grus

(GROOS) The Crane

I n his star atlas of 1603, Johann Bayer named this southern constellation Grus, the Crane—the bird which served as the symbol of astronomers in ancient Egypt. This group of stars—variously seen as a stork, a flamingo, and a fishing rod—has very little to offer the skywatcher who is using a small telescope, although some faint galaxies provide suitable targets for telescopes of 8 inch (200 mm) aperture or larger.

*It is interesting to compare the Crane depicted by Bode in* Uranographia *(1801) with our chart of Grus.*

*The cross-like shape of Grus is easily found just south of bright Fomalhaut.*

Grus has only three fairly bright stars, which can be used as a simple illustration of magnitude.
**Alpha (α) Gruis**, also known as Alnair, is a large, blue main-sequence star about 70 times as luminous as the Sun. Being only 100 light-years away, it is the brightest of the three stars only because it is relatively close to us.
**Beta (β) Gruis** is a much larger red giant star, some 800 times as luminous as the Sun, but its 170 light-year distance from us makes it appear fainter than Alpha (α) Gruis.
Finally, **Gamma (γ) Gruis**, a blue giant star, which is actually more luminous than either of the others, appears fainter than them since it is 200 light-years away.

313

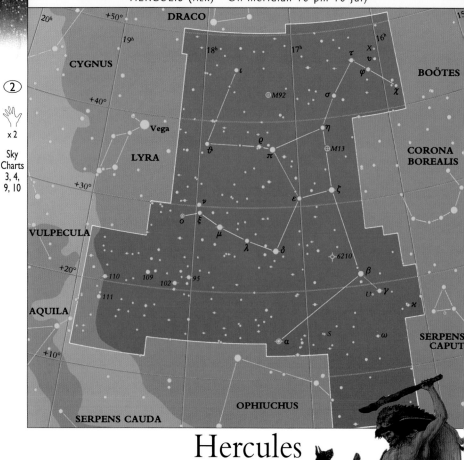

2

x 2

Sky
Charts
3, 4,
9, 10

# Hercules

(HER-kyu-leez) Hercules

F or northern observers,
Hercules, with its "key-
stone" of four stars—
**Epsilon (ε), Zeta (ζ), Eta (η),
and Pi (π)**—is one of the best
of the summer constellations.

One of the most famous of
all the classical heroes, Hercules
was immensely strong and was
revered throughout the Mediterranean. He was the half-
mortal son of Jupiter and was involved in many
noble exploits, the most famous being the
undertaking of the twelve labors. At the end of
his life, as a reward for his bravery, Jupiter made
him one of the gods, placing him in the sky.

👁 **The Hercules Cluster (M13):** The most
dramatic globular cluster in the northern sky, this
is faintly visible to the naked eye as a fuzzy spot,
but through a telescope it is a sight to behold.
The edges begin to resolve into stars in a 6 inch
(150 mm) telescope. When you view this cluster,
you are looking 25,000 years into the past.

*(Left) M13, the premier globular cluster of the northern sky.
(Right) Hercules and the Hydra, from a painting by Antonio
Pollaiuolo (1432–98).*

**M92:** M13's slightly smaller and fainter
cousin, this cluster of stars is some 27,000
light-years away.

👁 **Ras Algethi (Alpha [α] Herculis):** This
is a very red star, varying from magnitude 3.1 to
3.9. It is also a splendid colored double, with a
5th magnitude blue-green companion about
5 arcseconds away from an orange primary.

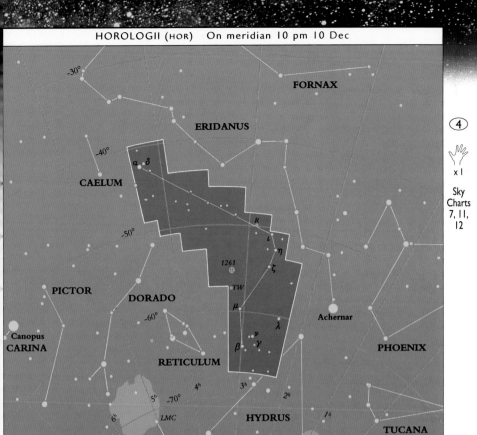

FORNAX

ERIDANUS

CAELUM

-30°

-40°

α  δ

-50°

R

ι

η

ζ

1261

TW

PICTOR

DORADO

-60°

μ

λ

Achernar

Canopus
CARINA

RETICULUM

ν
γ
β

PHOENIX

4ʰ

3ʰ

2ʰ

5ʰ  -70°

6ʰ

LMC

HYDRUS

1ʰ

TUCANA

VOLANS

MENSA

SMC

0ʰ

Wil Tirion

④

✋
x I

Sky
Charts
7, 11,
12

# Horologium

(hor-oh-LOH-jee-um)  The Clock

A small group of stars lying east of Archernar, this is one of the con-stellations mapped by Nicolas-Louis de Lacaille. Originally called Horologium Oscillatorium, it honors the invention of the pendulum clock by Dutch scientist Christiaan Huygens in 1656 or 1657.

*A 10 inch (250 mm) telescope captured this view of NGC 1261, seen at a distance of over 50,000 light-years.*

Huygens was a leader in Renaissance thinking. By apply-ing the law of the pendulum discovered by Galileo to clockmaking, he significantly increased the accuracy of timekeeping. Huygens' second great contribution to science was the discovery of Saturn's ring.

**R Horologii**: This long-period variable star was discovered from an observing station that Harvard University used to run in Peru. In 13½ months, it completes its cycle of variation from 5th to 14th magnitude and back.

**NGC 1261**: This 8th magnitude globular cluster is only 6 arcminutes across so is a target for a larger telescope.

## SKYWATCHING TIP

The vigil of meteor observation can be a more pleasurable experience when done in groups. The entire sky can be observed continuously over a longer period, with each person allocated a particular segment to monitor, and there is the added benefit of conversation to keep participants from falling asleep! Since effective meteor watching entails constant surveillance of the sky, sleeping bags and reclining chairs are recommended to prevent the inevitable stiff neck from setting in. Comfort, as well as patience, are necessary for producing useful results.

315

③

🖐
x 5

Sky
Charts
2, 7, 8,
9

+10°

13ʰ

ECLIPTIC    18° 12ʰ

19°

0°

14ʰ

15ʰ

VIRGO

20°

-10°

Spica

21°

CRA

220°

CORVUS

230°

-20°

R  γ  ψ

M68

LIBRA

π

58

M83

-30°

CENTAURUS

β

LUPUS

Wil Tirion

# Hydra

(HY-dra) The Sea Serpent

Hydra was the nine-headed serpent that Hercules had to kill as one of his twelve labors. Each time he lopped off one head, two others grew in its place. Hercules emerged from this nightmare by having his nephew burn the stump of each severed neck, preventing new heads from sprouting. In the midst of the struggle, Juno sent Cancer the crab to attack Hercules and distract him. The crab nipped Hercules, who then stepped on it and killed it. For its bravery, Juno rewarded the crab with a place in the sky (see Cancer p. 284).

As in the case of other large constellations, some mapmakers have tried to break up the snaking form of Hydra. In 1805, the French astronomer Joseph Lalande entertained himself by making up a constellation called Felis, the Cat. "I am very fond of cats," Lalande wrote.

"The starry sky has worried me quite enough in my life, so that now I can have my joke with it." Lalande formed his feline from stars of Hydra and Antlia, but it has not survived. Hydra remains, snaking a quarter of the way across the sky.

**R Hydrae**: One of the earliest known variables, this Mira star's light changes were first seen in the late 1600s. It varies over 13 months from a maximum as high as magnitude 4.5 to a minimum of 10.

**V Hydrae**: A rare example of a carbon star, this is a low-temperature red giant producing carbon. It is so deeply red that you can be sure you have found it merely by its color. The star varies somewhat erratically between magnitudes 6 and 12, with two superimposed periods—one about 18 months, and the other about 18 years.

*In Uranometria (1603), Bayer featured Hydra as a sea serpent rather than as the nine-headed serpent.*

316

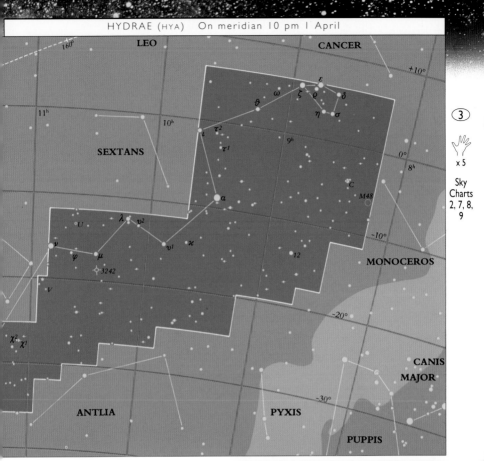

LEO

CANCER

+10°

160°

11ʰ

10ʰ

SEXTANS

9ʰ

0°

8ʰ

M48

τ²
ι
τ¹

ε
ω  ζ  θ
ϑ        δ
η  σ

C

α

λ  ν²

U

ν¹  κ

ν

φ  μ

12

–10°

MONOCEROS

3242

V

–20°

χ²
χ¹

CANIS
MAJOR

–30°

ANTLIA

PYXIS

PUPPIS

(3)

x 5

Sky
Charts
2, 7, 8,
9

**M48 (NGC 2548):** Long considered a
missing Messier object because he wrongly
reported its position, M48 is now considered to
be one and the same as NGC 2548—a large
open cluster best seen through binoculars or
a wide-field telescope.

**M83:** This is a strange-looking spiral
galaxy with three obvious spiral arms. At
8th magnitude, it is one of the brighter galaxies
visible in binoculars and will show more detail at
higher magnification in a telescope. Watch out
for "new" stars here, as M83 has produced four
supernovas in the last 60 years.

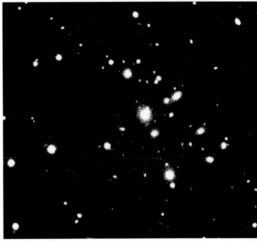

A CCD image of the distant Hydra cluster of galaxies. The
brightest galaxies here are magnitude 18!

**The Ghost of Jupiter Nebula
(NGC 3242):** This nebula is the brightest
planetary nebula in this part of the sky. It is
about 16 arcseconds across and shows its
structure well in 10 inch (250 mm) telescopes
and larger.

M83 is often classified as a barred spiral, and a bar across the
nucleus is apparent in this amateur photograph.

317

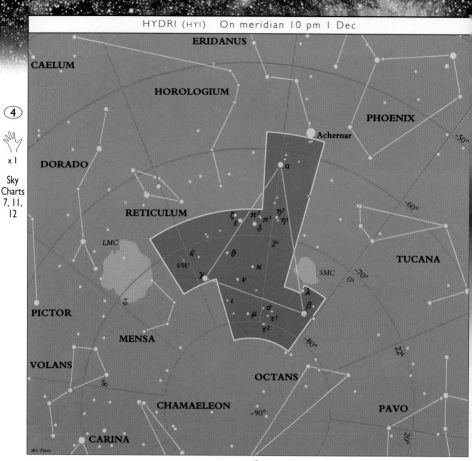

4

x 1

Sky
Charts
7, 11,
12

# Hydrus

(HY-drus) The Water Snake

Johann Bayer created this constellation, publishing it in his star atlas of 1603. He placed it near Achernar, the mouth of the River Eridanus and nestled between the Large and Small Magellanic Clouds. It is sometimes called the Male Water Snake, to avoid confusing it with Hydra.

**VW Hydri**: This star is the most popular cataclysmic variable with Southern Hemisphere observers. When in its usual state, it shines at a faint 13th magnitude, but when it goes into outburst, an event which occurs about once a month, it can become brighter than 8th magnitude in just a few hours.

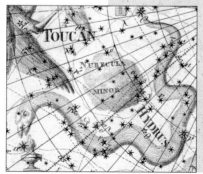

*In his Uranographia (1801), Bode shows Hydrus snaking past Nubecula Minor, the Small Magellanic Cloud in Tucana.*

## SKYWATCHING TIP

One frigid night in Flagstaff, Arizona, Clyde Tombaugh—the astronomer who discovered Pluto—was staring into the eyepiece of his telescope, guiding a one-hour exposure. Feeling sleepy, he struggled to keep from dozing off. When the hour was up, he realized that he had become so cold that he could hardly move. In great pain, he managed to close the telescope and move to a warm room, where he had to sit for some time beside a heater to thaw out.

Be careful not to become so absorbed in your observations that you are unaware of how cold you are. Walk around the telescope to stimulate your circulation, or go inside.

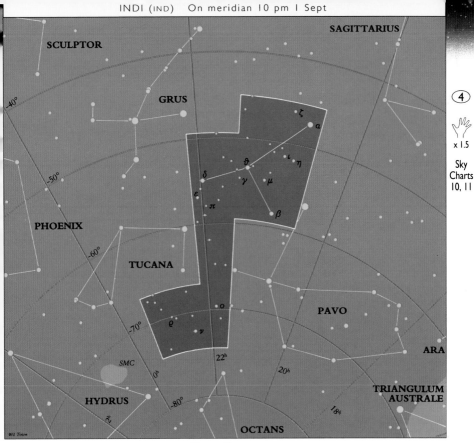

# Indus

(IN-dus) The Indian

This constellation was added to the southern sky by Johann Bayer to honor the Native Americans that European explorers encountered on their travels. The figure of Indus is positioned between three birds: Grus, the Crane; Tucana, the Toucan; and Pavo, the Peacock.

👁 **Epsilon (ε) Indi**: Only 11.8 light-years away, this is one of the closest stars to the Sun and is somewhat similar to it.

With four-fifths of the Sun's diameter and one-eighth its luminosity, scientists consider Epsilon (ε) Indi to be worth investigating for planets and for evidence of extraterrestrial intelligence, such as radio signals.

In the early 1960s, when Frank Drake began searching for signs of life elsewhere in the galaxy, he used this star as one of his targets. In 1972, the Copernicus Satellite searched unsuccessfully for laser signals from this star.

*Indus, standing between two of its feathered neighbors: Tucana (left) and Pavo (right).*

319

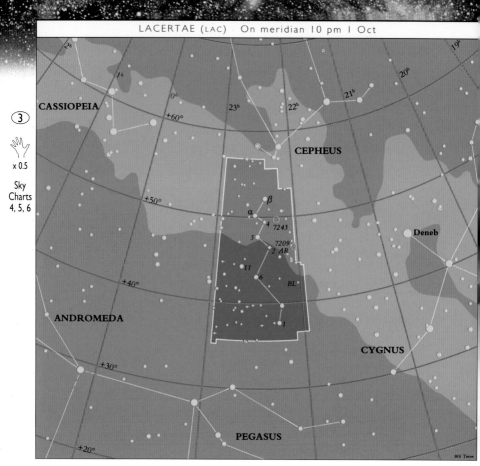

CASSIOPEIA

CEPHEUS

Deneb

ANDROMEDA

CYGNUS

PEGASUS

3

x 0.5

Sky
Charts
4, 5, 6

Wil Tirion

# Lacerta

(Iah-SIR-tah) The Lizard

Lacerta is far enough north to be circumpolar at the higher mid-northern latitudes. It lies south of Cepheus.

The German astronomer Johannes Hevelius suggested that this group of stars be named Lacerta in 1690, but a few revisions were needed before it evolved from a small, long-tailed mammal into a lizard. Other cartographers came up with names for the region to honor France's Louis XIV and Prussia's Frederick the Great, but these names were ignored.

**BL Lacertae**: Since this object varies from 13 to 16.1, it is invisible to any but the largest amateur telescopes. However, BL Lacertae is worth a look, since it is not a star at all but the nucleus of a distant elliptical galaxy. Some of this class of BL Lacertae-type (BL Lac) objects have been known to change by as much as two magnitudes in a single day. Recent theories suggest that

BL Lac objects, quasars, and other high-powered galaxies are all closely related "active galaxies." This powerful energy source at the center may be a black hole surrounded by a complex, swirling mass of gas and dust.

## SKYWATCHING TIP

As you become more involved in observing and develop connections with your local astronomy club, you might be invited to a star party sponsored by the club. Take full advantage of this wonderful opportunity to meet people, ask questions, share experiences, compare notes, and look through the various telescopes that are being used. Remember to take a red flashlight with you, so that you can operate effectively in the dark without inconveniencing anyone, and always point it downward, away from people's dark adapted eyes.

*Lacerta, the Lizard, as depicted in the constellation cards Urania's Mirror (1825).*

320

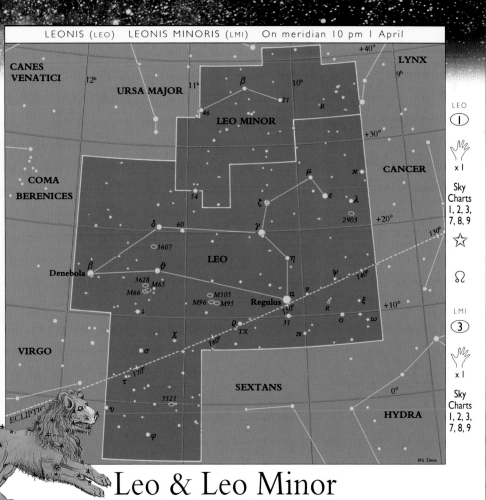

CANES VENATICI

URSA MAJOR

LEO MINOR

COMA BERENICES

LYNX

CANCER

VIRGO

SEXTANS

HYDRA

Denebola

LEO

Regulus

ECLIPTIC

LEO

①

×1

Sky Charts
1, 2, 3,
7, 8, 9

☆

♌

LMI

③

×1

Sky Charts
1, 2, 3,
7, 8, 9

Wil Tirion

# Leo & Leo Minor

(LEE-oh) The Lion    (LEE-oh MY-ner) The Little Lion

Unlike most of the zodiacal constellations, Leo, with its sickle (or backward question mark) tracing out a great head, really can be pictured as its namesake, a lion reclining not unlike the Egyptian Sphinx.

The Babylonians and other cultures of Southwest Asia associated Leo with the Sun, because the summer solstice occurred when the Sun was in that part of the sky.

Leo Minor is a recent addition to the constellations, introduced by Johannes Hevelius during the seventeenth century.

**Gamma (γ) Leonis**: This beautiful double star has orange-yellow components of 2nd and 3rd magnitude separated by 5 arcseconds.

**R Leonis**: This favorite Mira variable is easily found near **Regulus**. It ranges from magnitude 5.8 to 11 over about 10½ months.

**R Leonis Minoris**: Another Mira star, taking about a year to vary between magnitudes 7.1 and 12.6.

**M65 and M66**: These two spiral galaxies near **Theta (θ) Leonis** are visible in

binoculars but give a better view in a telescope. Other interesting galaxies in Leo are **NGC 3628**, **M95**, **M96**, **M105**, and **NGC 2903**.

**Leonids**: This meteor shower peaks annually on 17 November. In 1966, observers recorded up to 40 meteors per second at its peak.

*(Above, left) The Urania's Mirror (1825) Leo is leaping rather than reclining. (Below) This view of Leo is dominated by blue-white Regulus and orange Gamma (γ) Leonis in the sickle.*

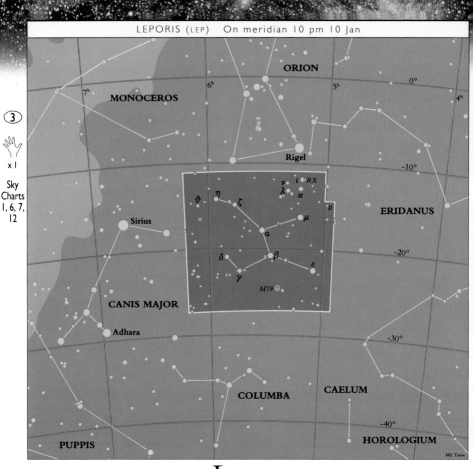

ORION

MONOCEROS

7ʰ

6ʰ

5ʰ

0°

4ʰ

Rigel

-10°

ν  ι  RX
λ  χ
R

η  ϑ

ζ

μ

ERIDANUS

Sirius

α

R

-20°

δ

β

γ

ε

M79 ⊕

CANIS MAJOR

Adhara

-30°

CAELUM

COLUMBA

-40°

PUPPIS

HOROLOGIUM

Wil Tirion

Sky
Charts
1, 6, 7,
12

3

x 1

# Lepus

(LEE-pus)  The Hare

A faint constellation, Lepus is nevertheless easy to find because it is directly south of Orion. In ancient times it was thought of as Orion's chair. Egyptian observers saw it as the Boat of Osiris. The Greeks and Romans gave it the name Lepus. Since Orion particularly liked hunting hares, it was appropriate to place one below his feet in the sky.

**Gamma (γ) Leporis**: Easy to separate in virtually any telescope, this wide double star with contrasting colors has a separation of 96 arcseconds. It is relatively close to Earth, at a distance of 29 light-years, and is part of the Ursa Major stream.

**Hind's Crimson Star**: Likened by some observers to a drop of blood in the sky, **R Leporis** is the variable that the nineteenth-century British

*Lepus, from the constellation cards Urania's Mirror (1825).*

*The globular cluster M79, pictured using a 12 inch (300 mm) telescope.*

astronomer J. Russell Hind called the Crimson Star. Over a period of 14 months, the star varies in magnitude from a maximum of as much as 5.5 to a minimum of 11.7. Its coloring is at its most striking when the sky is dark and the star is near maximum brightness.

**M79**: With the galactic center in Sagittarius half the sky away, this is a surprising place to find a globular cluster. Nevertheless, M79 is here to enchant you, especially if you have an 8 inch (200 mm) or larger telescope which will begin to resolve the stars around its edges.

322

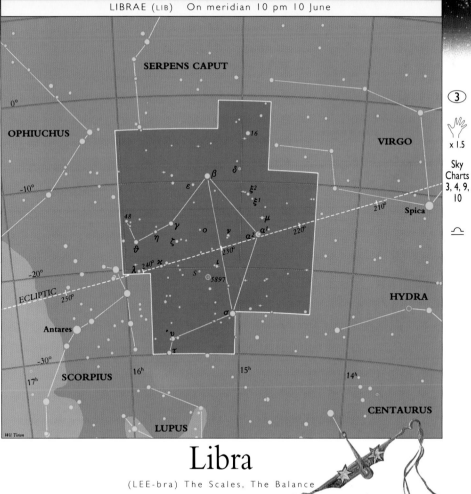

SERPENS CAPUT

OPHIUCHUS

VIRGO

HYDRA

SCORPIUS

CENTAURUS

LUPUS

ECLIPTIC

Antares

Spica

*Wil Tirion*

③

🖐 x 1.5

Sky Charts 3, 4, 9, 10

♎

# Libra

(LEE-bra) The Scales, The Balance

*A detail from a fresco at the Villa Farnese, in Italy, showing Libra, the Scales.*

ooking like a high-flying kite, Libra is easy to find by extending a line westward from Antares and its two bright neighbors in Scorpius. The line reaches a point between Alpha (α) and Beta (β) Librae.

Libra is one of the constellations of the zodiac and was associated with Themis, the Greek goddess of justice, whose attribute was a pair of scales. Originally these stars were thought of as part of Scorpius: the alpha (α) and beta (β) stars both carry Arabic names, the former being Zuben El Genubi, "southern claw" and the latter Zuben

*Libra leads the bright stars of Scorpius across the sky. The claws of the scorpion and brilliant red Antares dominate the lower corner of the image.*

Eschamali, "northern claw." Our understanding is that Libra became a separate constellation at the time of the ancient Romans.

👁 **Delta (δ) Librae**: Similar to Algol, this eclipsing variable star fades by about a magnitude every 2.3 days, from 4.9 to 5.9. The entire cycle is visible to the naked eye.

🔭 **S Librae**: A Mira star, S Librae varies from an 8.4 maximum to a 12 minimum over a period of a little more than six months.

323

ECLIPTIC

240°

230°

-20°

Antares

4

x 1

Sky
Charts
8. 9,
10

ξ χ
ψ1
ψ2   φ1
θ        φ2
RU   5986   ν
η

2

-30°

γ
ω        δ

β
o      1.4406
τ1
ε        λ            τ2
ν1 μ      π        α      ι
ν2 χ            ρ
σ
ζ              σ

5822

16ʰ        15ʰ                14ʰ
17ʰ

18ʰ

Hadar
Rigil Kentaurus

Mimosa

Acrux

-40°

-50°

-60°

13ʰ

12ʰ

Wil Tirion

# Lupus

(LOO-pus) The Wolf

South of Libra and east of Centaurus, Lupus, the Wolf, is a small constellation with some 2nd magnitude stars. It is almost joined with Centaurus, as if the Centaur is stroking the wolf like a pet. The ancient Greeks and Romans called this group of stars Therion— an unspecified wild animal.

Lying within the band of the Milky Way, this constellation is home to a number of open and globular clusters.

## SKYWATCHING TIP

Variable stars are named in a strange way. The first variable discovered in a constellation, say Lupus, is called R Lupi; the second S, then T, until Z. Then the names go RR Lupi, RS to RZ, then SS to SZ, and finally ZZ. In constellations with a great many variables, the list continues with AA to AZ, and on to QZ, but leaving out J. If more than these 334 variables are found, the system continues (more sensibly) with numbers, such as V 1500 Cygni for the 1975 nova in Cygnus.

**RU Lupi**: (Read this name out loud. After immersing yourself in constellation lore to this point, you possibly are!) RU Lupi is a faint nebular variable, with a maximum of only 9th magnitude. Its irregular variation is characteristic of young stars still involved with nebulosity.

*NGC 5986, a globular that is visible in binoculars, close to some 6th and 7th magnitude stars.*

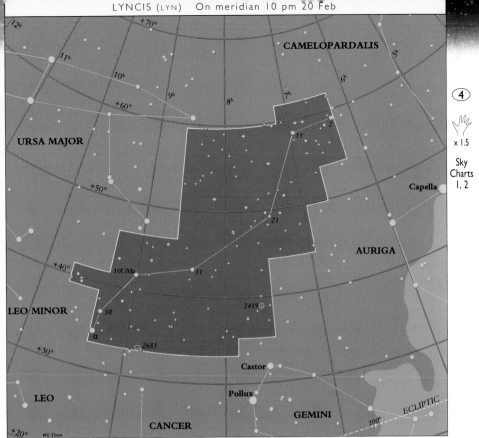

CAMELOPARDALIS

URSA MAJOR

Capella

AURIGA

LEO MINOR

Castor

Pollux

LEO

GEMINI

CANCER

ECLIPTIC

# Lynx

(LINKS) The Lynx

With only one 3rd magnitude star, Lynx is one of the hardest constellations to find. Johannes Hevelius charted this figure around 1690, apparently naming it Lynx because you need to have the eyes of a lynx to spot it. The same is true of its deep-sky objects.

**The Intergalactic Tramp (NGC 2419):** Lying some 7 degrees north of Castor, the brightest star in Gemini, this is a very faint and distant globular cluster. It is more than 60 degrees from any

*The lynx is a short-tailed cat of the Northern Hemisphere. The tail is rather long in this representation from Urania's Mirror (1825).*

*Almost edge-on, NGC 2683 is a 10th magnitude spiral galaxy just near the border with Cancer. Dust in the spiral arms is visible on the left side of the galaxy.*

other globular. At 210,000 light-years, it is more distant than the Large Magellanic Cloud and is so far away that it might escape the gravitational pull of our galaxy. It is for this reason that astronomer Harlow Shapley called it the Intergalactic Tramp. Through a 10 inch (250 mm) or larger telescope, NGC 2419 appears as a fuzzy knot of light.

325

DRACO

+50°   17ʰ

21ʰ

20ʰ

19ʰ

18ʰ

Deneb

CYGNUS

R

+40°

ε¹
ε²   XY   μ
η                α
ϑ               ζ   Vega
ι        δ²  δ¹
κ

HERCULES

+30°

β
γ   ν¹
λ   M57  ν²

M56

+20°

VULPECULA

SAGITTA

OPHIUCHUS

AQUILA

Wil Tirion

# Lyra

(LYE-rah) The Lyre

This beautiful constellation is dominated by Vega, one of the brightest stars in the sky. You can imagine the lyre strings stretched across the parallelogram of four stars that accompany it.

The lyre was given by Apollo to his son Orpheus. Orpheus played it so exquisitely that wild beasts and the mountains were enchanted. He passionately loved his wife Eurydice, and when she died he descended into the under-world to save her. He persuaded the gods to release her, on condition that he did not look at her on the journey. In his impatience, however, he glanced at Eurydice before they reached the upper world, and

*The Lyre, with dazzling Vega near the top, from Urania's Mirror (1825).*

she was swept back into Hades for ever. Inconsolable, Orpheus was torn to pieces by a group of young women after he ignored their advances. The lovers were then reunited and Orpheus's lyre was placed in the sky by Zeus.

**Epsilon (ε) Lyrae**: This is a "double double" star. The slightest optical aid shows two 5th magnitude stars—ε¹ and ε². Both are themselves doubles, with separations under 3 arcseconds. A 4 inch (100 mm) telescope operating at a magnification of 100 or more will split both of them.

**Beta (β) Lyrae**: This eclipsing variable ranges from magnitude 3.3 to 4.4 in 13 days.

**The Ring Nebula (M57)**: This famous planetary nebula lies midway between Beta (β) and **Gamma (γ) Lyrae**. Through a 3 inch (75 mm) or larger telescope it appears as a star out of focus at low magnification. Higher power will show its ring shape, about 2 arcminutes across.

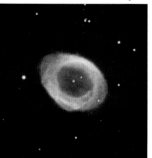

*A smoke ring in space? The famous Ring Nebula.*

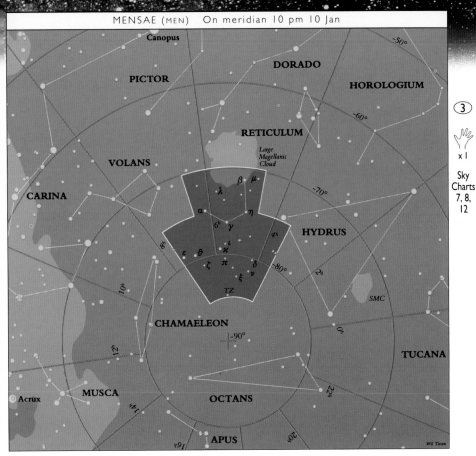

# Mensa

(MEN-sah) The Table, The Table Mountain

The only constellation that refers to a specific piece of real estate, Mensa was originally called Mons Mensae by Nicolas-Louis de Lacaille, after Table Mountain, south of Cape Town, South Africa, where he did a good deal of his work. He developed this small constellation from stars between the Large Magellanic Cloud and Octans.

Mensa, without any bright stars, offers little to draw your attention away from the glories of the Large Magellanic Cloud. The northernmost stars of the constellation, representing the summit of the mountain, are hidden in the Large Magellanic Cloud, in the same way that Table Mountain is often shrouded in clouds.

👁 **Alpha (α) Mensae:** This dwarf star has an apparent magnitude of 5.1. Alpha (α) Mensae lies comparatively close to us, its light taking only 33 years to reach Earth.

👁 **Beta (β) Mensae:** Lying very near the edge of the Large Magellanic Cloud, this faint star of magnitude 5.3 lies at a distance of 640 light-years away.

## SKYWATCHING TIP

Almost all of the variable stars we mention can be seen with the naked eye at their brightest. However, our charts mainly include only the brighter stars, so if a variable drops below about 6th magnitude, as many do, it will be difficult to identify. Mira stars, though, can still be recognized by their reddish color.

Finding a variable star is only half the challenge. The rest involves estimating the magnitude of the star using nearby stars for comparison. For example, if Delta (δ) Cephei (see p. 294) is a little fainter than the magnitude 3.5 star Zeta (ζ) and much brighter than the 4.4 Epsilon (ε), then Delta (δ) Cephei will be about magnitude 3.6 or 3.7.

But what if you do not know the magnitudes of the comparison stars, *a* and *b*? If the variable, *V*, were ¾ of the way from *a* to *b* in brightness, you could write your estimate as "*a*, 3, *V*, 1, *b*." This way you can monitor the changing brightness of *V* over time.

327

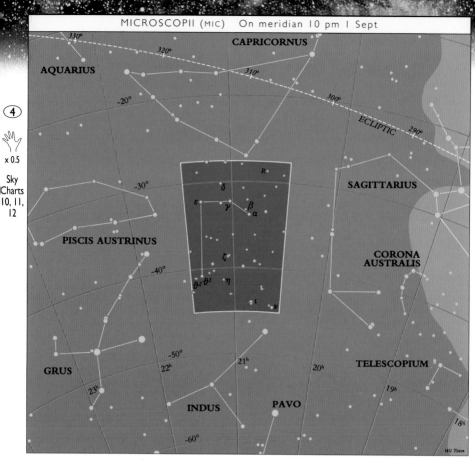

CAPRICORNUS

AQUARIUS

ECLIPTIC

SAGITTARIUS

PISCIS AUSTRINUS

CORONA
AUSTRALIS

GRUS

TELESCOPIUM

INDUS     PAVO

Wil Tirion

# Microscopium

(my-kro-SKO-pee-um) The Microscope

This small, faint constellation, which lies just south of Capricornus and east of Sagittarius, was created by Nicolas-Louis de Lacaille in about 1750. It commemorates the microscope, the invention of which is credited to the Dutch spectacle-maker Zacharias Janssen, around 1590, and to Galileo, among others.

## SKYWATCHING TIP

How good is your telescope? On a night when the atmosphere is steady, you can test its quality by trying to resolve double stars of equal magnitude. Beta (β) Monocerotis (opposite page), for example, is a triple system. Two stars, of magnitudes 4.7 and 5.2, form a pair separated by 2.8 arcseconds which should be split by a 2.4 inch (60 mm) telescope at 100x magnification. The third star, 7.3 arcseconds away, is even easier to split. Theta 2 (θ2) Microscopii is another triple system but the close pair, separated by 0.5 arcseconds, is a test for a 10 inch (250 mm) telescope in good seeing.

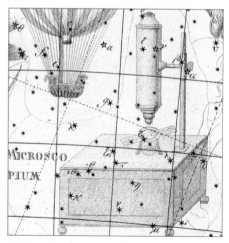

MICROSCO
PIUM

*Bode depicted a microscope typical of his time (1801).*

**R Microscopii**: This faint Mira variable has a rapid cycle lasting only 4½ months, during which it drops from magnitude 9.2 to 13.4 and climbs back again.

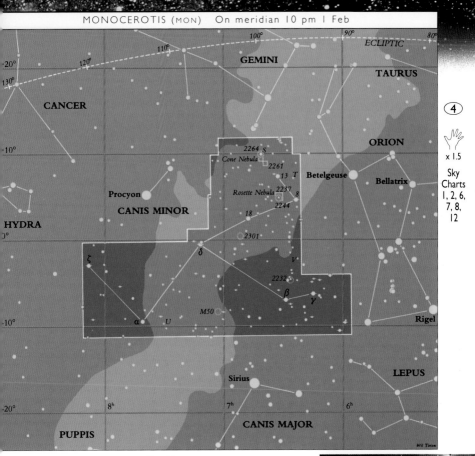

GEMINI

TAURUS

CANCER

ORION

2264 S
Cone Nebula
2261
13 T    Betelgeuse
Bellatrix
2237
Rosette Nebula
2244
18
Procyon
CANIS MINOR

HYDRA

2301

δ

ζ

ν

2232

β
γ

M50

α    U

Rigel

Sirius

LEPUS

8ʰ
7ʰ
6ʰ

CANIS MAJOR

PUPPIS

Wil Tirion

④

🖐
x 1.5

Sky
Charts
1, 2, 6,
7, 8,
12

# Monoceros

(moh-NO-ser-us) The Unicorn

This faint constellation was formed in about 1624 by the German astronomer Jakob Bartsch. Monoceros is the Latin form of a Greek word meaning "one-horned," and it seems that the mythical unicorn may have come into existence as a result of a confused description of a rhinoceros.

Anxious to create a winter equivalent to the Northern Hemisphere's summer triangle, some observers advocate a winter triangle, bounded by Betelgeuse (in Orion), Sirius (in Canis Major), and Procyon (in Canis Minor). Monoceros, the Unicorn, and the band of the Milky Way fill the space inside this triangle.

*The fabled unicorn, as seen in the constellation cards,* Urania's Mirror *(1825).*

*The blue-white stars of the Christmas Tree Cluster form a triangle which comes to a pinnacle at the top (south) of this picture.*

**M50**: This beautiful open cluster, lying slightly more than one-third of the way from Sirius to Procyon, is easy to find. Some of the cluster's stars are arranged in pretty arcs.

**The Rosette Nebula (NGC 2237)**: Through a 10 inch (250 mm) telescope, this ring-shaped nebula, and the open cluster it contains (**NGC 2244**), offer a scene of delicate beauty. Smaller telescopes and binoculars will reveal the nebula on very clear nights.

**The Christmas Tree Cluster (NGC 2264)**: This open cluster really does resemble a Christmas tree.

329

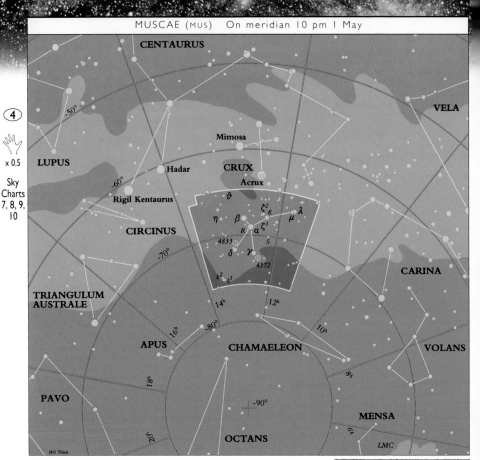

④

x 0.5

Sky
Charts
7, 8, 9,
10

CENTAURUS

VELA

LUPUS

Mimosa

CRUX

Hadar

Acrux

Rigil Kentaurus

CIRCINUS

$\vartheta$

$\zeta^2$ $\varepsilon$ $\lambda$

$\eta$ $\beta$ $\mu$

$R$ $\alpha$ $\zeta^1$

4833

$S$

$\delta$ $\gamma$

4372

$\iota^2$ $\iota^1$

CARINA

TRIANGULUM
AUSTRALE

$14^h$

$12^h$

APUS

CHAMAELEON

$10^h$

VOLANS

PAVO

-90°

MENSA

OCTANS

LMC

Wil Tirion

# Musca

(MUSS-kah) The Fly

M usca is an easy constellation to find, just to the south of the Southern Cross. It was originally described by Johann Bayer in his 1603 star atlas as Apis, the Bee. Later on, Edmond Halley called it Musca Apis, the Fly Bee, then Nicolas-Louis de Lacaille named it Musca Australis, the Southern Fly—to avoid it being confused with the fly on the back of Aries, the Ram. Now that this northern fly is no longer a constellation, the Southern Fly is known simply as Musca.

*Musca, the Fly, as featured in Johann Bode's Uranographia (1801).*

*The Southern Cross (Crux) and the Coal Sack overshadow the fainter stars of Musca, just to the south, on the edge of the glow of the Milky Way.*

**Beta (β) Muscae:** This elegant double star consists of two 4th magnitude stars that revolve around each other in a period that spans several hundred years. The pair is some 310 light-years from Earth. The separation of 1.6 arcseconds is very tight, presenting a challenge for a 4 inch (100 mm) telescope.

**NGC 4372:** This globular cluster is close to **Gamma (γ) Muscae** and has faint stars spread over 18 arcminutes.

**NGC 4833:** This is a large, faint globular cluster within 1 degree of **Delta (δ) Muscae.** A 4 inch (100 mm) or larger telescope is needed to begin to resolve the cluster into individual stars.

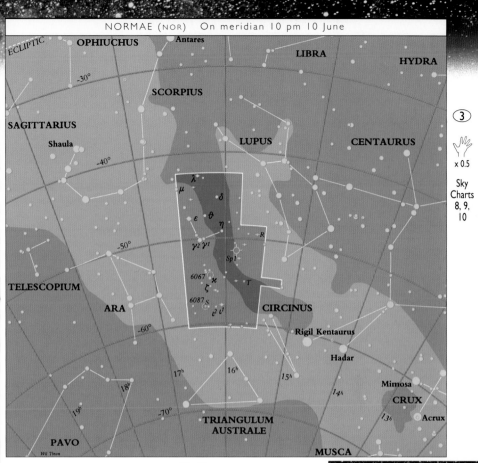

OPHIUCHUS
Antares
ECLIPTIC
-30°
SCORPIUS
LIBRA
HYDRA
SAGITTARIUS
Shaula
-40°
LUPUS
CENTAURUS
λ
μ
δ
ε  θ
η
R
γ² γ¹
Sp1
-50°
6067  κ
ζ
6087 S  T
ε¹ ι¹
TELESCOPIUM
ARA
CIRCINUS
-60°
Rigil Kentaurus
Hadar
17ʰ    16ʰ    15ʰ
Mimosa
18ʰ
14ʰ
CRUX
19ʰ
-70°
13ʰ  Acrux
TRIANGULUM
AUSTRALE
PAVO
MUSCA
Wil Tirion

③

× 0.5

Sky
Charts
8, 9,
10

# Norma

(NOR-muh) The Square

E ast of Centaurus and Lupus is a small
constellation called Norma, the Square.
When he named this group of stars,
Nicolas-Louis de Lacaille decided to call it
Norma et Regula, the Level and Square, after
a carpenter's tools. Since those days, however,

*NGC 6067 is a group
of 8th magnitude and
fainter stars overlying
the background of Milky
Way stars.*

the Regula has
been forgotten.
The constellation
lies alongside Circinus, the Drawing Compass,
which he named at the same time.

Set in the southern Milky Way, Norma
presents good fields for binoculars, with a
number of open clusters. For a small constellation, Norma has also been quite lucky with
the appearance of novas: there was one in 1893
and another in 1920.

**NGC 6067**: This is a small open cluster.
Large binoculars or a telescope reveal some
100 stars within a stunning field.

**NGC 6087**: This is another of Norma's
striking open clusters.

## SKYWATCHING TIP

J ust because the sky is clear does not necessarily
mean that the night is an ideal one for observing.
The best nights offer both good transparency and
good seeing.

A night without clouds, haze, or light pollution is
said to be transparent. However, if the atmosphere
above you is turbulent, the images you see through
a telescope will wave about like flags in a breeze.
When this happens, we say that the seeing is poor.
Under such conditions, your telescope will not see
the faintest stars, or be able to resolve the double
stars that it should. However, a crisp view of
planetary detail in good seeing is well worth the wait.

331

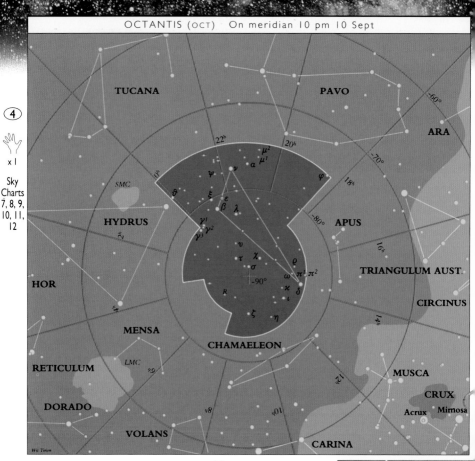

TUCANA

PAVO

ARA

-60°

22ʰ 20ʰ

μ²
μ¹
ν α
φ

SMC 18ʰ

ψ
ξ ε
ϑ β λ

γ¹
γ² APUS

HYDRUS

-70°

-80°

16ʰ

TRIANGULUM AUST.

υ
τ χ
σ
-90° ρ
ω π¹ π²
κ ι δ

R CIRCINUS

14ʰ

ζ η

MENSA

CHAMAELEON

LMC MUSCA

RETICULUM CRUX

Acrux Mimosa

DORADO

VOLANS

CARINA

HOR

Wil Tirion

Sky
Charts
7, 8, 9,
10, 11,
12

④

x I

# Octans

(OCK-tanz) The Octant

To honor John Hadley's invention of the octant in 1730, Nicolas-Louis de Lacaille formed this south polar constellation and called it Octans Hadleianus. The forerunner of the sextant, the octant was an instrument used for measuring the altitude of a celestial body—an essential device for navigators and astronomers.

*The Octant, as drawn by Bode in* Uranographia *(1801).*

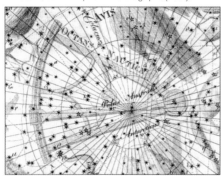

*Octans, with the SMC at top left and the stars of Apus to the lower right.*

**Sigma (σ) Octantis**: This is the south pole star. At magnitude 5.4, it is barely visible to the naked eye on a dark night, so while it does mark the pole, it is not as convenient a marker star as Polaris.

The celestial poles move with time as the axis of Earth precesses, or wobbles like a top, over some 26,000 years (see p. 113). Sigma (σ) Octantis was at its closest to the pole in about 1870, at just under ½ degree. It was just over 1 degree from the pole in the year 2000.

In about another 3,000 years, the pole will begin to move through Carina, and it will pass near **Delta (δ) Carinae** in about 7,000 years. At 2nd magnitude, this is the brightest south pole star Earth ever sees.

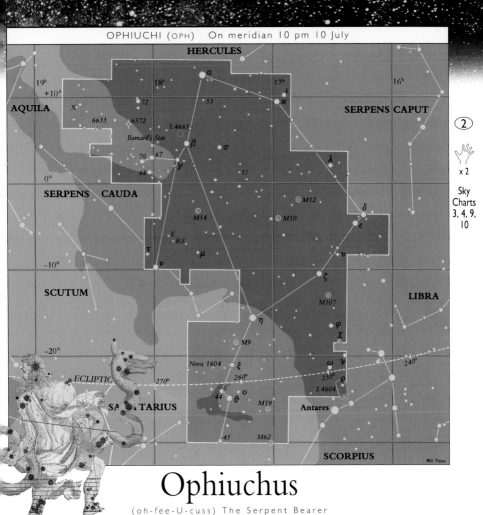

HERCULES

19h  +10°   18h       17h       16h
AQUILA    X

72

6633    6572        53        ι
                                κ
                              SERPENS CAPUT

I.4665
Barnard's Star

70  67   β      σ            λ
68      γ

0°                  • U

SERPENS    CAUDA                    ⊕ M12        δ

M14              ⊕ M10            ε
⊕                    ⊕
Υ R.S.                            υ

τ        μ

-10°   ν

ζ

SCUTUM                        M107
                             ⊕                    LIBRA

η        φ
χ

⊕ M9

-20°                          ω ψ      240°
Nova 1604   ξ
ECLIPTIC    270°   260°      230° ρ
                              I.4604
44  ϑ  o
SA  TARIUS        M19
                 ⊕        Antares

45     M62
        ⊕

SCORPIUS                     Wil Tirion

②
🤚
x 2

Sky
Charts
3, 4, 9,
10

# Ophiuchus

(oh-fee-U-cuss) The Serpent Bearer

Ophiuchus, entwined with the constellation of Serpens, covers a large expanse of sky and is filled with points of interest, including some of the Milky Way's richest star clouds. Greek for "serpent bearer," Ophiuchus is usually identified with Asclepius, the god of medicine.

In one legend, Asclepius learned about the healing power of plants from a snake. His medical skills were so great that he could even raise the dead, which was a cause of concern to Hades, god of the underworld. He therefore persuaded Zeus, his brother, to strike Asclepius dead with a thunderbolt. Zeus then placed Asclepius in the sky, in recognition of his healing skills, along with Serpens, his serpent.

On 9 October 1604, Ophiuchus hosted our galaxy's most recent supernova. Known as Kepler's Star, it outshone Jupiter for several weeks.

**M9, 10, 12, 14, 19, and 62**: These globular clusters provide a range of examples of different concentrations of stars. M9 and 14 are rich; M10 and 12 are looser; M19 is

oval; M62 is somewhat irregular in outline. All are visible in binoculars but require a 6 or 8 inch (150 or 200 mm) telescope to do them justice.

**RS Ophiuchi**: This recurrent nova had outbursts in 1898, 1933, 1958, 1967, and 1985. Its minimum is around magnitude 11.8, rising as high as 4.3 during outbursts.

**Barnard's Star**: Discovered by E. E. Barnard in 1916, this 9.5 magnitude red dwarf star has the greatest proper motion (apparent motion across the sky) of any known star. Only 6 light-years away, it is the nearest star after the Alpha (α) Centauri system.

*(Above, left) Ophiuchus is shown holding the Serpent, Serpens, in Bayer's Uranometria (1603). (Right) M62, one of Ophiuchus's globular clusters, as recorded by an 8 inch (200 mm) telescope.*

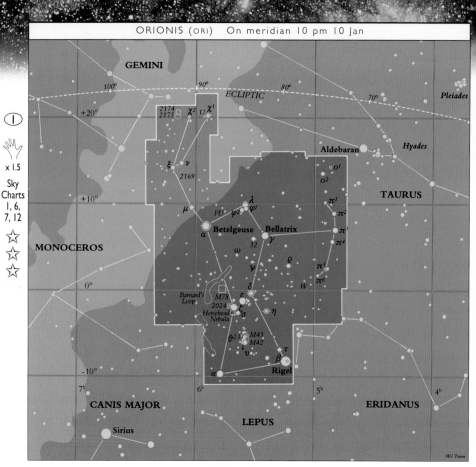

GEMINI

100°   90°   ECLIPTIC   80°   70°   Pleiades

+20°   2174 2175   χ²  U  χ¹

ξ  ν
2169

Aldebaran   Hyades

o¹
o²

TAURUS

+10°   λ   π¹
μ   FU  φ²  φ¹   π²
Betelgeuse   Bellatrix   π³
α   32   γ   π⁴
ω   ρ   π⁵
ψ   π⁶
δ   W
MONOCEROS   0°   Barnard's   ε   η
Loop   M78
2024   ζ  σ
Horsehead   θ²  θ¹  M43
Nebula   ι   M42
ν   τ
κ   β
-10°   Rigel

7ʰ   6ʰ   5ʰ   4ʰ

CANIS MAJOR   ERIDANUS

LEPUS

Sirius

*Wil Tirion*

x 1.5

Sky
Charts
1, 6,
7, 12

# Orion

(oh-RYE-un) The Hunter

Orion has been recognized as a distinctive group of stars for thousands of years. The Chaldeans knew it as Tammuz, named after the month in which the familiar belt stars first rose before sunrise. The Syrians called it Al Jabbar, the Giant. To the ancient Egyptians it was Sahu, the soul of Osiris.

In Greek mythology, Orion was a giant and a great hunter. In one story, Artemis, goddess of the Moon and of the hunt, fell in love with him and neglected

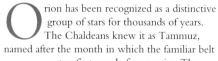

*Orion, with blue-white Rigel at his foot and red Betelgeuse at his shoulder.*

her task of lighting the night sky. Her twin brother, Apollo, seeing Orion swimming far out to sea, challenged his sister to hit what was no more than a dot among the waves. Not realizing that this was Orion, Artemis shot an arrow and killed him. Later, when Orion's body was washed up on the shore, she saw what she had done. Inconsolable, she placed his body in the sky, together with his hunting dogs. Her grief explains why the Moon looks so sad and cold.

Orion is a treasure, with Rigel, Betelgeuse, and its three belt stars in a row lighting up the sky from December to April. The stars are arranged so that it is easy to see the figure of a hunter, complete with lion's skin in his right hand and raised club in his left.

👁 **Betelgeuse (Alpha [α] Orionis):** Betelgeuse (pronounced BET-el-jooze but sometimes corrupted to BEETLE-juice) is fabulous. (Its name comes from the Arabic for "house of the twins," apparently because of the adjacent constellation of Gemini.) A variable star, it varies in magnitude from 0.3 to 1.2 over a

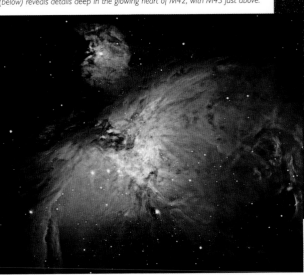

The Great Nebula in Orion (M42) is clearly visible (right) as the central star of Orion's sword, hanging from the three belt stars. The specially processed image (below) reveals details deep in the glowing heart of M42, with M43 just above.

period of about six years. However, the semi-regular nature of the variation means that it is often possible to detect changes over just a few weeks.

👁 **Rigel (Beta [β] Orionis):** The name Rigel is derived from the Arabic for "foot." This mighty supergiant, which is about 800 light-years away, is more than 40,000 times as luminous as the Sun.

👁 **The Great Nebula (M42):** This star nursery, one of the marvels of the night sky, is also known as the **Orion Nebula.** Plainly visible to the naked eye under a dark sky, it can be clearly seen through binoculars in the city. The swirls of nebulosity spread out from its core of four stars called the **Trapezium,** which power the nebula.

Photographs usually "burn-out" the inner region of the nebula and obscure the Trapezium stars. In 1880, using M42 as his subject, Henry Draper was the first person to successfully photograph a nebula.

**M43:** This is a small patch of nebulosity just north of the main body of the Great Nebula. In fact, the M42 complex is simply the brightest part of a gas cloud covering the constellation of Orion at a distance of some 1,500 light-years.

**The Horsehead Nebula (IC 434):** Also known as **Barnard 33**, this dark nebula is projected against a background of diffuse nebulosity, alongside the bright belt star **Zeta (ζ) Orionis**. It can be quite difficult to see, usually requiring a dark sky and at least an 8 inch (200 mm) telescope.

**NGC 2169:** This is a small, bright open cluster of around 30 stars.

The famous Horsehead Nebula, as recorded by the 150 inch (3.9 m) Anglo-Australian Telescope in Australia.

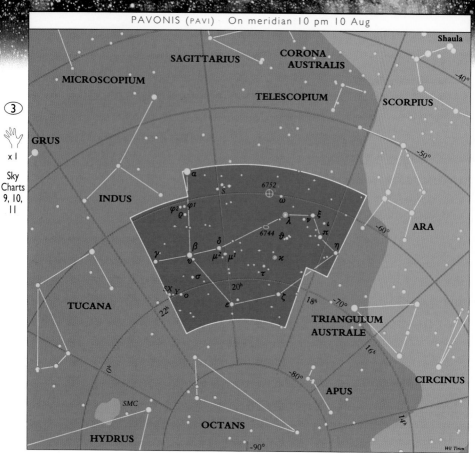

# Pavo

(PAH-voh) The Peacock

Pavo, the Peacock, lies not far from the south celestial pole, south of Sagittarius and Corona Australis. It is a modern constellation, devised by Johann Bayer and published in his star atlas of 1603, but he may have been thinking of the mythical peacock that was sacred to Hera in Greek mythology.

Hera suspected that her husband, Zeus, had fallen in love with the mortal Io and had changed his mistress into a white heifer as a disguise. Hera therefore asked Argus Panoptes— a giant with 100 eyes—to guard the heifer. Hermes was sent by Zeus to kill Argus, and when Hera heard of the giant's death she honored him by distributing his eyes over the tail of the peacock.

**The Peacock Star (Alpha (α) Pavonis):** This star is 180 light-years away. It is a binary system whose members orbit each other in less than two weeks, but the pair is too close to separate telescopically.

**NGC 6752:** A spectacular globular cluster at a relatively close distance of 13,000 light-years, this huge family of stars is the third largest globular cluster (in apparent size) after Omega (ω) Centauri and 47 Tucanae.

**NGC 6744:** This faint but beautiful galaxy is one of the largest known barred spirals. Smaller telescopes reveal only the nuclear regions, a 10 inch (250 mm) telescope being necessary to reveal more.

*Large and bright, NGC 6752 is one of the best globular clusters in the sky, but is little known because of its southerly location.*

336

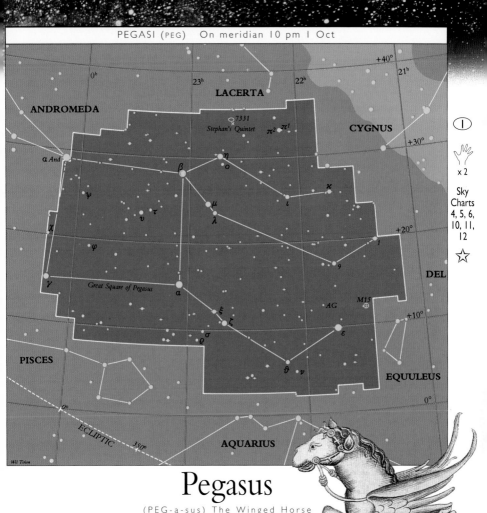

# Pegasus

(PEG-a-sus) The Winged Horse

Although this constellation has no really bright stars, it is easy to spot because its three brightest stars, with Alpha (α) Andromedae, form the Great Square of Pegasus.

The winged horse has been found on ancient tablets from the Euphrates Valley, and on Greek coins minted in the fourth century BC. According to Greek legend, when Perseus decapitated the Gorgon Medusa, Pegasus sprang up from her blood. When Pegasus was brought to Mount Helicon, one kick of his hoof caused the spring of Hippocrene to flow—source of inspiration for poets.

*Pegasus, as conceived by Domenico Bandini in his fifteenth-century encyclopedia of the universe, Fons Memorabilium Universi.*

*Four of the five galaxies of Stephan's Quintet appear to be interacting, distorting each other, and drawing out long streamers of stars.*

**M15**: One of the best of the northern sky globular clusters, M15 is 34,000 light-years away. Although it is visible through binoculars as a nebulous patch, in a telescope it is a real showpiece.

**NGC 7331**: This spiral galaxy is the brightest one in Pegasus, but is still only 9th magnitude.

**Stephan's Quintet**: This very faint group of galaxies lies ½ degree south of NGC 7331. Even though faint streamers of material seem to connect the largest of the galaxies to the others, detailed study indicates that this galaxy is probably closer to us than they are. These galaxies are not really targets for the beginner, as they need at least a 10 inch (250 mm) telescope to be seen clearly.

337

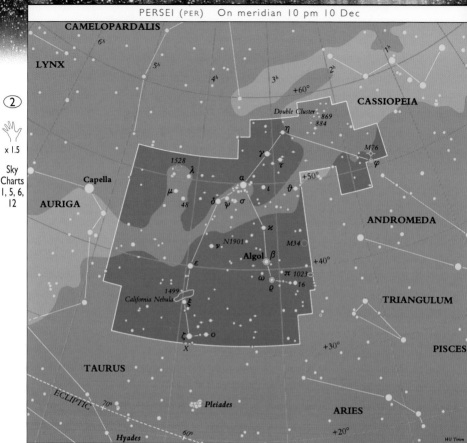

CAMELOPARDALIS

LYNX

6ʰ

5ʰ

4ʰ

3ʰ

2ʰ

1ʰ

+60°

CASSIOPEIA

Double Cluster
869
884

η

M76

γ

τ

φ

AURIGA

Capella

1528

λ

μ

α

ι

ϑ

+50°

48

δ

ψ

σ

ANDROMEDA

κ

ν

N1901

M34

Algol

β

+40°

ε

ω

π

1023

TRIANGULUM

1499
California Nebula

ξ

ϱ

16

ζ

o

χ

+30°

PISCES

TAURUS

ECLIPTIC

70°

Pleiades

ARIES

+20°

Hyades

60°

WIL Tirion

②

🖐
x 1.5

Sky
Charts
1, 5, 6,
12

# Perseus

(PURR-see-us) The Hero

*Perseus, carrying
Medusa's head, with
Pegasus, in a detail
from Rubens' Perseus
and Andromeda.*

A pretty constellation that straddles the Milky Way, Perseus is in the northern skies from July to March. Its stars arc from Capella, in Auriga, to Cassiopeia.

The son of Zeus and the mortal Danaë, Perseus's most famous exploit was to kill the Gorgon Medusa—one of three sisters that were so terrifyingly ugly that one glance of them would turn the viewer to stone. Using Athene's shield as a mirror, he severed Medusa's head, and the winged horse Pegasus sprang from her blood.

👁 **Algol**: The star that winks, this is the most famous of the eclipsing variables. Every 2 days, 20 hours, and 48 minutes, it begins to drop in brightness from magnitude 2.1 to 3.4 in an eclipse lasting 10 hours.

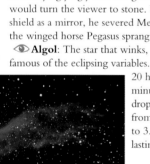

*NGC 1499, the California
Nebula, is not easy to see
as it is large and faint.*

**M34**: This bright open cluster sits in the middle of a rich field of stars. It is an interesting view through either binoculars or a telescope.

👁 **Double Cluster (NGC 869 and 884)**: Two of the finest examples of open clusters in the sky, NGC 869 and 884 (h Persei and Chi [χ] Persei respectively) are magnificent through binoculars or the low-power field of a small telescope.

👁 **Perseids**: One of the best meteor showers, these meteors, which come from Periodic Comet Swift-Tuttle, peak on 11 and 12 August.

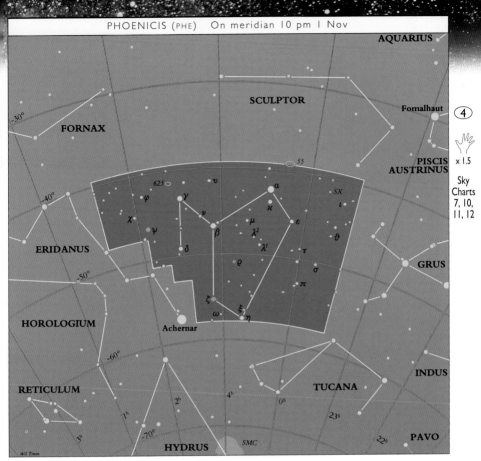

4

× 1.5

Sky
Charts
7, 10,
11, 12

# Phoenix

(FEE-nicks) The Phoenix

A fabulous symbol of rebirth, in mythology the Phoenix was a bird of great beauty that lived for 500 years. It would then build a nest of twigs and fragrant leaves which would be lit by the noontime rays of the Sun. The Phoenix would be consumed in the fire, but a small worm would wriggle out from the ashes, bask in the Sun, and quickly evolve into a brand new Phoenix. Depictions of this miraculous bird have been found in ancient Egyptian art and on Roman coins.

Although the constellation first appeared in Bayer's *Uranometria* of 1603, the idea of a Phoenix in the sky goes back to the ancient Chinese who visualized a firebird that was known as Ho-neaou.

**SX Phoenicis**: The best example of a "dwarf Cepheid" variable, this star changes from magnitude 7.1 to 7.5 and back again in only 79 minutes and 10 seconds! Cepheid periods are very exact. In this case, however, the range varies, with some maxima as bright as 6.7. The variation probably occurs because the star

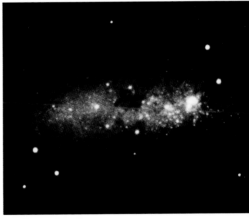

*NGC 625 is a faint (12th magnitude) irregular galaxy, lying about 13 million light-years away.*

has two different oscillations occurring at once. Such a small range in brightness can be difficult to monitor, requiring very careful comparison with neighboring stars.

339

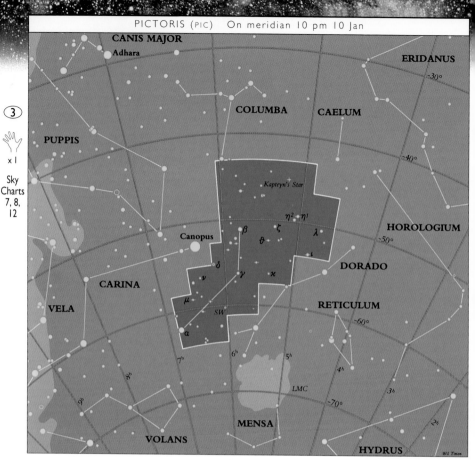

CANIS MAJOR
Adhara
ERIDANUS
-30°
COLUMBA
CAELUM
PUPPIS
-40°
Kapteyn's Star
η² η¹
β ζ λ HOROLOGIUM
θ -50°
Canopus ι
δ DORADO
γ κ
ν RETICULUM
CARINA μ -60°
VELA SW
α
7ʰ 6ʰ 5ʰ 4ʰ
8ʰ 3ʰ
9ʰ LMC -70° 2ʰ
MENSA
VOLANS HYDRUS
Wil Tirion

# Pictor

## (PIK-tor) The Painter's Easel

This southern constellation was originally named Equuleus Pictoris, the Painter's Easel, by Nicolas-Louis de Lacaille. Nowadays its shortened name refers solely to the painter. It is a dull group of stars lying south of Columba and alongside the brilliant star Canopus.

An unusual nova appeared here in 1925. Although it was a bright 2nd magnitude star at the time it was discovered, it continued to brighten until it was almost 1st magnitude. It then began to fade, but brightened again to a second maximum two months later.

👁 **Beta (β) Pictoris**: This 4th magnitude star is host to a disk of dust and ices which could be a planetary system in formation. The surrounding nebula is only visible using special techniques on large telescopes.

**Kapteyn's Star**: Only 12.8 light-years away, this star was discovered by the famous Dutch astronomer Jacobus Kapteyn in 1897. It moves quickly among the distant background stars, crossing 8.7 arcseconds of sky per year—the width of the Moon every two centuries. At magnitude 8.8, the star is visible through binoculars and small telescopes.

*We see the disk of dust and ices around Beta (β) Pictoris nearly edge on. The central circle and dark lines are artifacts of the imaging process.*

Sky Charts 7, 8, 12

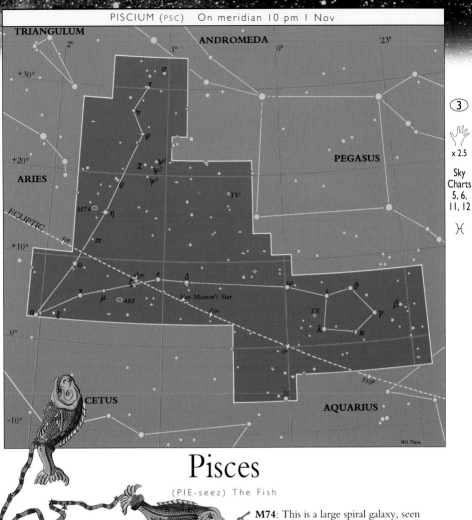

TRIANGULUM

ANDROMEDA

PEGASUS

ARIES

M74

ECLIPTIC

Van Maanen's Star

TX

CETUS

AQUARIUS

Wil Tirion

3

x 2.5

Sky
Charts
5, 6,
11, 12

♓

# Pisces

(PIE-seez)  The Fish

For thousands of years, this faint zodiacal constellation has been seen either as one fish or two. In Greco-Roman mythology, Aphrodite and her son Heros were at one time being pursued by the monster Typhon. To escape him, they turned themselves into fish and swam away, having tied their tails together to make sure that they would not be parted.

The ring of stars in the western fish, which is beneath Pegasus, is called the Circlet. The eastern fish is beneath Andromeda.

A triple conjunction, in which Jupiter and Saturn appeared close to one another three times in a single year, took place in Pisces starting in August of 7 BC. This rare event is one candidate for a celestial phenomenon which might have been the star of Bethlehem.

**Zeta (ζ) Piscium**: A beautiful double star of magnitudes 5.6 and 6.5, separated by 24 arcseconds.

**M74**: This is a large spiral galaxy, seen face-on, close to **Eta (η) Piscium**. While it is the brightest Pisces galaxy, it is still rather faint and requires a dark sky and an 8 inch (200 mm) telescope or larger to be seen.

**Van Maanen's Star**: This is a rare example of a white dwarf star that, at magnitude 12.2, can actually be identified in an 8 inch (200 mm) telescope.

*(Above, left) The Fish in Urania's Mirror (1825), with their tails tied together in accordance with the legend.*
*(Above) M74 is faint, but it is a classic example of a spiral galaxy with a prominent nucleus and well-developed spiral arms.*  341

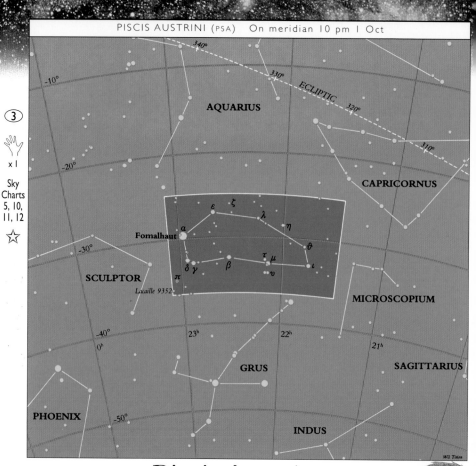

AQUARIUS

ECLIPTIC

CAPRICORNUS

Fomalhaut

SCULPTOR

*Lacaille 9352*

MICROSCOPIUM

GRUS

SAGITTARIUS

PHOENIX

INDUS

Wil Tirion

3

x 1

Sky
Charts
5, 10,
11, 12

☆

# Piscis Austrinus

(PIE-sis OSS-trih-nuss) The Southern Fish

Lying to the south of Aquarius and Capricornus, Piscis Austrinus, the Southern Fish, is relatively easy to spot because of its lone, bright star, Fomalhaut, which is often referred to as The Solitary One. For the Persians, 5,000 years ago, this was a Royal Star that had the privilege of being one of the guardians of heaven.

*Bayer's 1603 depiction of the Southern Fish looks more like a mythical sea monster than any fish you are likely to find in the southern oceans.*

Many early charts of the heavens show the Southern Fish drinking water that is being poured from Aquarius's jar.

**Fomalhaut:** At magnitude 1.2, this star is 25 light-years away—close by stellar standards. It is about twice as large as our Sun and has 14 times its luminosity.

Some 2 degrees of arc southward is a magnitude 6.5 dwarf star that seems to be sharing Fomalhaut's motion through space. They are so far apart that it is hard to call them a binary system. Maybe these two stars are all that is left of a cluster that dissipated long ago.

*Fomalhaut stands out as a beacon of Piscis Austrinus and the surrounding sky.*

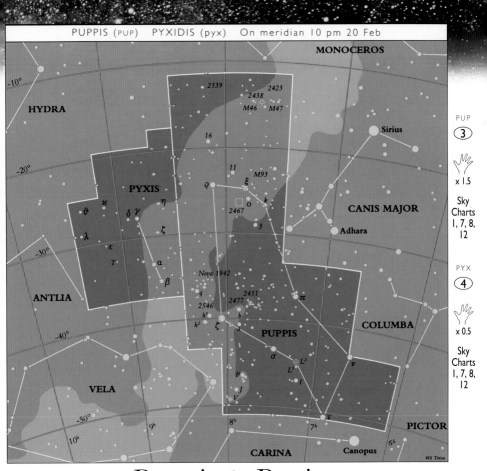

MONOCEROS

HYDRA

2539
2438   2423
M46   M47

Sirius

-10°

16

-20°

11
M93

PYXIS

ϱ

ξ
o   k
2467

M93

CANIS MAJOR

Adhara

η
δ γ
ζ
3

κ
ϑ
λ
ε
T

α
β

Nova 1942

ANTLIA

q
2546
h¹
h²   ζ
u

2451
2477

π

PUPPIS

σ
L²
L¹
i
P
J
V

ν

COLUMBA

VELA

PICTOR

-30°

-40°

-50°

10ʰ   9ʰ   8ʰ   7ʰ   6ʰ

CARINA   Canopus

Wil Tirion

PUP
③
🖐 x 1.5
Sky Charts
1, 7, 8, 12

PYX
④
🖐 x 0.5
Sky Charts
1, 7, 8, 12

# Puppis & Pyxis

(PUP-iss) The Stern   (PIK-sis) The Compass

Found just south of Canis Major, Puppis is the stern of the ship *Argo*. It is the northernmost of the constellations that formed the ship. With the Milky Way running along it, Puppis provides a feast of open clusters for binoculars or telescopes. Right alongside is the smaller and fainter constellation of Pyxis, which used to be Malus, *Argo*'s mast, before Lacaille made a ship's compass out of it.

👁 **Zeta (ζ) Puppis:** This blue supergiant sun is one of our galaxy's largest. About 1,400 light-years away, it shines at 2nd magnitude.

👁 **L² Puppis:** One of the brightest of the red variable stars, L² Puppis varies from magnitude 2.6 to 6.2 over a period of five months.

*Bode's illustration of the Compass, from Uranographia (1801).*

*Look for the faint red ring of the planetary nebula NGC 2438 in the north-east (upper left) quadrant of the cluster M46.*

 **M46:** A beautiful open star cluster, through small telescopes M46 is a circular cloud of faint stars

the apparent diameter of the Moon. A faint planetary nebula, **NGC 2438**, appears to be a part of the cluster but is not a true member. It is 11th magnitude and 1 arcminute across, needing an 8 inch (200 mm) or larger telescope to be seen well.

🔭 **T Pyxidis:** A recurrent nova with a faint 16th magnitude minimum, T Pyxidis sometimes reaches 7th magnitude during its outbursts, which occur at intervals of 12 to 25 years.

343

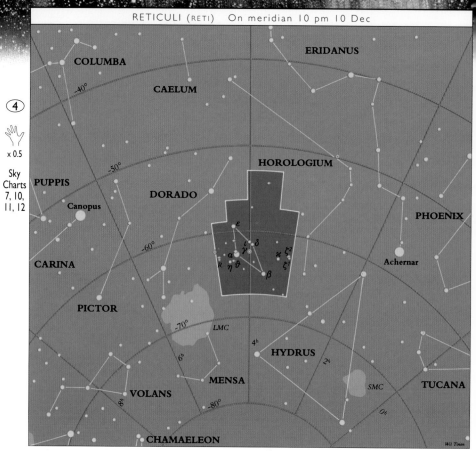

# Reticulum

(reh-TIK-u-lum) The Reticule

A small constellation of faint stars halfway between the bright star Achernar and Canopus, Reticulum was first set up as Rhombus by Isaak Habrecht of Strasburg. De Lacaille changed its name to Reticulum to honor the reticle—the grid of fine lines in a telescope

## SKYWATCHING TIP

The joy of skywatching is enriched when shared with young children. As you point out constellations to them, give them time to formulate questions, and don't be afraid to say if you don't know the answers. One eight-year-old girl once told me that the globular cluster M13 looked like a "little fluffy dog." It wasn't astrophysics, but it signaled the start of her personal relationship with the sky.

eyepiece that aids with centering and focusing. It is occasionally also known as the Net.

**R Reticuli**: This Mira star is quite red, and at maximum light it shines at about magnitude 7. Over a period of nine months, it drops to magnitude 14, then returns to maximum brightness.

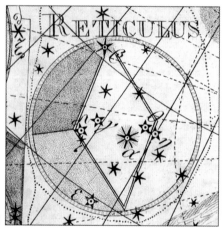

Bode's illustration reflects Reticulum's origin as a rhombus rather than its more recent identification as the scale sometimes found in a telescope eyepiece.

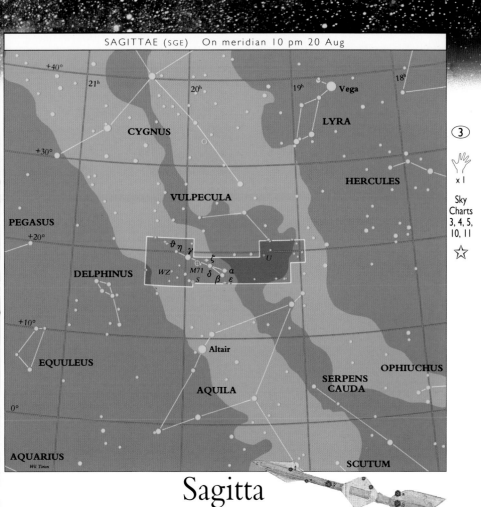

+40°
21ʰ        20ʰ              19ʰ    Vega      18ʰ

CYGNUS                              LYRA

+30°

HERCULES

VULPECULA

PEGASUS
+20°
ϑ η γ
ζ           U
DELPHINUS    WZ    M71  δ    α
S    β   ε

+10°

EQUULEUS              Altair              OPHIUCHUS

SERPENS
CAUDA
AQUILA

0°

AQUARIUS
Wil Tirion

SCUTUM

(3)

🖐
x I

Sky
Charts
3, 4, 5,
10, 11

☆

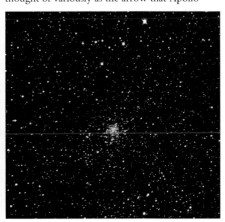

# Sagitta

(sa-JIT-ah)  The Arrow

Although only a small constellation, Sagitta is easy to find halfway between Altair in Aquila, and Albireo (Beta [β] Cygni). It really is true to its name—the ancient Hebrews, Persians, Arabs, Greeks, and Romans all saw this group of stars as an arrow. It has been thought of variously as the arrow that Apollo

*The Arrow, as depicted by Bayer in his famous star atlas Uranometria (1603).*

used to kill the Cyclops; one of the arrows shot by Hercules at the Stymphalian Birds; and as Cupid's dart.

**U Sagittae**: Every 3.4 days, this eclipsing binary drops from magnitude 6.5 to a minimum of 9.3.

**V Sagittae**: Although this star is faint, varying erratically from magnitude 8.6 to magnitude 13.9, it is interesting for it alters a little almost every night. It might have been a nova a long time ago.

**M71**: A little south of the midpoint of a line joining **Delta ($\delta$)** and **Gamma ($\gamma$) Sagittae**, M71 is a fertile cluster of faint stars. It is now generally regarded as a poor, uncondensed globular cluster, rather than as a rich open cluster.

*M71 is a very loose globular cluster, without the bright central core typical of its kind.*

345

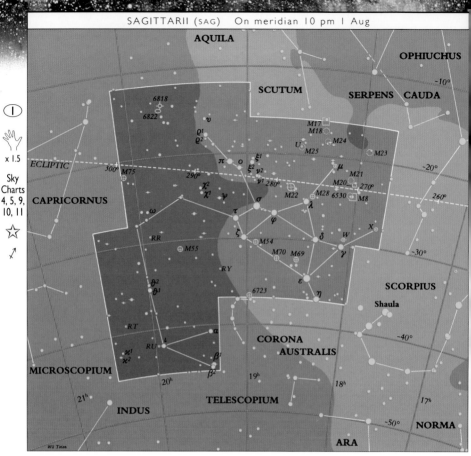

AQUILA

OPHIUCHUS

SCUTUM

SERPENS CAUDA

−10°

6818

6822

M17
M18

M24

M23

ν

θ¹
θ²

U

M25

M22

π

o

ξ¹
ξ²

ν²

ν¹

280°

M20

M28

6530

M21

M8

−20°

ECLIPTIC

300°

M75

290°

χ²
χ¹

ψ

μ

270°

λ

260°

CAPRICORNUS

ω

τ

σ

φ

−30°

RR

M55

ζ

M54

δ

W

X

RY

M70 M69

γ

SCORPIUS

θ²
θ¹

6723

ε

η

Shaula

RT

α

CORONA
AUSTRALIS

−40°

κ¹
κ²

RU

ι

β¹
β²

MICROSCOPIUM

20ʰ

19ʰ

18ʰ

17ʰ

21ʰ

TELESCOPIUM

INDUS

−50°

NORMA

ARA

Wil Tirion

# Sagittarius

(sadge-ih-TAIR-ee-us)   The Archer

This is one of the 12 constellations of the zodiac. The most distinctive aspect of Sagittarius is the group of stars within it that look like a teapot, complete with spout and handle. The charts mark the eastern set of Sagittarius's brighter stars as a quadrilateral figure, which is the teapot's handle. The western group, marked as a triangle, is the spout. The handle also stands by itself as the Milk Dipper.

Not knowing much about teapots, the ancient Arabs thought of the western triangle as a group of ostriches on their way to drink from the Milky Way, and the eastern quadrilateral as ostriches returning from their refreshment.

Sagittarius is generally thought to be a centaur—half man and half horse—and is usually considered to be Chiron, who is also identified with the constellation Centaurus. However, Sagittarius is seen as holding a drawn bow, which is not in character with Chiron, who was known for his wisdom and kindness. Some say that Chiron created the constellation to guide Jason and the Argonauts as they sailed on the *Argo*.

Sagittarius is located on the Milky Way in the direction of the center of the galaxy. Here the band of the Milky Way is at its broadest, although cut by dark bands of dust. It is a treasure trove of galactic and globular clusters, plus bright and dark nebulas.

*The Archer, as he appears in a sixteenth-century fresco at the Villa Farnese, in Italy.*

The bright western half of Sagittarius overlaps the star clouds of the Milky Way looking toward the galactic center. The pink glow of the M8 nebula is quite apparent, and is shown close up below. Several other bright features of Sagittarius and Scorpius can also be discerned.

**M22**: The Great Sagittarius star cluster is a very large globular—the best of the constellation's many globulars. At magnitude 5.1, it is an easy object to see in binoculars, but a telescope really brings out the cluster's beauty. Only 10,000 light-years away, it is one of the closest globulars, and with an 8 inch (200 mm) telescope you should be able to resolve it into seemingly countless stars.

**M23**: Just one of many galactic clusters in Sagittarius, M23 presents over 100 stars in an area the size of the Moon. It is a striking sight in binoculars or in a telescope at low magnification.

**The Lagoon Nebula (M8)**: This spectacular diffuse nebula envelops the cluster of stars called **NGC 6530**. On a dark night the nebula is visible to the naked eye just to the north of the richest part of the Sagittarius Milky Way. In photographs, the extensive nebula is marked by several tiny dark splotches. Bart Bok identified these as globules in which new stars are being formed.

**The Trifid Nebula (M20)**: Found only 1½ degrees to the northwest of the Lagoon Nebula, the Trifid Nebula is likely to be part of the same complex of nebulosity. It is known as the Trifid because three lanes of dark clouds divide the nebula in the most beautiful way. You should be able to detect these dark lanes with a 6 inch (150 mm) telescope under a good sky.

**The Omega Nebula (M17)**: Also called the Swan, the Horseshoe, or the Check-mark, this nebula can be seen quite clearly in binoculars and is a stunning sight through a large telescope.

(Above) M22 is the most striking of the many globular star clusters in Sagittarius.
(Left) The Omega Nebula (M17) on the border of Sagittarius and Serpens Cauda.

OPHIUCHUS

SERPENS CAPUT

SERPENS CAUDA

-10°

ScoX-1

ψ
χ  ξ

ν  β
240°
230°

-20°

260°
250°
ω2,1
M80  δ

o

ECLIPTIC
270°

Antares
σ

LIBRA

α  M4
π
τ

ρ

M62
RR

-30°

BM  M6

ε
N

M7
H

Shaula  6302

λ  ν
6641  G
μ²  μ¹

Q  6337

ι²
κ

SAGITTARIUS
-40°

ι¹
δ
η
ζ²  6231
ζ¹

CRA
6388

CENTAURUS

LUPUS

TELESCOPIUM
18ʰ
ARA
17ʰ
16ʰ
15ʰ

19ʰ
-50°
NORMA

Wil Tirion

① 

🖐 x 1.5

Sky Charts 3, 4, 9, 10, 11

☆
☆

♏

# Scorpius

## (SKOR-pee-us) The Scorpion

In Greek mythology, Scorpius is the scorpion that killed Orion. The two constellations are set at opposite sides of the sky, to avoid further trouble between them.

A beautiful constellation of the zodiac, filled with bright stars and rich star fields of the Milky Way, Scorpius really looks like a scorpion, complete with head and stinger. Near the northern end is a line of three bright stars, with red Antares (Greek for "rival of Mars") at its center.

Some 5,000 years ago, the Persians thought of Antares as one of the Royal Stars, a guardian of heaven. The ancient Chinese referred

A thirteenth-century Italian view of Scorpius, with Antares near the center of the Scorpion's body.

348

to Antares's red glow as the "Great Fire" at the heart of the Dragon of the East. Another Chinese legend refers to Antares and its two attendants as the Ming T'ang—the "Hall of Light" or the "Emperor's Council Hall."

**Antares**: The Romans called this mighty star Cor Scorpionis, meaning "heart of the scorpion," a title the French also use—Le Coeur de Scorpion. Thought to be about 600 light-years away, Antares is a red supergiant some 600 million miles (1,000 million km) across and is 9,000 times more luminous than the Sun. However, with a mass only 10 or 15 times that of the Sun, it is not very dense. Its insides might be like a very hot vacuum.

**Beta (β) Scorpii**: This is a double star whose 2.6 and 4.9 magnitude components are 13.7 arcseconds apart, making resolution possible in a 2 inch (50 mm) telescope.

**M4**: "There are several M4s," said Walter Scott Houston, a highly insightful observer who died in 1993 at the age of 81. What he meant was that this strange globular cluster has a

*This three-color composite image shows Antares enmeshed in nebulosity at lower left, with the globular cluster, M4, to its west (right). The Rho (ρ) Ophiuchi nebulosity (IC 4604) appears at the top.*

different appearance with each instrument you use. Binoculars show a fuzzy patch of light; a small telescope shows a large patch of mottled haze; and 4 or 6 inch (100 to 150 mm) instruments begin to show the individual stars. This is one of the best globulars for viewing in small telescopes.

**The Butterfly Cluster (M6)**: The stars of this large, bright open cluster really resemble a butterfly when they are viewed at high power.

**M7**: This large, bright open cluster, lying southeast of M6, needs to be seen through the large field of view of binoculars to be fully appreciated.

**NGC 6231**: Half a degree north of **Zeta (ζ) Scorpii**, this bright open cluster lies in a rich region of the Milky Way. It is best surveyed in binoculars or at very low power in a telescope.

**M80**: This small, bright globular cluster can be seen in binoculars but needs a 10 inch (250 mm) telescope to resolve its stars.

**Scorpius X-1**: This is a close binary star in which one star expels gas onto a dense neighbor which could be a white dwarf, a neutron star, or a black hole. It is a bright X-ray source, but appears visually as a faint 13th magnitude star.

*(Above) The relatively loose globular cluster M4, is easily found just over 1 degree west of Antares.*
*(Left) Scorpius, with splendid Antares, is one of the few constellations that looks like the creature after which it is named. Its tail overlies rich Milky Way star clouds, which are dimmed and cut by dark dust lanes.*

ECLIPTIC
340°
-10°

CETUS

AQUARIUS
-20°

③
×1.5

Sky
Charts
5, 6,
10, 11,
12

288    253
South
Galactic Pole
α
ε
κ²  κ¹
ι    δ
τ
σ
π    R    z    η    S    7793    μ
ϑ
γ

Fomalhaut
-30°

FORNAX

PISCIS
AUSTRINUS

-40°

ξ    λ²··λ¹    55
β

ERIDANUS

1ʰ    0ʰ    23ʰ
-50°    22ʰ
3ʰ    2ʰ

PHOENIX    GRUS

HOR

Wil Tirion

# Sculptor

(SKULP-tor) The Sculptor

*Johann Bode's depiction of Sculptor in Uranographia (1801) reflects the origins of the name rather than a sculptor.*

This constellation, originally named L'Atelier du Sculpteur (the Sculptor's Workshop) by Nicolas-Louis de Lacaille, lies to the south of Aquarius and Cetus. Since that time, its name has been shortened to Sculptor. It contains little of interest, its most significant feature being a small cluster of nearby spiral galaxies.

**NGC 253**: For a small telescope user, this galaxy is one of the most satisfying, especially for observers in the Southern Hemisphere. It is very large and is viewed almost edge-on. It

*NGC 253, at magnitude 7, is a large, bright spiral galaxy. Ten million light-years distant, it is the largest member of the Sculptor Group of galaxies.*

was discovered by Caroline Herschel one night in 1783, while she was searching for comets. It appears as a thick streak in binoculars and begins to show the texture evident in photographs when larger instruments are used.

**NGC 55**: This is another very fine edge-on galaxy, similar to NGC 253. It is distinctly brighter at one end than the other when seen through an 8 inch (200 mm) telescope.

Both NGC 55 and 253 are members of the Sculptor Group, an arrangement of several galaxies that might be our Local Group's nearest neighbor in the cosmos.

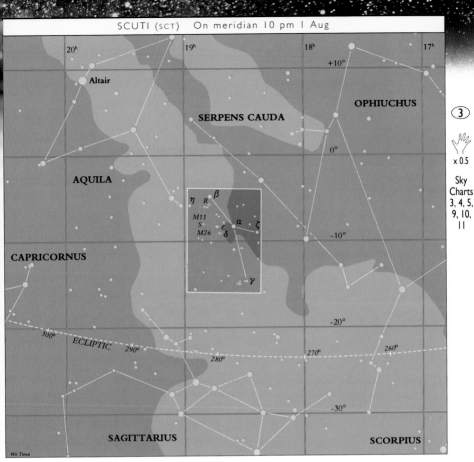

20ʰ   19ʰ   18ʰ   17ʰ

+10°

Altair

SERPENS CAUDA

OPHIUCHUS

③

✋
x 0.5

Sky
Charts
3, 4, 5,
9, 10,
11

AQUILA

0°

η  R  β
M11
S        α    ζ
M26   ε  δ

–10°

CAPRICORNUS

γ

–20°

300°   ECLIPTIC   290°

280°

270°

260°

SAGITTARIUS

–30°

SCORPIUS

Wil Tirion

# Scutum

(SKU-tum) The Shield

Although Scutum is not a large constellation and has no bright stars, it is not difficult to find in a dark sky because it is the home of one of the Milky Way's most dramatic clouds of stars. Johannes Hevelius created the constellation at the end of the seventeenth century, giving it the name Scutum Sobiescianum (Sobieski's Shield) in honor of King John Sobieski of Poland, after he had successfully fought off a Turkish invasion in 1683.

**R Scuti**: This is a semi-regular RV Tauri-type variable star. It changes from magnitude 5.7 to 8.4 and back over a period of about five months.

*The Wild Duck Cluster, M11, is quite a concentrated clutch of stars and presents a glorious view in almost any telescope.*

**The Wild Duck Cluster (M11)**: This spectacular open cluster is clearly visible in binoculars, rewarding in a small telescope, and stunning in an 8 inch (200 mm) one. One of the most compact of all the open clusters, the presence of a bright star in the foreground adds to its beauty.

③

🖐 x 3

Sky charts
3, 4, 9, 10

CORONA BOREALIS

19ʰ      18ʰ      17ʰ      16ʰ

HERCULES

+20°

AQUILA

+10°

ρ

π

κ

γ

β

R ν

φ

χ

τ

δ

OPHIUCHUS

λ

α

SERPENS
CAPUT

ε

ω

ψ

M5

I.4756

ϑ

0°

σ

μ

η

ζ

-10°

SERPENS
CAUDA

SCUTUM

ν

M16

o

LIBRA

M17

ξ

SCORPIUS

ECLIPTIC 23ⁱ⁰

Wil Tirion

# Serpens

(SIR-penz) The Serpent

T his is the only constellation that is divided into two parts. The head (Serpens Caput) and the tail (Serpens Cauda) are separated by the constellation of Ophiuchus, the Serpent Bearer. At one time both the Serpent and the Serpent Bearer formed a single constellation. Serpens was familiar in ancient times to the Hebrews, Arabs, Greeks, and Romans.

**R Serpentis**: A Mira star almost midway between **Beta (β)** and **Gamma (γ)** Serpentis, this variable has a bright maximum of 6.9. It fades to about 13.4, although it sometimes can become fainter. Its period is about one year.

The Eagle Nebula derives its name from the evocatively shaped dark dust features visible at the center of the photograph.

*(Above) Another of the coiling snakes in the sky, Serpens' full length is partially concealed by the figure of Ophiuchus. (Left) M5, a 5th magnitude globular cluster in Serpens Caput is one of the finest in the sky.*

**M5**: This very striking globular cluster is about 25,000 light-years away.

**The Eagle Nebula (M16)**: Through an 8 inch (200 mm) or larger telescope on a dark night, this combination of nebula and star cluster is quite stunning. But you can still enjoy the sight of the cluster in smaller telescopes.

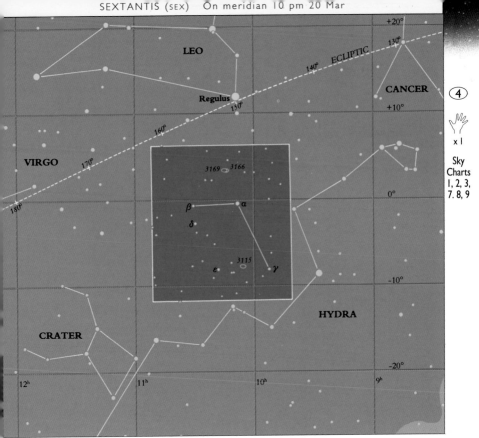

# Sextans

(SEX-tanz) The Sextant

Sextans Uraniae, now known simply as Sextans, was the creation of Johannes Hevelius. He chose this name for the constellation to commemorate the loss of the sextant he once used to measure the positions of the stars.

Along with all his other astronomical instruments, the sextant was destroyed in a fire that took place in September 1679. "Vulcan overcame Urania," Hevelius remarked sadly, commenting on the fire god having defeated astronomy's muse.

Placed between Leo and Hydra, Sextans' brightest star is barely visible to the unaided eye at magnitude 4.5. Despite this, in ancient times the Chinese chose one of the faintest stars in Sextans to represent Tien Seang, the Minister of State in Heaven.

*The galaxy NGC 3115 is a rarity, in that the way it appears to the eye in amateur telescopes is similar to the view that is provided in photographs, although perhaps the pointed ends cannot be seen so well.*

**The Spindle Galaxy (NGC 3115):** Because we see this 10th magnitude galaxy almost edge-on, it appears to be shaped like a lens. Unlike many faint galaxies, the Spindle Galaxy gives quite satisfying views at high power. In classification, it seems to be somewhere between an elliptical and a spiral.

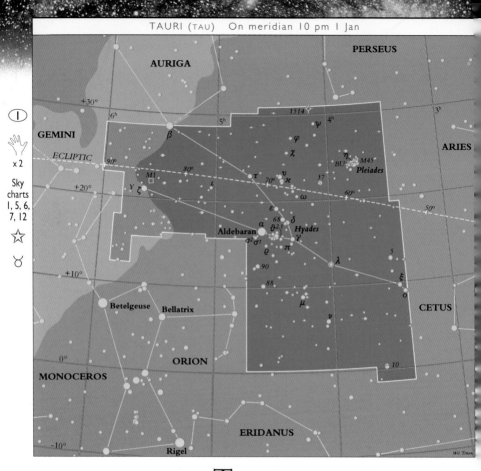

# Taurus

(TORR-us) The Bull

Just northwest of Orion, Taurus is a prominent northern constellation containing two of the largest star clusters we can see—the Hyades and the Pleiades. From the time of the Chaldeans, some 5,000 years ago, this constellation has been seen as a bull.

Bulls have been worshipped since ancient times as symbols of strength and fertility, and they appear in numerous myths. The ancient Egyptians worshipped Apis, the Bull of Memphis, a real bull that was thought to be an incarnation of Osiris. The Israelites worshipped the Golden Calf. Carvings of winged bulls stood at the gates of Assyrian palaces.

In classical times, the Greeks saw the constellation as Zeus disguised as a bull. The story goes that

Zeus fell in love with the beautiful Europa, daughter of Agenor, king of Phoenicia. One day, playing at the water's edge, Europa's attention was caught by a majestic white bull—Zeus in animal form—grazing peacefully among her father's herd. The bull knelt before her as she approached it, so she climbed on its back, wreathing flowers around its horns. Springing to its feet, the bull then took off into the sea and swam to Crete, where Zeus made Europa his mistress. One of their three sons, Minos, later became king of Crete. Only the forequarters of the bull are visible in the constellation, as it is emerging from the waves.

*A depiction of the zodiacal figure of Taurus from a fresco in the palace of Schifanoia in Ferraro, Italy, by Francesco del Cassa (1436–78).*

354

Mag  3  4  5  6  7  8  9

+25°

Asterope
Maia    Taygeta
    Celaeno
BU  η
Pleione        Electra    +24°
Atlas   Alcyone
        Merope

3ʰ 48ᵐ        3ʰ 44ᵐ   +23°

*The Seven Sisters of the Pleiades make a spectacular photograph, bathed in a blue reflection nebula. (Left) A chart of the Pleiades region.*

*Many a night I saw the*

*Pleiades, rising thro' the*

*mellow shade,*

*Glitter like a swarm of fire-flies*

*tangled in a silver braid.*

Locksley Hall, ALFRED LORD TENNYSON, (1809–92), English poet

They remained in the sky, staying close together. The seventh sister is hard to see because she really wants to go back to Earth, and her tears dim her luster.

On a reasonably dark night, you should be able to see at least six of the stars in the Pleiades with the naked eye; under good conditions, you might be able to see as many as nine. Containing more than 500 stars in all, the Pleiades is about 430 light-years away and covers an area four times the size of the Full Moon. It is best seen with binoculars.

**The Hyades:** Like the Pleiades, this is also an open cluster, but it is so close to us (only 150 light-years away) that even when viewed with the naked eye the stars appear to be spread out. The stars of the Hyades form the bull's head.

**Aldebaran (Alpha (α) Tauri):** This is an orange giant and is the brightest star in Taurus. Its name means "the follower" (of the Pleiades) in Arabic. Only 65 light-years away, it marks the eye of the bull.

**The Crab Nebula (M1):** This nebula marks the site of the supernova seen in 1054. It is clearly visible through a 4 inch (100 mm) telescope on a dark night as an oval glow, but this only hints at the complex structure that can be seen in high-magnification photographs.

*M1, the famous Crab Nebula, is an oval glow 5 arcminutes across in small telescopes.*

**The Pleiades (M45):** Also known as the Seven Sisters, this is the most famous open star cluster in the sky and forms the bull's shoulder. Greek legend tells that the sisters called for help from Zeus when they were being pursued by Orion. Zeus turned them into doves and placed them in the sky. Alcyone (Eta [η] Tauri) is the most dazzling sister. She is accompanied by Maia (20 Tauri); Asterope I and II (the double star 21 Tauri); Taygeta (19 Tauri); Celaeno (16 Tauri); and Electra (17 Tauri). Finally, there is Merope (23 Tauri), a star surrounded by a beautiful cloud of cosmic grains producing a blue reflection nebula. Atlas (or Pater Atlas, 27 Tauri) and Pleione (Mater Pleione, 28 Tauri) represent the girls' father and mother.

In a Native American tale, the Pleiades are seven youngsters who, on a walk through the sky, lost their way and never made it home.

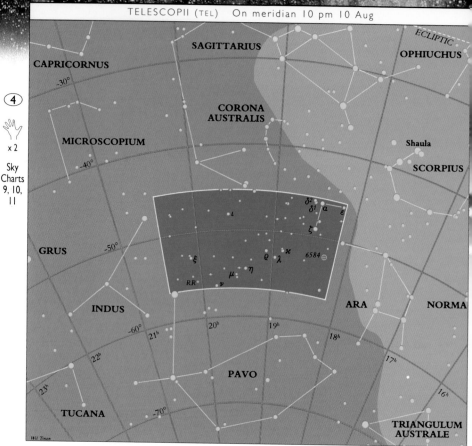

SAGITTARIUS
OPHIUCHUS
ECLIPTIC
CAPRICORNUS
-30°
CORONA
AUSTRALIS
MICROSCOPIUM
Shaula
-40°
SCORPIUS
GRUS
-50°
ι
δ² δ¹ α ε
ζ
ξ ϱ λ κ 6584
μ η
RR ν
INDUS
ARA
NORMA
-60°
21ʰ 20ʰ 19ʰ 18ʰ
22ʰ
17ʰ
PAVO
23ʰ
16ʰ
TUCANA
-70°
TRIANGULUM
AUSTRALE

Wil Tirion

# Telescopium

(tel-eh-SKO-pee-um) The Telescope

Originally bearing the name Tubus Telescopium, this constellation was created by Nicolas-Louis de Lacaille during the eighteenth century to honor the invention of the telescope. It was the only large telescope in space until the launching of the Hubble Space Telescope in 1990. Telescopium is surrounded by Sagittarius, Ophiuchus, Corona Australis, and Scorpius, and de Lacaille "borrowed" a number of stars from these constellations in order to create it.

**RR Telescopii**: Although this star is normally too faint for small telescopes, it is one of the most interesting novas on record. Before 1944, this star varied over about 13 months between 12.5 and 15th magnitude, but in that year it began a rise to magnitude 6.5 that took some five years. As the nova declined in the following years,

A simple refracting telescope honors the importance of the astronomical instrument in Urania's Mirror (1825).

NGC 6584 is a 9th magnitude globular cluster about 6 arc-minutes across. This image was recorded using a 10 inch (250 mm) telescope.

it still displayed its original 13-month period. It is thought that the star may be a binary system, in which a large red star is responsible for the minor variations that take place, and a smaller, hotter star puts on the nova part of the performance.

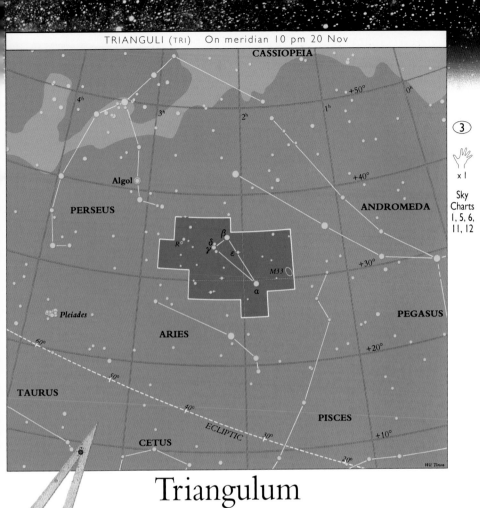

③

✋ x I

Sky
Charts
1, 5, 6,
11, 12

# Triangulum

(tri-ANG-gyu-lum)   The Triangle

Triangulum is a small, faint constellation extending just south of Andromeda, near Beta (β) and Gamma (γ) Andromedae. Despite its lack of distinction, the group of stars was known to the ancients, and because of its similarity to the Greek letter delta (Δ) it was sometimes called Delta or Deltotum. It has been associated with the delta of the River Nile and has also been connected with the island of Sicily, which is shaped like a triangle. The ancient Hebrews gave it the name of a triangular musical instrument.

👁 **The Pinwheel Galaxy (M33):** This galaxy is one of the brightest and biggest members of our Local Group and we have a front row view because it appears face-on. The galaxy is listed at magnitude 5.5 but its light is spread out over such a large area that it is notoriously difficult to see. Although it can be seen by the naked eye on very clear nights, you need a dark sky and binoculars to see a fuzzy glow larger than the apparent diameter of the

Moon. A telescope with a wide field of view will also show the galaxy, but one with a narrow field will show nothing at all.

*(Above, left) The Triangle was represented in a simple form by Bayer in* Uranometria *(1603).*

*(Below) M33 is a loosely wound spiral and the next largest galaxy in the Local Group of galaxies after the Milky Way and Andromeda (M31) galaxies.*

357

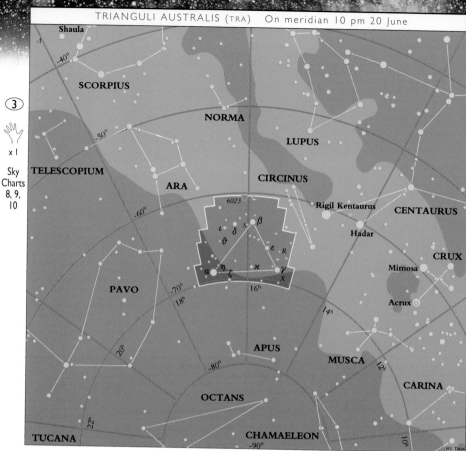

Sky Charts 8, 9, 10

# Triangulum Australe

(tri-ANG-gyu-lum os-TRAH-lee) The Southern Triangle

A simple three-sided figure deep in the southern sky, Triangulum Australe first appeared in Johann Bayer's great atlas *Uranometria* in 1603. It lies just south of Norma, the Level, and east of Circinus, the Drawing Compass—tools used by woodworkers and navigators on early expeditions to the Southern Hemisphere.

**R Trianguli Australis**: One of several Cepheids in the constellation, this interesting variable alters by about a magnitude—from 6 to 6.8. Because it is a Cepheid variable, we know its period precisely, which is 3.389 days. For Cepheids with this rapid variation, magnitude estimates at least once a night are worthwhile.

**S Trianguli Australis**: Another bright Cepheid variable, S varies from magnitude 6.1 to 6.7 and back over a period of 6.323 days.

**NGC 6025**: This is a small open cluster of about 30 stars of 9th magnitude, with fainter background stars.

*A small, but easily identified constellation, Triangulum Australe is the southern counterpart of Triangulum. Alpha ($\alpha$) Centauri, in the lower right of the picture, far outshines the constellation's three brightest stars.*

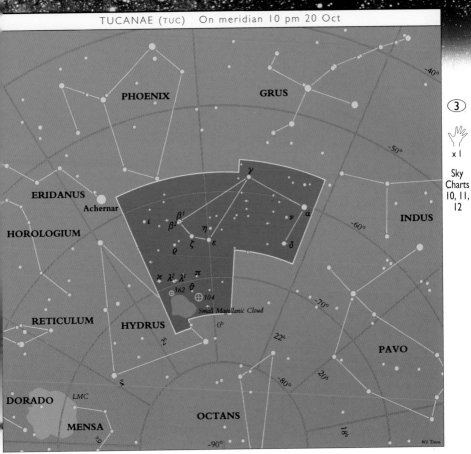

PHOENIX

GRUS

ERIDANUS

Achernar

HOROLOGIUM

γ

β¹
β²
ι
η
ν
α
INDUS
ϱ
ζ
ε
δ

-40°

-50°

-60°

κ λ² λ¹ π
362
104
Small Magellanic Cloud

RETICULUM

HYDRUS

0ʰ

-70°

22ʰ

PAVO

DORADO   LMC

MENSA

OCTANS

-80°

20ʰ

-90°

18ʰ

Wil Tirion

# Tucana

(too-KAN-ah)   The Toucan

Johann Bayer first published this constellation as Toucan, which in time became Tucana, the Latin form. Toucans are larger members of the genus *Ramphastos*—brightly colored, large-billed birds related to woodpeckers that are found in tropical America. From the earliest drawings, Tucana sat on the Small Magellanic Cloud, one of the two closest galaxies to the Milky Way, tending it like an egg.

*The Small Magellanic Cloud (SMC), with the spectacular globular cluster 47 Tucanae alongside. NGC 362, a much smaller and fainter globular, is just north (above) of the SMC.*

👁 **47 Tucanae (NGC 104):** From its perch 15,000 light-years away, this glorious globular cluster shines brightly at magnitude 4.5. Although it is a

*The Toucan, as seen in Johann Bode's Uranographia (1801).*

naked-eye object under dark conditions, a 4 inch (100 mm) or larger telescope really brings out the best in this cluster, which competes with Omega (ω) Centauri for the title of the most splendid globular cluster in the entire sky. It is more centrally condensed than its rival in Centaurus.

👁 **The Small Magellanic Cloud (SMC):** A member of our Local Group, this galaxy is visible to the naked eye on a good night, with 47 Tucanae alongside. Lying about 120,000 light-years away, the cloud is some 30,000 light-years wide.

359

CAMELOPARDALIS

DRACO

DRACO

+70°

15ʰ

14ʰ

13ʰ

12ʰ

11ʰ

10ʰ

9ʰ

8ʰ

M82

M81

σ²

σ¹

π¹

π²

τ

23

ο

ν

α

4605

+60°

τ

M101

80

ζ

ε

δ

Big Dipper

M108

β

M97
Owl Nebula

φ

26

ϑ

15

2841

ι

η

M109

γ

+50°

κ

+50°

χ

ST

ψ

λ

ω

μ

3184

LYNX

+40°

CANES VENATICI

LEO MINOR

+30°

ν

COMA
BERENICES

ξ

LEO

Wil Tirion

Sky
Charts
1, 2, 3,
4

x 2.5

# Ursa Major

(ER-suh MAY-jer) The Great Bear

One of the oldest of the constellations, Ursa Major, The Great Bear, is also perhaps the best known, and numerous legends have been associated with it. Particularly famous is the group of seven stars that make up what is commonly known as the Big Dipper, or the Plough.

In a Cherokee legend, the handle of the Big Dipper represents a team of hunters pursuing the bear from the time he is high in the sky in spring

The Big Dipper, forming the tail and back of Ursa Major, is only a part of the constellation shown in Urania's Mirror.

until he sets on fall evenings. At the start of each evening, the bear, with his hunters, has moved a little farther west in the sky.

The Iroquois of Canada's St Lawrence River Valley and the Micmacs of Nova Scotia have a more elaborate story. Represented by the Dipper bowl, the bear is hunted by seven warriors. Each spring the hunt begins when the bear leaves Corona Borealis, his den. The bear isn't killed until fall, and the skeleton remains in the sky until the following spring. Then a new bear emerges from Corona Borealis and the hunt begins again. Instead of a bear, the Sioux of central North America see a long-tailed skunk.

According to one Chinese legend, the stars of the Big Dipper form a bushel measure to deliver food in fair amounts to the population in times of famine. The ancient Hebrews also saw a bushel measure.

The early Britons saw the Big Dipper forming King Arthur's chariot, whereas in Germany people pictured the group of stars as a wagon and three horses. The Romans viewed

(Left) M101, a loosely wound spiral galaxy, is one
of the largest and brightest galaxies in the sky.
(Below) Ursa Major is the third largest constellation in the sky,
but the Big Dipper is clearly the most prominent part.

it as a team of seven oxen, harnessed to the pole
and driven by Arcturus.

In the Greek legend, Zeus and Callisto, a
mortal, had a son called Arcas. Hera, Zeus's
jealous wife, turned Callisto into a bear, and one
day, while out hunting, her son, not knowing
that the bear was his mother, almost killed her.
Zeus rescued Callisto, placing both her and her
son, whom he also turned into a bear, in the
sky together. Callisto is Ursa Major and Arcas is
Ursa Minor.

**Mizar (Zeta [ζ] Ursae Majoris) and
Alcor**: This is the famous apparent double star
in the middle of the Dipper's handle. It is
separated by 12 arcminutes and is thus possible
to see as a pair by naked eye. Mizar is itself a
true binary star, separated by 14 arcseconds.

**M81**: This spiral galaxy can be easily seen
through binoculars, even when observing
in the city, and it is dramatic when observed
under good conditions. The oval disk becomes
more apparent with increasing telescope size.

**M82**: This is a long, thin peculiar galaxy,
just ½ degree from M81. It appears as
a thin, gray nebulosity in a 4 inch (100 mm)
telescope, but begins to show some detail in an
8 inch (200 mm) or larger one. Even in large
telescopes or photographs, however, it is not
clear what type of galaxy this is.

**M101**: This large, spread-out spiral galaxy
is visible through small telescopes if the sky
is dark enough. It needs a wide
field and a low-power eyepiece.
At around 27 million light-years,
it is one of the closer spiral
galaxies to the Milky Way.

**The Owl Nebula (M97)**:
This is an oval planetary
nebula that takes the shape of an
owl when it is seen in a 12 inch
(300 mm) telescope. It is large
and therefore dim, and a 4 inch
(100 mm) or larger telescope is
needed to find it.

M81—a spectacular galaxy— is probably
a fair representation of how the Milky Way
galaxy would look from the outside.

361

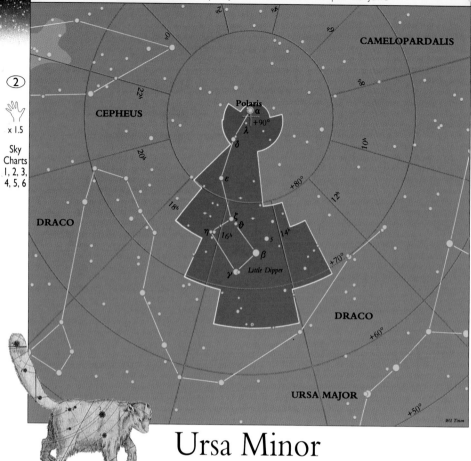

CAMELOPARDALIS

CEPHEUS

Polaris
α
+90°

λ

δ

ε

DRACO

18ʰ

η   16ʰ

γ

β

Little Dipper

DRACO

+80°

+70°

+60°

URSA MAJOR

+50°

Wil Tirion

# Ursa Minor

(ER-suh MY-ner) The Little Bear

Also known as the Little Dipper, Ursa Minor looks somewhat like a spoon whose handle has been bent back by a playful child. This group of stars was recognized as a constellation in 600 BC by the Greek astronomer Thales.

The Little Bear, according to Greek legend, is Arcas, son of Callisto—Ursa Major, the Great Bear. Placed in the heavens by Zeus, he and his mother follow each other endlessly around the north celestial pole.

👁 **Polaris (Alpha [α] Ursae Minoris):** The pole star for the Northern Hemisphere, this Cepheid variable is currently almost 1 degree from the exact pole. Precession of Earth's axis will carry the pole to within about 27 arcminutes of Polaris around the year 2100, and then it will start to move away again.

Polaris is 430 light-years away, with a 9th magnitude companion some 18½ arcseconds away. Splitting this pair is an interesting test for a 3 inch (75 mm) telescope.

While touring England's Lake District in

1819, a gravely ill John Keats was thinking of Polaris when he wrote these lines:

> Bright Star, would I were steadfast as thou art—
> Not in lone splendor hung aloft the night
> And watching, with eternal lids apart,
> Like Nature's patient, sleepless Eremite.

*(Above, left) The Little Bear, as seen by Bayer in Uranometria (1603), the first "modern" star atlas.*
*(Below) The relatively faint stars of Ursa Minor swing around Polaris as the sky rotates during the night.*

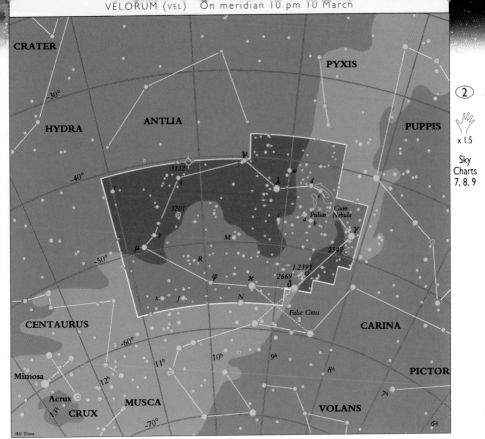

CRATER
PYXIS
HYDRA
ANTLIA
PUPPIS
-30°
-40°
3132
λ
ω
d
e
3201
Gum
Pulsar Nebula
-50°
M
a   b
γ
ι
R
2547
φ
χ
1.2391
N
2669
δ
J
False Cross
CENTAURUS
CARINA
-60°
11ʰ
10ʰ
9ʰ
8ʰ
PICTOR
Mimosa
12ʰ
7ʰ
Acrux
MUSCA
VOLANS
13ʰ
CRUX
-70°
6ʰ
Wil Tirion

②

x 1.5

Sky Charts 7, 8, 9

# Vela

(VEE-lah) The Sail (of Argo)

This constellation, along with Carina (the Keel), Puppis (the Stern), and Pyxis (the Compass), once formed part of a huge group of stars in the southern skies known as Argo Navis, the Ship Argo. This was the vessel that Jason and the Argonauts sailed in on their search for the Golden Fleece. Argo Navis was divided up by Nicolas-Louis de Lacaille in the 1750s, sharing its stars between the four resulting constellations. This left Vela with no stars designated alpha (α) or beta (β).

**The False Cross: Delta (δ) and Kappa [κ] Velorum**, together with **Epsilon (ε)** and **Iota (ι) Carinae**, make up a larger but fainter version of the Southern Cross which is known as the False Cross.

**Gamma (γ) Velorum**: This double star is resolvable in a steady pair of binoculars. The primary is a Wolf-Rayet star—very hot and luminous.

**NGC 3132**: This bright planetary nebula accompanies the many clusters in Vela, but lies right on the border with Antlia. Being 8th

*A satellite trail cuts across a small portion of the extensive nebulosity forming the Vela Supernova remnant.*

magnitude and almost 1 arcminute across, it is considered the southern version of Lyra's Ring Nebula, but with a much brighter central star.

363

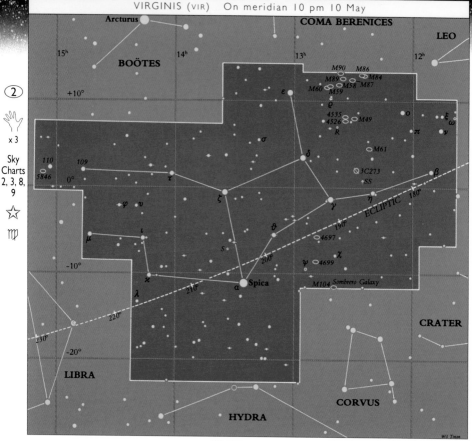

COMA BERENICES

LEO

Arcturus

BOÖTES

M90  M86
M89  M84
M60  M58  M87
M59
ρ
4535
4526  M49
R
M61
3C273
SS

LIBRA

HYDRA

CORVUS

CRATER

Spica

ECLIPTIC

Wil Tirion

# Virgo

(VER-go) The Maiden, The Virgin

Virgo is the only female figure among the constellations of the zodiac and has been thought to represent a great array of deities since the beginning of recorded history. Among others, she has been identified with the Babylonian fertility goddess Ishtar; Astraea, the Roman goddess of justice; and Demeter, the Greek goddess of the harvest (Roman, Ceres). Virgo is usually shown either holding an ear of wheat or carrying the scales of Libra, the adjoining constellation.

*The lovely Virgo, as she is featured in Urania's Mirror of 1825.*

**Spica (Alpha [α] Virginis)**: A bright white star, Spica is the ear of wheat Virgo is holding. The star is almost exactly 1st magnitude, although it has a slight variation. It is 260 light-years away and has more than 2,000 times the luminosity of the Sun.

**Porrima (Gamma [γ] Virginis)**: This is one of the best double stars in the sky, each component shining at magnitude 3.7. At 3 arcseconds separation, the pair is still easy to separate, but by the year 2017 they will appear much closer together.

## THE REALM OF THE GALAXIES

Scattered throughout Virgo and Coma Berenices are more than 13,000 galaxies. Known as the Virgo Cluster or Coma–Virgo Cluster, this mighty club of distant systems of stars repays sweeping with a small, wide-field telescope on a dark night. To see much detail usually requires an 8 inch (200 mm) or larger telescope. Listed here are some of the galaxies, but you are likely to see many more.

**M49**: This elliptical galaxy is one of the brightest galaxies in the Virgo cluster. It is slightly bigger and brighter than M87.

**M84 and 86**: These two elliptical galaxies are close enough to be seen in the same low-power telescope field. On a dark night, an 8 inch (200 mm) telescope will show several smaller galaxies in the same view.

**M87**: This elliptical galaxy is one of the mightiest galaxies we know. Through a small telescope it appears as a bright patch of fuzzy light about a magnitude brighter than M84 and 86. Interestingly, larger telescopes don't show a great deal more. In the professional size range, however, more details do emerge. With a 60 inch (1.5 m) telescope, for example, you can see a jet emerging from the galaxy's center, and photographs with a 200 inch (5 m) telescope have shown more than 4,000 globular clusters on the galaxy's outskirts.

**The Sombrero Galaxy (M104)**: Although this galaxy is quite a distance south of the main concentration of galaxies, it seems to be gravitationally attracted to the swarm and so is thought to be a part of it. The brightest of the Virgo galaxies, a dark lane cuts along its equator, making it look a little like a Sombrero hat in an 8 inch (200 mm) telescope.

*The Sombrero Galaxy presents a bright, 8th magnitude glow, just 8 arcminutes across, but readily seen in smaller telescopes.*

*Some of the brighter members of the stunning Virgo Cluster, dominated in this view by M86 (center) and M84 (right), both of them elliptical galaxies. The majority of galaxies in the Cluster are about 65 million light-years from Earth.*

**3C273 Virginis**: This is the brightest known quasar, but being only 13th magnitude, an 8 inch (200 mm) telescope is needed to identify it. Lying 3 billion light-years away, it is the most distant object amateurs are likely to see in their telescopes.

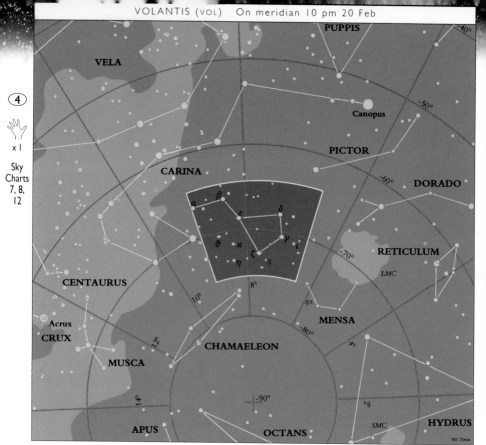

VELA

PUPPIS

Canopus

PICTOR

CARINA

DORADO

α   β

ε

δ

ϑ   χ   ζ   γ   ι

η      s

RETICULUM

LMC

CENTAURUS

8ʰ

6ʰ

MENSA

Acrux
CRUX

10ʰ

MUSCA

CHAMAELEON

12ʰ

APUS

OCTANS

-90°

SMC

HYDRUS

4
x 1
Sky
Charts
7, 8,
12

# Volans

(VOH-lanz) The Flying Fish

The constellation of Piscis Volans, the Flying Fish, lies south of Canopus, and was introduced by Johann Bayer in his *Uranometria* of 1603. It is now known only as Volans. Sailors in the south seas had reported seeing schools of flying fish, which may have been the inspiration for the name. The pectoral fins of these fish are as large as the wings of birds and they glide across the water for distances of up to ¼ mile (400 m).

## SKYWATCHING TIP

For long exposure photography of stars and planets using an equatorial mount, it is important to align the telescope mounting more precisely than for visual observing. This will save you major guiding headaches later. Computerized "Go To" mounts on some modern telescopes remove the need for manual alignment. Remember that an equatorial mount is still needed to avoid rotation of the field (see p. 91).

*NGC 2442 is an 11th magnitude barred spiral galaxy, seen nearly face-on. Its faintness and 6 arcminute size make it a target for a 12 inch (300 mm) telescope.*

**S Volantis**: A Mira star, S Volantis usually has a maximum magnitude of 8.6, but it has occasionally risen to 7.7. Its faint minimum averages 13.6. The star completes its cycle in a little less than 14 months.

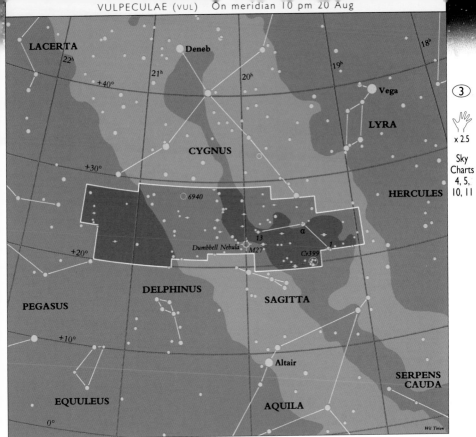

# Vulpecula

(vul-PECK-you-lah) The Fox

This constellation, invented by Johannes Hevelius in 1690, is without an exciting story or a moral tale. Hevelius's name for it was Vulpecula cum Anser, the Fox with the Goose, but now the constellation is simply referred to as the Fox.

**The Dumbbell Nebula (M27)**: This is one of the finest planetary nebulas in the sky and is well suited to small telescopes. Bright and large, it is easy to find just north of **Gamma (γ) Sagittae**. Being 7th magnitude, it can be found through binoculars, but it appears only as a faint nebulous spot. If you use a small telescope, you can make out its odd shape. A larger telescope will reveal its 13th magnitude central star.

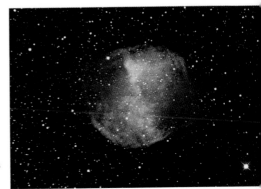

The Dumbbell Nebula is a sight for any telescope. The larger the instrument that is used, the greater the details revealed in the nebula's 5 arcminute disk.

Urania's Mirror (1825). shows Vulpecula, the Fox, with the goose which he has since lost!

Although the gases in this nebula are expanding at the rate of some 17 miles (27 km) per second, there will be no noticeable change in the nebula's appearance within a human lifetime.

367

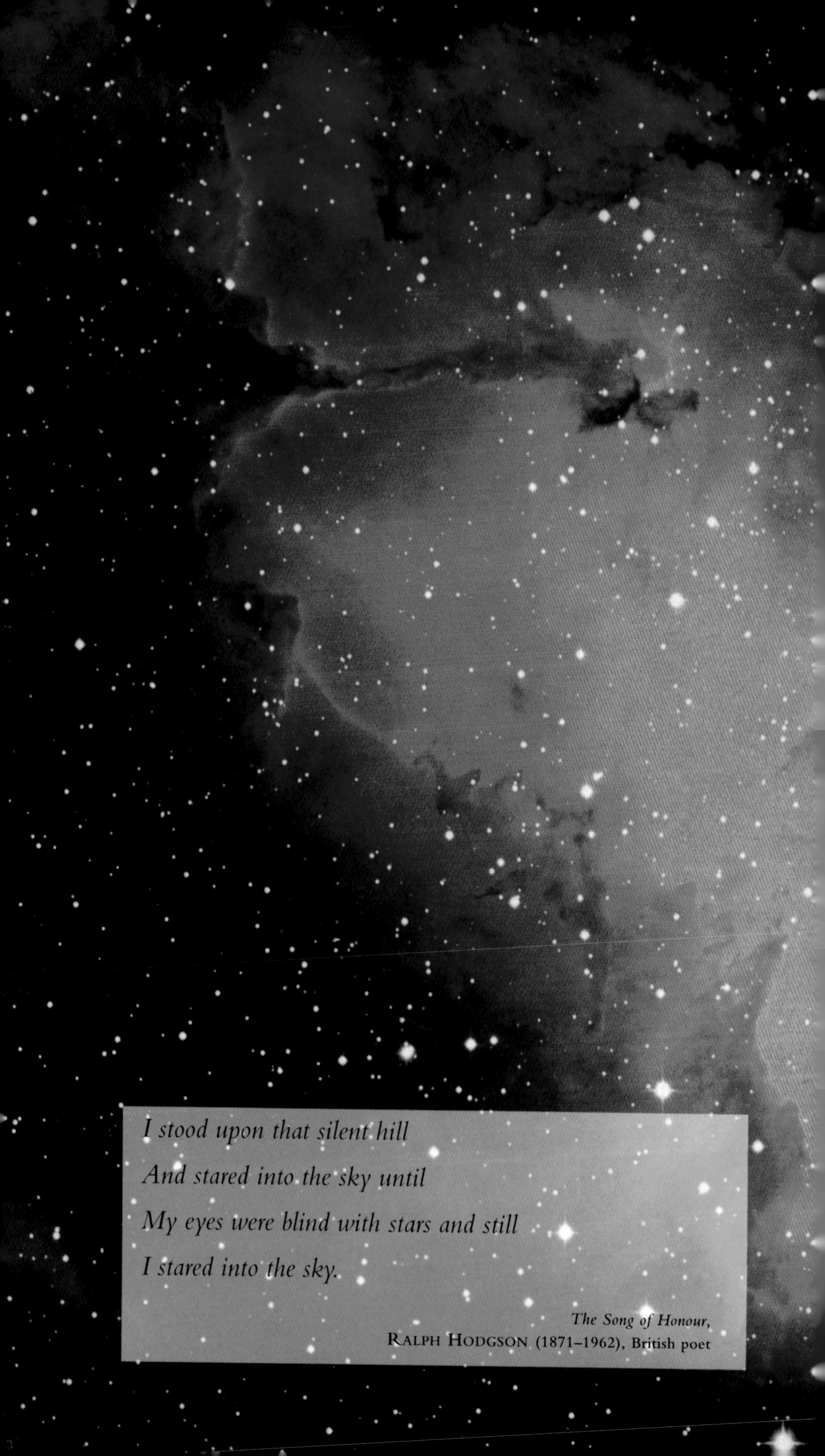

I stood upon that silent hill

And stared into the sky until

My eyes were blind with stars and still

I stared into the sky.

*The Song of Honour,*
RALPH HODGSON (1871–1962), British poet

CHAPTER NINE

# STARHOPPING
# GUIDE

# USING *the*
# STARHOPPING GUIDE

*For a closer look at some highlights of the sky,*

*try a guided tour.*

The 20 starhops on the following pages cover a selection of some of the best-known areas across the night sky. They are intended as an introduction to the technique of starhopping, with the hope that you will move on from here to discover for yourself some of the multitude of other amazing objects appearing elsewhere in the sky.

The **starting point** for each starhop is an easy-to-find bright star.

TOWARD THE HEART OF OUR GALAXY

**Color photographs** supplement the text by showing some of the objects described on the starhop.

Each **object** on the starhop is numbered to correspond with the 1 degree telescope fields of view (circles) on the main sky chart.

The **starhopping text** takes you on a guided tour of the night sky and is frequently cross-referenced to other relevant chapters in the book. It provides important observer-based information about each of the objects on the starhop, along with tips for finding fainter objects and some interesting historical facts.

The illustrated **feature box** takes a closer look at one of the more interesting objects on the starhop, elaborating on aspects of the science behind it.

M17, the Omega Nebula (above) lies ...

**⑬ M25 (IC 4725)** ⊂⊙⊃ Move 5 degrees north and 1 degree west of M22 to take a look at M25. This scattered open cluster, about ½ degree across, is an easy object for binoculars and small telescopes. Near the center of M25 is the yellowish Cepheid variable **U Sagittarii**, which varies from magnitude 6.3 to 7.1 over 6.75 days.

### THE STELLAR NURSERY IN M16

In 1995, the Hubble Space Telescope returned an eerie picture of the nebulosity of M16, showing huge columns of gas that are several light-years in length (below). A close look at the surfaces of the giant pillars shows what scientists have called evaporating gaseous globules, or EGGs. Ultraviolet light from nearby stars is stripping some of the gas from the columns and exposing the EGGs, in which stars are forming. We see the globules because they are more dense than the rest of the nebula, and so are not "blown away" as easily. However, the gradual loss of material feeding the developing stars is thought to play an important role in limiting their size. We can only wonder how many more stars are forming deep within these huge columns, gathering material in private and becoming larger and larger.

**⑭ M24** 👁 To locate the ...hop, scan your eye 3 degree... is unique among Messier ... huge star cloud about 1 d... grees long. Sweeping y... reveal a very rich regio... The cluster is visible ... scope, but can be re...

**⑮ The Omega N...** M17, NGC 6618... 2½ degrees farth... comes into view... see in binoculars, app... light. In a 4 inch (100 mm) sco... shape looks a bit like the figure 2 with an ... baseline. Because of this pattern, M17 is called ... both the Swan Nebula and the Omega Nebul...

**⑯ The Eagle Nebula (M16, NGC 6611)** ⊂⊙⊃ Across the border into Serpens, about 2½ degrees north of M17, the combination ... hazy star cluster and nebula, M16, appears. T... cluster is obvious enough with binoculars, bu... need a 6 inch (150 mm) scope and a dark sk... clearly see that it is actually immersed in a ... of nebulosity called the Eagle Nebula (see ... In larger scopes, the whole field is beautif...

**⑰ M23 (NGC 6494)** ⊂⊙⊃ Head 7½ ... southwest of M16, back into Sagittarius, ... view of M23. This open cluster is nearly ... gree across. It is a fine sight in a 4 inch ...

Being at a m... this region living in the... However, northern...

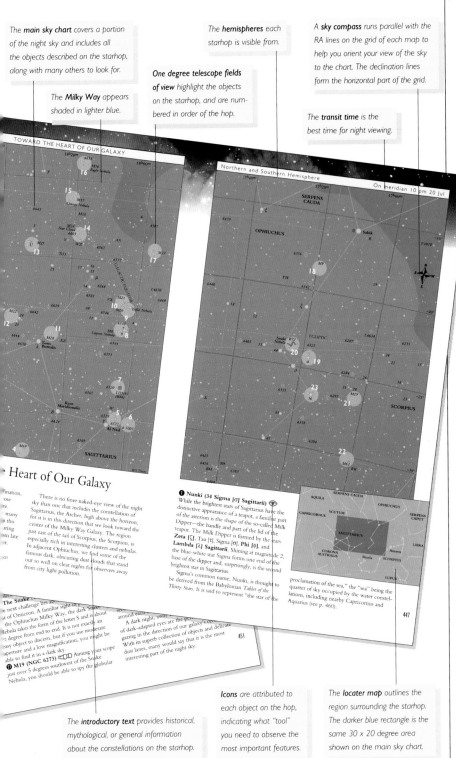

*The **main sky chart** covers a portion of the night sky and includes all the objects described on the starhop, along with many others to look for.*

*The **hemispheres** each starhop is visible from.*

*A **sky compass** runs parallel with the RA lines on the grid of each map to help you orient your view of the sky to the chart. The declination lines form the horizontal part of the grid.*

*One degree telescope fields of view highlight the objects on the starhop, and are numbered in order of the hop.*

*The **Milky Way** appears shaded in lighter blue.*

*The **transit time** is the best time for night viewing.*

*The **introductory text** provides historical, mythological, or general information about the constellations on the starhop.*

*Icons are attributed to each object on the hop, indicating what "tool" you need to observe the most important features.*

*The **locater map** outlines the region surrounding the starhop. The darker blue rectangle is the same 30 x 20 degree area shown on the main sky chart.*

371

# HOW *to* STARHOP

*For novice stargazers, navigating your way among the stars and searching for faint deep-sky objects can be daunting, until you learn how to starhop.*

asically, starhopping is a technique that uses bright stars as a guide to finding fainter objects. All you need to starhop is an idea of how much sky you can see (the field of view) in your finderscope and telescope eyepieces, and how that view will compare to the same amount of sky on your star chart.

Even if you have a computerized telescope to take you from one object to the next, the techniques of starhopping will help you pick out a faint deep-sky object from a field of stars. The 20 starhops on the following pages are a great way to get started.

## HOW MUCH SKY CAN YOU SEE?

To start out, you will need a variety of templates of some sort to represent the fields of view of your binoculars or finderscope, and of your telescope eyepieces. Illustrators' plastic circle templates work well, as they come in a wide variety of sizes to choose from and are available from most artists' and office supply stores.

To find out what diameter circle represents your finderscope field of view on the sky charts, center your finder on an easy-to-locate star, such as Mizar (Zeta [ζ] Ursae Majoris) (see p. 361). Southern Hemisphere skywatchers can use brilliant Alpha [α] Centauri (see p. 292). Look at the fainter stars around Mizar or Alpha Centauri, and study the view. Notice which stars you can see along the edges of the field, then find those same stars on your chart. Center your template circle on the same bright star on the map and try out different diameter circles until you find the one that closely matches your finderscope view.

To locate fainter objects, it helps to have an idea of how large the object will appear to be in your telescope eyepiece. The apparent size of an object is its angular diameter, usually measured in minutes or seconds of arc. Measure the field of view of your eyepieces by using a bright star located close to the celestial equator, such as Mintaka (Delta [δ] Orionis), and follow these simple steps: place the star on the eastern edge of your view; time, in seconds, how long it takes for the star to drift across the center of the field and exit on the opposite (western) side; divide this time by four. The result represents the angular diameter of your field of view in arcminutes.

You should multiply the answer by 60 to obtain your field diameter in arcseconds—many faint objects and the separations of most double stars are smaller than an arcminute across. Measure the field of view for all of your

**THE TRIFID NEBULA** *appears as a small, faint patch of light in a binocular-field view (left), but is revealed as a colorful nebula in the photograph below.*

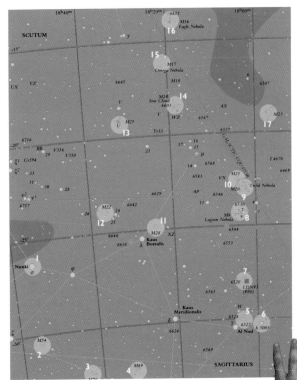

**A RED FLASHLIGHT** (left) is essential
for reading star charts under a dark sky.
The key to symbols (below) applies to
all the main sky charts in this chapter.

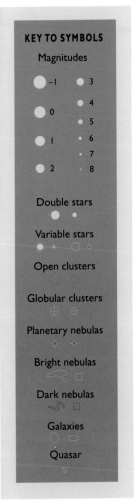

### KEY TO SYMBOLS

**Magnitudes**

| | | | |
|---|---|---|---|
| –1 | | 3 | |
| 0 | | 4 | |
| 1 | | 5 | |
| | | 6 | |
| | | 7 | |
| 2 | | 8 | |

**Double stars**

**Variable stars**

**Open clusters**

**Globular clusters**

**Planetary nebulas**

**Bright nebulas**

**Dark nebulas**

**Galaxies**

**Quasar**

eyepieces, along with the finderscope, and record this information in your observation logbook.

USING THE SKY CHARTS
It is important to have a feel for the size of your chart in relation to what you are looking at. All the main sky charts in the Starhopping Guide are to the same scale (about 30 x 20 degrees). The span of your outstretched hand at arm's length measures close to 20 degrees—equivalent to the area represented by the main sky charts' vertical edge. Look for the brightest stars on your chart, then match the chart orientation to the same arrangement of stars appearing in the sky.

Starhopping is like moving down the link of a chain, with each link represented by your field-of-view circle. On your chart, each selected object is centered in a 1 degree field—about the same size as a low-power eyepiece—but remember that each eyepiece's field size will be different. Estimate how many fields to move your telescope in making each hop, using your template circle. As you move the telescope, look out for the patterns of fainter stars along the hop.

One of the most difficult aspects of starhopping is being sure to move your telescope

**COMPARE** sky distances, relative to your chart, using your hand as a rough guide.

in the right direction, as the view of the sky through a telescope is often upside down or back to front (see p. 76). The easiest way to orient yourself is to nudge your scope toward the north (or south) celestial pole to see where stars enter the northern (or southern) edge of the field. Nudge your scope at right angles to this to find east and west. A compass symbol is also marked on the grid of each main sky chart to help keep you on track.

The stars on the star charts are marked from magnitude –1 down to magnitude 8. The deep-sky objects are marked to as faint as magnitude 12.5.

Now that you know how to starhop, you are ready to embark on any of the guided sky tours featured.

# KEY MAPS

*Finding some constellations can be difficult, depending on your location, but a map of the whole sky is a great way to orient yourself to the overall picture.*

It is easy to get around the night sky if you have some knowledge of the constellations and their position in relation to each

**DRACO,** *the Dragon, is one of the Northern Hemisphere's circumpolar constellations.*

other. A naked-eye view of the sky is the best way to learn the positions of the constellations, regardless of whether you own a telescope or not. When you are familiar with the constellations, you can then move on to particular regions, such as those in the starhops.

The key maps below represent the entire night sky. The projection used here halves the celestial sphere from top to bottom (celestial north to south). The north celestial and south celestial poles are at +90° and –90°, respectively. To avoid the

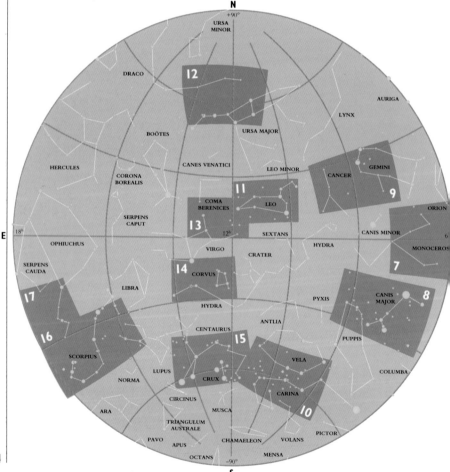

## THE STARHOPS

distortion of the constellation patterns that occurs as the sphere narrows toward the two poles, the key maps have been presented in stereographic projection. The only "distortion" from this projection comes from an increase in scale toward the edge of each map,

giving them a more rounded appearance, making the star patterns easier to recognize.

The starhops are represented by the darker blue patches on the key maps, and are numbered from 1 to 20. They are ordered west to east across the sky according to the times that

they come into view in the evening throughout the year. These regions have been chosen for their variety of objects as well as their location. One or two starhops might be too far north or south for you to observe but most of them can be enjoyed by everyone.

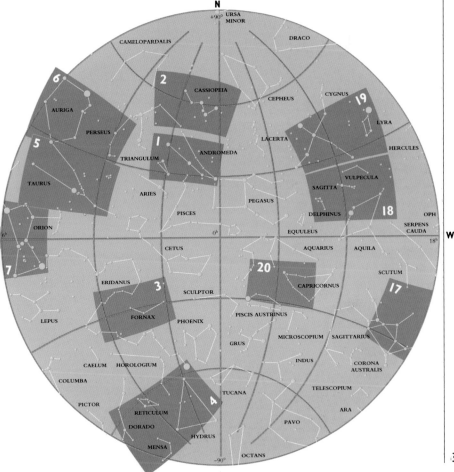

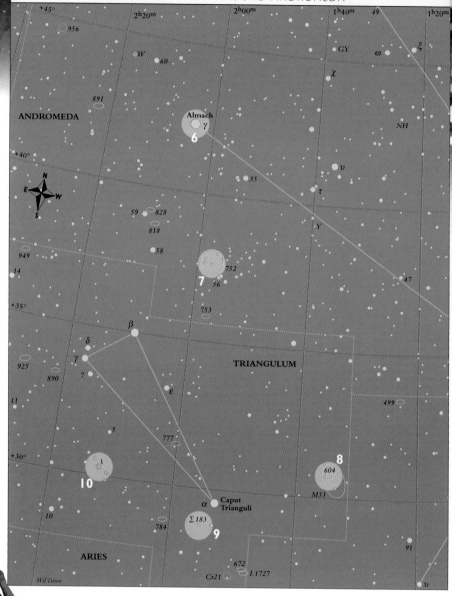

# A Galaxy Hunt around Andromeda

Visible from both hemispheres, this star-hop offers skywatchers plenty of celestial treasures. The tour highlights are the Andromeda and Pinwheel galaxies, providing unforgettable views in a clear, dark sky.

*Perseus rescuing Andromeda, the Chained Maiden, by Pierre Puget (1620–94).*

The Andromeda Galaxy, in the constellation of Andromeda, is one of the brightest and most accessible galaxies for observers, and has been well known since ancient times. The earliest extant drawing of it is on a sky chart published in AD 964 by the Persian astronomer Abderrahaman al-Sufi. The Pinwheel Galaxy lies in the neighboring constellation of Triangulum, the Triangle.

Keep in mind, though, that these galaxies are not the sum total of the area. It is also rich in many other interesting deep-sky objects.

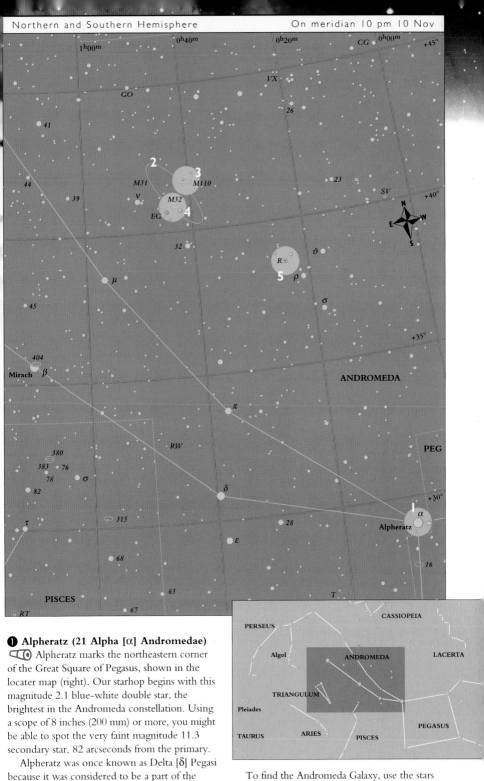

❶ **Alpheratz (21 Alpha [α] Andromedae)**
Alpheratz marks the northeastern corner
of the Great Square of Pegasus, shown in the
locater map (right). Our starhop begins with this
magnitude 2.1 blue-white double star, the
brightest in the Andromeda constellation. Using
a scope of 8 inches (200 mm) or more, you might
be able to spot the very faint magnitude 11.3
secondary star, 82 arcseconds from the primary.

Alpheratz was once known as Delta [δ] Pegasi
because it was considered to be a part of the
constellation of Pegasus before being renamed
and allocated to Andromeda. You might see it
labeled as Delta [δ] Peg on some sky charts.

To find the Andromeda Galaxy, use the stars
**Mirach (43 Beta [β] Andromedae)** and
**37 Mu [μ] Andromedae**, located in the twin
chains of stars northeast of Alpheratz, as pointers.

**2 The Andromeda Galaxy (M31, NGC 224)** 👁 The spectacular Andromeda Galaxy is the most distant object we can see with the naked eye. As you gaze at the soft glow of the closest spiral galaxy to the Milky Way, you are looking across about 2,900,000 light-years. M31 is a large, bright object measuring 160 x 50 arcminutes. In a dark sky it is clearly visible to the unaided eye, appearing as a broad, hazy-white cloud. However, city observers need binoculars to catch a satisfying view. In a 4 inch (100 mm) telescope, the core of M31 seems flattened, flowing evenly into the disk of the galaxy. A telescope fitted with a low-power, wide-field eyepiece is an advantage when observing M31 because of its sheer size. A higher power eyepiece will reveal only its nucleus.

In 1924, Edwin Hubble proved that M31 was not a part of the Milky Way, helping to determine the size of the universe. He did this by comparing the brightness of Cepheid variables in M31 and the Milky Way, finding that they were dimmer in M31 because they were so much farther away.

**3 M110 (NGC 205)** Within ½ degree of M31 is one of its two companion galaxies, M110, glowing at magnitude 8. In a 4 inch (100 mm) scope, M110 appears as an elongated patch of light, brightening toward its core.

M110 is the last object appearing on Charles Messier's list, having been added in 1967.

*M31 (above) is vastly more distant than the open cluster NGC 752 (left), its neighbor in the Andromeda constellation.*

**4 M32 (NGC 221)** M32 is a compact elliptical galaxy about 1 degree southeast of M110. Also a companion galaxy of M31, M32 appears as an oval patch of light in a 4 inch (100 mm) scope, but is best viewed through larger telescopes.

**EG Andromedae** Half a degree farther southeast is the cataclysmic variable red giant EG Andromedae. Its magnitude ranges from 7.1 down to 7.8 in a complicated cycle.

**5 R Andromedae** Southwest of EG is R Andromedae, a long-period Mira variable well known for its broad-ranging cycle, moving from magnitude 5.3 down to 15 over a period of about 409 days. It is sometimes visible in binoculars, but at minimum, is a challenge for an 8 inch (200 mm) telescope. The stars **24 Theta [θ] Andromedae**, **25 Sigma [σ] Andromedae**, and **27 Rho [ρ] Andromedae** form a triangle west of R, that helps locate the star when it is faint.

**6 Almach (57 Gamma [γ] Andromedae)** Sweep your scope about 12½ degrees east to Almach, a fine, color-contrast close double star—one of the best doubles for small-telescope observers. The magnitude 2.3 golden primary contrasts with its magnitude 5.1 greenish blue companion, lying 10 arcseconds away.

**❼ NGC 752**  Shift about 5 degrees south of Almach to NGC 752. The 60 stars in this open cluster can be seen with the unaided eye from a dark site. With a finderscope or binoculars, you may be able to see chains and knots of stars forming a twisted X within this rich cluster.

**❽ The Pinwheel Galaxy (M33, NGC 598)** About 7 degrees southwest of NGC 752 is the beautiful Pinwheel Galaxy, across the border in Triangulum. Although bright, at magnitude 5.5, this spiral galaxy is hard to see because it appears face-on and is spread over a large patch of sky. It appears as a fuzzy glow through binoculars in a dark sky, but you will need a scope of 8 inches (200 mm) and a wide-field eyepiece to see more of the galaxy, including its diffuse, twisting arms. **NGC 604** The splotch of nebulosity sitting prominently on the north-eastern tip of the Pinwheel Galaxy is NGC 604, shining more brightly than other parts of M33's spiral arms. NGC 604 is an oval-shaped emission nebula of glowing hydrogen gas where new stars are born.

**❾ Struve [Σ] 183** About 4 degrees southeast of M33 is the bright yellow multiple star **Caput Trianguli (2 Alpha [α] Tri)**. Use it as a pointer to locate the much dimmer double star Struve 183, about 1 degree southeast. A fine close double, Struve 183 has a magnitude 7.7 light yellow primary and a magnitude 8.4 blue secondary, 6 arcseconds away.

*M33, the Pinwheel Galaxy, is the biggest and brightest galaxy in the Local Group, after the Milky Way and the Andromeda Galaxy.*

**❿ 6 Iota [ι] Tri** Four degrees northeast of Struve 183 is the stunning color-contrast double 6 Iota [ι] Tri. The primary is a magnitude 5 yellow star and the secondary is a magnitude 6.5 pale blue star, shining 3.8 arcseconds away.

Take the time to re-explore the highlights of our tour, along with the many other deep-sky objects surrounding them in this area.

## RADIO MAPPING OF M31

In 1912, American astronomer V. M. Slipher made the first spectrogram of M31. The blue-shifts and redshifts of the spectrum lines of different parts of the galaxy showed that it was rotating. This image (left), based on the 21 centimeter wavelength by the Westerbork Synthesis Radio Telescope, also shows that M31 is rotating and that there is a high concentration of cold hydrogen gas in the galaxy's outer arms. The color enhancement of the radio image uses yellow and red to indicate the areas that are rotating away from us, while green and blue indicate areas that are in systematic motion toward us. From our vantage point, the Andromeda Galaxy is rotating in a clockwise direction.

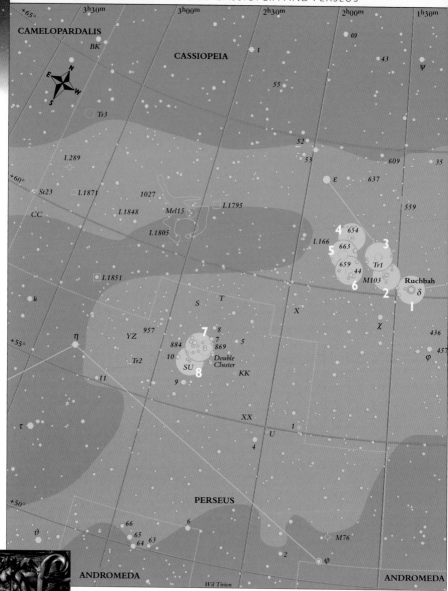

Wil Tirion

# Clusters around Cassiopeia and Perseus

Unfortunately, the spectacular open clusters in this part of the Milky Way are not visible to stargazers in mid-southern latitudes. For northern observers, however, Cassiopeia is a circumpolar constellation that never sets. The best months to explore this region are September through February. This area of the night sky is especially rewarding for binocular observers.

Cassiopeia, the Queen of Ethiopia, and Perseus, the Hero, are central figures in the lore connected to several other mythological characters for whom nearby constellations are named. Cassiopeia sits on her throne in the form of an easy-to-locate W-shaped asterism of bright northern stars. To her southeast is the Y-shaped figure of her son-in-law Perseus.

*The figure of Perseus in a detail from* The Doom Fulfilled, *a painting by English artist Edward Burne-Jones (1833–98).*

**380**

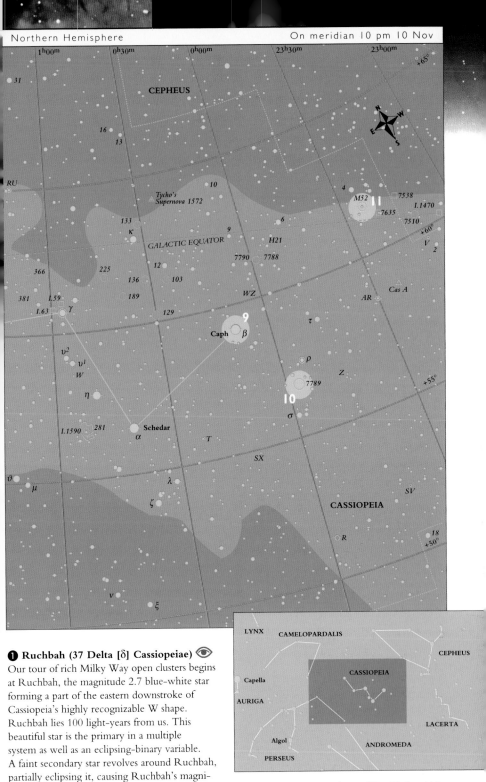

**① Ruchbah (37 Delta [δ] Cassiopeiae) 👁**
Our tour of rich Milky Way open clusters begins
at Ruchbah, the magnitude 2.7 blue-white star
forming a part of the eastern downstroke of
Cassiopeia's highly recognizable W shape.
Ruchbah lies 100 light-years from us. This
beautiful star is the primary in a multiple
system as well as an eclipsing-binary variable.
A faint secondary star revolves around Ruchbah,
partially eclipsing it, causing Ruchbah's magnitude to dim very slightly over a 760-day period.

Ruchbah's name originates from the Arabic
*Al Rukah*, in relation to its position as "the knee"
of the woman in the chair.

About ½ degree northeast of Ruchbah, within
the same binocular or finder field, is the U-shaped
gathering of open clusters: M103, Trumpler 1,
NGC 654, NGC 663, and NGC 659.

**❷ M103 (NGC 581)** Shift your scope about 1 degree northeast of Ruchbah to locate M103. Pierre Méchain discovered this loose, magnitude 6.2 open star cluster in 1781. The brighter stars within the cluster seem to form the shape of an arrowhead. A 4 inch (100 mm)

*The Double Star Cluster (above)—NGC 869, at right, and NGC 884, at left—is a favorite target for Northern Hemisphere observers. Appearing close in the sky, they are not really associated.*

scope can resolve many of the fainter stars, some of which are colored. M103 may not actually be a star group that is bound by gravity, but simply a grouping of about 60 stars that appear as a scattered cluster from our vantage point.

**❸ Trumpler 1** Looking ½ degree farther northeast from M103, you should be able to spot Trumpler 1. This poor open cluster is a little difficult to pick out from the rich Milky Way star field it appears to be a part of. The small clump of stars seems to have a bright streak running through it that cannot be properly resolved, even with a 6 inch (150 mm) telescope.

**❹ NGC 654** About 1 degree northeast from Trumpler 1 is NGC 654. This cluster consists of about 60 stars in a loose association, but is still an easy object to see with binoculars, appearing as a hazy glow of tiny, faintly resolved stars.

**❺ NGC 663** Nudge ½ degree south of NGC 654 to the open cluster NGC 663. The 80 stars in this cluster give it a roundish look, somewhat like a cross between an open and a globular cluster, through a 6 inch (150 mm) telescope.

### SUPERNOVA REMNANT CASSIOPEIA A

When Cassiopeia A (3C 461) exploded as a supernova some 9,700 years ago, a gigantic circular shell of gas raced out into space. Today, this still-expanding supernova remnant is too faint to see with amateur equipment, but its powerful energy is detectable at radio wavelengths. The radio output emission is created by high-speed electrons spiraling around magnetic field lines as the expanding cloud collides with thin gas between the stars.

Radio images of the cloud (right), the brightest radio source outside of the Solar System, show gas racing away from the spot where the star exploded. By calculating this speed and the distance traveled by the gas since the explosion, astronomers estimate that light from the explosion reached Earth around 1680, creating a 5th magnitude star. No record exists of anyone noticing this short-lived supernova in Cassiopeia.

*In the rich star field (above), the Bubble Nebula (NGC 7635) appears at lower right to the open cluster M52, seen at upper left. NGC 7789 (right) is a rich open cluster, slightly larger than M52.*

**6 NGC 659** Some ½ degree southwest of NGC 663 lies NGC 659, an X-shaped cluster of about 40 stars. The magnitude 5.8 golden-yellow double star **44 Cassiopeiae**, which is probably not a member of the cluster, lies close by.

**7 The Double Star Cluster (NGC 884 {Chi [χ] Persei} and NGC 869 {h Persei})** Return to Ruchbah, then hop about 8 degrees southeast to the Double Star Cluster in Perseus. This magnificent conglomeration of bright stars has been well known since ancient times. Because of their size, these two 100-plus member clusters are best viewed with binoculars or a wide-field telescope eyepiece. Careful study of each cluster will reveal the many double-, multiple-, and variable-star systems they contain.

**8 SU Persei** Some ½ degree southeast of the center of NGC 884 is the semi-regular variable SU Persei, which is actually a member of NGC 884. This pulsating red supergiant varies from magnitude 9.4 down to 10.8 during a cycle that lasts about 533 days.

**9 Caph (Beta [β] Cassiopeiae)** Sweep some 17 degrees west and slightly north of SU Persei to Caph, a wide double star and a short-period pulsating variable of a rare type known as Delta Scuti. Its variability is hard to detect visually because its magnitude of 2.2 fluctuates very slightly during a cycle that lasts a couple of hours.

**10 NGC 7789** Hop 4 degrees southwest to the broad open cluster NGC 7789. This large cluster is located between the yellow semi-regular variable **7 Rho [ρ] Cassiopeiae** and the wide, white multiple-star system **8 Sigma [σ] Cassiopeiae**. Use binoculars or a wide-field eyepiece to view the cluster, because it covers an area roughly the apparent size of the Moon.

**11 M52 (NGC 754)** Located about 6 degrees northwest from NGC 7789 is M52, an open star cluster. The 200 members of this cluster are located on the western border of Cassiopeia. A fine string of stars extending like an arm in an east-west direction across the cluster is a good test for 4 inch (100 mm) scopes or larger.

In areas filled with relatively easy binocular targets, such as this, it is a pleasure to amble around the many "families" of open clusters.

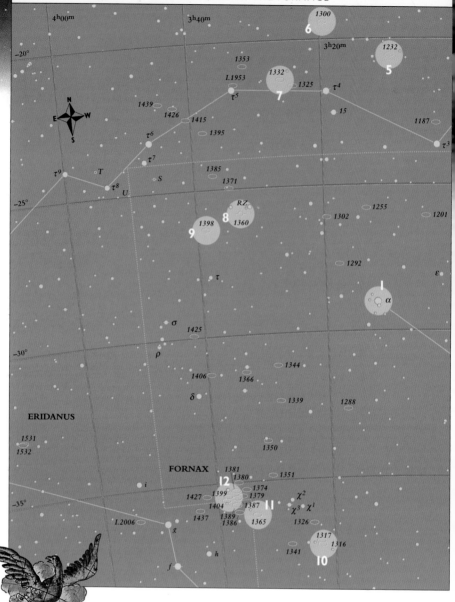

# A Galaxy Feast in the Furnace

Although it can be seen from both hemispheres, this part of the sky is much
more accessible to Southern Hemisphere
skywatchers. Northerners are best advised to
view this region during winter evenings, when it
is at its highest point in the sky.

The constellation of Fornax was introduced
by the eighteenth-century French astronomer

*The flaming Furnace burns beside the Phoenix and the River
in Carel Allard's eighteenth-century Celestial Planisphere.*

Nicolas Louis de Lacaille. He originally named it
Fornax Chemica, the Chemical Furnace, but
today it is simply known as the Furnace.

Although this patch of sky may appear fairly
barren at first, throughout Fornax and in parts of
neighboring Eridanus, the River, you are looking
toward a region rich in galaxies. The Fornax
Galaxy Cluster, a relatively nearby cluster of
galaxies, is one of the highlights of this starhop.
You need at least a 4 inch (100 mm) telescope
to see many of the star systems in this area.

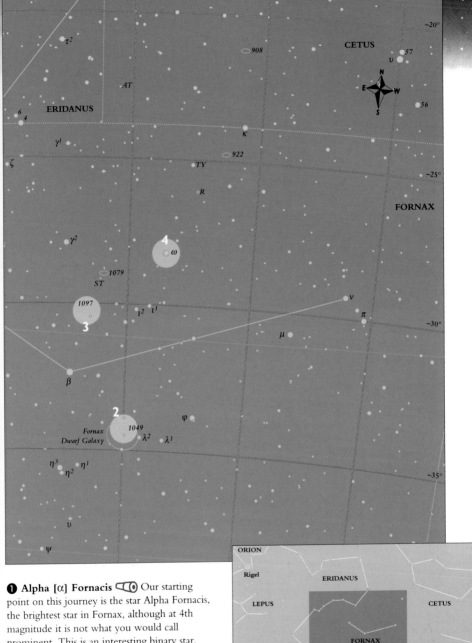

**❶ Alpha [α] Fornacis** Our starting point on this journey is the star Alpha Fornacis, the brightest star in Fornax, although at 4th magnitude it is not what you would call prominent. This is an interesting binary star, with its magnitude 7 companion, also a deep yellow color, shining about 5 arcseconds away. The exact period during which the stars orbit each other is not altogether certain, but the currently accepted value is about 314 years.

The difference in brightness between the two components makes this a difficult object for small scopes. On a steady night with good seeing, a 3 inch (75 mm) scope should show the secondary fairly easily. The secondary star is also suspected of being variable, possibly fading to as low as magnitude 8 at minimum, when you may have some difficulty seeing it.

385

**❷ NGC 1049** ⊂⊏◉ About 9 degrees southwest from Alpha, you will find the star **Beta [β] Fornacis.** Use it as a pointer to help locate NGC 1049, 3 degrees farther southwest. This globular cluster calls for a larger aperture because it shines at only 13th magnitude. An 8 inch (200 mm) scope may just show it as a fuzzy blob about 20 arcseconds across. It appears faint because it is not a part of our galaxy, belonging instead to the dim Fornax System, a 1 degree wide dwarf galaxy that looms so large it cannot be recognized visually, even though it is within the Local Group of galaxies.

**❸ NGC 1097** ⊂⊏◉ Shift 4 degrees north and a little east to NGC 1097, a barred spiral galaxy that is easy enough to see with a small scope. It shines at magnitude 9.3 and has a very bright nucleus. Its elongated form should be revealed through a 4 inch (100 mm) scope or larger.

**❹ Omega [ω] Fornacis** ⊂⊏◉ Just over 3 degrees northwest of NGC 1097 is Omega Fornacis, a double star and an easy object for any small telescope. Its 5th and 7th magnitude components glow 11 arcseconds apart.

**❺ NGC 1232** ⊂⊏◉ To find NGC 1232, hop back to Alpha Fornacis, then move just over 8 degrees north and a little west, across the border into Eridanus. This large but faint spiral galaxy is not an easy object for small scopes and its total

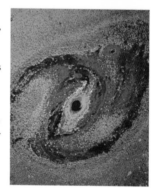

*A portion of the Fornax Cluster of galaxies (above). The computer-enhanced image of NGC 1097 (left) shows its spiral structure.*

magnitude of 9.9 is somewhat deceptive. As with many objects on this tour, an 8 inch (200 mm) telescope will reveal more of the galaxy.

**❻ NGC 1300** ⊂⊏◉ To locate NGC 1300, nudge your scope 2½ degrees east and a little south of NGC 1232, where you will find the 4th magnitude star **16 Tau⁴ [τ⁴] Eridani.** Just over 2 degrees north of Tau⁴ is the 10th magnitude barred spiral galaxy NGC 1300. This beautiful galaxy has a bright core and is easy to see through a 4 inch (100 mm) scope. In a dark sky, a hint of the arms can be detected in 12 inch (300 mm) telescopes.

**❼ NGC 1332** ⊂⊏◉ Return to Tau⁴ before moving 1½ degrees east and a little north to the elongated elliptical galaxy NGC 1332. Shining at magnitude 10.3, this fuzzy patch has a fairly bright nucleus and is not hard to see through a 4 inch (100 mm) telescope.

**❽ NGC 1360** ⊂⊏◉ About 4½ degrees south and a little east, back in Fornax, is NGC 1360. This planetary nebula, about 6 arcminutes across, offers an interesting change from observing the many galaxies in this region. Its magnitude 11 central star can be distracting, but the nebula is not a difficult object in a 6 inch (150 mm) scope.

*The elliptical shape of NGC 1360 (above) becomes obvious to the observer only when it is seen through a large telescope.*

**❾ NGC 1398** ⊂⊏◉ Gently guide your scope 1 degree southeast of NGC 1360 to NGC 1398, glowing at magnitude 9.7. This barred spiral galaxy is not difficult to spot, but is best suited to 6 inch (150 mm) scopes and larger.

**❿ NGC 1316** ⊂⊏◉ The next hop is quite some distance away. First, shift 10 degrees south to the little triangle of stars **Chi**[1,2,3] **[χ¹, χ², and χ³] Fornacis**. About 2 degrees southwest of this triangle is the 9th magnitude spiral NGC 1316, the brightest galaxy in the Fornax Cluster. A 3 inch (75 mm) telescope shows it clearly as a fuzzy patch of light. NGC 1316 is also known as the radio source Fornax A. See if you can spot the 11th magnitude galaxy **NGC 1317**, just 6 arcminutes north of NGC 1316.

**⓫ The Great Barred Spiral (NGC 1365)** ⊂⊏◉ Retrace your steps to the Chi[1,2,3] triangle, then hop 1 degree east to find the magnificent NGC 1365, a favorite among deep-sky observers. This 9th magnitude galaxy is the best example of a barred spiral in the southern sky. Easily found in 4 inch (100 mm) telescopes because of its prominent central region, NGC 1365 has a bright bar that is visible with an 8 inch (200 mm) telescope.

**⓬ NGC 1399** ⊂⊏◉ Using a wide-field eyepiece, nudge 1 degree northeast of NGC 1365 to the heart of the Fornax Cluster. With a 4 inch (100 mm) scope you should easily be able to see quite a few faint galaxies, the brightest two being NGC 1399 and **NGC 1404**, both ellipticals, lying only a few arcminutes apart.

The Fornax region has the effect of making you feel very small. Several of the galaxies we have seen on this tour belong to the Fornax Cluster, our close neighbor in the universe.

## BARRED SPIRAL GALAXIES

In a barred spiral galaxy, the spiral arms seem to originate from the ends of a "bar," composed of stars, gas, and dust, that crosses the nucleus. The bar relates to dynamic conditions within a galaxy. On close examination, many spiral galaxies— including the Milky Way—show the trace of a bar. The ratio of the mass of the faint halo surrounding a galaxy to that of its disk may play a role, as calculations suggest that galaxies with less massive haloes form bars more quickly. Many barred spirals, such as NGC 1365 (left), have well-defined arms, and there seems little doubt that an explanation of the bar is linked to our understanding of spiral structure.

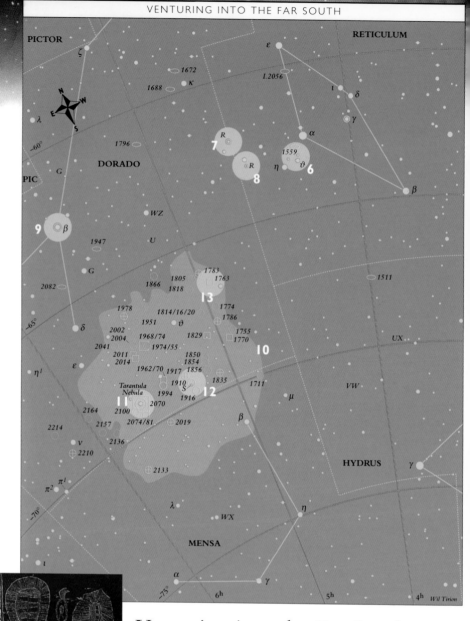

*An Australian Aboriginal bark painting depicting the Large and*
**388** *Small Magellanic Clouds, from Groote Eylandt, circa 1954.*

# Venturing into the Far South

This is our southernmost starhop, so unfortunately this region of the sky is probably not visible for skywatchers in the Northern Hemisphere. Even if you were observing within 15 degrees of the equator, you could still only expect to see a few of the objects.

The highlights of this area are the Magellanic Clouds, originally known as the Cape Clouds because they were associated with being as far south as the Cape of Good Hope. Portuguese seamen saw these two large patches of sky during voyages made in the fifteenth century. The explorer Ferdinand Magellan later described them and they were duly named in his honor, although no one knew what it was they were looking at.

Four centuries later, astronomers finally recognized the existence of galaxies other than our own, realizing that the Magellanic Clouds were two of the nearest galaxies to the Milky Way.

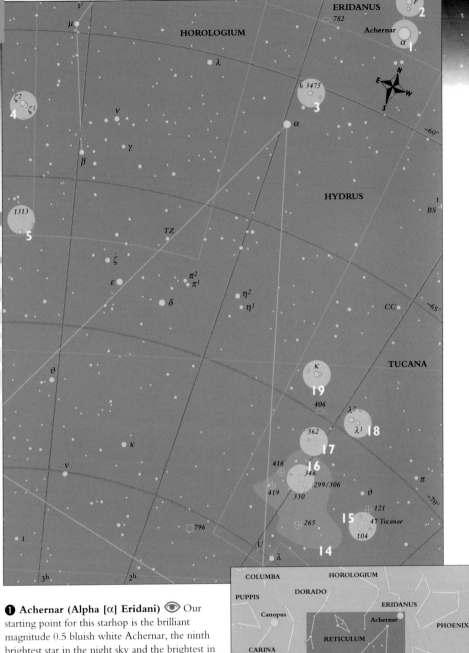

**❶ Achernar (Alpha [α] Eridani)** 👁 Our starting point for this starhop is the brilliant magnitude 0.5 bluish white Achernar, the ninth brightest star in the night sky and the brightest in the constellation of Eridanus, the River.

Achernar is the most southerly of the very bright stellar beacons, standing out in an otherwise barren patch of sky. Finding it is easy—it is about as far from the south celestial pole as the Southern Cross is on the opposite side.

The name Achernar comes from the Arabic meaning "the end of the river," and was originally attributed to Theta [θ] Eridani. Theta was so named because it appeared to mark the end of the River for ancient astronomers, whose view of the sky was limited by their Northern Hemisphere locations. We now know Theta as Acamar, which is a corruption of its former name.

389

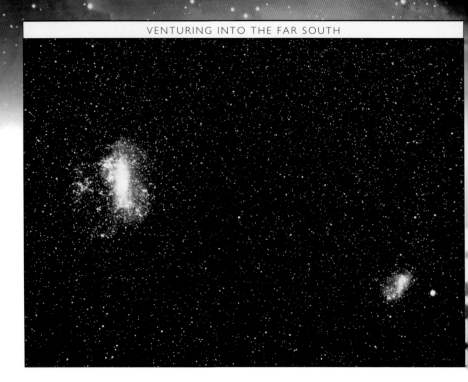

**❷ p Eridani (Dunlop 5)** 🔭 Aim your scope toward Achernar, before nudging it just over 1 degree north to locate the double star p Eridani, upstream from Achernar. This is a wonderful double for any size telescope, and a favorite among many observers. Shining at magnitude 5.8, its yellow-white components are equal in brightness, lying 10.4 arcseconds apart.

**❸ h 3475** 🔭 Return to Achernar, then move to a point about two-thirds of the way to **Alpha [α] Hydri** to find h 3475. The stars of this double are much closer together than those of p Eridani, with its two 7th magnitude components being 2.4 arcseconds apart. It is a good test for a 3 inch (75 mm) scope on a still night. This is one of the many double stars discovered by John Herschel during his stay in South Africa in the 1830s.

**❹ Zeta [ζ] Reticuli** 🔭 To find Zeta Reticuli, hop to Alpha Hydri, then shift west about 9 degrees along the way to **Alpha [α] Reticuli.** Zeta is a wide double star that is easy to separate with binoculars, its two 5th magnitude component stars, Zeta$^1$ and Zeta$^2$, lying 5 arcminutes apart. Each is a magnitude 4.7 yellowish star. These stars are remarkably similar to the Sun, a fact that has prompted speculation about the possibility of life existing on orbiting planets, although this is not known for sure.

**❺ NGC 1313** 🔭 Moving poleward 4 degrees will bring NGC 1313 into view. This is a well-known barred spiral galaxy near the corner of the constellation of Reticulum, the Reticule. It has a total magnitude of 9.4, but appears quite large and faint. An 8 inch (200 mm) scope or larger is recommended for viewing this galaxy, but even with a 12 inch (300 mm) telescope, the bar is only vaguely visible.

**❻ Theta [θ] Reticuli** 🔭 To the northeast of NGC 1313 and just southeast of Alpha Reticuli, Theta Reticuli shines faintly to the unaided eye. Theta is a pleasing double star comprising 6th and 8th magnitude components, separated by about 4 arcseconds. With a 3 inch (75 mm) scope, you should have little difficulty separating them, but the higher the magnification, the better the view.

**NGC 1559** 🔭 In the same field, about ½ degree north of Theta, is the barred spiral galaxy

*The LMC and the SMC (top) are classified as irregular galaxies. Despite its chaotic-looking appearance, the fainter NGC 1313 (above) has sufficient order to be classified as a barred spiral.*

NGC 1559. Although it is about a magnitude fainter in total brightness than NGC 1313, this galaxy is more compact, appearing as a broad, fuzzy line. A 4 inch (100 mm) scope, or possibly even a smaller one, should clearly show the galaxy. The brightening toward its core is easier to see with larger scopes.

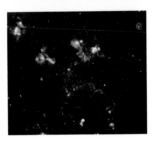

*The pink, glowing clouds (left) are just some of the many star-forming regions scattered throughout the LMC (above).*

**7 R Doradus** Moving toward the border with Dorado, 2 degrees to the northeast of NGC 1559, is R Doradus, just inside the constellation of Dorado, the Swordfish. R Doradus is a vivid orange-red color, not unlike a fainter view of the planet Mars. It is a semi-regular variable star with a period that fluctuates from cycle to cycle, averaging out at about 338 days. R Doradus is an easy binocular target because it only varies between magnitudes 4.8 and 6.6.

**8 R Reticuli** Moving back about 1 degree southwest is R Reticuli. Unlike R Doradus, this Mira-type variable can drop to magnitude 14 at minimum, so you will not see it with binoculars then. However, at maximum, around every 278 days, it can reach magnitude 7, which makes checking to see just how bright it is a worthwhile effort. It has a ruddy look that is obvious through binoculars, and it lies 12 arc-minutes south of a 6th magnitude star.

**9 Beta [β] Doradus** Swing 7 degrees east to the Cepheid variable Beta Doradus, which fluctuates between magnitude 3.5 and 4.1 over a period of 9.84 days. It is thought to lie 1,700 light-years away.

**10 The Large Magellanic Cloud (LMC)** Far more distant, but seen just a few degrees poleward of Beta, lies the Large Magellanic Cloud. Easily visible to the unaided eye as a "cloud" of light several degrees across, the LMC is a fine sight in binoculars, which help show the elongated, patchy appearance of this irregular galaxy. Lying about 160,000 light-years from the Milky Way, the LMC is the second closest galaxy to our own—the closest known being a dwarf galaxy in Sagittarius that was detected in 1994.

**11 The Tarantula Nebula (NGC 2070)** Near the eastern end of the the LMC is the wonderful Tarantula Nebula. Being the brightest object in the LMC, the Tarantula Nebula is visible to the unaided eye—a remarkable fact, given that it is in another galaxy. Even through a 3 inch (75 mm) scope, some of its intricate spider-like structure can be seen, while the view in an 8 inch (200 mm) is stunning. Within the nebula's dazzling heart is a dense and mysterious cluster of brightly glowing supergiants known as R136. NGC 2070 is more than 1,000 light-years across, making it the largest known diffuse nebula, some 30 times the size of the Great Nebula in Orion.

**⑫ NGC 1910** ⬭🔘 Sweep your scope through the LMC, heading slowly west for 3 degrees, until you come across NGC 1910. One of the many clusters in the LMC, this compact group contains the bluish white variable star **S Doradus**, which is the prototype of its class. S Doradus

*The wispy tendrils of the Tarantula Nebula are clearly seen in this image (above), taken just after nearby SN1987A blazed into view.*

---

### SUPERNOVA 1987A

Astronomers around the world were excited when, on 24 February 1987, a new naked-eye star suddenly appeared within the Large Magellanic Cloud. It was the first supernova seen with the unaided eye since 1604, reaching magnitude 2.9 at its peak and remaining visible for several months. The object was originally a massive blue star called Sanduleak −69 202.

The star actually exploded some 165,000 years ago, the light having taken that long to reach us. The light was accompanied by a burst of neutrinos—tiny, elusive particles. Scientists had predicted that these would be produced in very large numbers during a supernova explosion.

The study of SN1987A continues today, with rings of light, shown in the illustration of the Hubble Space Telescope image (right), being the most recent surprises in the ongoing tale of the supernova's discovery.

---

varies irregularly between magnitude 9 and 11, the variations being the result of the periodic shedding of the star's outer layers.

**⑬ NGC 1763** ⬭🔘🔘 To spot the bright nebula NGC 1763, head northwest, to the edge of the Large Magellanic Cloud. NGC 1763 is easily seen with binoculars, and a 3 inch (75 mm) scope reveals two bright, seemingly separate areas.

**⑭ The Small Magellanic Cloud (SMC)**
👁 Leaving the LMC and heading some 20 degrees west, you will come across its smaller counterpart—the SMC. Slightly more distant than the LMC and somewhat smaller, the SMC, also an irregular galaxy, is nevertheless an easy naked-eye target. Binoculars will not show nearly as much of the SMC as the LMC. This lesser galaxy has a more irregular form, appearing as a large blob of light, about 3½ degrees across, that fits neatly into a typical binocular field.

**⑮ 47 Tucanae (NGC 104)**
👁 A chief item of telescopic interest in this region is the naked-eye spot of light about 3 degrees west of the SMC, 47 Tucanae. A 3 inch (75 mm) telescope shows a "granular" appearance, while a 6 inch (150 mm) reveals a multitude of stars. It is widely

*In the sky, the fabulous globular cluster 47 Tucanae (right) appears to lie close to the much more distant SMC (above).*

accepted that 47 Tucanae is the most brilliant globular after Omega Centauri, although it has a condensed core, giving it quite a different appearance. This is the astronomer Johann Bode's celebrated "ball of suns," also having been referred to as "a stupendous object" by John Herschel. Seemingly leading the SMC around the sky, 47 Tucanae is a superb globular belonging to the Milky Way Galaxy.

**⑯ NGC 346** Move 3 degrees back to the southeast, into the SMC, to find NGC 346, an open cluster that lies within the SMC. About 5 arcminutes across, it is embedded in a region of nebulosity and appears as an easy-to-see fuzzy spot in a 3 inch (75 mm) scope. Its total brightness is close to that of a 10th magnitude star. See if you can pick out the nearby cluster **NGC 330**, which is visible in the same low-power field, about ½ degree southwest of NGC 346.

**⑰ NGC 362** Edge your scope 1 degree north to NGC 362. Seen as a round patch of light through binoculars, this is a conspicuous globular cluster. It is hard to resolve through a 4 inch (100 mm) telescope, which shows only a hint of its starry nature, but a 6 inch (150 mm) will show it reasonably well.

**⑱ Lambda¹ [λ¹] Tucanae (Dunlop-2)** One and a half degrees farther north lies a little pair of naked-eye stars. The fainter of the two is Lambda¹ Tucanae. This star is a wide double that

is easy to resolve in any size telescope. The 6th and 8th magnitude yellow-white components of Lambda¹ are separated by 20 arcseconds.

**⑲ Kappa [κ] Tucanae** The final object on our hop is another double star, Kappa Tucanae, 2 degrees east and a little north of Lambda¹. Kappa is an easy pair in a 3 inch (75 mm) scope, especially on high power. Its 5th and 7th magnitude stars lie 5.4 arcseconds apart. Look for another 7th magnitude star, shining just a few arcminutes away in the same field.

Southern Hemisphere observers really are the envy of northerners for having the Magellanic Clouds all to themselves. It is a rewarding pastime simply to spend an evening browsing around this region with anything from the unaided eye to a large telescope.

393

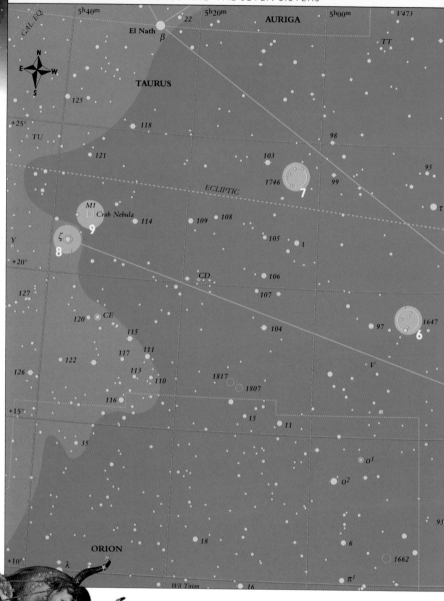

El Nath β
TT
125    TAURUS
+25°    118    98
TU
121    103    95
1746    99
ECLIPTIC    7
M1
Crab Nebula    114    109    108    τ
9    105    ι
Y    ζ
8    106    +20°
CD    107
127
120    CE    104    1647
115    97    6
117    111    V
122
126    113    110    1817
116    1807
+15°    15    11
35
o1
o2
93
18    6
ORION    +10°    λ    π1    1662
Wil Tirion    16

# Taurus and the Seven Sisters

The zodiac constellation of Taurus, the Bull, is clearly visible from both the Northern and Southern hemispheres, from February through April.

According to legend, the red eye of mighty Taurus glares at Orion, the Hunter, as the bull guards the Pleiades, the Seven Sisters, from Orion's advances. Orion's heart is set on making Merope, one of the Pleiades, his wife. Her parents, the Titan Atlas and the Oceanid Pleione, watch closely from the edge of the Pleiades Cluster. Of the seven young sapphire-blue Pleiade sisters, only Merope married a mortal, the King of Corinth, so she hides her shame behind a wispy reflection nebula.

The Pleiades' half-sisters form the nearby Hyades Cluster, another fantastic naked-eye and binocular object on this starhop.

*The well-known zodiacal figure of Taurus, the Bull, from a sixteenth-century fresco in the Villa Farnese, Caprarola, Italy.*

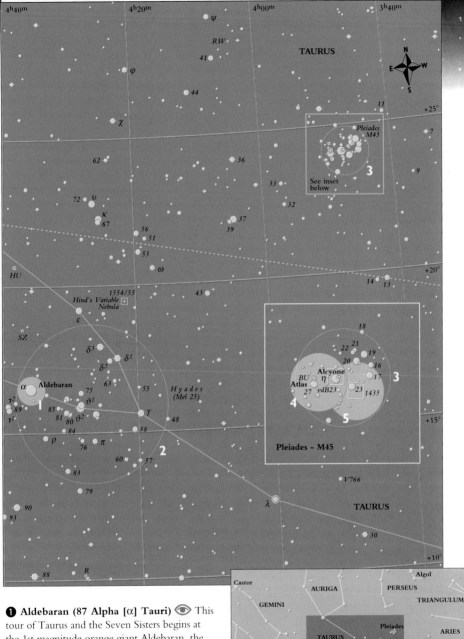

**❶ Aldebaran (87 Alpha [α] Tauri)** 👁 This tour of Taurus and the Seven Sisters begins at the 1st magnitude orange giant Aldebaran, the 13th brightest star in the sky. Although it seems to form the eastern tip of the Hyades Cluster, Aldebaran, just 65 light-years distant, is actually much closer to us. Like many other orange giants, it is possible that this star is also slightly variable. Aldebaran has five close companions, but they are faint and extremely difficult to observe through amateur equipment.

At about 40 times larger than the Sun and intrinsically some 125 times brighter, Aldebaran would fill most of the area inside Earth's orbit.

The name Aldebaran is derived from the Arabic *Al Dabaran*, meaning "the follower," probably because this star follows the Pleiades through the night sky.

395

**❷ The Hyades (Mel 25)** 👁 This large (more than 4 degrees in diameter), rich, distinctly V-shaped open star cluster consists of a mixture of about 100 bright stars and many faint ones, including dozens of double and variable stars. The cluster, which is 150 light-years away, marks the head of the bull. It is named for the half-sisters of the Pleiades, being the daughters of Atlas and Aethra. The Hyades' distance to us provides a crucial step in measuring distances in the universe, as it is one of the closest clusters to us.

*The Pleiades, at upper right, and the Hyades, beside Aldebaran at left (above), are nearby clusters appearing large in the sky.*

**❸ The Pleiades (M45, Mel 22)** 👁 About 14 degrees northwest of Aldebaran is M45—the Seven Sisters—the brightest and most famous star cluster in the sky. M45 is a young open cluster dominated by youthful hot blue stars and enveloping nebulosity. The brightest part of the nebula surrounding the Pleiades is around the magnitude 4.1 star **Merope (23 Tauri)**. The Pleiades star nursery has been known since ancient times. Seven bright stars are visible to the naked eye from a dark site, although the cluster shows best through binoculars or a low-power telescope eyepiece, since it covers an area about four times the size of the Full Moon. It is essential to use high power when studying the fainter stars and patches of nebulosity.

**❹ Alcyone (25 Eta [ε] Tauri)** 👁 The brightest member of the Pleiades is magnitude 2.9 Alcyone, a wide quadruple-star system embedded within the reflection nebula **van den Bergh 23**.

The individual components of Alcyone's system are easy to separate with a small telescope. **Atlas (27 Tauri)** 👁 This young blue giant star is a close double that usually requires a telescope of at least 10 inches (250 mm) and good seeing conditions to separate its magnitude 3.6 and 6.8 components,

### M45, THE SEVEN SISTERS

The ancient Greeks were not the only people to develop myths and legends about the origins of the cluster they named the Pleiades. The Australian Aborigines also integrated the skies they observed into a heritage of stories, known as the Dreamtime. The bark painting (below), by Wongu, a member of the Yolŋu tribe in Arnhem Land, depicts the Seven Sisters inside a canoe. The three stars of Orion's belt are shown in a line at the head of the canoe, to the right of the image, with the seven prominent stars in the middle representing the Pleiades, their wives. The lone fish inside the canoe, at center top, represents the nearby Hyades Cluster, while the fish swimming in the water are bright stars in the Milky Way.

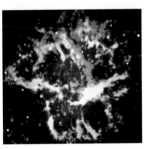

The Pleiades (above) and the Crab Nebula (left) are highlights in Taurus.

which lie only 0.4 arcseconds away from each other.

**❺ Tempel's Nebula (NGC 1435)** ⊂⊃⊙ This is the reflection nebulosity in which blue-white Merope is embedded. The nebula appears to be nearly transparent so is quite hard to detect, but once you "see" it you might think that it looks as if someone has smeared white shoe polish on a pane of glass. This nebula is part of a series of small nebulas extending over much of the western side of the Pleiades.

**❻ NGC 1647 (Mel 26)** ⊂⊙⊙ Return to Aldebaran, then hop about 3½ degrees northeast to NGC 1647. This open star cluster contains about 25 uniformly bright 8th magnitude stars gathered in a loose grouping along the line of the western horn of the bull. More than 200 additional fainter stars are part of this rich cluster. As you study the individual stars, try to pick out some of the many close double stars.

**❼ NGC 1746 (Mel 28)** ⊂⊙⊙ Look for the open star cluster NGC 1746 about 5½ degrees northeast of NGC 1647 and just 1 degree southwest of the magnitude 5.5 blue-white double star **103 Tauri**. NGC 1746 has a dense central region of about 20 stars, along with about 30 more in knots scattered about the cluster's core.

**❽ 123 Zeta [ζ] Tauri** 👁 Shift your scope about 8 degrees southeast to Zeta Tauri, the last bright star in the southern horn of the bull. Zeta

is an extremely close binary star whose components cannot be separated with amateur scopes. It is also a variable, belonging to the rare Gamma Cassiopeiae-type class. Zeta lies about 420 light-years away from us.

**❾ The Crab Nebula (M1, NGC 1952)** ⊂⊙⊙ Nudge about 1 degree northwest of Zeta to the famous Crab Nebula, lying about 6,500 light-years from us. The nebula is visible with small telescopes but can be somewhat disappointing. Details in the cloud can be detected in 10 inch (250 mm) scopes or larger.

First discovered by British amateur astronomer John Bevis in 1731, M1 is the gaseous supernova remnant of the explosion of a star witnessed and recorded by Chinese astronomers in 1054 (see also p. 223). The explosion was so bright that the star was visible in daylight for a period of 23 days.

In the core of the Crab Nebula, a tiny spinning neutron star flashes a beam of energy on and off 30 times a second. This "star," called a pulsar, is all that remains of the original star that exploded as a supernova so long ago. This was the first pulsar to be detected visually.

This patch of sky has intrigued and entranced skywatchers since ancient times because its highlights, the Pleiades and Hyades, are so big and bright, as well as being rich in mythology.

397

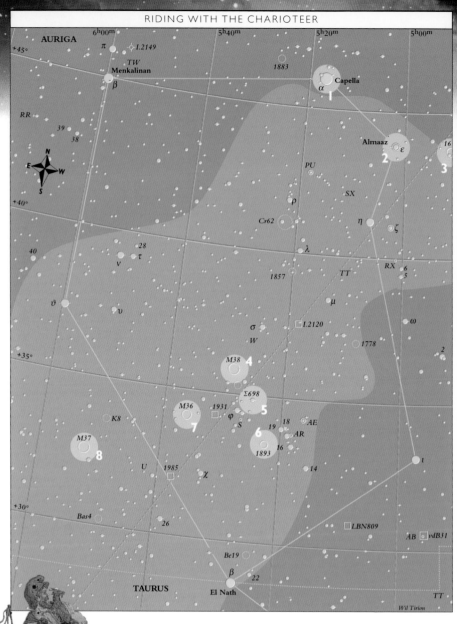

# Riding with the Charioteer

The rich star fields of the Milky Way in the pentagon-shaped constellation of Auriga, the Charioteer, hug the galactic equator, forming the centerpiece of this area of sky. It is visible to skywatchers in both the Northern and Southern hemispheres.

The route of this hop includes some of the finest open star clusters in the northern half of the sky. Cutting diagonally across Auriga is a string of open star clusters, including M36, M37, and M38. In addition to these highlights, many other star clusters lie among a variety of objects.

To the ancient Greeks and Romans, the constellation of Auriga represented either a charioteer or a herd of goats. In the herd, bright, white Capella is the mother of the flock and her three starry "kids" graze close by.

*Auriga, the Charioteer, carrying radiant Capella on his back, from Johann Bayer's seventeenth-century star atlas,* Uranometria.

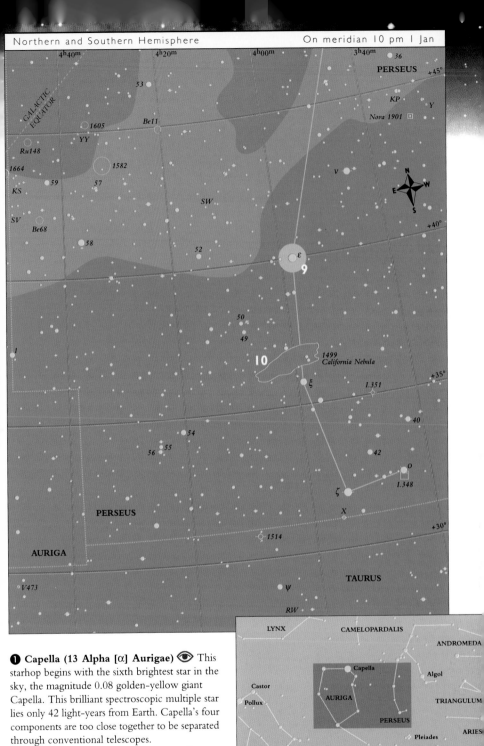

**❶ Capella (13 Alpha [α] Aurigae)** 👁 This starhop begins with the sixth brightest star in the sky, the magnitude 0.08 golden-yellow giant Capella. This brilliant spectroscopic multiple star lies only 42 light-years from Earth. Capella's four components are too close together to be separated through conventional telescopes.

Capella A and Capella B are binary yellow companions that form one of the star systems making up Capella. The A and B stars are estimated to be 90 and 70 times more luminous than the Sun, respectively. The other binary star completing the multiple system is Capella H, consisting of a pair of red dwarf stars.

In many ancient myths, Capella, the Goat Star, is represented as a she-goat, carried over the shoulder of the charioteer. It is also sometimes referred to as the Little She-goat.

399

**❷ Almaaz (Epsilon [ε] Aurigae)** 👁 About 3½ degrees southwest of Capella is Almaaz, an eclipsing binary. Intrinsically, this magnitude 3.8 white star is one of the most luminous individual stars we know of. It is partially eclipsed by a mysterious companion object every 27 years, the eclipse itself lasting about 2 years. The next eclipse is due to start in 2009. During the eclipse, Almaaz's brightness drops by about ²/₃ magnitude. Many theories have been posed as to the nature of the eclipsing object. One theory holds that it is an almost transparent infrared supergiant star that is so large it would fill most of the area inside the orbit of Saturn. It is also possible that it may be simply a cloud of dust and gas.

*M37, M36, and M38 (left to right, above) lie within the Milky Way in Auriga (left).*

**❸ NGC 1664** ⊂〇 Nudge 2 degrees west of Almaaz to the faint open star cluster NGC 1664. Use low power for this rather sparse 30-member cluster, which appears like a string of tiny jewels on a chain, or an elongated hollow diamond. The brightest star, of about 7th magnitude, is located near the southern end of the chain.

**❹ M38 (NGC 1912)** ⊂〇〇 Scanning some 10 degrees to the southeast along the galactic equator you will find M38, the dimmest of the three Messier objects in the center of Auriga. This cross-like open cluster, consisting of about

160 stars of magnitude 10 or fainter, was discovered by the Italian astronomer Giovanni Hodierna in the seventeenth century. He also discovered the open clusters M36 and M37.

**❺ Struve [Σ] 698** ⊂〇 The double star Struve 698 lies about 1 degree southwest of M38. This beautiful little double star consists of two red stars of magnitudes 6.6 and 8.7, separated by 31 arcseconds. They were first measured by astronomer F. G. W. Struve in 1831. The components should be easy enough to resolve through a 4 inch (100 mm) telescope.

**❻ NGC 1893** ⊂〇〇 Another 1½ degrees farther southwest is NGC 1893, an elongated open star cluster comprised of about 40 to 60 members. Most of the component stars of the cluster are faint, but they still stand out in the foreground of the surrounding star fields.

**❼ M36 (NGC 1960)** ⊂〇〇 Sweeping some 3 degrees east of NGC 1893 is M36, an open star cluster of about 60 young blue and white stars, with a slight condensation of stars near its center. Search through the cluster to find the numerous double stars within it.

**❽ M37 (NGC 2099)** ⊂〇〇 Moving on about 4 degrees southeast from M36 lies M37. This

*The California Nebula (above) is a large, faint cloud of gas amid the stars of Perseus, including bright Minkib at center bottom.*

very rich, magnitude 6.2 open star cluster, the best in Auriga, contains about 170 bright stars, as well as hundreds of fainter stars scattered throughout. The majority of the stars are young blue-white giants and supergiants. The brighter ones form a rough trapezoid shape with a belt of fainter stars cutting across the geometric figure.

**❾ 45 Epsilon [ε] Persei** ◀▣◑ Return to M38, then sweep about 17 degrees west and a little north to Epsilon Persei. The primary for this double-star system is a magnitude 2.9 blue-white giant. The greenish magnitude 8.1 secondary is 8.8 arcseconds away and can be seen through scopes of 6 inches (150 mm) or larger. The difference in brightness between the two components makes them quite difficult to separate.

**❿ The California Nebula (NGC 1499)** ◀▣◑ Hop 4 degrees due south from Epsilon Persei to **Minkib (46 Xi [ξ] Persei)**, the 4th magnitude blue-white star located on the southwestern fringe of the California Nebula. Minkib is the illuminating power for the nebula, the last object on our hop. This elusive and faint reflection nebula appears as a huge, elongated patch of grayish wispy reflection nebulosity. It is named for its vague

resemblance to the state of California, USA, but you need to use your imagination to see that. The nebula has a very low surface brightness, and is best seen with a nebula filter (see p. 84). Using low power and a wide-field eyepiece at a dark site is also a great help to observers.

Ride with the Charioteer, scanning the depths of the Milky Way Galaxy for the many other star clusters and double stars to be found.

## STELLAR EVOLUTION: CAPELLA AND THE SUN

The life of a star can last from a few million to more than a hundred billion Earth years. Its life span is directly related to the amount of hydrogen fuel it contains and the rate at which it converts this fuel to helium. Part of our Sun's life cycle is shown in the illustration below. Currently about five billion years old (below, top), the Sun will grow to become a yellow giant (below, right), Capella's present size. The Sun is expected to use up the hydrogen fuel in its core in about another five billion years, before swelling up as a red giant (below, left) then fading away to become a white dwarf star.

Capella is 13 times larger than the Sun. It is already well on the way to becoming a red giant star.

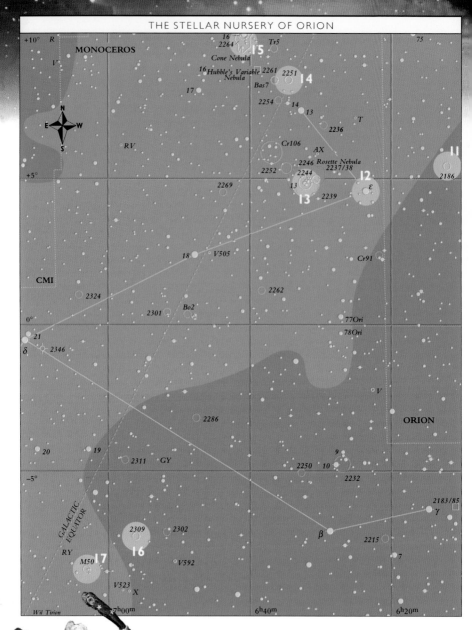

+10°   R

MONOCEROS

16
2264
15
Cone Nebula
Tr5
75

V

16 Hubble's Variable 2261 2251
Nebula
Bas7
17
2254   14
13
2236
T

N
E   W
S

RV

Cr106
AX
2246 Rosette Nebula
2252   2244   2237/38
13
2239
ε

II
2186

+5°

2269

13
13

12
ε

18   V505

Cr91

CMI

2324

2301   Bo2

2262

0°

21
δ   2346

77Ori
78Ori

V

2286

ORION

20   19
2311   GY

9
2250   10
2232

–5°

2183/85
γ

GALACTIC
EQUATOR

2309   2302

β
2215

7

RY
M50   17
16
V592

V523   X

Wil Tirion   7h00m   6h40m   6h20m

Orion, the Hunter, as depicted in a sixteenth-century fresco
painted on the walls of the Villa Farnese, Caprarola, Italy.

# The Stellar Nursery of Orion

Orion straddles the celestial equator, making its star nursery visible to observers in both the Northern and Southern hemispheres. With his mighty club raised, Orion, the Hunter, dominates the night sky early in the year as he prepares to battle Taurus, the Bull. Behind Orion's back is the mystical Monoceros, the Unicorn, partly obscured by a maze of Milky Way star fields.

The Hunter's starry sword dangles from his three-star belt. The sword harbors the Great Nebula in Orion, where gas covering the sky throughout the constellation is unmasked by the blazing light of young, hot blue-white stars. Even younger stars lie hidden within the gas clouds around it, revealed by their infrared glow.

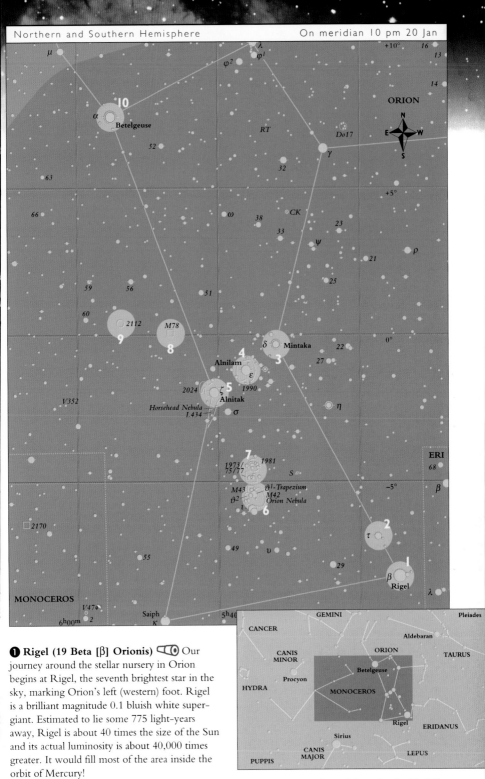

**ORION**

**❶ Rigel (19 Beta [β] Orionis)** ⊂⓪ Our journey around the stellar nursery in Orion begins at Rigel, the seventh brightest star in the sky, marking Orion's left (western) foot. Rigel is a brilliant magnitude 0.1 bluish white super-giant. Estimated to lie some 775 light-years away, Rigel is about 40 times the size of the Sun and its actual luminosity is about 40,000 times greater. It would fill most of the area inside the orbit of Mercury!

Although Rigel is clearly visible to the naked eye, a 6 inch (150 mm) telescope is needed to separate the magnitude 6.7 bluish companion from the primary star.

Rigel's name is derived from the Arabic *Rijl Jauzah al Yusra,* meaning "left leg of the giant." It has been said that honors and splendor would befall those who were born under this star.

403

**❷ 20 Tau [τ] Orionis** 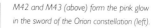 About 2 degrees northeast of Rigel is the magnitude 3.6 double star Tau Orionis. This is an easy-to-observe, wide double, with its magnitude 3.6 blue primary shining 35 arcseconds from the much fainter secondary.

**❸ Mintaka (34 Delta [δ] Orionis)** Follow Orion's left leg north to the westernmost of the three bright belt stars— Mintaka. This is a magnitude 2.2 white double star shining just south of the celestial equator. In telescopes of 4 inches (100 mm) or more, Mintaka's bluish magnitude 6.7 companion is visible 52 arcseconds directly north of it.

**❹ Alnilam (46 Epsilon [ε] Orionis)** Follow the line of the belt east from Mintaka to Alnilam, enveloped in the faint emission/reflection nebula **NGC 1990**. Alnilam is a young magnitude 1.7 blue-white supergiant with an intrinsic luminosity of about 40,000 times that of the Sun. Alnilam's name comes from the Arabic *Al Nitham*, meaning "the string of pearls."

**❺ Alnitak (50 Zeta [ζ] Orionis)** The third of the belt stars, Alnitak is a double star, shining at magnitude 1.8 and illuminating much of this region. The elusive Horsehead Nebula (see Box) is within the long stretch of nebulosity IC 434, located about ½ degree south of Alnitak.

*M42 and M43 (above) form the pink glow in the sword of the Orion constellation (left).*

**❻ The Great Nebula (M42, NGC 1976)** Shift about 4 degrees southward, following the stars that mark Orion's sword, to discover the wonderful Great Nebula in Orion. This glowing, irregular cloud is one of the most impressive sights for skywatchers, as more astonishing details unfold with every increase in aperture. The Great Nebula was first seen telescopically by Italian astronomer Nicholas Peiresc in 1611.

M42 is an emission nebula visible to the naked eye in dark skies, appearing as a soft fuzzy spot in the middle of Orion's sword. It is about 30 light-years in diameter, and estimated to be some 1,600 to 1,900 light-years away from us. There are dozens of variable stars within it.

The vibrant red colors of the nebula show up beautifully in photographs, but are too faint to be seen with telescopes. However, some observers have reported seeing a slight pale-green tint to it through an 8 inch (200 mm) scope.

The turbulent appearance of M42 is borne out by spectroscopic studies of the cloud, which show that gas is racing in different directions within it, indicating the localized areas where gas is condensing into new stars.

*The glimmering nebula NGC 1977 (above) is another illuminated portion of the gas that envelops Orion.*

M42 is considered to be one of the most beautiful objects in the heavens, full of wreaths of swirling gas. It is a sight you will undoubtedly want to return to many times.

**The Trapezium (41 Theta¹ [θ] Orionis)**

⊂⊂⊙ The Trapezium is a group of four young, bright, hot white stars at magnitudes 5.1, 6.7, 6.7, and 8.0. Theta¹ is the main source of the strong ultraviolet light that causes the abundant hydrogen gas in M42 to glow. It is only possible to separate Theta's four additional faint, magnitude 11 to 16 components in telescopes of 15 inches (380 mm) or larger.

**M43 (NGC 1982)** ⊂⊂⊙ Slightly northeast of the Trapezium is the diffuse nebula M43. This circular emission nebula appears to be attached to the northern side of M42, although it is not actually a part of it. An 8th magnitude star is the source of the light that makes M43 visible to us. The dark patch lying between M42 and M43 is known as the Fishmouth.

❼ **NGC 1973, 1975, and 1977** ⊂⊂⊙ About ¹/₂ degree north of M43 is the triple nebula of NGC 1973, 1975, and 1977. This glowing triangle of nebulosity consists of three separate emission/reflection nebulas that appear to be connected. Most of the energy exciting these wispy nebulas to glow is sourced from the multiple star 42 Orionis.

**NGC 1981** ⊂⊂⊙ Move on about 25 arcminutes north from NGC 1973, 1975, and 1977 to find the star cluster NGC 1981. This loose open star cluster consists of about twenty magnitude 8 to 10 stars. Most of the members of the cluster are young white stars.

### THE HORSEHEAD NEBULA (B33)

An interesting, but exceedingly difficult, object to observe is the famous Horsehead Nebula (below). The Horsehead is a thick black cloud of dust and gas that can be seen only because it blots out some of the light coming from the faintly glowing streamers of the diffuse emission nebula IC 434. The Horsehead is estimated to be 1 light-year across and composed of a thin haze of dust in non-luminous gas.

You need a clear dark sky to have any chance of observing this elusive object—probably the most challenging you will encounter. To locate the nebula, scan the area about halfway between Alnitak and the 11th and 12th magnitude stars directly south of it. An H-beta filter that screws into your eyepiece may help you to observe it.

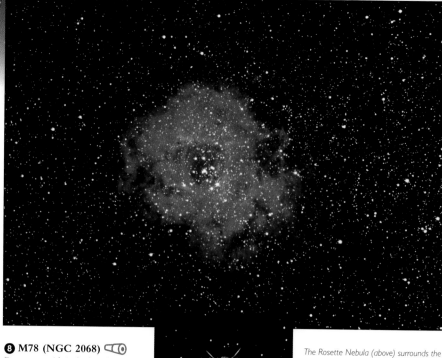

**8 M78 (NGC 2068)**
Return to Alnitak then hop about 2¹/₂ degrees northeast to M78. This small, circular nebula softly glows at magnitude 8.3 in reflected light from two young, hot blue-white stars embedded within it. M78 is a small, illuminated portion of a large, dark nebula that partly encircles much of Orion's belt and sword area. Other illuminated patches form an arc known as Barnard's Loop, after the American astronomer Edward Emerson Barnard (see p. 87).

**9 NGC 2112** Two degrees east of M78 is the small open star cluster NGC 2112. The 50 members of this compressed open cluster shine at a combined magnitude of 9.1. It is a challenge to pick out from the surrounding star fields.

**10 Betelgeuse (58 Alpha [α] Orionis)**
This cool red supergiant lies 7 degrees north of NGC 2112, and is quite possibly the largest star in our part of the Milky Way. Betelgeuse is the only 1st magnitude star known to vary significantly in brightness, with an irregular period lasting about 6 years. Estimates indicate that at its normal size it would fill the area within the orbit of Mars, and that when it swells up its diameter would equal Jupiter's orbit. Betelgeuse is also the only star, besides the Sun, with surface features that have been seen by astronomers.

**11 NGC 2186** The open star cluster NGC 2186 lies about 4 degrees southeast of Betelgeuse. This is a large, loose open star cluster

*The Rosette Nebula (above) surrounds the star cluster NGC 2244 (below). They lie near the red supergiant Betelgeuse (left).*

enveloped in a rich Milky Way star field. Most of the 30 members of the cluster are 9th to 11th magnitude and can be seen with small scopes.

**12 8 Epsilon [ε] Monocerotis** The 4th magnitude star about 3 degrees east of NGC 2186 is Epsilon Monocerotis, our next target. Epsilon is a triple-star system in which the three pale yellow-white to bluish-white components are easy to separate through a small telescope. The A star shines at magnitude 4.3, the B star at magnitude 6.7, 13 arcseconds away, and the C star at magnitude 12.7.

**13 The Rosette Nebula (NGC 2237)**
The faint circular mass of gas 2 degrees east of Epsilon is our next stop. This doughnut-like emission nebula includes the open star cluster NGC 2244. The nebula is very large at about

80 arcminutes in diameter. Because it is so diffuse, it can be difficult to locate in telescopes smaller than 8 inches (200 mm). The Rosette is estimated to be at a distance of about 2,600 light-years from us. This would make it some 55 light-years in diameter—almost twice the size of M42.

**NGC 2244** This open star cluster, which is enveloped by the Rosette Nebula, appears in the hole of the doughnut. Its brightest apparent member is the 6th magnitude yellow giant 12 Monocerotis. Some astronomers think that this star is a foreground object and not actually part of the cluster, since the other members appear to be mostly young white stars.

**⓮ NGC 2251** Hop about 4 degrees north to NGC 2251. This elongated open star cluster contains about 30 stars located mostly in a twisted string. The cluster shines with a combined magnitude of about 7.3, but most of the stars are fainter than magnitude 12.

**⓯ The Christmas Tree Cluster (NGC 2264)** Slide about 2 degrees northeast to the open star cluster NGC 2264. The triangular shape of this open star cluster strongly suggests its common name, given by Lowell Observatory astronomer Carl Otto Lampland. There is a faint nebula surrounding the cluster, and the black **Cone Nebula** intrudes visually into this at the southern end.

This effect is too faint to be seen in telescopes smaller than 12 inches (300 mm). The brightest star in the cluster, S Monocerotis, is an intensely luminous, magnitude 4.6 blue-white double.

**⓰ NGC 2309** Take a long hop, about 17 degrees south, to NGC 2309. About 40 stars make up this magnitude 10.5 compact open star cluster. It appears to be centered on a magnitude 11.5 star, in front of a very attractive field.

**⓱ M50 (NGC 2323)** About 2 degrees southeast of NGC 2309 is the last object on our starhop, the easy-to-find, rich open star cluster M50. This beautiful magnitude 6.3 cluster contains about 50 stars, appearing as a mottled patch of light barely visible to the unaided eye in dark skies. The cluster is roughly diamond shaped and is best viewed with binoculars or a low-power telescope eyepiece. A bright yellow-orange star holds center stage slightly southwest of the core. Two parallel arms of stars extend out from the core.

The Orion area is another patch of the night sky that you could spend years exploring and yet still not see all of the objects that are to be found. Our tour has only touched on some of the highlights.

*M50 (left) lies in Monoceros, along with the Christmas Tree Cluster, which extends to the north of the Cone Nebula (above).*

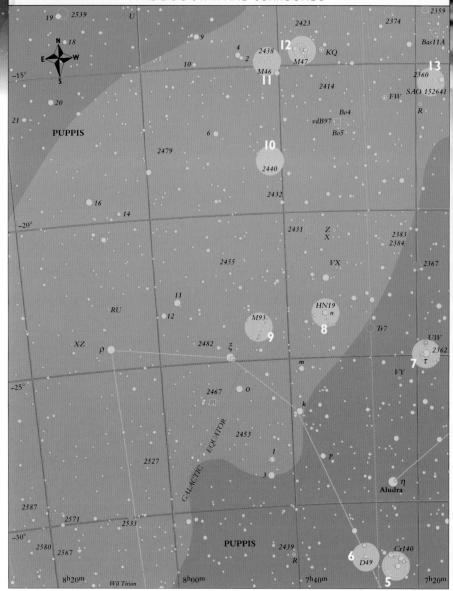

# The Dog Star and Surrounds

The best views of the constellation of Canis Major, the Great Dog, are to be had during the evenings of the early months of the year. This prominent, easy-to-find constellation lies on the southwestern side of the Milky Way.

*Canis Major, the Great Dog, is one of the most striking of all the constellations.*

Although visible from both hemispheres, southern observers will have a better view of it.

Farther north of this sky chart is the constellation of Canis Minor—which, as you might suspect, is the Little Dog. In classical mythology, the two canines are said to have attended Orion. Moving into the band of the Milky Way near Canis Major, we cross the border into neighboring Puppis, the stern of the ship Argo. This region contains many interesting objects for binoculars and small telescopes.

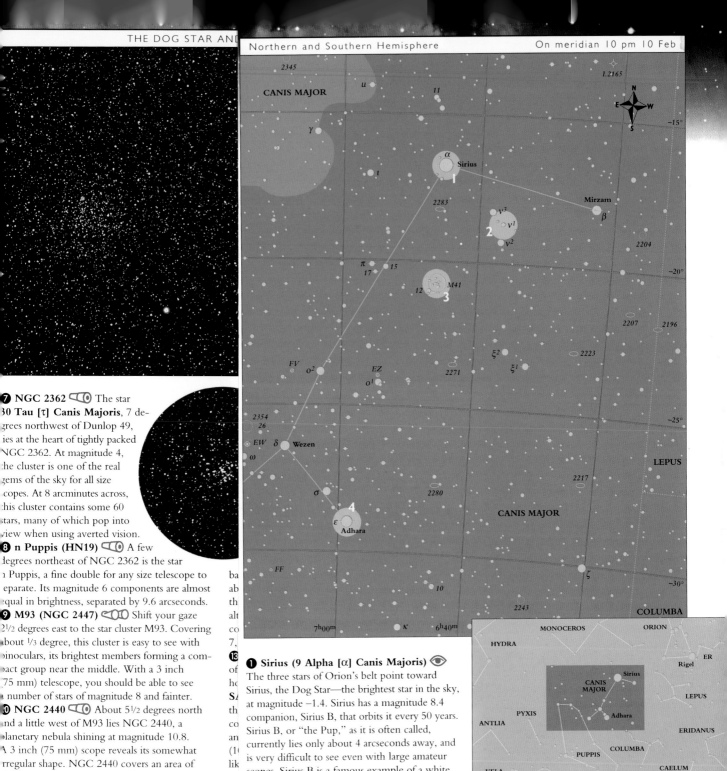

**7** **NGC 2362** 🔭 The star
**30 Tau [τ] Canis Majoris**, 7 degrees northwest of Dunlop 49,
lies at the heart of tightly packed
NGC 2362. At magnitude 4,
the cluster is one of the real
gems of the sky for all size
scopes. At 8 arcminutes across,
this cluster contains some 60
stars, many of which pop into
view when using averted vision.

**8** **n Puppis (HN19)** 🔭 A few
degrees northeast of NGC 2362 is the star
n Puppis, a fine double for any size telescope to
separate. Its magnitude 6 components are almost
equal in brightness, separated by 9.6 arcseconds.

**9** **M93 (NGC 2447)** 🔭 Shift your gaze
2½ degrees east to the star cluster M93. Covering
about ⅓ degree, this cluster is easy to see with
binoculars, its brightest members forming a compact group near the middle. With a 3 inch
(75 mm) telescope, you should be able to see
a number of stars of magnitude 8 and fainter.

**10** **NGC 2440** 🔭 About 5½ degrees north
and a little west of M93 lies NGC 2440, a
planetary nebula shining at magnitude 10.8.
A 3 inch (75 mm) scope reveals its somewhat
irregular shape. NGC 2440 covers an area of
about 14 x 32 arcseconds. Its bluish color is
quite obvious, especially in larger telescopes.

**11** **M46 (NGC 2437)** 🔭 Hop 3½ degrees
farther north to the open star cluster M46.
Discovered by Messier in 1771, M46 contains a

**1** **Sirius (9 Alpha [α] Canis Majoris)** 👁
The three stars of Orion's belt point toward
Sirius, the Dog Star—the brightest star in the sky,
at magnitude −1.4. Sirius has a magnitude 8.4
companion, Sirius B, that orbits it every 50 years.
Sirius B, or "the Pup," as it is often called,
currently lies only about 4 arcseconds away, and
is very difficult to see even with large amateur
scopes. Sirius B is a famous example of a white
dwarf star, which is only a little bigger than
Earth and extremely dense. A cubic inch of
its material would weigh about 2 tons!

The brilliance of Sirius has given rise to
countless stories through the ages. In ancient

Egypt, Sirius was known as the Nile Star because
at the time of year when it rose just before
dawn, it heralded the annual flooding of the
Nile River, an important event in Egyptian life. **409**

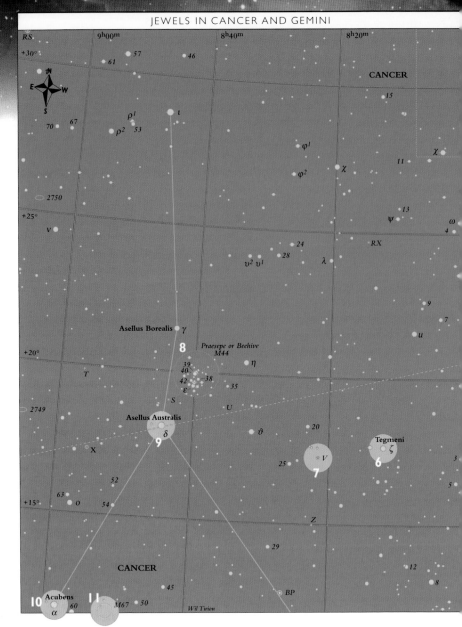

**❷ 6 Nu¹ [ν¹] Canis Majoris** 🔭 To find
Nu¹ Canis Majoris, nudge your scope 3 degree
southwest from Sirius. Nu¹ is the middle of a
group of three stars—the others, predictably, a
Nu² and Nu³. Nu¹ is a fine double that is easy
separate in any size scope. Its components, at
magnitudes 5.8 and 8.5, lie 17.5 arcseconds apar
**❸ M41 (NGC 2287)** 👁 A 3 degree hop
southeast brings us to the star cluster M41. Th

### WHITE DWARF STARS

Sirius B, Sirius A's companion (below), was the
first white dwarf to be identified. Incredibly,
Sirius B has a mass almost equal to that of our
Sun, but a diameter of only about 19,000 miles
(30,500 km)—less than three times that of Earth.

We now know that a white dwarf is the
collapsed core of a star—the last
stage in the evolution of a star
that was originally up to about
eight times as massive as
the Sun. In 1931, it was
discovered that a white
dwarf star cannot contain
more than 1.4 solar
masses. However,
stars lose a great deal of
mass during their lives by
shedding material, most
visibly in the form of planetary
nebulas late in their life cycle.

# Jewels in Cancer and Gemini

The constellations on this hop along the
zodiac are clearly visible from both the
Northern and Southern hemispheres.
The legends of
Cancer, the Crab,
and Gemini, the
Twins, come from
Greek mythology. The

*The Twins, from a sixteenth-
century Turkish literary text.*

goddess Hera sent Cancer to kill her enemy
Hercules, who crushed the crab with his mighty
club. Saddened by the crab's demise, Hera
placed it among the stars to honor its valiant bu
ill-fated efforts. Gemini represents the twin sons
of Leda, the Queen of Sparta. One was fathered
by her husband, King Tyndareus, and the other
by Zeus while in the form of a swan (represented
by Cygnus). The twins, Castor and Pollux,
served aboard Jason's ship, the Argo, in the
legendary voyage of the Argonauts.

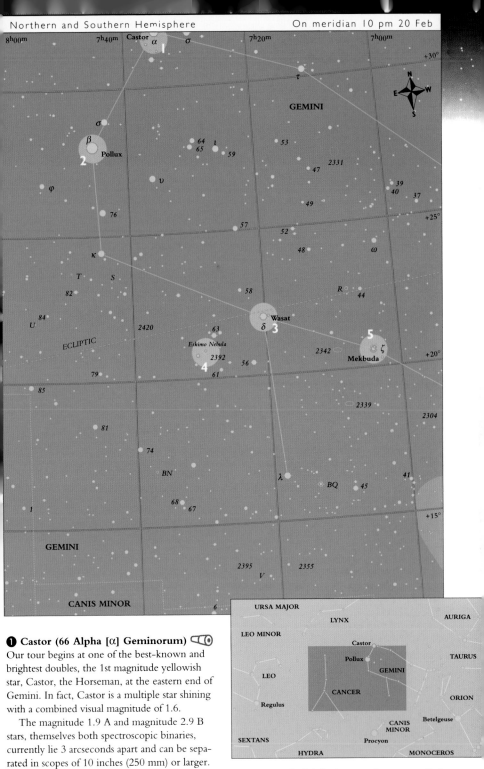

8h00m          7h40m   Castor          7h20m          7h00m

GEMINI

+30°

Castor α   σ

τ

σ

β

Pollux

2

φ

64  ι
65
59

53

47  2331

49

39
40
37

+25°

υ

76

57
52

48

ω

κ

58

R
44

T   S

82

84
U

2420

63

Eskimo Nebula
2392

56

2342

Mekbuda

ζ

5

+20°

ECLIPTIC

Wasat
δ  3

79

61

85

2339

2304

81

74

BN

λ

BQ

45

41

68
67

+15°

1

GEMINI

2395

2355

CANIS MINOR

6

V

**❶ Castor (66 Alpha [α] Geminorum)** 📷
Our tour begins at one of the best-known and
brightest doubles, the 1st magnitude yellowish
star, Castor, the Horseman, at the eastern end of
Gemini. In fact, Castor is a multiple star shining
with a combined visual magnitude of 1.6.

The magnitude 1.9 A and magnitude 2.9 B
stars, themselves both spectroscopic binaries,
currently lie 3 arcseconds apart and can be separated in scopes of 10 inches (250 mm) or larger.
The B star orbits the A star over a period of
about 400 years. The faint red dwarf C star
(designated YY Geminorum), also a spectroscopic
binary, completes the entire Castor system.

URSA MAJOR

LYNX

AURIGA

LEO MINOR

Castor

LEO

Pollux

GEMINI

TAURUS

CANCER

ORION

Regulus

CANIS
MINOR

Betelgeuse

SEXTANS

Procyon

HYDRA

MONOCEROS

A spectacular bright shower of meteors, called
the Geminid meteors, emanates from a point
near Castor each year, reaching its maximum
around 14 December (see also p. 200).      **413**

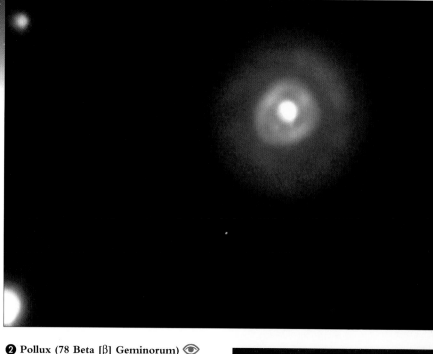

**❷ Pollux (78 Beta [β] Geminorum)** 👁
Four and a half degrees southeast of Castor is the other twin, Pollux. This is the sky's 17th brightest star, shining as a magnitude 1.2 golden-yellow giant. Pollux is about 34 light-years from Earth and intrinsically 35 times brighter than the Sun.

**❸ Wasat (55 Delta [δ] Geminorum)** 🔭
Move your scope about 8 degrees southwest to the double star Wasat, a magnitude 3.5 yellow star with a faint red dwarf companion at magnitude 8.2. This star system lies some 60 light-years away with an orbital period believed to be more than 1,000 years long. In 1930, Clyde Tombaugh discovered Pluto near this binary system.

**❹ The Eskimo Nebula (NGC 2392)** 🔭
About 2 degrees southeast of Wasat is the bright planetary nebula NGC 2392, discovered by William Herschel in 1787. It is intense and compact, looking somewhat like a large, fuzzy bluish green star, making an attractive contrast to the magnitude 8.3 orange star nearby, which has a similar brightness. As with most planetary nebulas, high magnification is essential for good viewing. You can only expect to see the eskimo's "face" in photographs taken through large scopes.

**❺ Mekbuda (43 Zeta [ζ] Geminorum)** 👁
Sweep some 6 degrees west of NGC 2392 to Mekbuda, one of the brightest variable stars. This yellow pulsating Cepheid variable has a period of 10.16 days during which its magnitude fluctuates between 3.6 and 4.2. With binoculars, you might also note a magnitude 7.6 companion star nearby, which is actually unrelated to Zeta.

*The Eskimo Nebula (top) and the Praesepe (above) are opposites in terms of the level of magnification that reveals them best.*

**❻ Tegmeni (16 Zeta [ζ] Cancri)** 🔭 Hop about 16 degrees east and slightly south of Mekbuda to the binary star Tegmeni, across the border in Cancer. The B component of this system orbits the A star over a period of 59.6 years. Both components are yellowish main-sequence stars with nearly equal magnitudes of 5.6 and 5.9.

**❼ V Cancri** 🔭 This red giant Mira-type variable lies 2 degrees east of Tegmeni. This is one of many variables whose full cycle can be followed in scopes of less than 8 inches (200 mm). The normal magnitude range of V Cancri is 7.9 to 12.8, during a period of 125 days. At maximum, it would fill the volume of Mars's orbit.

**❽ The Praesepe/Beehive Cluster (M44, NGC 2632)** 👁 Shift your scope 5 degrees northeast to the Praesepe, a large, naked-eye open

he open star cluster M67 (above) is often overlooked in favor of
s larger cousin, the Praesepe.

...ar cluster that has been known since ancient
...mes. It covers about 1¹/₂ degrees of sky and is
...ome 520 light-years distant. The Praesepe's
...everal hundred scattered stars show best through
...inoculars or a finderscope because the cluster is
...uite broad. At higher magnification, it is quite
...isappointing because the view is too close.

**⦿ Asellus Australis (47 Delta [δ] Cancri)**
👁 Nudge 2 degrees southeast of the center of
...he Praesepe to Asellus Australis, a magnitude 4.3
...ellow optical double star. The Romans gave
...his star its name, the Southern Donkey. Asellus
...ustralis and **Asellus Borealis (43 Gamma [γ]
...ancri)** are known collectively as the Donkeys.

**⦿ Acubens (65 Alpha [α] Cancri)** 👁
...bout 6 degrees southeast of Asellus Australis
...s the magnitude 4.3 white wide double star
...cubens, about 170 light-years away. Its name
...s derived from the Arabic *Al Zubanah,* meaning
..."the claw," because it marks one of the crab's
...laws. The secondary star in this system shines
...t magnitude 11.8 and is visible through 3 inch
...75 mm) telescopes or larger.

**⦿ M67 (NGC 2682)** 👁 Use the bright
...tars Asellus Australis and Acubens to help locate
...467, about 2 degrees west of Acubens. This
...luster, the last object on our tour, was
...iscovered by the German astronomer Johann
...Gottfried Koehler between 1772 and 1779. In
...ark skies, this densely packed magnitude 6.1

open cluster can be seen with the naked eye, but
it needs a small scope or binoculars to really
appreciate it. M67 contains some 500 stars of
magnitudes 10 to 16 and many more fainter
ones. M67 is estimated to be four to five billion
years old—one of the oldest open clusters known.

Cancer and Gemini include some interesting
star fields, beckoning you to mine their collection
of clusters and double and variable stars.

### EXTRASOLAR PLANETS

Only recently have astronomers detected
clear evidence of planets orbiting distant Sun-
like stars. One of the first of a handful of extrasolar
planets to be discovered was the planet-size companion of 55 Rho¹ [ρ] Cancri, found by a team
from the Universities of California and San
Francisco. The illustration below compares the
orbit of Mercury around the Sun (top) on a similar
scale to Rho¹ Cancri and its planet (bottom). Rho¹
is a magnitude 5.9 yellow star with a planetary
companion, about 0.8 times the mass of Jupiter,
that orbits Rho¹ over just 14.7 days.

Extrasolar planets are far too faint to be seen
visually, even by the HST,
but they reveal
their presence
by the tiny
wobbles their
gravity induces in
the motions of each star.

The Great Ship Argo from a
fresco in the Villa Farnese.

# Gems in the Great Ship Argo

Northern Hemisphere skywatchers are disadvantaged here because many objects in this patch of sky are not seen at all from latitudes above about 30 degrees north. Carina, the Keel, and Vela, the Sails, were two of the four constellations formed when the huge constellation of Argo

Navis was subdivided by the French astronomer Nicolas Louis de Lacaille in the eighteenth century. The other two are Puppis, the Stern; and Pyxis, the Compass. Straddling the constellations of Carina and Vela is the so-called False Cross, which newcomers to the southern skies can easily confuse with the Southern Cross.

With binoculars or a telescope, you can expect to see many fine objects within this part of the Milky Way Galaxy, including several of the best star clusters in the entire night sky.

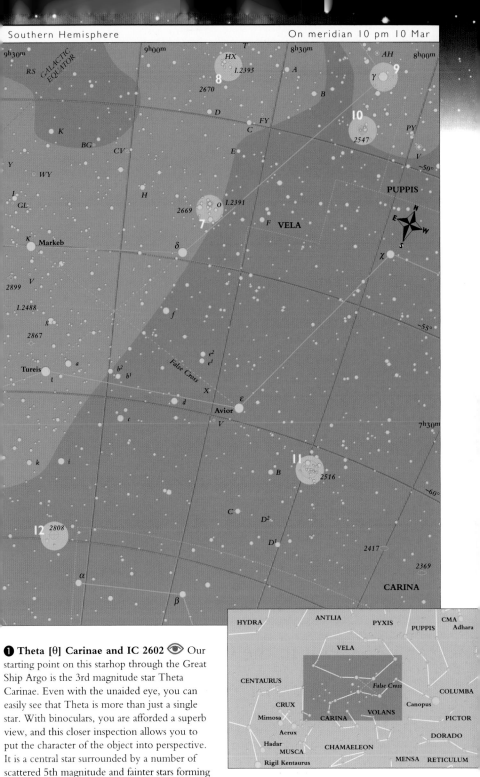

❶ **Theta [θ] Carinae and IC 2602** 👁 Our
starting point on this starhop through the Great
Ship Argo is the 3rd magnitude star Theta
Carinae. Even with the unaided eye, you can
easily see that Theta is more than just a single
star. With binoculars, you are afforded a superb
view, and this closer inspection allows you to
put the character of the object into perspective.
It is a central star surrounded by a number of
scattered 5th magnitude and fainter stars forming
the open star cluster IC 2602.

Theta Carinae, which covers a full degree of
sky, has also been called the Southern Pleiades
by some Southern Hemisphere observers.

To find the next stop on this starhop, the star
Eta Carinae, shift your scope about 4½ degrees
due north of IC 2602. Eta is enveloped in the
dramatic and beautiful Eta Carinae Nebula.

417

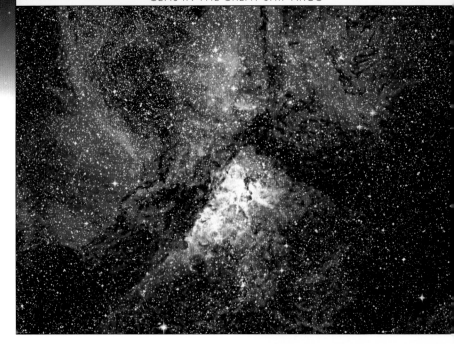

**2** **Eta [η] Carinae** ⟨⊙⊙ This is the brightest star in a field that is truly beautiful, even in scopes of less than 3 inches (75 mm). The star appears orange-red and is embedded in the superb **Eta Carinae Nebula.** Eta is a variable star—one of the most luminous in the Milky Way. In 1843, it outshone every star in the night sky except Sirius. A 4 inch (100 mm) scope will show Eta surrounded by a small "blob" of red light—the nebulous patch called the Homunculus (see Box).

**3** **NGC 3532** 👁 Turning about 3 degrees east and slightly north of Eta, the open cluster NGC 3532 is revealed. John Herschel, who cataloged thousands of celestial objects when observing from South Africa in the 1830s, considered NGC 3532 to be the finest cluster he had ever seen. It appears as a fuzzy patch to the naked eye and is a superb object with binoculars. Any size scope will resolve a large number of stars, which are best viewed using low power.

**4** **NGC 3293** ⟨⊙⊙ The compact cluster NGC 3293 is 4 degrees west and a little north of NGC 3532. Through binoculars, this cluster appears as a tiny bright spot, but a telescope view reveals a stunning group of several dozen stars of different colors. A 3 inch (75 mm) scope with moderate magnification gives a fantastic view. An orange star near one edge of the cluster is especially attractive.

**5** **IC 2581** ⟨⊙⟩ Nudge your scope about 1 degree west and slightly north to bring IC 2581 into view. Gathered around quite a distracting magnitude 5 star, IC 2581 is a cluster of some

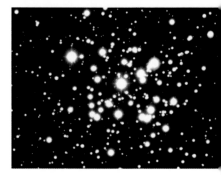

*Eta Carinae Nebula (top) and NGC 3293 (above) are two of the highlights of the Milky Way in Carina.*

25 stars that are much more obvious with high magnification. You should be able to resolve thi cluster well with a 4 inch (100 mm) telescope.

**6** **NGC 3114** ⟨⊙⊙ Shift your focus 4 degrees southeast to locate the next cluster, NGC 3114. Binoculars do a fine job of revealing this magnificent cluster of 7th magnitude stars and fainter covering an area of more than ½ degree. To the naked eye, it appears as a 4th magnitude patch o light. Low magnification is essential when viewing this cluster through a telescope.

**7** **IC 2391** ⟨⊙⊙ Almost 2 degrees north and a little west of **Delta [δ] Velorum** is IC 2391, also known as the Omicron Velorum cluster. Its members are loosely packed around the magnitude 3 bluish white **Omicron [o] Velorum**. is a better subject for binoculars than for a scope.

NGC 3532 (above) is one of the finest open clusters in the sky, lying about 1,300 light-years away from us.

**8 IC 2395** ⬤⬤ Head north about 5 degrees from IC 2391 to IC 2395. You can spot this cluster in binoculars, but it is better suited to a telescope. It is not exactly a rich group, but still one worthy of attention, and a 3 inch (75 mm) scope, or even smaller, provides a good view.

**9 Gamma [γ] Velorum** ⬤⬤⬤ A little more than 6 degrees west of IC 2395 is Gamma Velorum. This is a spectacular double star whose magnitude 2 and 4 components are 41 arcseconds apart and can be separated with binoculars. Any small scope will show two other stars, of magnitudes 8 and 9, about 90 arcseconds away.

**10 NGC 2547** ⬤⬤⬤ This attractive open cluster, about 2 degrees south of Gamma, contains stars with a moderate range in brightness. A 4 inch (100 mm) scope provides a good view of the cluster, which covers about 1/3 degree.

**11 NGC 2516** ⬤ Eleven degrees farther south of NGC 2547 is NGC 2516. This cluster is a fine sight in binoculars. The long axis of the False Cross points toward it, as if to draw attention to it, and it is easily located a few degrees southwest of **Epsilon [ε] Carinae**. The 1/2 degree wide cluster contains a number of stars of magnitude 7 and fainter.

**12 NGC 2808** ⬤⬤⬤ A little south of the halfway point between NGC 2516 and Theta Carinae is NGC 2808. This globular cluster shows up well as a glowing spot of light in binoculars.

A 6 inch (150 mm) scope or larger is needed to start to resolve it well, but a hint of its stellar nature can be obtained with smaller telescopes.

Whenever you fix your gaze toward the starry sky, you will always remember the fine objects in this area—especially the great clusters.

## THE ETA CARINAE NEBULA (NGC 3372)

The Eta Carinae Nebula is one of the finest areas of nebulosity in the southern sky—a visually stunning object whose complex structure is an impressive sight. Photographs show that the nebula is very detailed and extensive, covering some 4 square degrees of sky. This nebula is carved into two glowing halves by a dark dust lane, one of many that seem to divide the nebulosity into a number of areas of glowing light. The brightest of these "islands" of light contains the dark, irregular, and elongated mass that is the Keyhole Nebula (below). It covers quite a small area and appears close to the star Eta.

The star Eta (seen at lower left) is surrounded by a bright patch of nebulosity known as the Homunculus. It gets its name from the fact that, close up, its peanut shape appears to vaguely resemble the body of a man. This gas was ejected during the 1843 outburst.

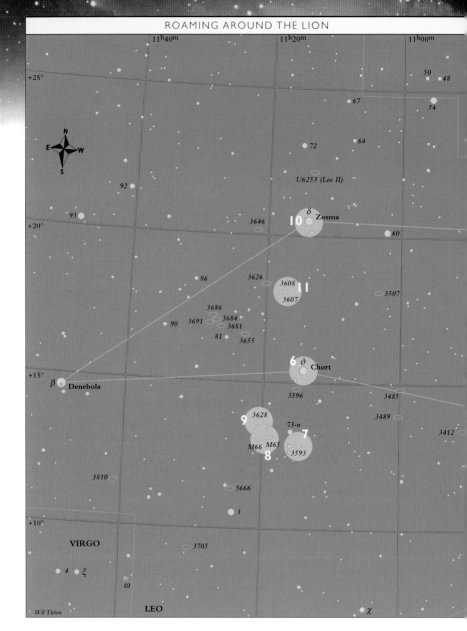

# Roaming around the Lion

Like all ancient zodiac constellations, Leo, the Lion, lies on the Sun's path through the sky, the ecliptic. It is easily visible from both the Northern and Southern hemispheres. There are plenty of bright galaxies, double stars, and variables in this region. Leo is easy

*Leo, from a fresco in the Villa Farnese, painted in 1575.*

to locate because this part of the sky is dominated by the backward-question-mark, or sickle, asterism that marks the lion's head and chest, and the triangle of stars that marks its hindquarters.

According to ancient Greek legend, Leo was the cave-dwelling Nemean Lion choked to death by the mighty Hercules. Because no weapon could penetrate Leo's skin, Hercules used the lion's own razor-sharp claw to remove its pelt. He made the pelt into a cloak to protect himself and carried it with him into the heavens.

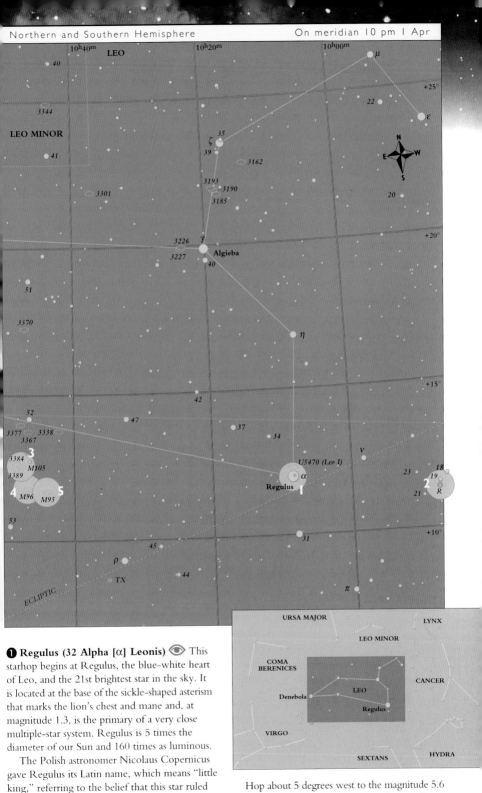

10h40m   **LEO**

40

3344

**LEO MINOR**

41

35

ζ

39

3162

3193

3190

3185

3301

+25°

22

ε

20

N

E W

S

+20°

3226 γ

3227   **Algieba**

40

51

3370

η

+15°

42

52

47

3377   3338

3367

3384

M105

3389

4 M96 M95 5

53

37

34

U5470 (Leo I)

α

**Regulus**

v

23

18

19

R

21

2

+10°

31

45

ρ

44

TX

π

**ECLIPTIC**

**URSA MAJOR**

**LYNX**

**LEO MINOR**

**COMA BERENICES**

**CANCER**

**LEO**

Denebola

Regulus

**VIRGO**

**SEXTANS**

**HYDRA**

❶ **Regulus (32 Alpha [α] Leonis)** 👁 This starhop begins at Regulus, the blue-white heart of Leo, and the 21st brightest star in the sky. It is located at the base of the sickle-shaped asterism that marks the lion's chest and mane and, at magnitude 1.3, is the primary of a very close multiple-star system. Regulus is 5 times the diameter of our Sun and 160 times as luminous.

The Polish astronomer Nicolaus Copernicus gave Regulus its Latin name, which means "little king," referring to the belief that this star ruled the heavens. This celestial imagery shares an apt symmetry with the commonly held notion of the lion as king of the beasts on Earth.

Hop about 5 degrees west to the magnitude 5.6 reddish star **18 Leonis** and magnitude 6.4 yellow-white **19 Leonis**. Use these two stars to locate R Leonis in the same low-power eyepiece view. **421**

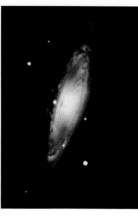

**❷ R Leonis**  In 1782, the Polish astronomer J. A. Koch first recorded the variability of this pulsating red giant Mira-type variable, one of the earliest to be discovered. The magnitude of R Leonis cycles from 5.8 down to fainter than 11th magnitude. Although it is normally a sharp red star, R Leonis can appear a deep shade of purple during its 312-day cycle.

**❸ M105 (NGC 3379)** Return to Regulus, then hop about 10 degrees east to find **52 Leonis**, a magnitude 5.4 yellowish star. About 1½ degrees south of 52 Leonis is a triangle of hazy light formed by three galaxies—M105 and NGC 3384 and 3389. At magnitude 9.2, M105 is the brightest and largest of the three. Typical of elliptical galaxies, M105 appears as a fuzzy ball that you cannot focus into a sharp point of light.

**NGC 3384** The second galaxy in the group is NGC 3384, also an elliptical, which appears as an elongated oval with a bright nucleus.

**NGC 3389** The third galaxy, NGC 3389, is fainter and harder to find. Appearing as a pale, thin streak of light, it is a magnitude 11.8 spiral that seems to glow evenly from edge to edge.

**❹ M96 (NGC 3368)** About 1 degree south of M105 is the spiral galaxy M96. This silver-gray beauty appears in a relatively starless

*Several galaxies in Messier's list, including M65 (left), are found in Leo (above).*

field, so it is easy to spot. The intense core of the galaxy is much sharper and brighter than the wispy wreath of its arms.

**❺ M95 (NGC 3351)** This magnitude 9.7 barred spiral galaxy, adjacent to M96, has a fairly bright core surrounded by wispy arms. Its nucleus appears almost stellar, with a fuzzy fringe around it. M95 is not as big or as bright as M96. Even in scopes larger than 10 inches (250 mm), it is difficult to see the faint arms of the galaxy, but a 4 inch (100 mm) shows the gradual brightening toward the core.

**❻ Chort (70 Theta [θ] Leonis)** Sweep about 8 degrees northeast of M95 to blue-white Chort. This stunning jewel marks the hips of the Lion and is the western apex of the triangle of stars marking the rear haunches of Leo.

**❼ NGC 3593** Nudge about 3 degrees south of Chort to find NGC 3593, a magnitude 11.3 spiral galaxy with a bright and elongated core. The outline of the galaxy's arms is quite diffuse, blending into the background at the edges

**❽ M65 (NGC 3623)** This beautiful magnitude 9.9 spiral galaxy is at least 200,000 light-years distant from M66, although the famous pair appear in the same low-power eyepiece view.

*The edge-on spiral galaxy NGC 3628 (above) is located quite close to M65 (left) and clearly shows a warp effect in its disk.*

**M66 (NGC 3727)** ⊂⦿ At magnitude 9.9, M66 is brighter and a bit shorter than M65. M66 is another spiral galaxy, with a finger of light extending from its southern tip. Many of its dark areas can be seen with a 4 inch (100 mm) scope.

**❾ NGC 3628** ⊂⦿ This elongated, pencil-slim spiral galaxy is at the northern apex of the richly studded triangle that includes M65 and M66. The edges are quite grainy on this magnitude 9.9 object. It is halfway between two magnitude 10 stars to the north and south, but because of its low surface brightness and being seen edge-on, it is not easy to locate.

**❿ Zosma (Delta [δ] Leonis)** ⊂⦿⦿ Sweep about 7 degrees northwest of NGC 3628 to blue-white Zosma, a magnitude 2.6 main-sequence star at the heart of a very open multiple system. These stars are easy to separate in any size scope because the magnitude 8.6 companion is located 191 arc-seconds from the primary.

**⓫ NGC 3607** ⊂⦿ Move back some 2½ degrees south of our last stop to find the elliptical galaxy NGC 3607, a magnitude 10.9 object that looks like an out-of-focus star through a telescope.

**NGC 3608** ⊂⦿ At about a magnitude fainter than NGC 3607, the elliptical galaxy NGC 3608 appears smaller, although astronomers believe that these galaxies are actually about the same size. NGC 3608 is roundish with a faint stellar core.

This is just a sampling of the night-sky wonders to be found in Leo. Additional faint galaxies, shown on the main sky chart, are within the body of the celestial king of beasts, waiting for you to explore them.

### THE MESSIER MARATHON

In the Northern Hemisphere, a window of opportunity opens around the time of the vernal equinox, every March, when all 110 Messier objects can be observed on one night.

French astronomer Charles Messier and his collaborator, Pierre Méchain, were engaged in a quest to discover comets and, in the process, listed many other wonderful celestial objects in order to avoid confusing them with comets.

To complete a Messier Marathon, begin with the spiral galaxy M74 in Pisces, in the western sky at dusk. During the night, sweep from one Messier object to the next, ending with M30 in Capricornus, shortly before the onset of morning twilight. The beautiful spiral galaxy M66 (left) is just one of the marathon objects to aim for while you are in Leo.

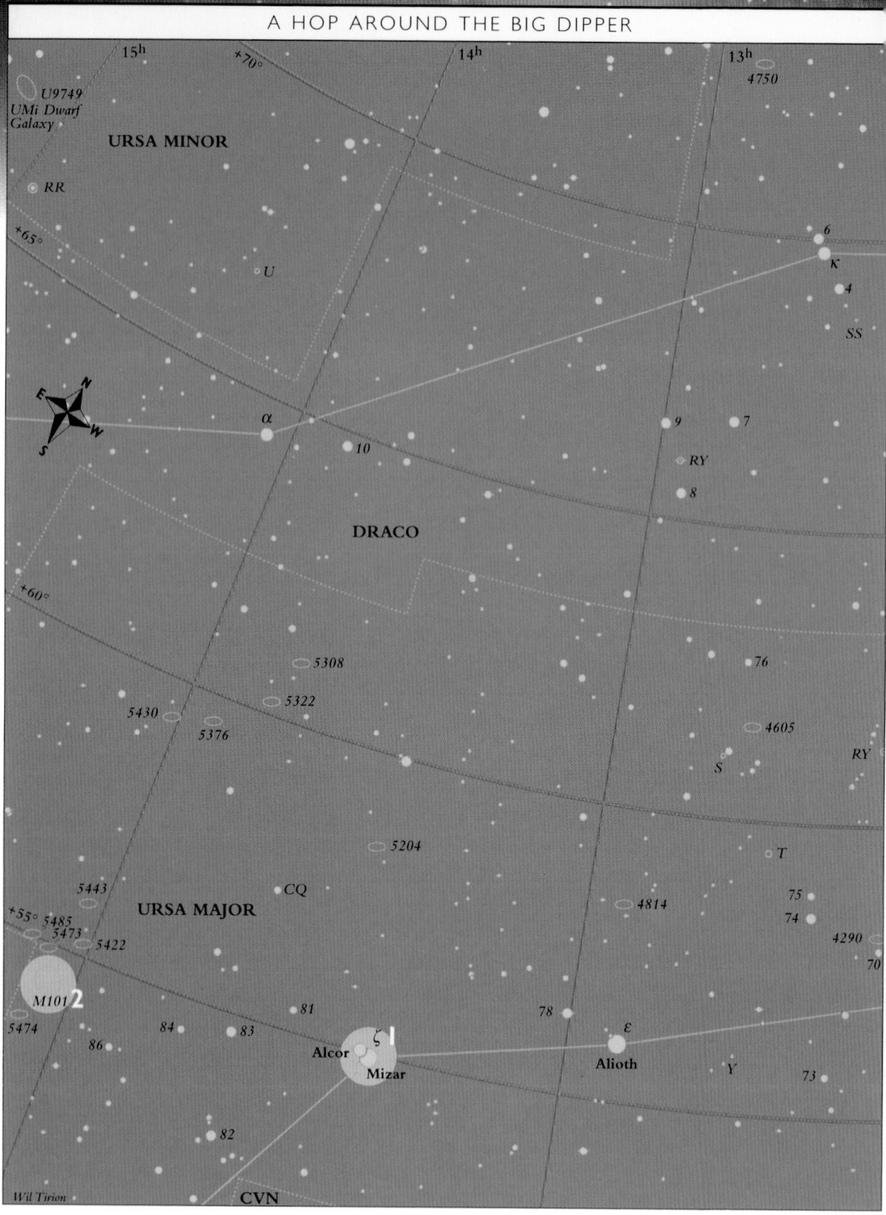

A HOP AROUND THE BIG DIPPER
15h
14h
13h
+70°
+65°
+60°
+55°
U9749
UMi Dwarf Galaxy
URSA MINOR
RR
U
4750
6
K
4
SS
N
E
W
S
α
10
9
7
RY
8
DRACO
5308
5322
5430
5376
76
4605
RY
S
5204
T
CQ
5443
URSA MAJOR
4814
75
74
4290
70
5485
5473
5422
M101
2
5474
81
78
ε
84
83
86
ζ
1
Alcor
Mizar
Alioth
Y
73
82
CVN
Wil Tirion

# A Hop around the Big Dipper

One of the first star patterns that Northern Hemisphere skywatchers usually learn to recognize is the Big Dipper, in the constellation of Ursa Major, the Great Bear. Unfortunately for observers in mid-southern latitudes, this famous asterism is not visible. The seven bright stars of the circumpolar bowl and handle are part of what is known as the Ursa Major Moving Group, the closest open star cluster to us.

The Dipper stars are rich in mythology and have been known by various names throughout history. The lore that many cultures created for this constellation is based on that of a bear. However, the ancient Egyptians considered the asterism to be either a hippopotamus or a Nile River boat for the god Osiris. Ancient Romans saw the Dipper as seven oxen pulling a plough.

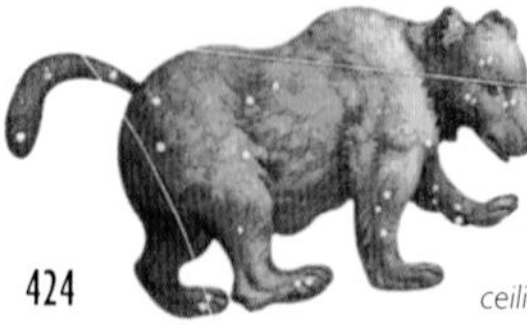

*The Great Bear, depicted in a ceiling fresco from the Vatican.*

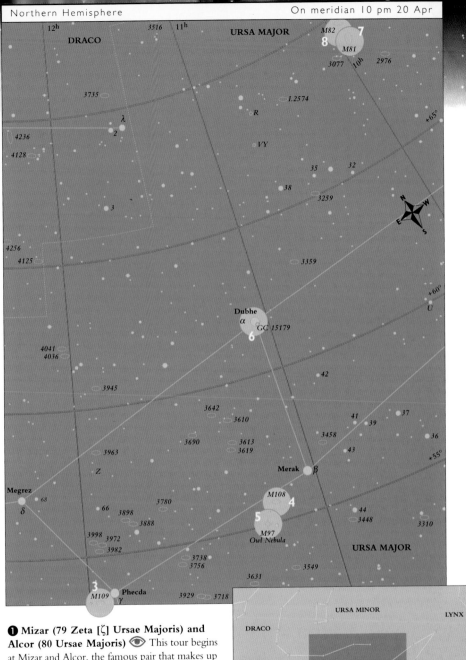

**❶ Mizar (79 Zeta [ζ] Ursae Majoris) and
Alcor (80 Ursae Majoris)** 👁 This tour begins
at Mizar and Alcor, the famous pair that makes up
the apparent wide double star in the middle of
the Dipper handle. If your eyesight is sharp, you
can separate the components with the naked eye,
since they are 12 arcminutes apart. A pair of
binoculars provides a great view of Mizar and
Alcor, and with a small telescope, you can see
that Mizar is actually a double star itself. Its 4th
magnitude companion, known as Mizar B, lies
about 14 arcseconds away. In 1650, Italian
astronomer Giovanni Riccioli identified Mizar's
two stars as the first true binary-star system.

M101, our next target on this starhop, is one
of three Messier objects commonly known as the
Pinwheel Galaxy. The other two "Pinwheels"
are M99 (see p. 430) and M33 (see p. 379). **425**

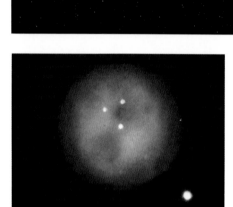

*The Big Dipper (above) is home to a variety of unique objects including the Owl Nebula (left), a very large planetary nebula.*

### ❷ The Pinwheel Galaxy (M101, NGC 5457)

This gem of a spiral galaxy lies about 5 degrees east of Mizar and Alcor. M101 is one of the largest and finest face-on spirals visible, and although it is fairly bright, at around magnitude 9, it is also diffuse, making it hard to spot, even with a 10 inch (250 mm) scope. It is best viewed in a dark sky using low magnification and a wide field in order to detect the knotty arms loosely extending outward from the core.

Pierre Méchain discovered M101 in 1781. He also found an object sometimes listed as M102, considered to be a missing Messier object. However, in a letter dated 1783, Méchain wrote that he thought M102 was just a re-sighting of M101.

### ❸ M109 (NGC 3992)

Hop over to the southeastern corner of the Dipper's bowl to **Phecda (64 Gamma [γ] Ursae Majoris)** and use this magnitude 2.4 white star, "the thigh of the Bear," to locate the barred spiral M109 about

¼ degree east of it. The arms, visible through the extended grayish nebulous haze of the galaxy, are best seen on a very dark night using a 10 inch (250 mm) telescope or larger.

### ❹ M108 (NGC 3556)

To find M108, hop first to the bright, white **Merak (48 Beta [β] Ursae Majoris)**. If you draw an imaginary line between Merak and Phecda, M108 is about 1½ degrees southeast from Merak along this line. M108 is a milky-white, edge-on spiral galaxy that appears flat, with no sign of the central bulge typical of most other spiral galaxies. M108 glows at magnitude 10.1. It has a mottled texture with about four bright spots visible within its arms. Visible through a 10 inch (250 mm) telescope, an elongated dust lane runs through the long axis of the galaxy.

### ❺ The Owl Nebula (M97, NGC 3587)

The Owl Nebula is found about ½ degree farther southeast from M108. This large, but diffuse, magnitude 12 planetary nebula was discovered by Pierre Méchain in March 1781. Although the nebula appears to us about the same size as Jupiter, it is estimated to be 3 light-years in diameter, making it one of the largest planetary nebulas. You should be able to see it through a 4 inch (100 mm) scope in a dark sky, but because of its low surface brightness you need a 10 inch (250 mm) telescope or larger to gain even a hint of the owl's dark eyes.

426

*This CCD image of M81 clearly reveals its classic spiral arms surrounding a bright central nucleus.*

**❻ Dubhe (50 Alpha [α] Ursae Majoris)**

To the Arabs, this magnitude 1.8 golden-yellow close double star was *Thar al Dubb al Akbar*, meaning "the back of the Great Bear." Dubhe and Merak form the western end of the Dipper's bowl and point the way to Polaris, the North Star. Dubhe also has a magnitude 7 bluish companion, **GC 15179**, lying 6.3 arcminutes from the primary, which is easily separated in small scopes. A line between Phecda and Dubhe will point you toward our next targets, the galaxies M81 and M82, which are located about 10 degrees northwest of Dubhe.

**❼ M81 (NGC 3031)** The bright core of M81 appears distinctly elliptical, at the center of a strikingly symmetrical overall structure that is easily visible with binoculars. Lurking within the bright core are most of the galaxy's 250 billion stars. The two spiral arms of the galaxy appear to be quite diffuse because they have fewer stars tracing them out.

M81 and M82 lie about 38 arcminutes apart, easily appearing in the same low-power telescope or binocular field. They are considered by many to be one of the finest pairs for observers.

**❽ M82 (NGC 3034)** Smaller and dimmer than M81, M82 is classified as a peculiar galaxy (see Box). In scopes smaller than 15 inches (380 mm), M82 looks like an unusually amorphous, edge-on spiral, although a spiral structure has not been confirmed, even in photographs taken at the largest Earth-bound telescopes or with the Hubble Space Telescope.

Sweep your scope in and around the bowl of the Big Dipper to observe and enjoy the many other faint galaxies waiting to be found.

### STAR OUTBURST IN M82

M82 is now thought to be a nearby example of a starburst galaxy. Astronomers attribute the burst of star formation in M82 to a possible encounter with M81, its companion spiral galaxy, approximately 100 million years ago. This may have severely disrupted M82's gas clouds, igniting the starburst that involved a mass of material equal to several million Suns. The massive young stars that formed within the galaxy's nucleus gave rise to raging winds of hot gas, which, combined, made a powerful galactic wind. This wind created the filamentary structures (left) that were once thought to be the products of an explosion.

427

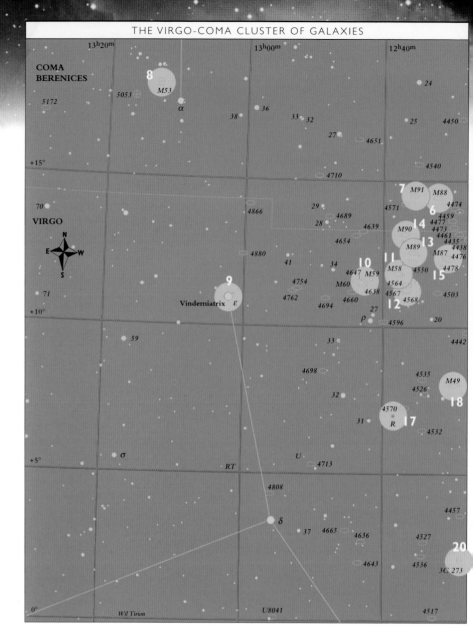

# The Virgo–Coma Cluster of Galaxies

From April through June, skywatchers are afforded the best opportunity to observe about 250 of the 3,000 "island universes" located within the Virgo Cluster of Galaxies. This cluster extends north and south through much of the constellations of Virgo, the Virgin, and Coma Berenices, the Hair

*Virgo, the Virgin, from a fresco in the Palazzo Schifanoia, Ferrara.*

of Berenice. This region is clearly visible from both the Northern and Southern hemispheres.

The majority of galaxies in the Virgo Cluster are some 65 million light-years from us. This means that the light from these distant multitudes of stars has been traveling, uninterrupted, through space for approximately 65 million years. Throughout this sky tour, you will be looking back into time and seeing what these galaxies were like about the time the dinosaurs died off on Earth. Amazing.

12ʰ20ᵐ           12ʰ00ᵐ           11ʰ40ᵐ

R

86

4394  4293         4147  4064

M85                                   3686

**5**   11                             3691  3684

                                  90     3681  3655

4350 4340     3                          81

4383          4152        WX

**4**                     95                    +15°

M100   4237

4419      M98               Denebola  β

4302  4262  6

4377 4298       **2**             **1**

M99  4212        **COMA**

**3**           **BERENICES**

4216  4189

M86        4168                        N

M84    4206                 E  W

**16**  4267      SU

4388                                   S

4371                     3810           3666

4429        4178       **LEO**

          12                          +10°

          4124                     3705

4417             **VIRGO**

         o   X

                 6      4  ξ

                              ω

4365      4235

                  π            v

4261     11

4324 4281  4273

     17                              +5°

**19**

M61                             80

                                  83

16                7             89    τ

10                β   ECLIPTIC

SS                     **VIRGO**

4385

**❶ Denebola (94 Beta [ß] Leonis)** 👁 To
locate the galaxies on the northwestern side of
Virgo and Coma Berenices, aim your scope
toward our starting point, the easy-to-locate
Denebola, a magnitude 2.1 blue-white giant on
the eastern edge of Leo. On this starhop, we will
be looking directly out of the disk of the Milky
Way Galaxy, where the lack of stars, dust, and
gas allows observation of objects located millions
of light-years beyond our galaxy's bounds.

To early Arab astronomers, Denebola was
called *Al Dhanab al Asad*, which denoted its
position as the tuft of hair on the lion's tail and
from which its modern name is derived.

BOÖTES             URSA MAJOR

            COMA
            BERENICES

Arcturus                    LEO

            Denebola        Regulus

VIRGO                      SEXTANS

Spica    CRATER    HYDRA

From Denebola, hop about 6½ degrees east
to the magnitude 5 whitish star **6 Comae**, to
help you find the spiral galaxies M98 and M99,
which all appear within the same finder view.  **429**

❷ **M98 (NGC 4192)** 🔭 On the evening of 15 March 1781, French astronomer and surveyor Pierre Méchain discovered M98, M99, and M100. The edge-on spiral galaxy M98 appears as a thin patch of light with a slight bulge visible near its nucleus. Although M98 may appear to observers as a member of the Virgo Cluster of Galaxies, it is actually located about halfway between the Sun and the cluster itself.

❸ **The Pinwheel Galaxy (M99, NGC 4254)** 🔭 Nudge your scope 1 degree southeast of 6 Comae to bring M99 into view. In 4 inch (100 mm) scopes or larger, the sweeping arms in this roundish face-on galaxy will become visible. At 50,000 light-years in diameter, M99 is about half the size of the Milky Way Galaxy.

❹ **M100 (NGC 4321)** 🔭 Return to 6 Comae, then hop about 2 degrees northeast. Two 5th magnitude stars heading in the same direction will point you toward M100—a large, face-on spiral galaxy glowing at magnitude 9.6. At 7 arcminutes across, it presents the largest apparent size of any galaxy in the Virgo Cluster. In 8 inch (200 mm) telescopes or smaller, it can resemble a dim globular cluster, appearing circular with very faint arms. It is relatively easy to see because of its brilliant star-like core.

❺ **M85 (NGC 4382)** 🔭 Almost 2 degrees north of M100 is the magnitude 4.7 yellow

*The center of the Virgo Cluster (above) includes M86 and M84. M100 (left) lies just a few degrees north of the center.*

double star, **11 Comae**. Use it to direct you to M85, about 1 degree northeast. A circular patch of light glowing at magnitude 9.2, M85 is estimated to be 100 to 400 billion times more massive than the Sun.

**NGC 4394** 🔭 In the same low-power telescope eyepiece as M85 is the magnitude 10.9 barred spiral galaxy NGC 4394. This small, round spiral galaxy is about 3.9 arcminutes across with a brightly shining nucleus that appears almost stellar.

❻ **M88 (NGC 4501)** 🔭 The easiest way to find M88 is to return to 6 Comae then hop back to M99. M88 and our next target, M91, are at the same declination as M99, so with an equatorially mounted telescope, center M99 in your eyepiece, then turn off the motor drive, if your scope is equipped with one, and wait 14 minutes or so for M88 to drift into your eyepiece field of view. M88 appears elongated through a telescope because we are looking at a spiral galaxy tilted to our line of sight. Its core seems almost stellar, shining brightly for such a distant object.

❼ **M91 (NGC 4548)** 🔭 Leave your scope fixed for three more minutes after sighting M88, and the barred spiral galaxy M91 will also make an appearance, slipping quietly into your 1 degree

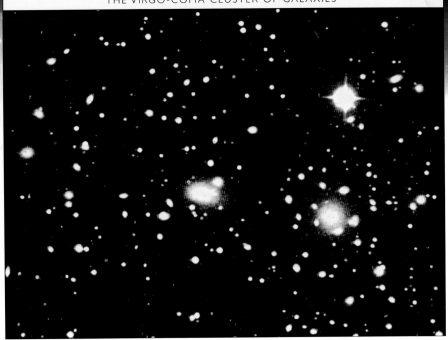

field. The slightly fainter M91 has low surface brightness, so it is more difficult to see than M88. The thin bar of its barred spiral structure is only visible in 12 inch (300 mm) telescopes or larger.

**8 M53 (NGC 5024)** Shift almost 9 degrees northeast of M91 to **42 Alpha [α] Comae.** One degree farther northeast is the magnitude 7.7 globular cluster M53, lying 65,000 light-years from us. In February 1775, German astronomer Johann Bode discovered this small, roundish globular cluster, describing it as "a new nebula, appearing through the telescope as round and pretty lively." At 3 arcminutes in diameter, M53 is a relatively easy target for any size telescope. The glowing halo of stars encircling the tightly compacted core of the cluster is a delight through a 4 inch (100 mm) telescope.

**9 Vindemiatrix (47 Epsilon [ε] Virginis)** About 8 degrees south of M53 lies the easy-to-locate magnitude 3 star Vindemiatrix. This yellow-white giant glows as a bright light in the night sky. From ancient Roman times, the early morning rising of Vindemiatrix in late August marked the time to commence harvesting the grapes. Its name is derived from the Latin word *Vindemitor*, which means "grape gatherer."

**10 M60 (NGC 4649)** Shift about 4½ degrees west and slightly north of Vindemiatrix to locate M60, one of the largest elliptical galaxies known. A halo seems to surround the bright core of this magnitude 8.8 galaxy. The mass of M60 is estimated to be equal to an amazing 1 trillion Suns, making it comparable in mass to M49.

**NGC 4647** This spiral galaxy and M60 appear to be interacting even though NGC 4647 is moving away from the Sun, while M60 is in approach. Although faint, at magnitude 12, NGC 4647 has a bright core and appears roundish and slightly smaller than M60.

**M59 (NGC 4621)** Also in the same low-power eyepiece view is M59, a magnitude 9.8 elliptical galaxy that appears oval with a bright core.

*The Coma Berenices Cluster (top) is much more distant than M99 (right) in the Virgo Cluster or M53 (center) in the Milky Way.*

431

**11 M58 (NGC 4579)** 〔📷〕 Swing about 1 degree west of M59 to M58, a magnitude 9.8 barred spiral galaxy. In an 8 inch (200 mm) scope, it appears elongated and faint, due to its low surface brightness. M58 is almost a twin, in terms of its size and mass, to the Milky Way Galaxy.

**12 The Siamese Twins (NGC 4567/68)** 〔📷〕 Nudge your scope about ½ degree southwest of M58 to the Siamese Twins. This unusual

*M87 (above), a giant elliptical galaxy surrounded by thousands of globular clusters, lies near the heart of the Virgo Cluster of Galaxies.*

pair consists of two seemingly attached galaxies, though at magnitude 10.8, NGC 4568 is slightly brighter than its "twin," and appears larger.

**13 M89 (NGC 4552)** 〔📷〕 M89 is an elliptical galaxy almost 1½ degrees north of the Siamese Twins. It appears as a round, glowing patch with a bright center through an 8 inch (200 mm) scope.

**14 M90 (NGC 4569)** 〔📷〕 Continuing about ½ degree northeast of M89 is the graceful M90, appearing as an elongated oval with a bright core. This beautiful magnitude 9.5 spiral galaxy is about 80,000 light-years in diameter. You should also be able to see M90's dust lanes through a 10 inch (250 mm) telescope or larger.

**15 M87 (NGC 4486, 3C 274)** 〔📷〕 Hop 1½ degrees southwest of M90 to bring M87 into view. This is a giant elliptical galaxy surrounded by more than 4,000 globular clusters. It is estimated to contain more than an incredible 1 trillion solar masses. We see it glow at magnitude 9.6, as a bright, fuzzy patch of light.

What you see is not all you get with this monster in the sky. Images taken by the Hubble Space Telescope reveal evidence for a black hole in the core of M87, surrounded by a mass of stars

## VIRGO'S BIG HEART, M87

Among more than 1,000 systems within the vast Virgo Cluster of Galaxies is the giant elliptical galaxy M87, one of the largest known. It covers an area of about 4 × 3 arcminutes and is surrounded by an extraordinary collection of more than 4,000 globular clusters. In a 6 inch (150 mm) scope, you will see it clearly as a magnitude 9 glow with a definite nucleus, but in photographs taken with professional equipment, it is possible to see a jet of very hot gas emanating from the nucleus. This was discovered by Heber Curtis at the Lick Observatory in 1918. The jet is about 4,000 light-years long and is now known to emit strong radio waves (right) and X-rays. Today, the jet is believed to be a high-speed beam of particles, probably associated with a central black hole.

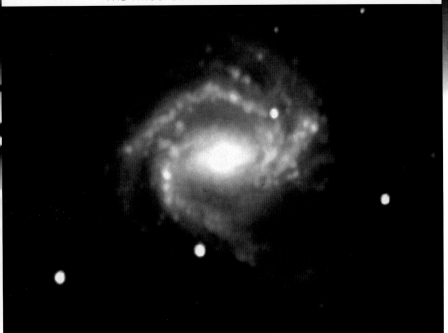

packed hundreds of times more tightly than in a normal elliptical galaxy. This galaxy is also known as Virgo A, because it is one of the earliest sources of radio waves discovered in the sky.

**16 M86 (NGC 4406)** ⊂◯◯ Located a little more than 1 degree northwest of M87 is M86, at magnitude 9.8. This slightly oval galaxy brightens a little toward its core. M86 may be as close to us as 20 million light-years, or as far away as 42 million. Despite the general expansion of the universe and unlike the majority of galaxies in the Virgo Cluster, which are moving away from us, M86 is moving toward us.

**M84 (NGC 4374)** ⊂◯◯ Within the same telescope-eyepiece view and lying directly west of M86 is another elliptical galaxy—M84. It may be located on the western edge of the core of the Virgo Cluster. At magnitude 9.9, M84 appears near-circular and, like M86, has a noticeable brightening toward its nucleus.

**17 R Virginis** ⊂◯◯ Swing your scope some 4½ degrees southeast of M86 and M84 toward the red giant Mira-type variable R Virginis. The magnitude of R Virginis ranges from 6.2 down to 12 over a period of about 145 days. It was discovered to be a variable star in 1809.

**18 M49 (NGC 4472)** ⊂◯◯ Our next stop is the giant elliptical galaxy M49, lying about 2 degrees northwest of R Virginis. It appears to us as a 9th magnitude oval with an almost even luminosity across most of its surface.

**19 M61 (NGC 4303)** ⊂◯ Moving on about 3 degrees southwest of M49, we come to the wide, yellow double star **17 Virginis**. About ½ degree farther south from 17 Virginis is the large, magnitude 9.7 face-on spiral galaxy M61. The arms and dust lanes of this galaxy are visible through a 10 inch (250 mm) telescope.

**20 3C 273** ⊂◯ Coming to the end of our journey through the Virgo Cluster is the really neat object, 3C 273, about 2½ degrees southeast of M61. This is the brightest known quasar and probably the most distant object visible in an 8 inch (200 mm) telescope. It appears as just one of many faint stars through a telescope eyepiece.

There are many more interesting galaxies to observe in the Virgo-Coma area, dozens more than we could possibly cover here, so, chart in hand, move on to enjoy discovering for yourself some of the other "island universes."

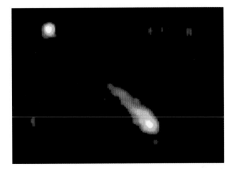

*The spiral galaxy M61 (top), in the Virgo Cluster, is nearer to Earth than 3C 273 (above). This radio image of the quasar, at top left, shows a jet of material, at center right, emerging from its core.* **433**

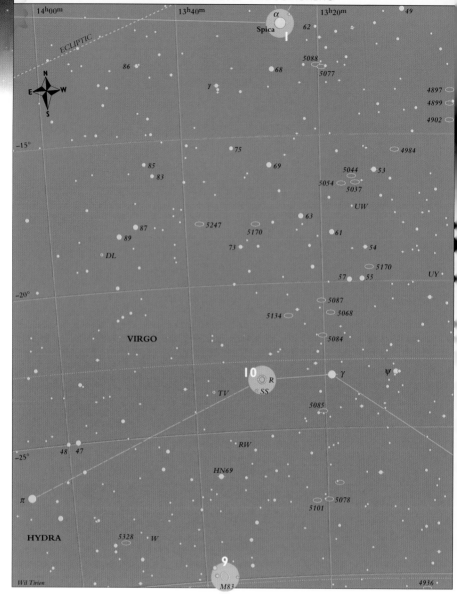

Wil Tirion

# A Stroll around the Sombrero

Although visible from both hemispheres, this part of the sky is in the southern half of the celestial sphere, so is a little better suited to Southern Hemisphere observers. However, fine views can be had in the evenings from April to June from most parts of the world. Some 30 degrees

*Corvus, the Crow, from Bayer's Uranometria (1639).*

south of the Virgo Cluster of Galaxies is the point where the Virgin meets the eastern end of the long constellation of Hydra, the Sea Serpent. Hydra and Virgo are the two largest constellations in the celestial sphere, but the third constellation on our chart for this starhop is one of the smallest—Corvus, the Crow. In mythology, Corvus was sent by Apollo to collect water with a cup. Instead, it returned with a watersnake. All three were banished to the sky, with the cup becoming the nearby constellation of Crater.

434

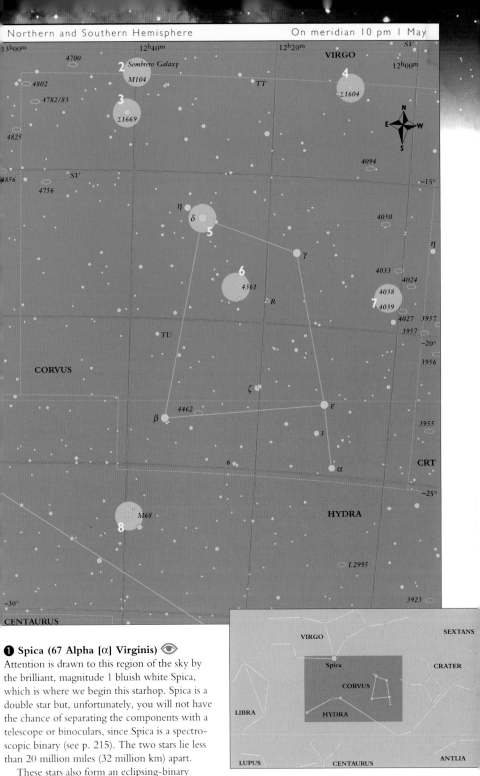

13h00m · 12h40m · 12h20m · **VIRGO** · 12h00m

4700
**2** *Sombrero Galaxy*
M104
4802
4782/83
**3**
Σ1669
4825
**4**
Σ1604
TT
4094
4856
SV
4756
δ
η
**5**
γ
4050
4033
4024
4038
**7** 4039
4027 3957
3957
6
4361
R
4094
TU
3956
CORVUS
ζ
4462
β
ε
3
α
CRT
6
M68
**8**
HYDRA
I.2995
CENTAURUS
3923

−15°
−20°
−25°
−30°

① **Spica (67 Alpha [α] Virginis)** 👁
Attention is drawn to this region of the sky by the brilliant, magnitude 1 bluish white Spica, which is where we begin this starhop. Spica is a double star but, unfortunately, you will not have the chance of separating the components with a telescope or binoculars, since Spica is a spectroscopic binary (see p. 215). The two stars lie less than 20 million miles (32 million km) apart.

These stars also form an eclipsing-binary system, in which one star slightly eclipses the other, resulting in small variations in the brightness of Spica. The two components revolve around each other every 4 days.

VIRGO · SEXTANS
Spica · CRATER
CORVUS
LIBRA
HYDRA
LUPUS · CENTAURUS · ANTLIA

Surprisingly, Spica does not form a part of the figure of Virgo. In some legends, Virgo becomes Demeter, the Corn Goddess, and Spica is depicted as an ear of wheat she is holding.

**435**

**❷ The Sombrero Galaxy (M104, NGC 4549)**
Aim your scope at Spica, before heading
11 degrees due west to find M104. This famous
8th magnitude spiral galaxy is easily visible,
appearing as a fuzzy patch of light through a
3 inch (75 mm) scope. With larger telescopes, it
is not difficult to see how it came by its common

*The Sombrero Galaxy (above) is the brightest of the Virgo galaxies,
with a distinctive dust band crossing its middle.*

name. The tightly coiled spiral arms result in a
strong resemblance to the brim of that famous
Mexican hat. Look for the prominent dark lane
that runs along the length of the galaxy.

**❸ Struve 1669 [Σ 1669]** Heading 1½ degrees to the south and just a little east of the
Sombrero, we come to Struve 1669, in Corvus.
This double star is listed in the catalog of Frederich
von Struve, a prolific discoverer of such objects
in the early 1800s. Its 6th magnitude yellowish
components are almost equal in brightness at
magnitudes 6 and 6.1, and, with a separation of
5 arcseconds, the object is a fine but fairly close
double for small scopes. A 10th magnitude star
can be seen about 1 arcminute from Struve 1669.

**❹ Struve 1604 [Σ 1604]** To find our
next target, Struve 1604, another of Struve's
doubles, first return to M104, then
shift your scope just over
7 degrees in a due westerly
direction. The brightest
component is of 7th
magnitude, with a 9th
magnitude attendant
9 arcseconds away.
There is another
9th magnitude star lying
some 19 arcseconds farther,
making the grouping an attractive sight for any small telescope.

## INTERACTING GALAXIES

As galaxies drift through space, there are times
when two or more of them can approach so
closely that they exert a strong gravitational pull
on each other, with the resulting tidal forces
causing interesting effects. The pair NGC 4038
and NGC 4039 is a superb example of this
(below). Named the Ring-Tail Galaxy because of
its shape, it is also known as the Antennae since
the material has been expelled in two directions
as a result of the interaction.

Many other examples of interacting
galaxies are known; for example,
in Canes Venatici, the galaxy
NGC 5195 appears to have
brushed past the Whirlpool
Galaxy (M51, NGC 5194).
Even our own Milky Way
Galaxy is suffering from
the tidal effects of an interacting system—the nearby
Magellanic Clouds have slightly
distorted its shape.

This patch of sky (above) shows Corvus at right, and Spica to the upper left. M83 (right) is a classic spiral galaxy lying in Hydra.

**❺ Algorab (7 Delta [δ] Corvi)** ⊂⟨◍⟩ Head almost 7 degrees southeast to the relatively bright star Delta Corvi. Shining at 3rd magnitude, Delta Corvi has a 9th magnitude companion that is fortunately 24 arcseconds away, making it easily visible in a 3 inch (75 mm) scope. Delta lies at the northeastern corner of the deformed rectangle that forms the main pattern in Corvus.

**❻ NGC 4361** ⊂⟨◍⟩ Try heading into the rectangle's interior, moving your scope southwest 2½ degrees, to find the 10th magnitude planetary nebula NGC 4361. This needs a 4 inch (100 mm) scope or larger to be seen well. It has a distinctly irregular disk shape and covers a total of about 45 x 110 arcseconds. NGC 4361 was discovered by William Herschel in 1785.

**❼ The Ring-Tail Galaxy (NGC4038/9)** ⊂⟨◍⟩ Cross northwest to **Gamma [γ] Corvi**. Use this magnitude 2.5 star as a midway point to help find NGC 4038/9 (see Box). Nudge your scope about 3½ degrees farther southwest from Gamma to find a patch of light in the field. These interacting galaxies shine at magnitude 10.7. The pair takes the form of the letter C. The best size scope for revealing detail is an 8 inch (200 mm).

**❽ M68 (NGC 4590)** ⊂⟨◍⟩ Moving on to **Beta [β] Corvi**, about 9 degrees southeast, head 3½ degrees farther south and slightly east to find M68, across the border into Hydra. This magnitude 8 globular cluster was discovered by

Messier in 1780. A 4 inch (100 mm) scope is necessary to begin to resolve the cluster into stars.

**❾ M83 (NGC 5236)** ⊂⟨◍⟩ Head 13 degrees southeast to locate M83, an 8th magnitude spiral galaxy—one of the easiest to see with a small scope and visible in 7 x 50 binoculars. A 6 inch (150 mm) scope reveals it as a fuzzy elliptical patch of light with a small, very bright core.

**❿ R Hydrae** ⊂⟨◍⟩ Moving 6½ degrees north and a little west we come to R Hydrae, just over 2 degrees east of **Gamma [γ] Hydrae**. This is one of the easiest long-period Mira-type variables to locate. It varies between magnitudes 4 and 10 over 390 days, making it fairly easy to see, even at minimum.

This starhop includes two of the best galaxies for large and small scopes—the Sombrero and M83—along with many other faint galaxies that seem to form a path between them.

**437**

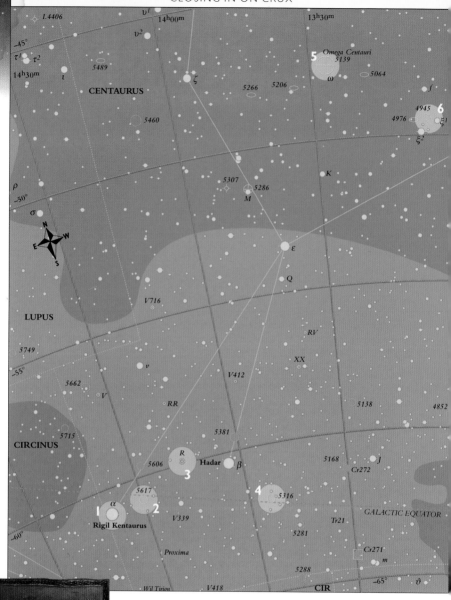

Wil Tirion

# Closing in on Crux

This part of the sky is far more accessible to observers in the Southern Hemisphere. However, those living south of about 20 degrees north can still catch a glimpse of these objects, low in the sky, during the evenings from May through July. Crux, the Southern Cross, and parts of nearby

*An Australian Aboriginal bark painting depicts the stars of the Southern Cross (a stingray) being chased by the Pointers (a shark).*

Centaurus, the Centaur, contain a number of interesting and visually stunning objects for one of our most southerly starhops.

In 1517, Andrea Corsali described a group of stars he saw in the southern sky as "a marveylous crosse in the myddest of fyve notable starres." Today, we know it as the Southern Cross, and although it is the smallest of all the constellations, it is a prominent group of stars. The Southern Cross is famous, among other things, for its appearance on the flags of several countries.

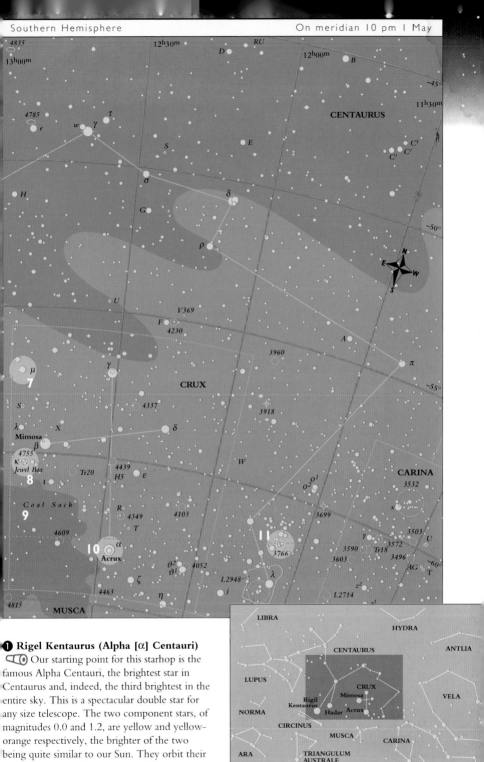

*(star chart labels)*

13h00m   4835   12h30m   RU   D   12h00m   B

−45°

11h30m

4785   e   w   γ   τ   **CENTAURUS**

S   E   C¹   C'

σ   C'   C²

H   G   δ

−50°

ρ

N

E   W

S

U   V369   F   4230   A   π

3960

−55°

μ   γ   **CRUX**

7   4337   3918

S   δ

λ   X   W   **CARINA**

**Mimosa**   3532

β   o¹

4755   o²

κ   **Jewel Box**   Tr20   4439   ε   3699

8   ι   H5   3503

*Coal Sack*   R   4349   4103   γ   3572   U

9   T   3590   Tr18   3496

4609   3766   3603   AG   −60°

**10**   α   T

**Acrux**   ϑ²   4052   **11**

ζ   ϑ¹   I.2948   λ   I.2714   z²

4463   η   j   z¹

4815   **MUSCA**

---

### ① Rigel Kentaurus (Alpha [α] Centauri)

Our starting point for this starhop is the famous Alpha Centauri, the brightest star in Centaurus and, indeed, the third brightest in the entire sky. This is a spectacular double star for any size telescope. The two component stars, of magnitudes 0.0 and 1.2, are yellow and yellow-orange respectively, the brighter of the two being quite similar to our Sun. They orbit their common center of gravity every 80.1 years, varying in apparent separation from about 2 to 22 arcseconds. Currently, they are 16 arcseconds apart. Just 4.4 light-years distant, Alpha Centauri is the nearest star system to the Sun.

*(locator map labels)*

LIBRA   HYDRA

CENTAURUS   ANTLIA

LUPUS   CRUX

Mimosa   VELA

Rigil   Kentaurus   Hadar   Acrux

NORMA   MUSCA   CARINA

CIRCINUS

ARA   TRIANGULUM   AUSTRALE

To make things really easy, the Southern Cross is clearly identified by the two Pointer stars, Alpha and **Hadar (Beta [β] Centauri)**, forming a line toward Crux in a westerly direction.   **439**

**❷ NGC 5617** ⊂◯◯ Move your scope just over 1 degree west of Alpha, in the direction of the blue giant Beta (β) Centauri, the other pointer, to locate NGC 5617. Shining at magnitude 6, this open cluster is visible with binoculars, and is also a good subject for small telescopes. A 4 inch (100 mm) scope on low power provides a good view of the cluster, which is about 10 arcminutes across.

**❸ R Centauri** ⊂◯◯ Aim your scope to a point two-thirds of the way from Alpha to Beta Centauri, and then shift its position ¹/₂ degree north, to locate R Centauri. This is a Mira-type variable star whose magnitude ranges between 5.3 and 11.8 over 546 days. At maximum, it can be seen with the unaided eye. A small scope will show the star's beautiful red color at its brightest.

**❹ NGC 5316** ⊂◯◯ Direct your scope toward Beta Centauri, before hopping 2 degrees southwest to NGC 5316. This open cluster looks similar to NGC 5617 in binoculars, appearing as a patch of light. A small scope provides a good view of its attractive field, with a 4 inch (100 mm) resolving quite a number of stars.

**❺ Omega [ω] Centauri (NGC 5139)** 👁 To find our next target, first follow a line about 7¹/₂ degrees from Beta to **Epsilon [ε] Centauri**. Extend it another 6¹/₂ degrees to locate Omega Centauri. This is the finest of all the globular clusters in the sky (see Box). It can be seen with

*Alpha Centauri (left) is one of the two stars pointing to Crux in the Milky Way (above).*

the unaided eye as a 4th magnitude spot of light, and stands out well in binoculars. A 3 inch (75 mm) scope begins to resolve it, and the view through an 8 inch (200 mm) is stunning.

**❻ NGC 4945** ⊂◯◯ From Omega, adjust your telescope's position just over 4 degrees to the southwest to locate NGC 4945. This magnitude 9 edge-on barred spiral galaxy appears as a slender streak of light about 20 arcminutes long, less than ¹/₂ degree east of the star **Xi¹ [ξ¹] Centauri**. A wide-field eyepiece will capture the galaxy, along with Xi¹ and the brighter **Xi² [ξ²] Centauri**, an easy double star. NGC 4945 is visible through a 4 inch (100 mm) scope, even though it has low surface brightness, but is better suited to larger telescopes.

**❼ Mu [μ] Crucis** ⊂◯◯ Swing your scope just over 7 degrees to the south and a little west, to close in on the star Mu Crucis, a showpiece wide double. Its bluish white 4th and 5th magnitude stars are separated by 35 arcseconds. Even a steadily held pair of 7 x 50 binoculars will separate them, but the view through a telescope is an opportunity not to be missed.

**❽ The Jewel Box (NGC 4755)** ⊂◯◯ Three degrees south of Mu is another showpiece object, the Jewel Box. A small scope will show many separate stars in the cluster. However, it is quite

*The Jewel Box (above) is justly famous for its attractive contrasting colors formed into a compact, A-shaped cluster.*

compact, so moderate magnification is a help. The cluster is shaped roughly like the letter A, with one of its brightest stars, a red supergiant, being very obvious in a field of contrasting colors.

**⑨ The Coal Sack** 👁 Continuing south a few more degrees, you will enter the south-eastern corner of Crux, where you should be able to see the Milky Way looking very dark. This is the area known as the Coal Sack. The most famous of the dark nebulas, the Coal Sack is a cloud of gas and dust, about 6 degrees across, which obscures the stars beyond it. Even in light-polluted skies, the Milky Way star field appears fairly barren in this area, but fine views of the Coal Sack can be obtained from dark, country sites where it stands out wonderfully.

**⑩ Acrux (Alpha [α] Crucis)** ⌐◯◯ Heading 2 degrees west of the Coal Sack, Alpha Crucis pops into view— the star at the foot of the Southern Cross. This is a celebrated double star, whose components, at magnitudes 1.4 and 1.9, lie 4.4 arcseconds apart and form a very long-period binary system. A 3 inch (75 mm) telescope with high magnification shows them well.

**⑪ NGC 3766** ⌐◯◯ Just over 6 degrees west of Alpha lies NGC 3766. In binoculars, this open cluster appears as an obvious fuzzy spot of light in a field very rich in stars. The view through a 4 inch (100 mm) scope is quite beautiful. Two of its brightest stars are a distinctive reddish orange.

There is little to outshine the view of this part of the night sky, especially on a moonless night away from bright city lights.

### A STAR ATTRACTION

Omega Centauri appeared as a star in Ptolemy's catalog nearly 2,000 years ago. We now know it as the most spectacular globular cluster in the entire sky, containing about a million stars of 11th magnitude and fainter. Omega Centauri's stars are packed so tightly that, near the center, they are typically only 1/10 of a light-year apart. At some 17,000 light-years away, it is also among the nearest globular clusters to Earth.

An interesting observation about Omega Centauri is that its stars vary significantly in their content of elements heavier than hydrogen and helium, suggesting that the stars differ in age and are not all about the same age, as was previously thought. The reasons for these variations remain unclear.

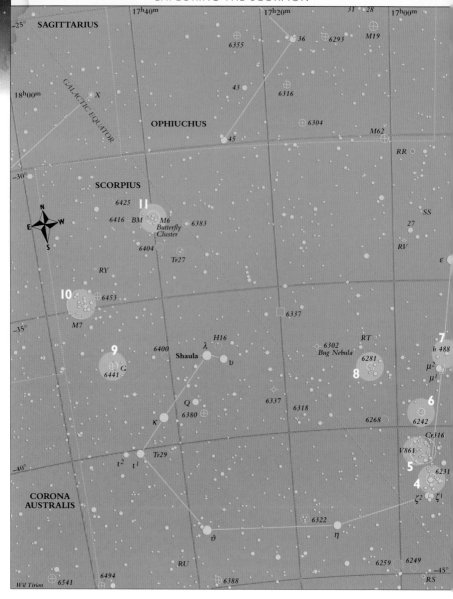

# Exploring the Scorpion

Scorpius is at its best when high in the sky during the evenings of June, July, and August. At this time, binocular and telescope observers are treated to a view of many fine celestial objects spotted throughout a very rich part of the Milky Way.

One of the ancient legends tells us that Orion, the Hunter, died after being stung by a scorpion, and that the two were turned into constellations on opposite sides of the sky. Orion must flee the scorpion for eternity, so as Scorpius rises in the east, Orion sets in the west.

For those less familiar with the night sky, Scorpius is one of the easiest constellations to locate. Simply look for the bright, reddish orange star Antares that marks the Scorpion's heart, and the distinctive curving shape of Scorpius's body.

*Scorpius, the Scorpion, as represented in the 1639 edition of Johann Bayer's illustrated star atlas, Uranometria.*

442

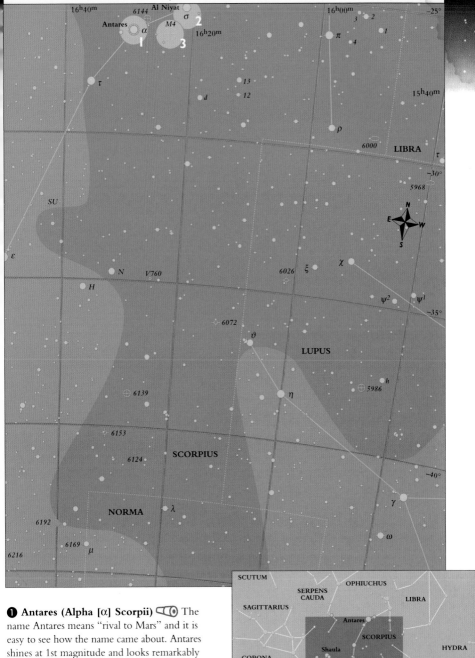

**❶ Antares (Alpha [α] Scorpii)** 🔭 The name Antares means "rival to Mars" and it is easy to see how the name came about. Antares shines at 1st magnitude and looks remarkably like a slightly fainter version of the planet Mars.

Antares is a red supergiant star at a very late stage of its evolution. It is immense, with a diameter close to 300 times that of our Sun and is enveloped in a haze of nebulosity (see p. 444). Antares is also a famous, but difficult to separate, double star. The magnitudes of the two components are 1.2 and 5.4, with the secondary often reported as appearing greenish. This is due to a contrast effect; it is actually a star of spectral class B, so that by itself it would appear bluish white. Antares's components lie almost 3 arcseconds apart, so you need at least a 6 inch (150 mm) telescope to separate them.

*A spectacular view of Scorpius (above). The nebulosity surrounding Antares (left).*

**❷ Al Niyat (20 Sigma [σ] Scorpii)** Move about 2 degrees northwest to look at 20 Sigma Scorpii, an easy but unequal double star, whose 3rd and 8th magnitude components are 20 arcseconds apart. They are easy to separate in a 3 inch (75 mm) telescope.

**❸ M4 (NGC 6121)** Shift 1 degree south and a little back toward Antares to bring M4 into view. This is one of the great globular clusters in the sky. Easily seen in binoculars, it is nearly ½ degree in diameter. A 4 inch (100 mm) scope begins to resolve individual stars in the cluster, although larger telescopes will reveal many more stars.

**❹ NGC 6231** The best way to exit the Antares area is by running southeast along the scorpion's body until you reach the bend in its tail, where NGC 6231 is located, some 17 degrees from M4. It contains a number of bluish white supergiant stars. The cluster covers about ¼ degree, but its brightest members form a more compact group. This is a fine open cluster for small scopes, being clearly visible to the naked eye as a bright spot of light just north of the star pair **Zeta¹·² [ζ¹, ζ²] Scorpii.** Zeta¹, an orange star, and bluish white Zeta² are not a physical pair but are worth a quick look for their contrasting colors.

**❺ v861 Scorpii** From NGC 6231 adjust your scope 1 degree north and about ½ degree east to encounter v861 Scorpii. Lying within the scattered 1 degree wide cluster **Collinder 316**, v861 Scorpii is a bluish star that varies in magnitude between 6.1 and 6.7 every 7.8 days. In mid-1978, the star was found to be a source of X-rays, and is now a well-known X-ray binary variable.

**❻ NGC 6242** Nudge your scope just under 1½ degrees north and a little west through this very rich field to find NGC 6242. This open cluster is a pleasing binocular target in addition to being a rewarding sight through a 3 inch (75 mm) telescope. Its brightest star, near the southern end of the cluster, is an attractive orange-red color.

**❼ h 4889** One and a half degrees northwest of NGC 6242 is the brilliant, wide pair **Mu¹·² [μ¹, μ²] Scorpii.** Another ½ degree farther is h 4889. This is a double whose 6th and 8th magnitude stars lie 7 arcseconds apart. Mu¹ and Mu² themselves are a fine naked-eye pair of stars that may be physically associated. In Polynesian legend, they were called "the inseparable ones."

**❽ NGC 6281** Sweep your telescope about 3 degrees east for a look at NGC 6281. Easily visible in binoculars, this is a 5th magnitude open cluster similar in size to NGC 6242,

*A number of impressive globular clusters (above), including M4, lie within the field around brilliant Antares, shown center left.*

but containing fewer stars. It is also well worth a look in any size telescope.

**9 NGC 6441** Hop just over 9 degrees east, past **Lambda [λ] Scorpii**, in search of NGC 6441. Although by no means bright, this globular cluster is quite easy to locate, being only a few arcminutes east of the yellow-orange 3rd magnitude star **G Scorpii**. The cluster requires quite a large scope to resolve its stars. A 10 inch (250 mm) scope reveals its granular appearance.

**10 M7 (NGC 6475)** Shift 2 degrees to the north and a little east to find M7. Easily seen as a 1 degree wide, hazy patch of light with the naked eye, M7 is one of the best open clusters for binoculars, which reveal many of its stars. A wide-field eyepiece is essential for a good view through a telescope.

**11 The Butterfly Cluster (M6, NGC 6405)** Four degrees northwest of M7 we come to another gem, the open cluster M6. It is just possible to see it with the naked eye, but binoculars or a small telescope provide a far better view. The view through a telescope reveals its main stars arranged in a pattern resembling the body and wings of a butterfly, hence its common name.

**BM Scorpii** Near the eastern end of M6, and marking the trailing end of the butterfly's left wing, lies the orange semi-regular variable star BM Scorpii. It varies between magnitude 6.8 and 8.7 over a period of about 850 days.

The grouping of stars and star clusters around Scorpius in a dark sky is an unforgettable sight. It is easy to lose track of time while sweeping your binoculars through this fascinating area.

## M7, AN ANCIENT OPEN CLUSTER

M7 is the southernmost object in Charles Messier's famous catalog of non-stellar objects. It is also one of the brightest. However, Messier observed it from Paris, where its maximum altitude of only about 6 degrees would have meant that he could not see it in its true splendor. The cluster contains about 80 stars scattered over more than 1 degree of sky. Studies of M7 show that it is about 220 million years old. This compares with only about 50 million years for its nearby companion M6 and 70 million years for the Pleiades in Taurus. There are few open clusters much older than M7, because as clusters age, they are dispersed by gravitational effects and encounters with interstellar clouds.

445

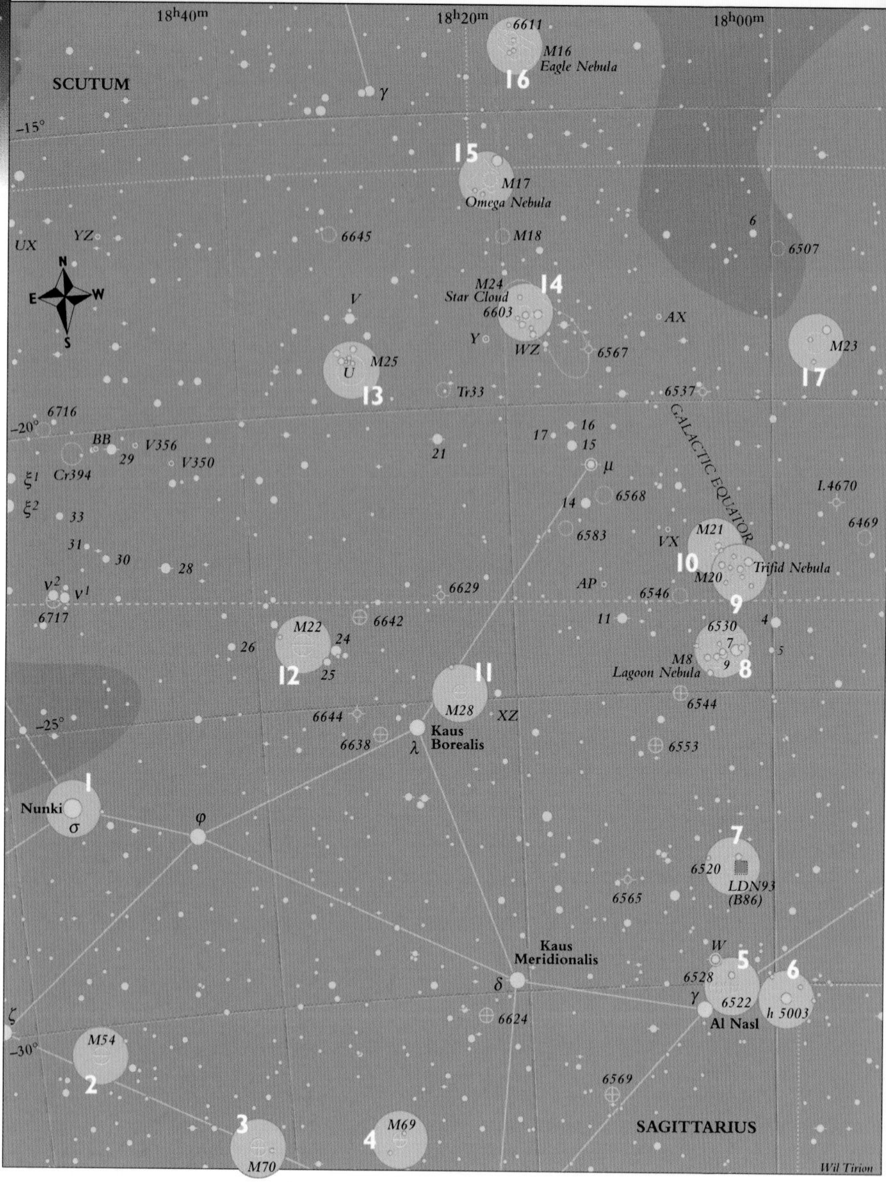

SCUTUM

SAGITTARIUS

# Toward the Heart of Our Galaxy

Being at a moderate southerly declination, this region passes overhead for those living in the Southern Hemisphere. However, northern skywatchers can spot many of the objects in this patch of sky during the evenings from late June to September.

*Sagittarius, the Archer, from an ancient Arabic manuscript.*

There is no finer naked-eye view of the night sky than one that includes the constellation of Sagittarius, the Archer, high above the horizon, for it is in this direction that we look toward the center of the Milky Way Galaxy. The region just east of the tail of Scorpius, the Scorpion, is especially rich in interesting clusters and nebulas. In adjacent Ophiuchus, we find some of the famous dark, obscuring dust clouds that stand out so well on clear nights for observers away from city light pollution.

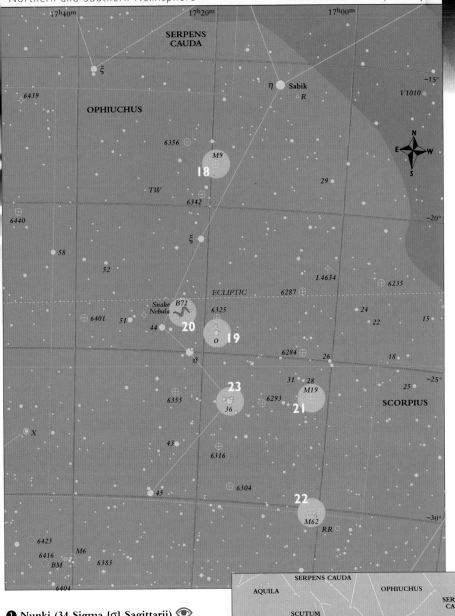

**❶ Nunki (34 Sigma [σ] Sagittarii)** 👁

While the brightest stars of Sagittarius have the distinctive appearance of a teapot, a familiar part of the asterism is the shape of the so-called Milk Dipper—the handle and part of the lid of the teapot. The Milk Dipper is formed by the stars **Zeta [ζ]**, Tau [τ], Sigma [σ], **Phi [φ]**, and **Lambda [λ] Sagittarii**. Shining at magnitude 2, the blue-white star Sigma forms one end of the base of the dipper and, surprisingly, is the second brightest star in Sagittarius.

Sigma's common name, Nunki, is thought to be derived from the Babylonian *Tablet of the Thirty Stars*. It is said to represent "the star of the proclamation of the sea," the "sea" being the quarter of sky occupied by the water constellations, including nearby Capricornus and Aquarius (see p. 460).

447

**❷ M54 (NGC 6715)** ◁◯◯
Center your telescope on Nunki
and head about 4 degrees due
south to M54, an 8th magnitude
globular cluster that seems to
have fallen out of the dipper's
bowl. It can be seen in binoculars, but is very difficult to
resolve. Even a 12 inch
(300 mm) telescope shows it
only as somewhat granular.

**❸ M70 (NGC 6681)** ◁◯◯
About 3 degrees southwest lies
another globular cluster, M70.
This is also an 8th magnitude
object, but it can be resolved a
little more easily than M54, with a 6 inch
(150 mm) scope showing some of its stars. M70
became famous in 1995 when astronomers Alan
Hale and Thomas Bopp, observing the cluster
from different locations, discovered Comet
Hale-Bopp (see p. 195), the Great Comet of
1997, in the same telescope field as the cluster.

**❹ M69 (NGC 6637)** ◁◯◯ The globular
cluster M69 lies 2½ degrees due west of M70
and appears virtually as its twin. It requires a
6 inch (150 mm) telescope to begin to resolve
it properly into individual stars. Unlike the
previous two globulars, which were discovered
by Charles Messier, M69 was first seen by
astronomer Nicolas Louis de Lacaille in 1752.

**❺ NGC 6522** ◯◯ Almost 6½ degrees north-
**448** west of M69 lies **10 Gamma [γ] Sagittarii**, a

*Sagittarius has many fine targets like the
globular cluster NGC 6522 (above), and
Barnard 86 and NGC 6520 (both left).*

bright star. Move across to it,
and place brilliant Gamma in
the southeastern corner of a
wide-field view. You should
also be able to see the 9th
magnitude globular cluster
NGC 6522. Some of the
cluster's stars are visible through
an 8 inch (200 mm) telescope.

**NGC 6528** ◯◯ The magnitude 10 globular cluster,
NGC 6528, is too difficult to
resolve even with a 12 inch (300 mm) telescope.

**❻ h 5003** ◯◯ Hop 1 degree west of our last
stop to h 5003. This delightful double star is a
great object for any size scope. A 3 inch (75 mm)
telescope, or even smaller, will show this orange
and yellow pair of stars, at magnitudes 5 and 7,
separated by 5.5 arcseconds.

**❼ NGC 6520** ◯◯◯ Return to NGC 6522
and NGC 6528, then shift 2 degrees due north to
locate NGC 6520. This compact open cluster,
visible in binoculars, is a pleasing sight in a 3 inch
(75 mm) scope or larger. Between NGC 6520
and a 7th magnitude star a few arcminutes northwest lies the dark nebula **Barnard 86**, also
known as LDN 93. A 4 inch (150 mm) telescope should reveal the nebula as an apparent
gap in the star field around it.

### 8 The Lagoon Nebula (M8, NGC 6523)

Ease your scope 3½ degrees north to find the wonderful Lagoon Nebula. One of the few bright nebulas visible to the unaided eye, this can be seen as a milky-white spot of light against the backdrop of the Milky Way. Although it is one of the finest nebulas for a small telescope, binoculars also provide a superb view.

### 9 The Trifid Nebula (M20, NGC 6514)

Try your hand at the Trifid Nebula, 1½ degrees north and just a little west of M8. This famous nebula is much fainter than the Lagoon Nebula and is best observed in a dark sky. There is always a feeling of delight to see the three dark lanes that give the Trifid Nebula its prominent shape. These dust lanes are visible in a 6 inch (150 mm) scope. It is helpful to use averted vision when viewing this nebula.

### 10 M21 (NGC 6531)

Nudge your scope just over ½ degree northeast of the Trifid to spot M21. Covering almost ¼ degree, M21 is easily resolved in a 3 inch (75 mm) telescope, with its brightest stars being of 8th magnitude.

### 11 M28 (NGC 6626)

Moving on about 6 degrees southeast, hop to the star Lambda [λ] Sagittarii, then backtrack 1 degree to pinpoint M28. Although visible in binoculars, it takes a 6 inch (150 mm) scope to begin to resolve the stars of this globular cluster. In smaller telescopes, M28 appears as a round, fuzzy glow.

### 12 M22 (NGC 6656)

About 3 degrees northeast of M28 lies one of the few globular clusters clearly visible with the unaided eye—M22, a spectacular object at magnitude 5.1. A 3 inch (75 mm) telescope will show some of its brightest members, and the view through a 6 inch (150 mm) is unforgettable. You may be able to detect its slightly elliptical shape. The cluster has a diameter of more than 20 arcminutes and, at a distance of only about 10,000 light-years, is one of the closest of its type to us.

*M20, the Trifid Nebula (above), lies at center bottom of the view of the Milky Way (right) which includes M17 at the top.*

449

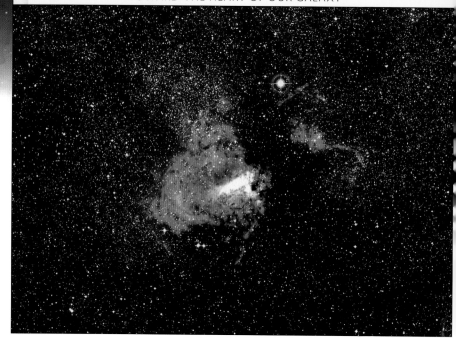

*M17, the Omega Nebula (above) lies on the border of Sagittarius.*

**⑬ M25 (IC 4725)** ⊂◯◯ Move 5 degrees north and 1 degree west of M22 to take a look at M25. This scattered open cluster, about ½ degree across, is an easy object for binoculars and small telescopes. Near the center of M25 is the yellowish Cepheid variable **U Sagittarii**, which varies from magnitude 6.3 to 7.1 over 6.75 days.

### THE STELLAR NURSERY IN M16

In 1995, the Hubble Space Telescope returned an eerie picture of the nebulosity of M16, showing huge columns of gas that are several light-years in length (below). A close look at the surfaces of the giant pillars shows what scientists have called evaporating gaseous globules, or EGGS. Ultraviolet light from nearby stars is stripping some of the gas from the columns and exposing the EGGS, in which stars are forming. We see the globules because they are more dense than the rest of the nebula, and so are not "blown away" as easily. However, the gradual loss of material feeding the developing stars is thought to play an important role in limiting their size. We can only wonder how many more stars are forming deep within these huge columns, gathering material in private and becoming larger and larger.

**⑭ M24** 👁 To locate the next object on our hop, scan your eye 3 degrees west of M25. M24 is unique among Messier objects because it is a huge star cloud about 1 degree wide and 2 degrees long. Sweeping your scope through it will reveal a very rich region, including the open cluster **NGC 6603** near the northeastern end. The cluster is visible with a 4 inch (100 mm) scope, but can be resolved in larger telescopes.

**⑮ The Omega Nebula (the Swan Nebula, M17, NGC 6618)** ⊂◯◯ If you look another 2½ degrees farther north and a little east, M17 comes into view. This bright nebula is easy to see in binoculars, appearing as a small streak of light. In a 4 inch (100 mm) scope, its unusual shape looks a bit like the figure 2 with an extended baseline. Because of this pattern, M17 is called both the Swan Nebula and the Omega Nebula.

**⑯ The Eagle Nebula (M16, NGC 6611)** ⊂◯◯ Across the border into Serpens, about 2½ degrees north of M17, the combination of hazy star cluster and nebula, M16, appears. The cluster is obvious enough with binoculars, but you need a 6 inch (150 mm) scope and a dark sky to clearly see that it is actually immersed in a region of nebulosity called the Eagle Nebula (see Box). In larger scopes, the whole field is beautiful.

**⑰ M23 (NGC 6494)** ⊂◯◯ Head 7½ degrees southwest of M16, back into Sagittarius, for a view of M23. This open cluster is nearly ½ degree across. It is a fine sight in a 4 inch (100 mm)

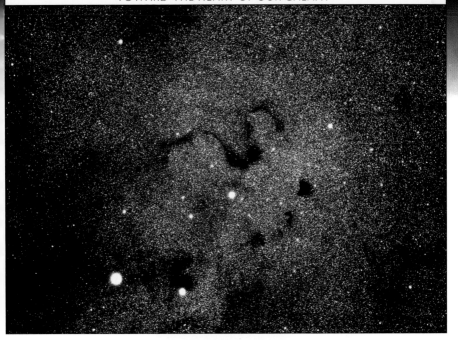

scope, which resolves quite a few stars. The straight and curved lines of stars within the cluster provide a delightful detail.

**⑱ M9 (NGC 6333)** ⊂⊃ Shift your scope 9 degrees farther west to find 8th magnitude M9, a globular cluster about 2½ degrees north of the star **Xi [ξ] Ophiuchi**. It appears fairly condensed and is visible in binoculars, although a 6 inch (150 mm) scope is needed to see some of its multitude of stars clearly.

**⑲ 39 Omicron [o] Ophiuchi** ⊂⊃ About 3 degrees south of Xi lies a small triangle of stars, the westernmost and faintest of the three being our next target. A beautiful double star for any small telescope, Omicron's colorful 5th and 7th magnitude orange and yellowish components are an easy 10 arcseconds apart, so should be no problem to visually separate with most telescopes.

**⑳ The Snake Nebula (Barnard 72)** ⊂⊃ The next challenge lies about 1½ degrees northeast of Omicron. A familiar sight in photographs of the Ophiuchus Milky Way, the dark Snake Nebula takes the form of the letter S and is about ½ degree from end to end. It is not exactly an easy object to discern, but if you use moderate aperture and a low magnification, you might be able to find it in a dark sky.

**㉑ M19 (NGC 6273)** ⊂⊃ Aiming your scope just over 5 degrees southwest of the Snake Nebula, you should be able to spy the globular

*Dark dust lanes create different and dramatic effects in B72, the Snake Nebula (above), and M16, the Eagle Nebula (left).*

cluster M19. At magnitude 7, M19 is an object that is easy to see with binoculars, appearing as a round, fuzzy spot. A 4 inch (100 mm) telescope will begin to show a granular appearance.

**㉒ M62 (NGC 6266)** ⊂⊃ Our next stop, the globular cluster M62, is about 4 degrees due south of M19. A little brighter but somewhat more difficult to resolve than M19, M62 appears as a more uniform haze of stars requiring an 8 inch (200 mm) telescope for a good view. However, a number of individual stars can just be resolved when using a 6 inch (150 mm) telescope or larger.

**㉓ 36 Ophiuchi** ⊂⊃ Returning to the northeast, two-thirds of the way to **Theta [θ] Ophiuchi** is the double star 36 Ophiuchi. Shining like a pair of distant orange headlights, the 5th magnitude stars of 36 Ophiuchi are equal in brightness and color, separated by about 5 arcseconds. The two form a binary system, moving around each other about every 550 years.

A dark night, away from city lights, and a pair of dark-adapted eyes are the perfect formula for gazing in the direction of our galaxy's center. With its superb collection of objects and delicate dust lanes, many would say that it is the most interesting part of the night sky.

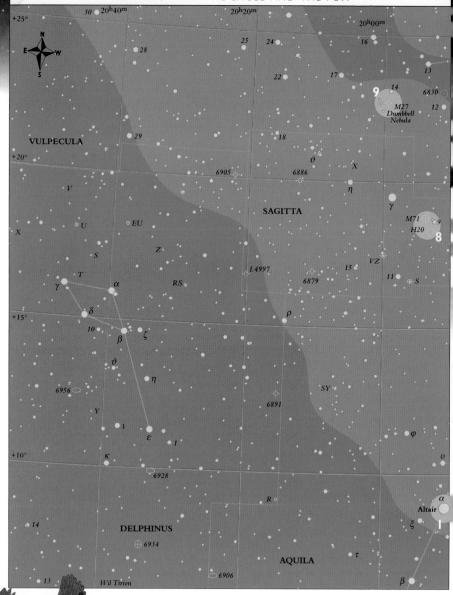

+25°   30  20h40m          20h20m                    20h00m

N
E—W
S

28

25   24                    16

22          17                    13

14           6830

9              M27        12
             Dunbbell
             Nebula

VULPECULA          29

18

+20°                                       ϑ          X

6905        6886

η

SAGITTA

V                                           γ

M71      9
H20
X        U          EU                               8

S            Z                                  VZ

T                                           15          11      S

γ         α       RS         I.4997                 6879

δ                                    ρ

+15°  10                           β  ζ

ϑ

η

6956                                   SY

6891

Y

ι

ε       1

φ

+10°   κ

o

6928

R                              α
                                            Altair
14                                       ξ

DELPHINUS
⊕ 6934                                   τ

AQUILA

13      Wil Tirion           6906                β

# Hunting with the Eagle and the Fox

This part of the night sky is visible to skywatchers in both hemispheres from July through September. The starhop zooms in on parts of the constellations of Aquila, Vulpecula, and Delphinus, the Dolphin, and includes the third smallest constellation of all—Sagitta, the Arrow. Aquila's name

is derived from the Arabic *Al Nasr al Tair*, meaning the Flying Eagle. Today, it is simply known as the Eagle, while Vulpecula is generally referred to as the Fox.

This region features rich fields containing thousands of sparkling bright stars, and coal-black dark nebulas. Wide swaths of dust clouds, known as the Great Rift, divide the Milky Way into two parallel bands of stars through the center of this starhop, as they do through nearby Cygnus, the Swan (see p. 456).

*In Bayer's seventeenth-century star atlas,* Uranometria, *he depicted Aquila as a bird belonging to the Greek God Zeus.*

452

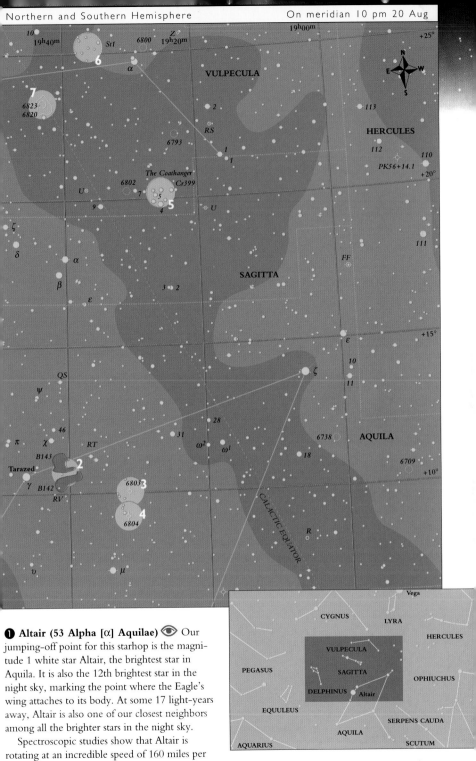

**19h00m**

+25°

10
19h40m
St1
6800
Z
19h20m

6
α

**VULPECULA**

7
6823
6820

2

113

**HERCULES**

RS

6793

112

110

PK56+14.1

+20°

*The Coathanger*

6802
U
7
Cr399

9
5
U

4

δ
α
ζ

111

FF

β

ε

**SAGITTA**

+15°

3 2

ε

+15°

QS

ζ

ψ

10

11

46

28

π
χ
RT
31
ω²
ω¹
6738
**AQUILA**

B143
2
18

Tarazed

6709

γ
B142

+10°

RV

6803

4

6804

*CALACTIC EQUATOR*

R

υ
μ

**❶ Altair (53 Alpha [α] Aquilae)** 👁 Our
jumping-off point for this starhop is the magnitude 1 white star Altair, the brightest star in
Aquila. It is also the 12th brightest star in the
night sky, marking the point where the Eagle's
wing attaches to its body. At some 17 light-years
away, Altair is also one of our closest neighbors
among all the brighter stars in the night sky.

Spectroscopic studies show that Altair is
rotating at an incredible speed of 160 miles per
second (258 km/s), a truly remarkable feature.
One Altair "day" is completed in approximately
6½ hours. By comparison, the Sun takes more
than 25 Earth days to complete one rotation.

Vega

CYGNUS
LYRA

HERCULES

VULPECULA

PEGASUS
SAGITTA

OPHIUCHUS

DELPHINUS
Altair

EQUULEUS

SERPENS CAUDA

AQUARIUS
AQUILA

SCUTUM

To find the next object on our starhop, first
nudge your scope about 2 degrees northwest of
Altair, to **Tarazed (Gamma [γ] Aquilae)**, then
hop 1½ degrees west to the Double Dark Nebula. **453**

**❷ The Double Dark Nebula
(B142 & B143)** 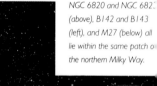 The U-
shaped nebula B143 and the
elongated oval nebula B142 are
made up of clouds of non-
luminous gas and dust and are
fine examples of dark nebulas.
These clouds completely blot
out the grainy star fields that
are thought to be located some
5,000 light-years beyond. An 8 inch (200 mm)
scope or larger and a dark sky are usually needed
to study the edges of these clouds in detail.

**❸ NGC 6803** About 2 degrees west and
a little south of the Double Dark Nebula is a
fine planetary nebula, NGC 6803, glowing at
magnitude 11. At magnitude 14, its central star is
visible in scopes larger than 16 inches (400 mm).
NGC 6803 is only 4 arcseconds in diameter, so it
is essential to use high power when observing it.

**❹ NGC 6804** Hop 1 degree south to
NGC 6804, another planetary nebula, making a
fine pair with NGC 6803. At 60 arcseconds in
diameter, NGC 6804 is an easier target to locate
than its close neighbor. It is dimmer overall, at
magnitude 13, but its central star is a little brighter.

Both NGC 6803 and NGC 6804 look like
smaller and much fainter versions of the Ring
Nebula (M57) in Lyra (see p. 458).

**❺ The Coathanger (Collinder 399)**
Sweep some 10 degrees north of NGC 6804 to
the spectacular open star cluster known as the
Coathanger. It includes about 40 stars forming

*NGC 6820 and NGC 682[?]
(above), B142 and B143
(left), and M27 (below) all
lie within the same patch o[f]
the northern Milky Way.*

an arc of 6 bright
stars running in an
east–west direction,
with a hook on the
southern side. The
Coathanger is best
viewed with binocu-
lars or a finderscope, because you can more
easily orient yourself within a wider field. The
magnitude 5.1 red giant at the bottom of the
hook is the multiple star **4 Vulpeculae**.

**❻ Stock 1** Hop 5 degrees northeast to
the large, very loose open star cluster Stock 1. It
contains about 40 stars that are scattered amor-
phously, making it a challenge to resolve individual stars from the surrounding Milky Way
star field. Stock 1 is about 1 degree in diameter,
so observe it with binoculars or a finderscope,
instead of with a telescope eyepiece.

**❼ NGC 6823** Swing your scope about
5 degrees southeast of Stock 1 to the open star
cluster NGC 6823, surrounded by the elusive
glow of the faint emission nebula **NGC 6820**.

M71 (above), in the constellation of Sagittae, is most often classified as a loosely packed globular cluster.

The 30-member cluster itself, with a combined magnitude of 7.1, is an easy binocular target.

**❽ M71 (NGC 6838)** Locate the 3rd magnitude stars **7 Delta [δ] Sagittae** and **12 Gamma [γ] Sagittae**, both red giant stars, about 4 degrees east of NGC 6823, to help find M71, lying midway between them. This compact cluster glows at magnitude 8.3, with many faint stars forming a triangular shape. Most astronomers consider M71 to be a loose globular rather than a compact open cluster because it contains red giant stars. However, no short-period variables, which are common to all globular clusters, have ever been detected in M71, leaving its true nature somewhat in doubt.

**Harvard 20** Only 30 arc-minutes southwest is the faint, sparse open cluster Harvard 20. An elongated box-like formation of 15 members, Harvard 20 is a challenge to pick out from the rich star field behind it.

**❾ The Dumbbell Nebula (NGC 6853, M27)** Return to Gamma Sagittae before hopping about 4 degrees north to the magnitude 7.6 Dumbbell Nebula. Considered by many to be one of the finest planetary nebulas, it is also

one of the closest to us at 1,000 light-years away. This softly glowing, greenish, hourglass-shaped planetary nebula is best viewed with a telescope, although, at 8 arcminutes across, it is also worth a look with binoculars or a finderscope.

There is so much more to see in this part of the night sky, so plan to return and slowly scan the wonderful Milky Way star fields and dark nebulas, again and again.

## PLANETARY NEBULAS

Seen with early scopes, planetary nebulas were described as somewhat like the disks of planets, when actually, they are the last stage of a star's life. When a star like the Sun gets old, it swells to become a red giant, soon running out of hydrogen and helium to burn in its core. Instead, it continues its nuclear reactions in shells around the core. As it does this, the star begins to pulsate, culminating in a shedding of its outer layers, which form a planetary nebula. This process reveals the hot central core which, in turn, pours ultraviolet radiation out into the expanding planetary shell, causing it to glow. This ghost-like shell survives for only about 50,000 years before it dissipates, leaving the core behind as a slowly cooling white dwarf star.

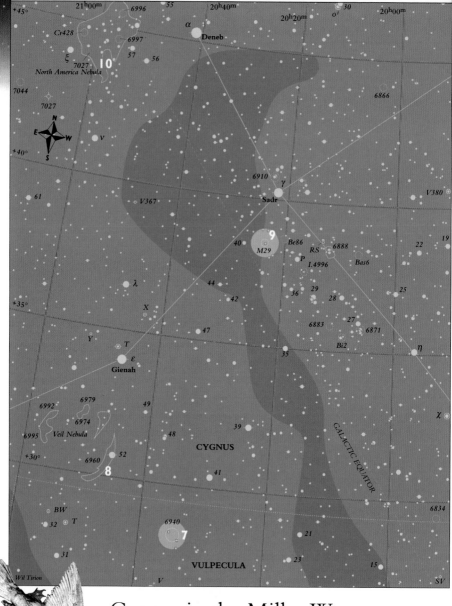

+45°  21ʰ00m    6996    55    20ʰ40m    20ʰ20m    o¹  30    20ʰ00m
Cr428
                6997
    ξ              57    56
      7027
North America Nebula  10
7044                                                        6866
    7027
            N
       W    E
            S
+40°        V

    61        V367            6910  γ                        V380
                              Sadr
                        40         9
                              M29   Be86  6888          22    19
                                        RS
                                    P  I.4996    Bas6
        λ          44              36   29   28        25
                        42
+35°        X                                27
        Y      T              47        6883      6871
           ε                            Bi2            η
        Gienah                   35

    6992   6979        49
        6974                                                  χ
6995   Veil Nebula                        39
                              48
+30°        6960        52            CYGNUS
        8                        41

        BW                    6940
    32  T                              21
    31                                              6834
                    VULPECULA      23        15
Wil Tirion            V                                SV

# Cygnus in the Milky Way

For mid-northern observers, the two constellations traversing this region of sky hover near the zenith on July and August nights, making these the best months for viewing. This particular part of the Milky Way is rich with a variety of celestial objects in the constellations of Cygnus, the Swan, and Lyra, the Lyre.

*This image of Cygnus, the Swan, was painted as part of a larger fresco on a ceiling of the Villa Farnese in Caprarola, Italy, in 1575.*

Cygnus forms a prominent cross shape that straddles the Milky Way at the point where the galaxy is split by a dark dust lane. It is also known as the Northern Cross—the Northern Hemisphere's answer to Crux, the Southern Cross (see p. 438). The constellation of Lyra appears as a small box lying to the west of Cygnus.

Vega in Lyra, Deneb in Cygnus, and Altair in Aquila (south of this area), are the three 1st magnitude stars that form what is known in the Northern Hemisphere as the Summer Triangle.

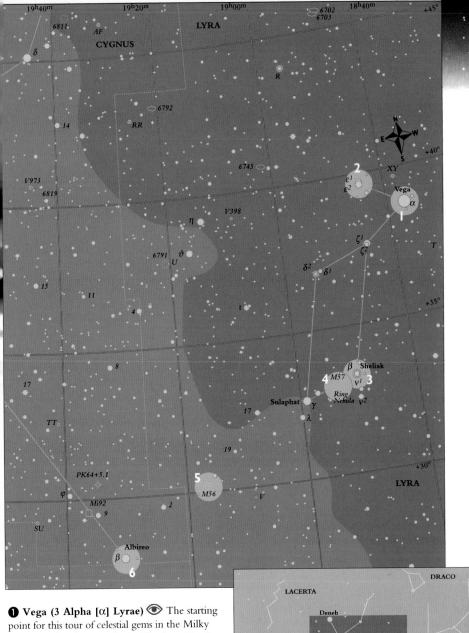

19h40m        19h20m         19h00m          18h40m

6702
6703

LYRA

6811    AF

CYGNUS

δ

R

6792

14      RR

6745

V973            V398

6819            η

6791  ϑ
U

15      11

4

8

17
TT

PK64+5.1
φ      Mi92
9
SU           2

Albireo
β

6

2       XY
ε¹
ε²           Vega
α
1

ζ¹
ζ²

δ²  δ¹

ι                                T

β  Sheliak
4  M57
ν¹  3
Ring
Sulaphat  γ  Nebula  ν²
17       λ

19

5
M56
V

LYRA

+45°

N
E   W
S

+40°

+35°

+30°

**❶ Vega (3 Alpha [α] Lyrae)** 👁 The starting point for this tour of celestial gems in the Milky Way is the brightest star in the Northern Hemisphere's summer sky—Vega. This bright, blue-white star is very close to zero magnitude, being listed at magnitude 0.03. Vega is the fifth brightest star in the entire sky, and is estimated to be intrinsically about 58 times brighter than the Sun. It is 25 light-years away from Earth.

Vega appears in many ancient legends. Its common name comes from the Arabic *Al Nasar al Waki*, meaning "the swooping eagle." In one Greek legend, it symbolizes the lyre of Hermes, so it is sometimes referred to as "the harp star."

DRACO
LACERTA
Deneb
Vega
LYRA
CYGNUS
HERCU
PEGASUS       VULPECULA
SAGITTA
OPHIUCHUS
DELPHINUS        AQUILA

Because Vega's magnitude does not vary significantly, the star is often used by astronomers when calibrating photometric equipment that is used to measure stellar brightness.

**❷ The Double Double Star
(4 Epsilon¹ & 5 Epsilon² [ε¹, ε²]
Lyrae)** The famous
"double double" star, lying about
1½ degrees northeast of Vega,
consists of four white stars. The
two primary stars are 208 arcseconds apart, and each is itself a
binary star. The magnitude 6
companion of Epsilon¹ is 2.8 arcseconds from its magnitude 5.1 primary. The
Epsilon² system is composed of two 5th magnitude stars lying 2.6 arcseconds apart.

**❸ Sheliak (10 Beta [β] Lyrae)** Hop
about 6 degrees south to Sheliak, a bright
eclipsing binary variable, whose magnitude fluctuates from 3.3 down to 4.4. The two unequally
bright stars in this system orbit about a common
center of gravity over a period of 12.9 days. The
stars seem to be constantly transferring massive
amounts of gas between their atmospheres.
Being only about 22 million miles (35 million km)
apart, the components of Sheliak have never
been visually separated.

**❹ The Ring Nebula (M57, NCG 6720)**
Located about midway between Sheliak
and **Sulaphat (14 Gamma [γ] Lyrae)** is the
famous bright planetary nebula M57. It is often
referred to as the "smoke ring" or "doughnut"
of the sky. M57 is the result of a star that blew
off a massive amount of gas thousands of years
ago. The shell of gas is easy to see through an
8 inch (200 mm) scope and has a slightly

*The Cross shape of Cygnus (above) with
Vega at center right. The Ring Nebula (left).*

greenish cast to it. From a dark
site, with a 15 inch (380 mm)
scope, you can glimpse the
magnitude 17 dwarf star from
which the gas was expelled.

**❺ M56 (NGC 6779)**
Some 4½ degrees southeast
from Sulaphat is the globular cluster M56. This
small, circular 8th magnitude mass of stars was
discovered by Messier in 1779. Many faint stars
around the edges of the cluster can be resolved
with scopes smaller than 8 inches (200 mm).

**❻ Albireo (6 Beta [β] Cygni)** Continue
southeast about 3½ degrees from M56 to Albireo.
This is probably the finest color-contrast double
star and a real showpiece at any star party. This
magnitude 3.1 star appears bright yellow or
topaz with a sapphire-blue companion shining
at magnitude 5.1. The stars are 34.3 arcseconds
apart, making them an easy pair to separate in any
size telescope. While near Albireo, scan the rich
star clouds with binoculars, looking for the
myriad clusters and dark nebulas.

**❼ NGC 6940** From Albireo shift about
14 degrees east to the open cluster NGC 6940,
across the border in Vulpecula. About 60 to
80 stars make up this large, scattered cluster.
Nine bright stars appear centered within the
cluster, while other stars seem to form chain-like
strings just north of the center of the cluster.

*The North American Nebula (above) is a mix of bright gas and dark dust lanes. The Pelican Nebula appears to its right (west).*

**❽ The Veil Nebula (NGC 6960)** ⬤⬤ Hop about 6 degrees northeast of NGC 6940 to magnitude 2.5 **Epsilon [ε] Cygni**. Three degrees due south of Epsilon is the magnitude 4.2 star **52 Cygni**, straddling the long filaments of the western side of the Veil Nebula. The nebula is the expanding shell of gas from a supernova explosion. Your best views will be through an Oxygen III or other high-contrast filter (see p. 84). To fully explore this intriguing object, trace the intertwining filaments from end to end with a 6 inch (150 mm) scope or larger.

**❾ M29 (NGC 6913)** ⬤⬤ Some 10 degrees northwest of the Veil Nebula is magnitude 2.1 **Sadr (37 Gamma [γ] Cygni)**. Use this star to find the open star cluster M29, about 2 degrees south. This attractive open cluster lies on the edge of one of the obscuring bands of the Great Rift dust clouds that seem to cut through the Milky Way in Cygnus. The seven brighter stars in M29 form a dipper-like asterism.

**❿ The North American Nebula (NGC 7000)** 👁 Hop about 6 degrees from Sadr to the bright magnitude 1.3 star **Deneb (50 Alpha [α] Cygni)**, "the tail of the swan." Another 3 degrees east of Deneb is NGC 7000. From a very

dark site, this giant, diffuse emission nebula can just be seen with the naked eye, glowing softly at magnitude 5.9. Its common name arises from its shape, which is very similar to that of the North American continent, including Mexico.

Take your time to fly with the Swan and seek out and explore the multitude of open clusters and double stars that are embedded in the star fields of Cygnus in the Milky Way.

## THE VISUAL MILKY WAY

One of the joys of stargazing is to sit out under a very dark sky and watch the broad swath of the faintly glowing Milky Way slowly pass overhead. Take the time to study the Milky Way without binoculars or a telescope. Look for the delicate interweaving of the bands of obscuring dust clouds and the hazy star fields. After observing the Milky Way with the naked-eye, scan the band with binoculars, stopping to study the rich star fields and coal-black dark nebulas throughout it.

The plane of the Milky Way Galaxy is made up of vast clouds of glowing gas and stars cut by dark lanes of dust (left). Billions of faint stars, many that are far too faint for our eyes to separate into individual points of light, merge like a mottled, grayish fog.

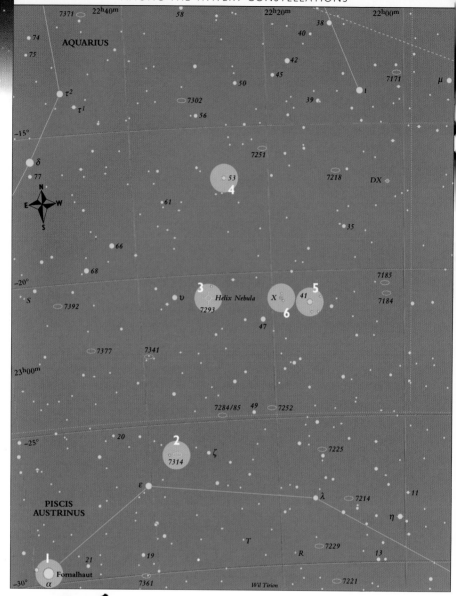

# Among the Watery Constellations

Visible from both hemispheres, this region follows the brilliant area of the galactic center across the sky. Of several water constellations, first comes Capricornus, the Sea Goat, which was depicted on charts of long ago as a goat with the tail of a fish. Following Capricornus is Aquarius, the Water Bearer, and Pisces, the

*Piscis Austrinus, the Southern Fish, is often depicted as a fish drinking the flow of water poured from the urn of Aquarius.*

Fishes, partly shown on the locater map (right). All three are zodiacal constellations, which the planets regularly pass. Also in this region are the constellations of Cetus, the Whale, and Piscis Austrinus, the Southern Fish—the parent of Pisces, according to ancient legend.

This starhop begins at Fomalhaut, near the southeastern corner of the main sky chart, before going on to browse around the celestial delights of parts of three watery constellations—Aquarius, Capricornus, and Piscis Austrinus.

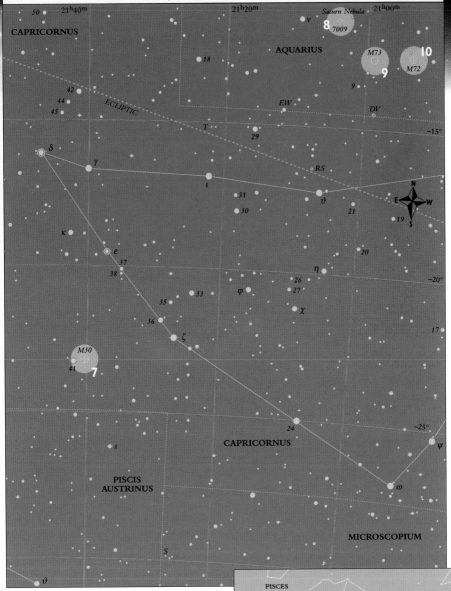

**❶ Fomalhaut (24 Alpha [α] Piscis Austrinus)** 👁 Our starting point for this star-hop is the brilliant magnitude 1.2 blue-white star Fomalhaut, the 18th brightest star in the night sky. This star lies in quite a barren region of the southern sky and because of this has become known as "The Solitary One." Its isolation makes it very easy to find. Fomalhaut lies some 25 light-years distant from Earth and is estimated to have about twice the diameter of the Sun and to be some 14 times more luminous.

Fomalhaut's name comes from the Arabic *Fum al Hut*, meaning "mouth of the fish," as it marks the position of the Fish's mouth in the constellation of Piscis Austrinus. The Greek poet Aratos described Fomalhaut as "One large and bright by the Pourer's feet," with the pourer, of course, being Aquarius, the Water Bearer.

461

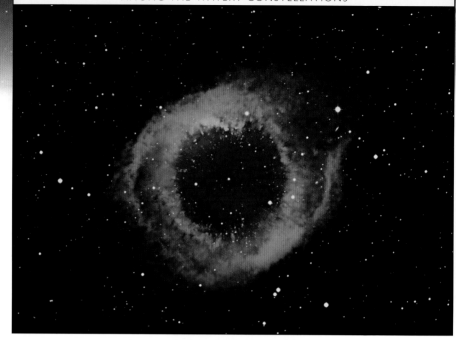

**❷ NGC 7314**  To find the next object on this hop, head just over 4 degrees north-west, to **18 Epsilon [ε] Piscis Austrius**. Another 1½ degrees farther northwest will bring you to NGC 7314. In a 4 inch (100 mm) scope, this 11th magni-tude spiral galaxy can be seen as a fuzzy patch of light with a small, but fairly bright, nucleus. A larger scope reveals considerably more detail surrounding its center.

**❸ The Helix Nebula (NGC 7293)** Cross the constellation border into Aquarius and aim your scope some 5 degrees north to find the star **59 Upsilon [υ] Aquarii**. About 1 degree to the west of Upsilon lies the famed Helix Nebula (see Box). The Helix is a planetary nebula—one of the best known of its type. You should just be able to capture a view of it in a good pair of binoculars on a clear, dark night, appearing as a circular patch of light about a ¼ degree in diameter. When using a scope, you need a low-power, wide-field eyepiece to really appreciate it. If you have an Oxygen III filter, you can treat yourself to a fantastic view.

**❹ 53 Aquarii** Continue swinging your scope northward, about 4 degrees, until you find 53 Aquarii. This pretty, yellowish pair of 6th magnitude stars was much easier to separate when measured early in the last century. They were then about 10 arcseconds apart, but now the distance has shrunk to a mere 3 arcseconds.

*The Helix Nebula (above) is a highlight in Aquarius. Fomalhaut a beacon of light (left).*

Even so, you should easily sepa-rate them with a 3 inch (75 mm) scope on a steady night.

**❺ 41 Aquarii** Direct your scope 5 degrees southwest of 53 Aquarii to the next double star, 41 Aquarii. The brighter of this pair, at magnitude 5.6, has a distinctive yellow color, while its 7th magnitude companion is yellow-white. At 5 arcseconds apart, they make a fine double for small scopes.

**❻ X Aquarii** Shift your telescope 1 de-gree east of 41 Aquarii, then nudge it just a few arcminutes north to view the field containing X Aquarii. A long-period Mira-type variable, X is also one of the rare S-type variables, which show zirconium oxide in their spectra. This orange-red star varies between magnitude 7.5 and 14.8 over a period of 312 days, so you need to catch it near maximum to see it clearly.

**❼ M30 (NGC 7099)** To find M30, make another border crossing, into Capricornus, 8 degrees west and 2 degrees south of X Aquarii, to find 5th magnitude **41 Capricorni**. About ½ degree west of it lies the small but attractive globular cluster M30, which is just visible with binoculars. A 3 inch (75 mm) telescope will begin to resolve some of its stars, but the view through a 6 inch (150 mm) telescope is without doubt the more attractive one.

*The globular cluster M30 (above) is the only bright deep-sky object to be found in the constellation of Capricornus.*

**❽ The Saturn Nebula (NGC 7009)** ⊂▯◉
Get ready to hop quite a distance to our next target. First, stop off at 4th magnitude **Zeta [ζ] Capricorni**, then turn northwest where, 7 degrees away, you will spot the less bright **Theta [θ] Capricorni**. Finally, cross back into Aquarius by moving 6 degrees farther north, to the Saturn Nebula. This delightful little 8th magnitude planetary nebula, a little more than 1 degree west of **Nu [ν] Aquarii**, glows a beautiful blue color. A 3 inch (75 mm) scope with moderate magnification shows the interesting elliptical shape of this object. It was given the name of the Saturn Nebula because of its almost ring-like projections that can be seen with 12 inch (300 mm) telescopes or larger.

**❾ M73 (NGC 6994)** ⊂▯◉ Aim your scope 2 degrees southwest of the Saturn Nebula to see M73. While not exactly a spectacular sight, it has historic signifi- cance, since Charles Messier included M73 in his catalog. It is merely a tight group of four faint stars between 10th and 12th magnitude.

**❿ M72 (NGC 6981)** ⊂▯◉
About 1 degree west of M73 is M72. Although it is moderately concentrated, 9th magnitude M72 is a

relatively dim globular cluster, and one which is, unfortunately, quite difficult to resolve. A 10 inch (250 mm) scope is needed to show a hint of some of its multitude of stars, and only observers with large scopes will obtain a good view of it.

Although this patch of sky is relatively sparse, our starhop around some of the watery constella- tions has shown that even here there are many fascinating objects waiting to be found.

## THE HELIX NEBULA

Lying at a distance of about 450 light-years from Earth, the Helix Nebula is the closest planetary nebula to us, and therefore appears as the largest in the sky. It has been the subject of a number of serious studies. In 1996, astronomers researching images of the Helix from the Hubble Space Telescope found thousands of comet-like gaseous fragments, especially near the inner edge of the nebula's ring. It is thought that these "knots" may have formed when new material from the star interacted with material that had been ejected earlier. One day, billions of years from now, the Sun will have a shell, probably not unlike the Helix Nebula, but by that time, the human race will probably have long since disappeared from Earth.

463

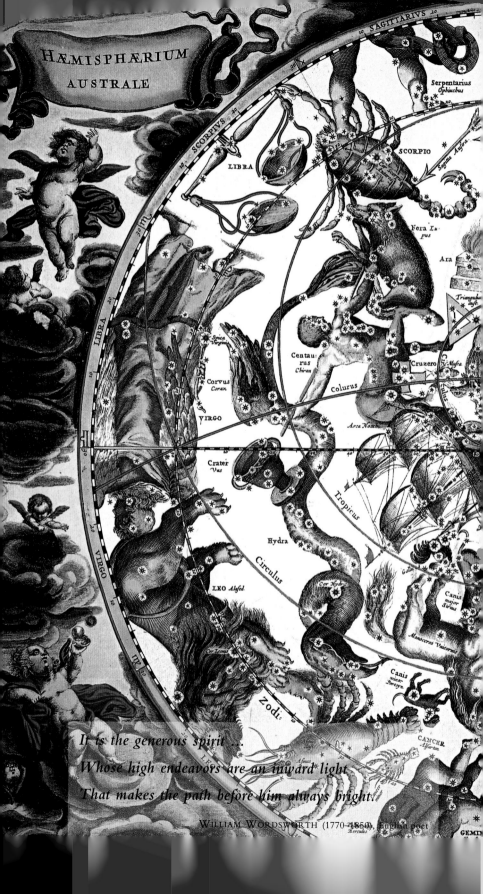

HÆMISPHÆRIUM
AUSTRALE

SAGITTARIVS

Serpentarius
Ophiuchus

SCORPIVS

SCORPIO

Sagitta Austra

LIBRA

Fera Lupus

Ara

Triangulus
aust.

LIBRA

Spica Virginis

Centaurus
Chiron

Cruzero C. Mufca

Corvus
Corax

Colurus

VIRGO

Arca Noachi

Crater
Vas

Tropicus

Hydra

Canis
major
Sirius

Circulus

Gallus

VIRGO

Monoceros Vnicornus

LEO Alfed.

Cor Hydr.

Cor Leonis

Canis
minor
Procyon

Zodi:

CANCER
Afarem

*It is the generous spirit ...*
*Whose high endeavors are an inward light*
*That makes the path before him always bright.*

WILLIAM WORDSWORTH (1770–1850), English poet

Hercules

GEMINI

# INDEX *and* GLOSSARY

# INDEX *and* GLOSSARY

In this combined index and glossary, **bold** page numbers indicate the main reference, and *italic* page numbers indicate illustrations and photographs.

## CAPTIONS

*Page 1:* A miniature sixteenth-century painting of astronomers observing the sky from Galata Tower, Constantinople.

*Page 2:* Tycho Brahe, Ptolemy, St Augustine, Copernicus, Galileo, and Andreas Cellarius with The Muse of Astronomy.

*Page 3:* Comet-like gaseous fragments are left behind as the Helix Nebula's ring expands out behind it.

*Pages 4–5:* The moon may look larger when near the horizon.

*Pages 6–7:* A long-exposure photograph showing Delicate Arch, Utah, silhouetted against star trails.

*Pages 8–9:* The observatory domes at Mauna Kea, Hawaii, USA.

*Pages 10–11:* A time-lapse montage of a partial solar eclipse in five stages.

*Pages 12–13: The School of Athens* by Raphael (1483–1520).

*Pages 22–23:* The space shuttle Atlantis launch, Florida, USA.

*Pages 50–1:* A group of observers viewing an eclipse of the Sun, Paris, 17 April 1912.

*Pages 96–7:* A spectacular aurora over Wasilla, Alaska.

*Pages 116–17:* A detail from Sun and Moon © Copyright Paul Klee (1879–1940). Reproduced by permission of BILD-KUNST (Germany) and VI$COPY Limited, 1997.

*Pages 146–7:* A detail of the surface of Jupiter, photographed by the Voyager 2 spacecraft.

*Pages 194–5:* A Hubble Space Telescope image of the interstellar gas pillars within the Eagle Nebula.

*Pages 240–1:* Constellation patterns depicted in a fresco in the Palazzo Farnese, Italy.

*Pages 368–9:* The Eagle Nebula (M16).

*Pages 464–5:* A detail from Andreas Cellarius's *Atlas Coelestis seu Harmonia Macrocosmica* (1660).

---

# PICTURE CREDITS

t = top, c = center, b = bottom, l = left, r = right, i = inset, co = chapter opener

A = Auscape International; AAO = David Malin/Anglo-Australian Observatory; AF = Akira Fujii; APL = Australian Picture Library; ASP = Astronomical Society of the Pacific; AV = Astro Visuals; BAL = The Bridgeman Art Library, London; ESA = European Space Agency; Granger = The Granger Collection, New York; IS = Image Select; LC = Lee C. Coombs; JPL = Jet Propulsion Laboratory; Mendillo = Mendillo Collection of Antiquarian Astronomical Prints; MEPL = Mary Evans Picture Library; MIC = Meade Instruments Corporation; NC = Newell Color; NOAO = National Optical Astronomy Observatories; NRAO = US National Radio Astronomy Observatory; NW = North Wind Picture Archives; OS = Oliver Strewe; PE = Planet Earth Pictures; RAS = Royal Astronomical Society; ROE = Royal Observatory, Edinburgh; SPL = Science Photo Library; SF = Space Frontiers Collection, London; STScI = Space Telescope Science Institute; TPL = The Photo Library, Sydney; TS = Tom Stack and Associates; UCO = University of California Observatory; USGS = US Geological Survey.

1 University Library Istanbul/ET Archive/APL. 2 Granger. 3 NC/NASA. 4–5 Franz X. Kohlhauf. 6–7 David Nunuk/SPL/TPL. 8–9 Chlaus Lotscher/Peter Arnold Inc./A. 10–11 Pekka Parviainen/SPL/TPL. 12–13 co Scala, Firenze (Stanze di Raffaello, Vaticano); i Photodisc. 14 tr Ancient Art and Architecture Collection, London; tl BAL (The British Museum, London); b TPL (Paul Thompson). 15 tr APL; tl Griffith Observatory (Lois Cohen); br National Anthropological Archives/Smithsonian Institution photo 1285; bl OS. 16 tl APL (John Miles); tr & b By permission of the British Library. 17 b BAL (Raymond O'Shea Gallery, London); tr Jean-Loup Charmet; tl The Museum of New Zealand Te Papa Tongarewa, Wellington. Photo no. B18727 (J. Nauta). 18 t Ann Ronan/IS; bl APL; br APL. 19 t MEPL (Explorer); b MEPL. 20 br Ann Ronan/IS; bl A (Explorer); t APL. 21 t APL; b BAL (Private Collection); c Scala, Firenze (Museo della Scienza, Firenze). 22 c NW; b SPL/TPL. 23 t Baum & Henbest/SPL/TPL; c Mt Wilson & Las Campanas Observatories/ASP slide set. "Astronomers of the Past." 24 t Granger; b NASA (Dana Berry/STScI). 25 t NASA (Dana Berry/STScI); b TPL (SPL/NASA). 26 b NASA (JPL); tl TPL (SPL). 27 t TPL (SPL/David Hardy). 28 t TPL (SPL/David Parker). 29 b NASA (Dana Berry/STScI). 30–31 co & i TSM-Mark M Lawrence/Stock Photos. 32 t Ann Ronan/IS; c Hulton-Deutsch/TPL; b NASA/TPL. 33 t SF/PE; c Max Planck Institute/SPL/TPL; b NC/NASA. 34 t NOAO/AV; cr ROE/SPL/TPL; b European Southern Observatory. 35 t Mary Lea Shane Archives, Lick Observatory/ASP slide set "Astronomers of the Past"; b Astronomy Group, School of Physics, University of Sydney. 36 cl Jodrell Bank/SPL/TPL; cr NRAO; b SPL/TPL. 37 t Ann Ronan/IS. 38 t Seth Shostak/SPL/TPL; b Dornier Space/SPL/TPL. 39 t SF/PE; c NASA; b NASA/SPL/TPL. 40 c NC/NASA; bl NASA; br C. R. O'Dell/Rice University/NASA. 41 t NASA and The Hubble Heritage Team (STScI/AURA); c & b NC/NASA. 42 tl NASA/AV; tr SPL/TPL; c Novosti Press Agency/SPL/TPL. 43 c NASA; b NASA/Peter Arnold Inc./A. 44 t NASA/Airworks/TS; cl IS; c SF/PE. 45 tl IS; tr, SF/PE; b SPL/NASA/TPL. 46 c SF/PE; b NASA/SPL/TPL. 47 t NASA. 48 tl Novosti/SPL/TPL; cl JPL/TSADO/TS; cr NASA. 49 tl & tr NASA; bl & br NASA/SPL/TPL. 50 c & b SF/PE. 51 tl & b SF/PE; tr David Parker/SPL/TPL. 52 tl Julian Baum/SPL/TPL; tr NASA/AV; c PE; b JPL/NASA. 53 t NASA/JPL/Malin Space Science Systems; b Seth Shostak/SPL/TPL. 54 c F. Espenak/SPL/TPL; b Ray Nelson/Phototake/Stock Photos. 55 t ESA; bl Ray Pfortner/Peter Arnold Inc./A; br ROE/SPL/TPL. 56–57 co & i Roger-Viollet. 58 t Scala, Firenze (Pinacoteca Vaticano); b OS. 59 David Miller. 60 t & b OS. 61 t AF; tr SPL/TPL; b Tony & Daphne Hallas/Astro Photo. 62 b International Dark Sky Association; t Gregg Thompson. 63 c International Dark Sky Association; b NOAO; t TPL (SPL/David Nunuk). 64 t & c Dave Parkhurst/The Aurora Collection; b NASA. 65 t Dave Parkhurst/The Aurora Collection; b John Sanford/SPL/TPL. 66 t Ann Ronan/IS; b Pekka Parviainen/SPL/TPL; c Pekka Parviainen/SPL/TPL; b David Miller. 67 t Dave Parkhurst/The Aurora Collection; c Pekka Parviainen/SPL/TPL; b David Miller. 68 t TPL (TSI); b OS. 69 OS. 70 t Hulton-Deutsch/TPL; bl Bill and Sally Fletcher; br Alan Dyer. 71 t & c OS; b Tenmon Guide, Japan. 72 OS. 73 t OS; b Alan Dyer. 74 t Museo della Scienza, Firenze/Scala; b MIC. 75 t & b OS. 77 t Don Whiteman; c AF; b Yerkes Observatory. 78 tl Roger-Viollet; c OS; b Alan Dyer. 79 cr & b

OS. 80 t & b MIC. 81 t AAO; c European Southern Observatory; b OS. 82 tl & c OS. 83 b OS. 84 tl & cr MIC. 85 t Tele Vue Optics Inc. 86 t OS; cl MIC; cr Luke Dodd. 87 t & b Yerkes Observatory. 88 t OS; cl & cr David Miller. 89 t Pekka Parviainen/SPL/TPL; b Alan Dyer. 90 t Alan Dyer; b Luke Dodd. 91 t Luke Dodd. 92 t MIC; c AF; cr C. Arsidi/MIC. 93 b MIC; br Tele Vue Optics Inc. 94 t OS; b Gregory Terrance. 95 t MIC; c Jack Newton/MIC; b Jack Newton. 96 t Mendillo; c Corbis/Bettman/APL. 97 t OS; b NW. 98 t OS; bl Maris Multimedia; br MIC. 99 t NASA/STScI (http://oposite.stsci.edu/pubinfo/Pictures.html); c NASA/JPL (www.mars.jpl.nasa.gov). 100 t TPL (SPL/John Sanford); bl Sky & Telescope Magazine (Robert E. Cox). 101 t TPL (SPL/Gordon Garrad); b OS. 102 t Richard Keen; b David H. Levy. 103 tr Gordon Hudson; tl Gregg Thompson; b Alan Dyer. 104–105 co AF & i Corbis. 106 br BAL (Roy Miles Gallery); t Granger. 109 Franz X. Kohlhauf. 110 t BAL (Roy Miles Gallery). 111 tr Warren Cannon; tl The Image Bank (Murray Alcosser). 112 t Dennis di Cicco. 114 tl Ann Ronan/IS; tr Mendillo; b TPL (SPL). 115 b By permission of Owen Gingerich; t RAS. 116 The Image Bank (Bernard Van Berg). 117 br AAO. 118 t BAL (Derby Museum and Art Gallery). 119 t Geoff Chester, Albert Einstein Planetarium, Smithsonian Institution; b NASA (JPL). 120–121 co & i Sun and Moon © Copyright Paul Klee/Repro-duced by permission of BILD-KUNST (Germany) and VI$COPY Ltd 1997/BAL. 122 t Egyptian Museum, Cairo/Werner Forman Archive; c F. Espenak; b Tele Vue Optics Inc. 123 t F. Espenak; b SPL/TPL. 124 t AKG, London. 125 t NASA/TSADO/TS; c JPL/TSADO/TS. 126 t MEPL; c NASA; b Photri Inc./APL. 127 t & c NOAO; b Dr. Dan Gezari/NASA/SPL/TPL. 128 t NRAO/AUI. 129 c Sky & Telescope Magazine; b Guildhall Library, London/BAL. 130 bl Francisco Arias/Sipa Press/TPL; br F. Espenak. 131 tr F. Espenak. 132 t Pacific Stock/APL; c F. Espenak; b Roger-Viollet. 133 t Tony and Daphne Hallas/Astro Photo; b F. Espenak. 134 t Granger; bl Rev. Ronald Royer/SPL/ TPL; br AF. 136 t MEPL; cl Chad Ehlers/TPL; cr F. Espenak. 137 t Ann Ronan/IS. 138 t Wallace & Gromit/Aardman Animations Ltd. 1989; c NASA; b NASA/Airworks/TS. 139 t & c NASA. 140 t Zuber et al./SPL/TPL; c Granger; b NASA/JPL/TSADO/TS. 141 t UCO/Lick Observatory; b NC/NASA. 142–149 UCO/Lick Observatory. 150–151 co & i NASA/SPL/TPL. 152 tl Chris Burrows (STScI), John Krist (STScI), Karl Stapelfeldt (JPL) and colleagues, the WFPC2 Science Team and NASA. 154 t BAL. 156 t Erich Lessing/AKG, London; c NASA/SPL/TPL; b JPL/TSADO/TS. 157 t NASA/SPL/TPL; b AF. 158 t Museum of Mankind, London/BAL; c & b NASA. 160 t & c NASA; b Image Library/State Library of NSW. 161 t NASA; c & b NASA/SPL/TPL. 162 t Archaeological Museum, Aleppo/ET Archive/APL. 163 tl Arthur Pengelly/Sipa Press/TPL; tr Sylvain Grandadam/TPL; bl Adam Jones/TPL; br Peter Knowles/TPL. 164 tl NASA/SPL/TPL; tr Worldsat/Knighton/SPL/TPL. 165 t NASA; b Penny Gentieu/TPL. 166 t Prado, Madrid/BAL; b USGS/TSADO/TS. 167 t NASA; b Mary Lea Shane Archives, Lick Observatory/ASP slide "Astronomers of the Past." 168 tl USGS/TSADO/TS; tr USGS/SPL/TPL; b NASA. 169 br Don Parker. 170 t NASA; c SPL/TPL; b NASA. 171 tl NASA/SPL/TPL; tr NASA; b David Hardy/SPL/TPL. 172 t NW; b NASA/SPL/TPL. 173 tl & tr NASA/SPL/TPL. 174 t Don Parker. 175 t NRAO/AUI; r1, r2, & r3 LC. 176 t & br SF/PE; bl NW. 177 t NASA/SPL/TPL; tr photolibrary.com/Science Photo Library; cr PE; b SPL/TPL. 178 c Palazzo Vecchio, Florence/BAL; b SF/PE. 179 tl NASA/SPL/TPL; b NW. 180 t & c NASA/SPL/TPL; bl Peter H. Smith/University of Arizona Lunar and Planetary Laboratory/NASA; br Hermitage, St Petersburg/BAL. 181 t Don Parker. 182 t Granger; c NASA. 183 cc NASA/SPL/TPL; tr NASA; b USGS/SPL/TPL. 184 tl Giraudon/BAL; tr NASA; bl NASA/SPL/TPL; bp MEPL. 185 t SF/PE; c NASA/SPL/TPL. 186 t Palazzo Vecchio, Florence/BAL; c NC/NASA; cr Corbis/Bettman/APL. 187 t NASA; b NASA/SPL/TPL. 188 t NASA (JPL). 189 t WSF/R. Helin. 188–189 b Lynette R. Cook. 190 t Ann Ronan/IS; b JHUAPL/NASA. 191 t R.S. Hudson and S. J. Ostro/NASA; c NASA/SPL/TPL; b Photofest/Retna/APL. 192 t Musee de la Tapisserie, Bayeux/BAL; cl AF; cr SPL/TPL; b ESA/SPL/TPL. 193 b Gordon Garradd/SPL/TPL. 194 c AF; b SPL/TPL. 195 t Index Stock/Phototake/Stock Photos; c Alan Dyer; b Observatoire de Paris/Bulloz. 196 br Hubble Space Telescope Comet Team/STScI/NASA; b Dr. H. A. Weaver and T. E. Smith/STScI/NASA. 197 t NASA/SPL/TPL; c Jonathan Blair/Woodfin Camp/APL. 198 t Boyer–Viollet; c & b Mineralogy Dept/Nature Focus/Aust-

## ILLUSTRATION CREDITS